# THE MOLTEN STATE OF MATTER

## Melting and Crystal Structure

# THE MOLTEN STATE
# OF
# MATTER

## Melting and Crystal Structure

A. R. UBBELOHDE

*Department of Chemical Engineering and Chemical Technology,
Imperial College, London University*

*A Wiley–Interscience Publication*

JOHN WILEY & SONS

Chichester · New York · Brisbane · Toronto

*Library of Congress Cataloging in Publication Data*:

Ubbelohde, Alfred René.
  The molten state of matter.

  'A Wiley–Interscience publication.'
  Includes bibliographical references.
  1.  Crystallography.     I.  Title.
QD 921.U2   1978      548'.5      77-28300

ISBN 0 471 99626 2

Printed in Great Britain by
J. W. Arrowsmith Ltd., Bristol.

# ACKNOWLEDGEMENTS

Figures for which permission to reproduce was gratefully acknowledged in an earlier book on melting by the author have in a number of instances been also used in the present text. Permission is also acknowledged to reproduce.

Thanks are likewise due to many colleagues and former students whose discussions have added zest to the labours of preparing this edition. Invaluable help in preparing the text has also been received from Miss G. M. Greene and Miss S. Hartman and from Mrs. G. Munday and Hon. Mrs. J. Lewis, and in preparing the figures to Mr. L. Moulder.

# CONTENTS

PREFACE                                                                    XV

## 1. GENERAL INTRODUCTION                                                  1
1.1.  Some historical aspects of melting                                   1
1.2.  Structual aspects of melting                                         1
1.3.  Melting leading to specific types of liquids                         2
1.4.  Theories of melting                                                  3
1.5.  Precursor anomalies in the solid–liquid transition                  4

REFERENCES                                                                 5

## 2. PHENOMENOLOGICAL THERMODYNAMICS OF MELTING                            6
2.1.  Classical thermodynamics of the solid–liquid phase change            6
   2.1.1.  Frozen-in microparameters                        6
   2.1.2.  General thermodynamic equations                   6
2.2.  Equilibrium between phases                                           7
2.3.  Derived entropy and volume changes on fusion                        10
2.4.  Effects of moderate pressures on melting                            11
   2.4.1.  Volume changes on fusion                         11
   2.4.2.  Benzene derivatives                              13
2.5.  Melting at very high pressures                                       14
   2.5.1.  Pressure/melting point relationships             15
2.6.  Maxima on pressure/melting point curves                             17
   2.6.1.  Ligand modification at very high pressures       18
   2.6.2.  Interpretation of the Simon equation             18
2.7.  Melting at very high temperatures                                    20
2.8.  Range of existence of condensed phases—sublimation                  20
2.9.  Thermodynamic criteria of continuous transition from crystal to melt  25
   2.9.1.  Critical melting—other criteria                  29
2.10. Melting of very small portions of solid (homomolecular effects)      30
   2.10.1.  Melting at grain boundaries                     31
2.11. Heteromolecular effects—melting of sorbed layers on surfaces         32
   2.11.1.  Regelation of ice                               34
2.12. Melting in capillaries                                               35
2.13. Melting of microcrystallites in gels                                 37
   2.13.1.  Effect of capillaries on nucleation            38
2.14. Effects of impurity on melting                                       39
   2.14.1.  Impurities not soluble in the crystalline phase  39
2.15. Melting of solid solutions                                           43
   2.15.1.  Melting of crystals of isotopic molecules       45
   2.15.2.  Melting of helium                               46
2.16. Melting of polymorphs: crystals yielding the same melt               48
REFERENCES                                                                 50

3. PHENOMENOLOGICAL THEORIES OF FUSION                                54

   3.1.   Some general aspects of phenomenological theories of fusion   54
      3.1.1. Corresponding state calculations                          55
   3.2.   Mechanical theories of melting                               58
   3.3.   Vibrational theories of melting: one-phase models            61
   3.4.   Vibrational melting of a line solid                          66
   3.5.   Vibrational properties in relation to thermodynamic behaviour of
          solids                                                       67
   3.6.   Tests of the Lindemann melting formula                       68
      3.6.1. Vibrational melting of helium and other inert gases       70
   3.7.   Vibrational solid–solid transformations: two-phase theories  72
   3.8.   Vibrational softening leading to transformations in crystals 73
   3.9.   Changes of vibrational parameters on melting                 75
      3.9.1. Communal entropy                                          78
   3.10.  Vibrational properties of melts in relation to the crystalline state  78
      3.10.1. Specific heat                                            79
      3.10.2. Thermal expansion                                        79
   3.11.  Transport coefficients of melts as a diagnostic guide to their structure  80
      3.11.1. Viscosity                                                80
      3.11.2. Thermal conductivity                                     83
      3.11.3. Temperature coefficients of electrical resistivity of metals  84
   REFERENCES                                                          84

4. SOLID–SOLID TRANSFORMATIONS RELATED TO
   FUSION                                                              86

   4.1.   Structural interpretations of phase transformations in crystals  86
      4.1.1. Some mechanisms of entropy increase in crystal transformations  87
   4.2.   Positional randomization in solid–solid transformations      88
   4.3.   Orientational randomization in crystals with quasi-spherical mole-
          cules                                                        90
      4.3.1. Orientational transformations in crystals with quasicylindrical
             molecules                                                 99
      4.3.2. Sensitiveness of orientational transformations to crystal lattice
             interactions                                              100
   4.4.   Thermodynamic transformations of 'higher order'              101
      4.4.1. Structural interpretation of transformations of 'higher order'  102
   4.5.   Hybrid crystals in a solid–solid transformation             103
      4.5.1. Persistence of crystal axes in a cyclic lambda transformation  104
      4.5.2. Appearance of new polymorphic forms                       105
      4.5.3. Independent nucleation                                    105
      4.5.4. Transformation by way of hybrid crystals                  106
   4.6.   Hysteresis in lambda transformations                         108
   4.7.   Coexistence of related transforms in hybrid crystals as origin of
          hysteresis                                                   110
      4.7.1. Acoustic detection of coexistence                         111
   4.8.   Generalized thermodynamics of hysteresis                     112
   4.9.   Storage of defect energy in exhaustive thermal cycles        114
   4.10.  Polymorphic transitions in condensed microphases             115
      4.10.1. Effects of impurities on transformation temperatures     115
   REFERENCES                                                          116

5. STRUCTURAL MELTING OF CRYSTALS. INTRODUCTION
OF POSITIONAL DISORDER                                    119
  5.1.  Melts as disordered versions of their crystalline counterparts: X-ray
and other physical evidence                               119
    5.1.1. Melts as quasi-crystalline lattices: the role of positional defects   122
  5.2.  Positional disordering on melting and quasicrystalline melts   124
    5.2.1. Positional melting                                124
      5.2.1.1. Cell models for liquids                      125
    5.2.2. Classification of types of liquids                 125
    5.2.3. Similitude theories of positional melting          125
  REFERENCES                                               128

6. STRUCTURAL MELTING. MOLECULAR CRYSTALS OF
RIGID MOLECULES                                           129
  6.1.  Positional melting of inert gases and related crystals   129
    6.1.1. Orientational disordering of molecules in crystals   130
  6.2.  Orientational disordering on melting                 130
  6.3.  Repulsion envelopes in the melting of molecular crystals with rigid
molecules                                                 132
    6.3.1. Effects of pressure on rotational disordering      133
  6.4.  Coexistence of related structures in rotator transitions in crystals   134
  6.5.  Trends in melting parameters showing similitude for structurally
related molecules                                         134
    6.5.1. Trends in melting points with increasing molecular size   135
    6.5.2. Melting of globular molecules                      136
    6.5.3. Plastic crystals                                   142
  6.6.  Melting of rigid linear molecules                    143
    6.6.1. Quantitative calculations of entropy increases on orientational
melting                                                   147
    6.6.2. Melting of rigid planar molecules—vibrational entropy contribu-
tions to fusion                                           148
    6.6.3. Room to rotate—orientational correlations in melts   149
  REFERENCES                                               152

7. MELTING OF FLEXIBLE MOLECULES                             154
  7.1.  Increases of molecular configurational entropy on melting   154
    7.1.1. Melting of n-alkane hydrocarbons                   154
    7.1.2. Melting of other flexible molecules                160
  7.2.  Vibrational melting of flexible molecules            162
  7.3.  Melting of 'loaded' polymethylene chains             163
    7.3.1. Effects of point of loading of chain molecules on $T_f$   166
  7.4.  Configurational melting—non-thermodynamic evidence   166
  7.5.  Melting of homologous flexible molecules loaded at several points   168
  7.6.  Crystals with mixed flexible tautomers in the melt   168
  7.7.  Configurational disordering giving anticrystalline melts   170
  REFERENCES                                               170

8. MELTING OF IONIC CRYSTALS                                 172
  8.1.  Introduction                                         172
  8.2.  Positional melting in ionic crystals—ions of inert gas type   172

8.3.   Ion defect formation in simple ionic crystals                    173
    8.3.1.  Interpenetrating sublattices                               174
    8.3.2.  Hydrogen in ionic lattices                                 174
    8.3.3.  Other ions of inert gas type                               175
8.4.   Effects of positional disordering on electrostatic compensation  176
8.5.   Melting parameters of ionic halides with low polarizabilities    177
    8.5.1.  Halides of high valence cations (lanthanides)              178
    8.5.2.  Ionic analogue for the iodide anion                        179
8.6.   Structural studies on ionic melts                                179
    8.6.1.  Diffraction studies on ionic melts—polarizability effects  180
8.7.   Precursor effects associated with specific ion crystal types     182
8.8.   Melting of salts with inert gas type ions—effects of moderate pressure  184
8.9.   Cryoscopy with molten salts                                      186
8.10.  Similitude rules for melting ionic crystals                      189
8.11.  Simple ions forming strongly polarizable systems                 192
8.12.  Formation of ionic melts by autocomplexing—simple structures     193
    8.12.1. Complexing in mixed melt environments                     196
    8.12.2. Autocomplexing leading to ion formation in other melts    197
8.13.  Extended interactions in ionic melts                             197
    8.13.1. Melts of zinc compounds                                   198
    8.13.2. Grotthus conduction in network ionic melts               199
8.14.  Melts of crystals containing polyatomic ions                     200
    8.14.1. 'Stable' polyatomic ions                                  201
    8.14.2. Melting of sulphates                                      203
    8.14.3. Melting of carbonates                                     203
8.15.  Formation of association complexes on melting                     205
    8.15.1. Association by closer packing into anticrystalline clusters  206
8.16.  Optical studies of complexing local ionic environments           210
    8.16.1. Infrared measurements on phase transitions in ionic crystals  213
8.17.  Atom transfer defects on melting ionic crystals                  215
    8.17.1. Anion dissociation                                        215
    8.17.2. Cation dissociation                                       216
    8.17.3. Valence switch defect formation                          216
8.18.  Other methods of investigating dissociation on melting           217
8.19.  Change from 'ionic' to 'metallic' structures on melting          217
8.20.  Sublattice melting of ionic lattices                             217
    8.20.1. Melting of the cation sublattice                          218
    8.20.2. Melting of anion sublattices—ionic crystals carrying doubly
            charged cations and singly charged anions                219
8.21.  Molten organic salts—problems of chemical stability of organic ionic
       melts                                                            220
8.22.  Thermodynamic parameters in the melting of alkali carboxylates   222
    8.22.1. Other organic anions                                      229
8.23.  Melting of salts with organic cations                            229
REFERENCES                                                              232

9. MELTING OF METALS                                                    237
9.1.   Metallic crystals as models of melting                           237
9.2.   Vibrational effects in the melting of metals—solid–solid trans-
       formations                                                       237
    9.2.1.  Melting parameters of rarer metals                        240

9.2.2. Role of crystal similitude 241
9.2.3. Correlation of vibrational properties with $T_f$ 242
9.3. Thermodynamic parameters of melting of metals 243
9.3.1. Specific heats of metals near $T_f$ 243
9.3.2. Entropies of fusion 244
9.4. Positional defects in melting of metals 245
9.5. Melting of metals at high pressures and temperatures 249
9.6. General similitude rules in the melting of metals 250
9.7. Structural information about molten metals 253
9.8. Melting of metals with non-equivalent bonds 255
9.9. Changes of electronic properties of metals on melting 256
9.9.1. Changes of electrical resistivity 256
9.9.2. Thermo-electric power of molten metals 260
9.9.3. Hall effect of liquid metals 261
9.9.4. Changes in magnetic susceptibility on melting 262
9.9.5. Changes in X-ray band edges on melting 264
9.9.6. Changes in thermal conductivity on melting 264
9.10. Melting of semiconductors 265
9.10.1. Melting of semiconducting crystal compounds 266
9.10.2. Environmental collapse and melting 267
9.10.3. Changes of electrical conductance on melting class II semiconductors 267
9.10.4. Effects of pressure 268
9.11. Melting of intermetallic compounds $AB_n$ 268
REFERENCES 270

10. NETWORK MELTING 274
10.1. Independent defects leading to network melting 274
10.2. Formation of non-cooperative defects in three-dimensional networks 276
10.2.1. Network melting in elements 278
10.3. Binary oxides and halides 278
10.2.1. Three-dimensional oxide networks 278
10.3.2. Three-dimensional halide networks 280
10.3.3. Two-dimensional oxide networks 282
10.4. Water as a network melt 282
REFERENCES 283

11. STATISTICAL THEORIES OF MELTING AND CRYSTAL STRUCTURE 285
11.1. Melting of crystals with simple units of structure 285
11.1.1. Communal melting entropy 285
11.2. Disorder theories of fusion—positional melting 287
11.3. Melting of inert gas crystals to quasicrystalline melts. The Lennard-Jones–Devonshire model 288
11.4. Solid-liquid transitions in hard sphere assemblies 295
11.5. Cooperative positional disorder leading to melting 296
11.5.1. Dislocation melting (elementary model) 296
11.5.1.1. Energy arising from dislocations 297
11.5.1.2. Entropy increase resulting from dislocations 298

11.5.2.   Dislocation melting (refined models)                           299
11.6.     Positional disorder theories with more than one mechanism of
          melting                                                        300
11.7.     Conglomerate models of melting                                 304
11.7.1.   'Swimming domain' models                                       305
11.7.2.   Conglomerate melting of hydrogen-bonded crystals               305
REFERENCES                                                               307


12. PREMELTING IN CRYSTALS                                               309

12.1.     Precursor effects in the transition from crystal to melt       309
12.2.     Theoretical possibilities for incipient disorder of a crystal lattice on
          approaching $T_f$                                              310
12.2.1.   Two-phase effects (trivial premelting)                         310
12.2.2.   Homophase premelting                                           310
12.2.3.   Phonon premelting                                              312
12.2.4.   Defect premelting                                              313
12.2.5.   Fluctuation premelting                                         314
12.2.6.   Corresponding state theories for precursor effects             316
12.3.     Surface melting of single crystals                             317
12.4.     Superheating of solids above $T_f$                             317
12.5.     Trivial two-phase premelting—the Raoult test                   318
12.6.     Homophase premelting in different types of crystal lattice     320
12.6.1.   Homophase premelting in molecular crystals                     321
12.6.2.   Homophase premelting in ionic crystals                         327
12.6.3.   Homophase premelting in network crystals                       333
12.6.4.   Homophase premelting in metallic crystals                      334
12.6.5.   Thermal macroscopic $vs.$ lattice thermal expansion            334
12.6.6.   Specific heats                                                 334
12.7.     X-ray diffraction                                              335
12.8.     Electromagnetic effects                                        336
12.9.     Thermal conductance                                            338
12.10.    Mössbauer effect                                               338
REFERENCES                                                               339


13. PREFREEZING PHENOMENA IN LIQUIDS                                     342

13.1.     Asymmetry of precursor phenomena on either side of $T_f$       342
13.1.1.   Non-inversion of prefreezing with premelting                   342
13.2.     Crystallizable and anticrystalline clusters in relation to theories of
          cooperative fluctuations                                       343
13.3.     Effects of cluster formation on enthalpy and specific volume   345
13.4.     The 'blocked volume' model for viscosity anomalies             348
13.5.     Cluster formation and the glassy state                         349
13.5.1.   Centrifugal fields and clustering                              350
13.6.     Properties indicating prefreezing in melts of molecular crystals 351
13.6.1.   Viscosity anomalies in polyphenyl melts                        351
13.6.2.   Clustering as a result of association forces                   353
13.6.3.   Prefreezing in melts of long-chain molecules                   354
13.6.4.   Rotational independence $vs.$ clustering of molecules in melts  355
13.6.5.   Correlation of neighbours in melts of inorganic molecules      358

13.6.6. Surface effects in prefreezing of melts of polar molecules 358
13.6.7. Rotokinetic effects 359
13.7. Prefreezing in water 359
13.7.1. Domain or cluster structure of molten ice 359
13.8. Melts of sulphur, selenium, and tellurium 360
13.9. Melts of ionic crystals 361
13.10. Molten metals 362
13.10.1. Structural measurements on molten metals 362
13.10.2. Viscosity and mass transport in metal melts 363
13.10.3. Other properties of molten metals near $T_f$ 367
13.11. Melts of semiconductors 367
REFERENCES 369

14. LIQUID CRYSTALS 373
14.1. Assemblies of partly ordered molecules in melts 373
14.1.1. Ionic mesophase melts 375
14.2. Melting in stages 377
14.2.1. Cholesteric type of partially ordered melts 378
14.3. Some structural relationships in molecular anisotropic melts 379
14.4. Ionic 'liquid crystals' 383
14.5. Disordering in stages above $T_f$—general thermodynamic studies 384
14.6. Domain thermodynamics in liquid crystals 385
14.6.1. Two-component liquid crystals 387
14.6.2 Compound liquid crystals 387
14.7. Transport studies 388
14.7.1. Flow properties in liquid crystals 388
14.8. Precursor effects in oriented melts 389
14.8.1. Precursor phenomena above $T_{cl}$ 391
REFERENCES 392

15. RATE PROCESSES IN THE SOLID–LIQUID TRANSITION 395
15.1. Nucleation and crystal growth 395
15.2. Classical kinetics at the solid–liquid interface 396
15.2.1. Rates of advancement of the crystal-melt interface 398
15.2.2. Dielectric charge accumulation 400
15.2.3. Formation of stepped and spiral structures at interfaces 400
15.3. Rythmic growth from melts 400
15.4. Dendritic growth 401
15.5. Spontaneous nucleation in melts 401
15.5.1. Heteronucleation vs. homonucleation 401
15.6. Melts which fail to nucleate spontaneously 404
15.7. Crystallization of anticrystalline melts through reconstructive changes 405
15.7.1. Some experimental aspects of nucleation from anticrystalline melts 407
15.8. Inverse crystallization temperatures of glasses 409
15.9. Homonucleation catalysts 409
15.9.1. The crystal-melt interface 410
REFERENCES 411

## 16. MELTS AND GLASSES                                               413

16.1.   Glasses as congealed melts                                    413
16.2.   Viscosity criteria for congelation                            413
  16.2.1.   Empirical correlations for viscosity of melts passing into glasses   414
  16.2.2.   Types of glass from different types of melt                415
16.3.   Thermodynamic criteria for glass formation                    418
16.4.   Changes of configurational energy and entropy on forming a glass   423
16.5.   Frozen-in configurational arrangements in glasses             424
16.6.   Compressibility changes accompanying glass formation          426
16.7.   Theoretical estimates of 'ideal' glass temperatures           427
16.8.   Vitrification in relation to molecular structures             428
REFERENCES                                                            430

## 17. MELTING AND CRYSTALLIZATION IN POLYMER SYSTEMS                  432

17.1.   Partly crystallized flexible macromolecules                   432
17.2.   Effects of past history on the melting of polymers            435
17.3.   Melting and molecular structure                               436
  17.3.1.   Effects of the molecular units on the melting of macromolecules   438
  17.3.2.   Transformations in polymers below $T_f$                    439
  17.3.3.   Degree of crystallization                                 440
  17.3.4.   Single crystals of macromolecules                         441
17.4.   Changes of molecular association on melting                   441
17.5.   Melting of macromolecules with mixed sequence of units        442
  17.5.1.   Melting of protein chain molecules                        442
17.6.   Role of configurational entropy in the melting of macromolecules—glassy polymers   443
17.7.   Rates of nucleation and crystal growth in high polymers       444
17.8.   Melting of polymers under pressure                            445
REFERENCES                                                            447

INDEX                                                                 449

# PREFACE

A classical route to the description of liquids by analogy with gases was first explored by van der Waals in 1873. Its mathematical developments still offer tempting opportunities of comparative precision, and of simplicity. However, this quasi-gaseous model is fully effective only for molecules with simple force fields, and not too far from the critical temperature. Very few liquids amongst the many thousands known to chemical physicists and chemical engineers can be adequately described by quasi-gaseous models.

Near the freezing point, distinctive differences between different types of molecules have special influences on their liquid state, which often resemble their influence on the properties of the crystal structures formed on freezing. Furthermore, information about many other physical properties of individual molecules often depends on precise studies of crystal structure. Determinations of structure began with the work of von Laue and of W. H. and W. L. Bragg in 1914, and by now an enormous diversity of substances has been studied. The approach to the study of liquids through the crystalline state of matter and through consideration of what happens to it on melting has become one of the main highways of modern molecular physics.

In order to emphasize basic foundations when developing this approach, it is helpful to refer to the 'molten state' of matter, as a more correct term than either the 'liquid' or the 'fluid' state. The range of factors that can be considered about the molten state of matter turns out to be much greater than in the quasi-gaseous treatment of liquids originated by van der Waals, though of course the two approaches are not in conflict.

In this book on the 'molten state of matter' my aim has been to include what may become effective lines of research in the future, at least by some reference, as well as to describe more fully the well-established present developments.

My own interest in melting in relation to crystal structure has stretched over many years. It began through discussions with Tamman and Eucken in Göttingen, and was strengthened by the personal enthusiasm of W. H. Bragg at the Royal Institution for fresh developments of crystallography. It is pleasant to record the stimulating and tenacious collaboration of numerous students and colleagues.

# 1. GENERAL INTRODUCTION

## 1.1 Some historical aspects of melting

Changes from solid to liquid states of matter are so obvious in nature that even ancient speculative physics might be expected to discuss them. More or less confused descriptions of heat as a mode of motion can be found in general terms in the records of Greek thought (e.g. Plato, Theaetetus). Such notions recurred from time to time in premetrical natural philosophy, whenever atomistic concepts were in fashion. However, as was also true for the rest of physics, definitive advances had to await suitable methods of measurement of the phenomena that accompany melting; in particular, with reference to the volume and heat changes. Black's discovery of the 'latent' heat taken up by water on changing into steam (1762) was followed by his calorimetric measurements of the change from solid to liquid water.

Humphry Davy's experiments to show that the heat generated by rubbing ice against itself sufficed to melt it were important because they could easily be grasped conceptually, in demonstrating the ancient idea that heat was not a fluid (caloric) limited in quantity, but a mode of motion. Michael Faraday devised various elegant demonstrations to show that increased pressure lowered the melting point of ice (1850). Measurements with reasonable precision of the heat changes accompanying melting came fairly late in the development of calorimetry. Bunsen's ice calorimeter gave close correlation between the volume change and the latent heat (1870).

The first applications of thermodynamic arguments to the phenomenon of melting emerged at about the same time as definitive formulations of the Second Law of Thermodynamics by Clausius and W. Thomson (Lord Kelvin) in 1851. These arguments appear to have described the first 'heat cycle' which permitted the application of an engineering law to a wide variety of physical phenomena (J. Thomson, 1849). Though conceptually clumsy, analogous procedures based on engineering concepts were used with considerable virtuosity in various ways in the second half of the nineteenth century, even by originators such as Nernst. However, at the present time the engineering approach is largely superseded. All the results of applying thermodynamics to melting can be compactly presented in mathematical form, as discussed in Chapter 2.

## 1.2 Structural aspects of melting

As is well known, the great power and generality of classical thermodynamics lies in the fact that its applications do not require any knowledge about the structure of the states of matter treated. In fact, some of the most important classical conclusions about melting and freezing were reached before even the atomic structure of matter was universally accepted. At the time of publication of Gustav Tammann's classic work *'Krystallisieren und Schmelzen'* (1903) the relative unimportance of structural considerations is shown by the fact that molecular theory had by no means become generally accepted. Until really powerful methods became available for studying the molecular structure of matter, the thermodynamics of phase changes made remarkable progress by simply ignoring it. Like many later authors, Tammann had as one aim to find 'laws' of melting of the greatest possible generality, in particular with reference to the effects of pressure and temperature.

More refined considerations discussed in this book show, however, that melting is a function of the detailed structure of the crystalline state, and that diverse laws of melting must be looked for because of the diversity of crystal chemistry. Determinations of detailed crystal structure, for example by X-ray crystallography, are of particular relevance for the present book. One of the originators of solid state science, the late Sir William Bragg, once told the author how astonished pioneer X-ray crystallographers had been to find that practically all 'solids' were in fact crystalline. This discovery has gradually introduced a fundamental change in discussions of melting. It has laid renewed stress on the 'ordered' nature of crystals, by contrast with disordered or 'amorphous' states of matter such as liquids, and thus has given additional meaning to investigations of melting in relation to crystal structures. It is indeed no longer profitable in discussing changes from one condensed 'phase' to another to ignore the detailed structure of matter, as was generally done in classical thermodynamics. It should be added that X-ray crystallography, despite its enormous power and precise formulations, is now in its turn proving to be inadequate to elucidate all the structural features of molten states of matter. Examples are discussed in the following pages of melts in which some kind of molecular cluster formation seems evident and whose domain topology needs to be discussed. For adequate understanding of liquids, new experimental and theoretical methods of study are now in great demand.

## 1.3 Melting leading to specific types of liquids

Classification of crystals into broad types according to their crystal chemistry has become more than a mere convenience. With the growth of knowledge, a broad classification of molten states of matter on corresponding lines is essential. After dealing with some of the more general aspects of melting in relation to crystal structure, successive chapters of this book deal with crystals that can be broadly described as

MOLECULAR
IONIC
METALLIC, and
GIANT NETWORK.

Each of these gives rise to its own type of liquid, with subtypes. Structural features that are being progressively elucidated are part of the general theory of the liquid state.

Liquids near their freezing point $T_f$ are conveniently referred to as *melts* to mark the distinction from the same liquids near their critical temperatures $T_{crit}$ for their liquid–gas transition. This distinction though often forgotten can be helpful. Thus many liquids show a gradual change from 'quasicrystalline' behaviour near the melting point, to 'quasigaseous' properties near the critical point. With more detailed experimental knowledge about parameters such as specific heat and thermal expansion the change should become easier to follow.

## 1.4 Theories of melting

A study of melting in relation to crystal structure must aim as far as possible to describe the structures of melts. In an important number of cases such structures can be conveniently referred to a single-crystal lattice as the point of departure, and such melts are often termed 'quasicrystalline'. More complex 'hybrid' or 'conglomerate' examples are also known which can be referred to more than one crystal lattice. By contrast, much of the structure of yet other important classes of melts cannot be referred to any crystalline lattice arrangement. The term 'anticrystalline' is convenient in referring to such instances.

Theories of melting need to discuss the various way in which the entropy of a crystal can be increased on passing to its more randomized melt in terms of the structure of both condensed phases. Although different modes in the total entropy increase $\Delta(S_{\text{Liquid}} - S_{\text{Crystal}})$ obviously interact with one another to some extent, it is convenient to discuss them separately; to a rough approximation the total entropy of fusion $S_f$ can be treated as the sum of the separate contributions

$$S_f = \Delta S_1 + \Delta S_2 + \Delta S_3 \ldots$$

each of which corresponds to a different kind of disordering.

On melting, positional disordering is found for all types of crystals. Orientational disordering arises in many molecular crystals with non-spherical units of structure, as well as in crystals with polyatomic ions. Specific types of crystals show yet other mechanisms of fusion which can be discussed in terms of the structure and ordering in solid and liquid phases. In all, a major part of this book thus provides essential foundations for general theories of the liquid state. It should be added, however, that only the molten state is discussed at all fully in this volume. Properties of liquids near the critical point $T_{crit}$ are often better discussed in terms of quasi-gaseous models with which this book is not primarily concerned.

### 1.5  Precursor anomalies in the solid–liquid transition

With the accumulation of precise experimental data, a number of peculiarities in relation to the transition from one condensed phase (solid) to another (liquid) have become increasingly well defined. Interpretation of these can only be made in terms of accurate knowledge about the structure of both phases. An important group of examples referred to in this book includes precursor effects on either side of the freezing point $T_f$.

There is an important asymmetry of the kinds of theoretical discussions that are most appropriate. Premelting in crystals can be treated in terms of various types of crystal defect. For example, general theories of cooperative thermodynamic fluctuations which have been developed in relation to premelting effects below $T_f$ can be given more or less precise consideration in terms of crystal structure. On the other hand, prefreezing effects above $T_f$ must be interpreted in terms of the formation of more than one kind of domain or 'cluster' in the *melt*, when this is progressively cooled down to and below its freezing point $T_f$. Clusters that are 'crystallizable' are one type. These are the well-known nuclei for spontaneous crystallization that become operative for many melts below a temperature $T_N$ of critical supercooling, characteristic of the structure of any melt. Various aspects of nucleation and growth of crystals from melts are reviewed in Chapter 15. However, in many kinds of melts 'anticrystalline' clusters may also be formed. If their concentration becomes appreciable their presence can lead to prominent prefreezing anomalies, and also promotes easy passage from a melt to a glass. These effects are discussed in Chapters 13 and 16.

Recognition that melting may involve progressive changes in a crystal was probably first made through the discovery of 'liquid crystals', often described as 'mesophases'.

Existence of this group of condensed phases depends on favourable molecular properties which are comparatively very rare. Skilful organic synthesis has promoted the investigation of many anomalous properties of molecular liquid crystals, some of which are reviewed in Chapter 14. An interesting subvariant is presented by ionic liquid crystals formed by melting organic salts, as discussed in Chapter 8.

Understanding of the various kinds of anomalies of melting has led to extensive developments in theories of the structure of solid polymers. Aspects of the melting of polymers closely related to melting processes in other solids are discussed in Chapter 17.

## REFERENCES

J. Black, in *Lectures on the Elements of Chemistry* **1,** 116, Edinburgh 1803. The announcement was made in a lecture given on 23rd April 1762 to the Philosophical Society, University of Glasgow.

Bunsen (1870) *Poggendorf's Ann. Physik* **141,** 1.

M. Faraday (1850) 17 June, Discourse at the Royal Institution. See *Experimental Researches in Chemistry and Physics* (1859), p. 372, Taylor and Francis, London.

G. Tamman (1903) *Krystallisieren und Schmelzen* appeared in various editions. An English translation was published of the 2nd German edition of the States of Aggregation in 1925, published by Van Nostrand, translated by R. F. Mehl.

James Thomson (1849) *Trans. Roy. Soc. Edin.* **16**(5), 575.

A large number of review articles and monographs on the liquid state have been published in recent years. In consequence, the complexity and diversity of the molten state of matter is becoming better understood theoretically. At the same time, few authors yet take the vital step of referring any liquid under discussion to the parent crystal as the systematic starting point. In fact, it is as yet not common to consider liquids generally as 'melts'.

References to specific classes of melts are given in the chapters which follow as appropriate. Some general texts are listed below.

A Munster (1965) *Theory of the Liquid State*, North-Holland Amsterdam.

Faraday Society discussion (1967) *Structure and Properties of Liquids.*

P. A. Eglestaff (1967) *Introduction to the Liquid State*, Academic Press, New York.

G. J. Janz (1967) *Molten Salts Handbook*, Academic Press, New York.

Liquid Metals (1967) *Advances in Physics, Vol. 16*, Taylor and Francis, London.

C. J. Pings (1968) *Physics of Simple Liquids*, North-Holland, Amsterdam.

A. Bondi (1968) *Properties of Molecules, Crystals, Liquids and Glasses*, Wiley, New York.

J. S. Rowlinson (1969) *Liquids and Liquid Mixtures*, 2nd edn., Butterworth.

H. Eyring and J. M. Shik (1969) *Significant Liquid Structures*, Wiley, New York.

Molten Salts (1969) ed. G. Mamantov, Marcel Dekker, New York.

Edgar F. Westrum (1972) *Thermodynamics of Crystals*, Vol. 10 of Physical Chemistry Series, ed. H. A. Skinner, Butterworth.

Non-simple Liquids **31** (1975) ed. I. Prigogine and S. A. Rice, Wiley (England).

This list should not be regarded as exhaustive.

# 2. PHENOMENOLOGICAL THERMODYNAMICS OF MELTING

## 2.1 Classical thermodynamics of the solid–liquid phase change

Quantities which can in principle be readily measured for any state of matter are the specific volume $V$ occupied by unit mass at a specified temperature $T$ and pressure $P$, and the thermodynamic heat content or enthalpy $H$ per unit mass under the same conditions. $H$ must be referred to a suitable zero, e.g. the solid at $0°$ K, and unit mass is often conveniently chosen to be 1 mole of the substance. Provided equilibrium states can be assured, a physical equation can be described for any 'state' of matter, solid, liquid, or gaseous, relating $V$ with $P$ and $T$.

This reservation about equilibrium is most likely to be significant for solid states of matter. One may accept the definition that a solid is able to withstand shear stresses reversibly up to a limit, without flow of matter. However internal stresses can arise in a substance, e.g. as a result of thermal operations to which it has been subjected. When the energy stored as a result is of a magnitude comparable with the other parameters in the equation of state, additional variables have to be introduced (see p. 106). Experience shows that rates of change at the melting point are generally sufficiently rapid to ensure thermodynamic equilibrium (see also Chapter 17) with respect to mechanical stresses.

### 2.1.1. *Frozen-in microparameters*

Pressure and volume are conventional macroparameters which do not need to specify molecular composition. However, when this can vary, 'frozen-in' statistical distributions of microparameters can also affect melting, particularly when the operation of crystallization imposes a degree of selection as to which of the alternative kinds of molecules are packed into the crystal. This could in principle occur, for example, in freezing liquid hydrogen when all the molecules have not attained the *para*-state, or in the freezing of organic melts containing a tautomeric mixture, or mixture of enantiomorphic *d–l* molecules, in cases when the microequilibrium is sluggish.

### 2.1.2. *General thermodynamic equations*

In order to develop relationships between *different* states of the same substance, some thermodynamic characterization must be introduced.

Various thermodynamic functions can be utilized in principle; modern practice has given much attention to the Gibbs free energy $G$ per unit mass. As is discussed in thermodynamic treatises, it is possible to evaluate the free energy of the substance to be characterized for any state of matter in physical (external and internal) equilibrium, using various experimental procedures and appropriate theoretical calculations. Classical equations relating $G$ with other parameters include the well-known relationships which may be particularized for example for the solid state (subscript S):

$$G = H - TS; \quad G_S = f_S(P, T)$$

$$\left(\frac{\partial G}{\partial P}\right)_T = V_S; \quad \left(\frac{\partial G}{\partial T}\right)_P = -S_S; \quad \left(\frac{\partial H}{\partial T}\right)_P = C_{P_S} \qquad (2.1)$$

where

$$\alpha_S = \frac{1}{V_S}\left(\frac{\partial V_S}{\partial T}\right)_P; \quad \beta_S = \frac{-1}{V_S}\left(\frac{\partial V_S}{\partial P}\right)_T$$

refer to the thermal expansion $\alpha_S$ and the isothermal compressibility $\beta_S$ of the solid under discussion. Corresponding equations apply for the melt (subscript L).

Under suitable conditions the same substance may be found in several condensed states. For example, a solid element such as tin can be prepared in polymorphic crystalline forms, tetragonal and cubic respectively, and in a molten state. In principle, classical thermodynamics treats each of these states as quite independent of any other. On this basis an independent set of thermodynamic equations can be written down for each state, and can be evaluated by ignoring the behaviour of all other states.

## 2.2 Equilibrium between phases

For thermodynamic equilibrium between two states of matter, well-known theorems show that the free energy per unit mass must be the same for each state. In mathematical terms, the relationships, for example, between the solid and liquid phases coexisting in equilibrium are given by the condition $G_S = G_L$. Various thermodynamic theorems of importance for melting can be derived from this condition.

For a single substance constituting a one-component system, according to the Phase Rule, the solid, liquid, and vapour phases can coexist at one temperature and pressure only, the triple point. Many, but by no means all, crystals have very low vapour pressures at the triple point, so that their triple point temperatures $T_{trip}$ are practically the same as the melting points $T_f$

under one atmosphere external pressure. For example, for ice

$$T_f = 0.000°C \text{ (by definition)}^*$$

$$T_{trip} = 0.010°C$$

At pressures above the triple point pressure, the vapour phase in a one-component system collapses and the melting point depends on the total pressure applied.

Thermodynamic parameters of predominant practical importance for the interpretation of melting are the volume change $\Delta V_f$ and the entropy change $S_f$ per unit mass. In the geometrical presentation of the free energy surfaces for the solid and liquid states, these quantities represent the intersection of tangent slopes to the free energy surfaces

$$\Delta V_f = \left(\frac{\partial G_L}{\partial P}\right)_T - \left(\frac{\partial G_S}{\partial P}\right)_T \qquad (2.2)$$

and

$$-S_f = \left(\frac{\partial G_L}{\partial T}\right)_P - \left(\frac{\partial G_S}{\partial T}\right)_P \qquad (2.3)$$

In classical phase theory, these free energy surfaces are wholly independent of one another. It is convenient to discuss sections of these surfaces by planes of constant pressure, such as $P = 1$ atm, or constant temperature, e.g. 25°C. Geometrically, equations (2.2) and (2.3) imply a discontinuous difference of angle between the tangents to the solid and liquid free energy curves if they intersect, as illustrated in Fig. 2.1. In principle, if appearance of the alternative phase can be prevented, the free energy of either liquid or solid can be investigated even in ranges of $T$ and $P$ when the other phase is more stable, so that the intersection of the curves can be studied in full experimental detail. In practice, numerous examples of melts supercooled below $T_f$ have been studied, but it is very difficult to superheat solids above $T_f$. Even the superheating of ice 0·3°C above $T_f$ is noteworthy (Käss and Magun, 1961). Reasons for this non-commutative relationship between solid and liquid involve the kinetics of melting and freezing and are discussed in Chapter 15.

---

* The current official definitions of the Celsius scale of temperature are published in 'The International Practical Temperature Scale of 1968' (H.M.S.O., 1969), and are also to be found in the French version given in the 'Comptes Rendus des Séances de la Treizième Conférence Générale des Poids et Mesures, 1967, Annexe 2'.

The International Practical Celsius temperature ($t_{68}$) is defined as

$$t_{68} = T_{68} - 273·15 \text{ K}$$

where $T_{68}$ is the International Practical Kelvin Temperature. The Kelvin is defined as the fraction $1/273·16$ of the thermodynamic temperature of the triple point of water, itself defined as 273·16 K.

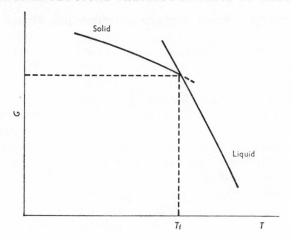

FIG. 2.1. Melting at constant pressure

The free energy surfaces of the solid and liquid separately intersect the free energy surface for the vapour. If the solid and vapour surfaces intersect in a range of temperatures at free energies below the intersection between free energy surfaces for solid and liquid, on raising the temperature direct sublimation occurs in this range without prior appearance of any liquid. Such behaviour implies a relative lack of 'stability' or relatively high free energy of the liquid phase, and is associated with certain types of liquid structures.

Despite their fundamental importance, direct experimental measurements of the entropy change, and more especially of the volume change on melting, are by no means abundant in relation to the great variety of crystal structures now known. Entropies of fusion are most usually calculated from calorimetric measurements of the heat uptake on passing from solid to liquid. Such determinations are made in the course of systematic evaluation of standard entropies. $S_f$ has thus become available for all those substances for which standard thermodynamic functions have been established. Entropies of fusion are discussed in relation to specific types of crystal structures in the chapters which follow.

Particularly for solids with $T_f$ below about 500 K, calorimetry using electrical heat input has yielded information about the change of heat content with temperature, in great detail. With regard to calorimetry, one practical difficulty of relevance to the present chapter arises when there are rate processes (usually in the solid state) involving heat uptake whose relaxation time is comparable with or longer than the duration of the experiment. This may happen particularly in connection with structural solid–solid transformations, which involve movements of the atoms to new positions. 'Drop methods' of calorimetry to some extent avoid difficulties which arise from the slow attainment of equilibrium in condensed phases (e.g. Plester, Rogers, and Ubbelohde, 1956). At temperatures above 300 K drop methods can probably be made as accurate as electrical heat input methods, though unavoidably they furnish less detail (see also Goodkin, Solomons, and Janz, 1957).

Two further general comments may be noted here about entropy changes. When a transformation is accompanied by *hysteresis* (Chapter 4) special considerations have to be introduced in evaluating thermodynamic functions. When marked *premelting or other precursor* phenomena accompany a change of phase, it is advisable to specify whether these should be included or be treated as 'excess' effects in the overall evaluation of quantities such as $S_f$ or $\Delta V_f$ (see Chapters 12 and 13).

Direct evaluations of the volume change are often even less widely available than evaluations of entropies of melting, since they do not at present have the same general practical importance, and also since their determination can involve experimental difficulties. Experimental values of specific volumes of melts are frequently known from other studies on liquids. In favourable cases, $V_L$ is known down to the freezing point, and even below it for supercooled liquids. The thermal expansion of most liquids is in any case a smooth function of temperature. Precise values of $(V_L)_{T_f}$ can thus often be calculated by extrapolation as well as by direct observation. On the other hand, experimental determinations of $V_S$ are not often available up to the melting point. Fortunately, provided marked premelting phenomena can be excluded, known values of the thermal expansion can be extrapolated with moderate confidence up to the melting point to give $(V_S)_{T_f}$. Unfortunately, in certain solids whose melting is of particular interest, premelting by defect formation can vitiate this extrapolation, particularly near the melting point (see Chapter 12). For this reason values of $(V_S)_{T_f}$ determined using fairly extensive extrapolation tend to be too low, except in very special cases. This means that estimated values of $\Delta V_f$ from the difference

$$\Delta V_f = (V_L)_{T_f} - (V_S)_{T_f}$$

then tend to be too large unless precursor anomalies are included by definition.

## 2.3  Derived entropy and volume changes on fusion

Indirect determinations of the *entropy change* on fusion can often be made by adding a second component and using the measured depression of freezing point to calculate heats of fusion. Various difficulties arise, and cryoscopic entropies are not always fully reliable. Examples are quoted in relation to specific crystal types in the chapters which follow.

Indirect determinations of the *volume change* on fusion can be made using the observed dependence of the melting temperature $T_f$ of any suitable substance on the applied pressure, provided a reliable value of the entropy of fusion $S_f$ is available. These are discussed in the next section.

## 2.4 Effects of moderate pressures on melting

### 2.4.1. Volume changes on fusion

A basic thermodynamic relation for correlating phase changes is the Clausius–Clapeyron equation. Applied to the free energy surfaces for solid and liquid, a standard geometrical argument about tangents to surfaces yields the Clausius–Clapeyron equation:

$$\frac{S_f}{(V_L - V_S)} = \frac{dP}{dT} \tag{2.4}$$

Each major advance in experimental techniques for working at higher pressures (e.g. Bridgman, 1941, 1945, 1946) has led to a fresh group of studies of phase changes and melting essentially related to the Clausius–Clapeyron equation. When the pressures used are small and extrapolate down to $P = 1$ atm, $\Delta V_f$ and $S_f$ can be treated as independent of pressure except in comparatively rare instances where there is marked premelting (Chapter 12). Investigations of the melting curve at fairly low pressures have thus provided one of the most general means of calculating the volume change, from the much more generally known entropy change. Unfortunately, pressure coefficients of $T_f$ are somewhat scattered in publications, which often do not even recalculate the data in units that make the volume change readily available (e.g. Deffet, 1951).

For reasons connected with mathematical models of melting processes, it is most convenient to record the fractional change $\Delta V_f / V_S$ rather than the absolute change in specific volume on passing from crystal to melt. For many organic molecules whose non-spherical repulsion envelopes could lead to unusual behaviour on either side of the melting point (see Chapter 6) and that have medium- or long-range intermolecular force fields, this significant melting parameter has unfortunately hardly been explored in relation to crystal structure. Much information about liquids with unusual characteristics thus still waits to be gathered. When the crystal structure is known, the change in volume often gives important clues about the structure of the melt into which it transforms at $T_f$. As theories of the liquid state become more sophisticated, there can be no doubt that this parameter must be made much more widely available and become more widely used, as a guide to the structure of melts.

By way of illustration only, some directly observed values of $\Delta V_f / V_S$ are recorded in Table 2.1 for a homologous series of organic molecules, in which the crystal structures are closely related. A second series of molecules whose molecular structure is related, but whose crystal structure may not be, is recorded in Table 2.2.

**Table 2.1.** *Volume changes on melting for homologous series of long-chain poly-methylene derivatives crystallizing in closely similar structures (Sackmann and Sauerwald, 1950; Sackmann and Venker, 1952)*

| n-Aliphatic paraffins | | n-Aliphatic alcohols | | n-Aliphatic acids | |
|---|---|---|---|---|---|
| $T_f$ (°C) | $\Delta V_f / V_s$ (%) | $T_f$ (°C) | $\Delta V_f / V_s$ (%) | $T_f$ (°C) | $\Delta V_f / V_s$ (%) |
| $C_1$  −184 | 8·69 | $C_1$  −96 | 8·47 | $C_1$  31·3 | 18·96 |
| $C_4$  −139 | 9·10 | $C_2$  −114·2 | 11·14 | $C_2$  16·58 | 19·93 |
| $C_5$  −130·5 | 9·80 | $C_4$  −84 | 12·50 | $C_3$  −22·4 | 12·19 |
| $C_6$  −99·5 | 11·77 | $C_5$  −78·5 | 8·17 | $C_4$  −5·6 | 12·08 |
| $C_7$  −90·5 | 11·47 | $C_6$  −50·5 | 12·79 | $C_5$  −34·5 | 12·14 |
| $C_8$  −57 | 14·77 | $C_7$  −35·5 | 8·44 | $C_6$  −3·6 | 10·61 |
| $C_{20}$  36·5 | 11·4 | $C_8$  −14 | 11·34 | $C_7$  −7·5 | 8·32 |
| $C_{21}$  40·2 | 11·0 | $C_9$  −5 | 9·00 | $C_8$  16·32 | 12·24 |
| $C_{22}$  44·3 | 11·3 | $C_{10}$  4·7 | 12·12 | $C_9$  12·35 | 9·11 |
| $C_{23}$  47·5 | 11·2 | $C_{11}$  24·3 | 9·44 | $C_{10}$  31·5 | 13·16 |
| $C_{24}$  51·0 | 11·2 | $C_{12}$  23·8 | 11·85 | $C_{11}$  30 | 10·72 |
| | | | | $C_{12}$  44 | 14·36 |
| | | | | $C_{14}$  53·8 | 15·01 |
| | | | | $C_{18}$  71·5 | 13·34 |

**Table 2.2.** *Molar volumes at $T_f$ for long-chain molecules*

| Number of C atoms | Paraffins | | Alcohols | | Acids | |
|---|---|---|---|---|---|---|
| | $V_L$ | $V_S$ | $V_L$ | $V_S$ | $V_L$ | $V_S$ |
| $C_1$ | 33·63 | 30·94 | 34·07 | 31·49 | 37·28 | 31·34 |
| $C_2$ | — | — | 50·82 | 45·72 | 57·02 | 47·55 |
| $C_3$ | 59·86 | — | 64·91 | — | 71·23 | 63·49 |
| $C_4$ | — | — | 82·46 | 73·31 | 89·56 | 79·91 |
| $C_5$ | 93·72 | 85·36 | 99·28 | 91·77 | 103·07 | 91·91 |
| $C_6$ | 113·30 | 101·38 | 117·14 | 104·14 | 122·48 | 110·73 |
| $C_7$ | 129·24 | 115·95 | 134·84 | 124·33 | 138·20 | 127·60 |
| $C_8$ | — | — | 153·35 | 137·74 | 157·54 | 140·37 |
| $C_9$ | — | — | 170·94 | 156·81 | 173·51 | 159·03 |
| $C_{10}$ | — | — | 188·69 | 168·29 | 192·40 | 170·02 |
| $C_{11}$ | 201·7 | — | 206·24 | 188·44 | 208·76 | 188·55 |
| $C_{12}$ | 220·2 | — | 224·35 | 200·61 | 228·41 | 199·73 |
| $C_{13}$ | 237·6 | — | — | — | — | — |
| $C_{14}$ | 255·8 | — | — | — | — | — |
| $C_{15}$ | 273·7 | — | — | — | — | — |
| $C_{16}$ | 291·9 | — | — | — | — | — |
| $C_{17}$ | 309·5 | — | — | — | — | — |
| $C_{18}$ | 327·5 | — | — | — | — | — |
| $C_{19}$ | 345·3 | — | — | — | — | — |

It is interesting to note (see also Chapter 7) that in some series of n-alcohols and n-acids both crystals and melts show an alternation of molar volume, as is shown in Table 2.2. The data illustrate a typical problem in attempting to make detailed comparisons of thermodynamic properties in terms of molecular rather than crystal structure. For melting processes very often the appropriate temperature at which to make comparisons is not known.

Further calculation shows (Sackmann and Venker, 1952) that the alternation in the melts is predominantly due to the fact that the temperature of fusion itself shows alternation (Chapter 7). Comparisons of $\Delta V_f / V_S$ thus do not refer to properly chosen 'corresponding temperatures' for successive values of $n$, since the melts with larger $V_L$ also have the larger $T_f$. If the values of $V_L$ for $n$-odd are extrapolated to the virtual melting points given by the mean of $T_f$ for $(n-1)$ and $(n+1)$, the alternation in $V_L$ with $n$ vanishes, whereas the alternation in $V_S$ increases. The phenomenon of alternation can thus be related with structural effects in the crystals (see Chapter 7).

## 2.4.2. Benzene derivatives

For this large group of compounds organic chemists have from time to time considered that melting parameters should show similarities based on the

Table 2.3. *Volume changes on melting derivatives of benzene*

| Molecule | $\Delta V_f / V_s$ (%) | $T_f$ (°C) | $S_f$ (cal mol$^{-1}$ deg$^{-1}$) |
|---|---|---|---|
| Benzene | 13·32 | 5·4 | 8·4 |
| Nitrobenzene | 10·93 | 5·66 | — |
| Cyclohexane | 5·16 | 6·5 | 2·3 |
| $C_6H_5F$ | 12·8 | −41·9 | — |
| $C_6H_5Cl$ | 7·1 | −45 | 7·89 |
| $C_6H_5Br$ | 7·9 | −30·6 | 8·24 |
| $C_6H_5I$ | 7·9 | −31·4 | — |
| $C_6H_5CH_3$ | 11·3 | −95 | 14·80 |
| $C_6H_5NH_2$ | 10·8 | −6·2 | 12·85 |
| $C_6H_5OH$ | 10·9 | −41 | 9·00 |
| $o$-$C_6H_4(CH_3)_2$ | 8·0 | −27 | 12·55 |
| $m$-$C_6H_4(CH_3)_2$ | 9·6 | −47 | 12·47 |
| $p$-$C_6H_4(CH_3)_2$ | 22·1 | 13·2 | 14·11 |
| $\sigma$-$C_6H_4(NO_2)_2$ | 9·45 | 118 | 14·00 |
| $m$-$C_6H_4(NO_2)_2$ | 10·6 | 89·6 | 11·42 |
| $p$-$C_6H_4(NO_2)_2$ | 18·7 | 174 | 15·04 |
| $\sigma$-$C_6H_4(OH)CH_3$ | 9·6 | 30 | 10·99 |
| $m$-$C_6H_4(OH)CH_3$ | 9·8 | 12 | 7·89 |
| $p$-$C_6H_4(OH)CH_3$ | 11·8 | 36 | 9·19 |
| $\sigma$-$C_6H_4(OH)_2$ | 13·3 | — | 14·39 |
| $m$-$C_6H_4(OH)_2$ | 10·0 | 110 | 13·26 |
| $p$-$C_6H_4(OH)_2$ | 13·5 | 173 | 14·61 |
| $\sigma$-$C_6H_4(NO_2)NH_2$ | 12·1 | 71·5 | 11·62 |
| $m$-$C_6H_4(NO_2)NH_2$ | 12·8 | 111·8 | 14·37 |
| $p$-$C_6H_4(NO_2)NH_2$ | 15·1 | 147·5 | 11·56 |

obvious close similarities of *molecular* structure rather than on (often unverified) correlations between *crystal* structures. However this is fallacious. Not surprisingly, Table 2.3 illustrates that any expectation of similarity is only erratically fulfilled. At present it is not possible to determine whether even the similarities that do appear in Table 2.3 also correspond with similarities in crystal structure, but the incomplete correlation between $\Delta V_f$ and $S_f$ makes this doubtful (see Schinke and Sauerwald, 1961).

## 2.5 Melting at very high pressures

With the development of practicable means for applying very high pressures, melting curves have been determined for quite a number of substances up to $10^3$–$10^4$ atm, and for some up to $10^5$ atm or even higher. Theoretical preoccupations and practical considerations have up to the present directed much of the work at the highest pressures to substances with very simple crystal structures, such as the inert gases or metals. It should be remembered that since $T_f$ normally increases with increasing pressure, investigation is necessarily restricted to substances which remain stable up to the temperatures required. Chemical instability can obstruct experimentation with some crystals of particular interest at high pressures (see section 2.9). Melting at high pressures as observed from shock wave data (Horie, 1967) occurs in short time intervals, partly avoiding problems of chemical instability, but it does not as yet give data very consistent with static methods of measurement.

A number of organic molecules are likely to behave reversibly in their melting, even at high pressures, particularly if their repulsion envelope is roughly ellipsoidal and provided they are free from strongly polar groups. Thus tetrahydronaphthalene freezes reversibly up to at least 1100 atm (Bassett, 1939). $T_f = 18°C$ at 3200 kg cm$^{-2}$ where $\Delta V_f / V_S = 0.041$ and $S_f = 1\cdot4$ e.u. approximately. As the density of solid and liquid increase, both $\Delta V_f / V_S$ and $S_f$ show trends with increasing pressure. These depend on the crystal type as well as on the nature of the molecular force fields. No wide generalizations can be conveniently made at the present time. It is interesting to note that the effect of pressure on the melting of polyethylene has been observed up to about 200 kbar (Maeda and Kanetsuna, 1975).

### 2.5.1. *Empirical pressure/melting point relationships*

For crystals with simple structures, it is sometimes useful to express the findings in relation to the semiempirical equation of melting of Simon. This can be written in the form

$$(P - P_0)/a = (T/T_0)^c - 1 \qquad (2.5)$$

in which $P_0$ and $T_0$ are the coordinates of the triple point of the solid, and $a$

and $c$ are constants characteristic of the substance. For most solids $P \gg P_0$ and a simpler equation

$$P/a = (T/T_0)^c - 1 \qquad (2.5.1)$$

may be used. A logarithmic form of this equation

$$\log (P + a) = c \log T + b \qquad (2.5.2)$$

can be used for simple computations of smoothed curves from accurate experimental data. More accurate direct computation of the parameters $a$ and $c$ involves the solution of awkward transcendental equations. Programmed computing has been carried out for over thirty elements and numerous compounds (Babb, 1963). For inert gases a better fit can be obtained using a modified Simon equation

$$P = A(T - T_0)^c - B \qquad (2.5.3)$$

though this introduces an additional empirical parameter (Hardy, Crawford, and Daniels, 1971).

At the origin, the slope of the Simon equation must conform to the Clausius–Clapeyron equation, i.e.

$$\frac{dP}{dT} = \frac{S_f}{\Delta V} = \frac{c(a + P)}{T}$$

Table 2.4 illustrates some of the data for some simple crystals of different types. In principle, it might be expected that, at very high pressures, the effects of forcing the molecules together would depend very much on the class of liquid. For inert gases, or other *molecular liquids*, in the limit, overlap of closed molecular orbits must ultimately lead to electron delocalization with metallic conduction and other metallic properties appearing; though for many substances the pressures required are still out of reach experimentally. For

**Table 2.4. Parameters in Simon's melting equation (Babb, 1963)**

| Crystal | $T_0$ (K) | $a$ (bar) | $c$ |
|---------|-----------|-----------|-----|
| $^4$He | 2·046 | 50·96 | 1·5602 |
| $^3$He | 3·252 | 117·60 | 1·5178 |
| Kr | 115·745 | 2376 | 1·6169 |
| Xe | 161·364 | 2610 | 1·5892 |
| In | 429·76 | 35800 | 2·30 |
| Sn(I) | 505·05 | 57000 | 3·4 |
| Sn(II) | 591 | 14700 | 5·2 |
| Ni | 1726 | 1020000 | 2·2 |
| Pt | 2046 | 1020000 | 2·0 |
| Rh | 2253 | 500000 | 1·30 |
| Fe | 1805 | 1070000 | 1·76 |

substances that are already *metallic* even at $P = 1$ atm, and with simple orbits such as sodium or lithium, the jump in volume on melting steadily decreases. For sodium this decrease in $\Delta V_f / V_S$ has been observed up to 25 000 kg cm$^{-2}$ and probably $\Delta V_f / V_S \rightarrow 0$ at still higher pressures (Ivanov, Makarenko, and Stishov, 1970). It should be stressed, with reference to the possibility of critical melting (see below) that at high pressures and temperatures $S_f$ asymptotes not to zero but to about 80% of the value at $P = 1$ atm. Plots of $S_f$ against $\Delta V_f / V_S$ show no possibility that vanishing will occur simultaneously at high pressures. A closely similar asymptote is claimed for argon as for sodium (Stishov, Makarenko, Ivanov, and Nikolaenko, 1973). A further 'pressure ionization' of the ion cores of alkali metals is anticipated around 600 kbar (Stocks and Young, 1969).

A useful generalization is found when $T_f$ is plotted against $\Delta V_f / V_0$. For many metals the Kraut–Kennedy equation

$$T_f^P / T_f^0 = 1 + c(\Delta V_f / V_0) \tag{2.6}$$

gives a good representation of the linear relationship (Kraut and Kennedy, 1966a,b). For molecular melts the plot is concave upwards (away from the volume axis) and for ionic melts it is concave downwards (Kennedy and Vaidya, 1970) who also give references to correlations of this relationship with other equations describing melting under high pressures).

A transformation of the Kraut–Kennedy relation can also be obtained (Libby and Thomas, 1969) in the form

$$c + \ln(V_L - V_S)_f = \alpha (S_L - S_S)_f \tag{2.6.1}$$

if it is combined with an empirical rule (Stishov, 1967)

$$\frac{d(\ln V_f)}{dT_f} = \text{constant}$$

A logarithmic plot, $\log(T_f^P / T_f^0)$, against $\log V_0 / V$ remains linear over a much wider range of pressures than the Kraut–Kennedy equation. By combining the Lindemann equation $T_f = CM\theta^2 V^{\frac{2}{3}}$ with the Gruneisen law $d\theta / dV = 2(Y - \frac{1}{3}) T_f / V$ the logarithmic equation has been derived in the form (Gopal, Vaidya, Gambhir, and Govindarajan, 1967)

$$\log(T_f^P / T_f) = 2(Y - \tfrac{1}{3}) \log \frac{V_0}{V} \tag{2.6.2}$$

which simplifies into the Kraut–Kennedy equation when $\Delta V_f / V_0$ is small. It may be noted that, to test these and other semiempirical pressure equations, experimental data of high precision are requisite (Babb, 1966). Piezothermal analysis (Ter Minassian and Pruzan, 1977) using high-pressure calorimetry (Pruzan, Ter Minassian, Figuiere, and Swarc, 1976) may permit valuable developments in this direction.

## 2.6 Maxima on pressure/melting point curves

Quite generally, $S_f$ is always finite and positive on melting. The Clausius–Clapeyron equation expressed in the form

$$dT_f/dP = (V_L - V_S)/S_f$$

indicates that if the compressibility of $V_L$ is greater than that of $V_S$ the melting curve will be concave towards the pressure axis and may pass through a maximum.

Although some crystals of simple structure follow equation (2.5) to within about 1%, others depart from it and can show quite complex behaviour at higher pressures. Both $^3$He and $^4$He show minima on the $P$ vs. $T$ melting curves (Straty and Adams, 1966; Grilly, 1971). On the other hand caesium shows two maxima on the melting curve between 0 and 50 000 atm (Kennedy, Jayaraman, and Newton, 1962). At these maxima, since $dT_f/dP = 0$, volumes of crystal and melt are identical, though these two condensed phases have different detailed structures and different entropies.

Various discussions of this interesting behaviour have been proposed. It seems likely that maxima arise for more than one reason, depending on ligand relationships in the melts; this makes it advisable to discuss them in relation to the different classes of liquids separately, but a brief summary can conveniently be given here. By far the most extensive data have been obtained for molten alkali metals and other molten metals. Whereas caesium shows a definite maximum, the metals Li, Na, K, show only a flattening of the rise in $T_f$ vs. $P$ above 60 kbar. Changes in the electrical resistance of the melts in the same region of pressures suggest that a different packing of the ion cores is assumed at the higher pressures (Luedemann and Kennedy, 1968). For other classes of melts, there may also be other reasons for the relatively greater compressibility of the molten state than of the crystals, which appears to be necessary for the appearance of a maximum in the melting curve (Stishov, 1969; Weir, 1971; Yoshida and Kamakura, 1972). Elements that show this effect include Rb, Cs, Ca, Sr, Ba, C, As, Sb, Se, Te, S, and Eu (Kawai and Inokuti, 1968). The elements Ce and Ge melt with contraction in volume and in these melts, at increasing pressures, minima are found instead of maxima (for Ce, see Jayaraman, 1965; for Ge see Vaidya, Akella, and Kennedy, 1969). Maxima have also been calculated from data on certain molten salts including $KNO_2$, $Sb_2S_3$, $Sb_2Te_3$, $Bi_2Te_3$, $NaClO_3$ and $Li_2CrO_4$ (Kawai and Inokuti, 1970) as well as for $CdAs_2$ (Clark and Pistorius, 1974). No detailed discussion has yet been proposed. In the case of molten graphite, interpretation of the greater compressibility of the melt compared with the crystals, which is necessary to account for the maximum observed in the pressure/melting point curve, is further discussed in Chapter 10 (see also Babb, 1963). The peak on the pressure/melting point curve of graphite resembles that for other solids (Korsunskaya, Kamenetskaya, and Aptokar, 1972; Kawai and Inokuti, 1970). The exceptionally high compressibility calculated for liquid carbon ($5 \times 10^{-3}$ l kbar$^{-1}$) appears to be many times higher than that for most other materials (Jayaraman, Klement, and Kennedy, 1963). For the sequence of elements As, Sb, Bi (Klement, Jayaraman, and Kennedy, 1963), the coordination number (which is $3 + 3$ at zero pressure) tends to increase at higher pressures.

### 2.6.1. *Ligand modification at very high pressures*

When a melt is compressed, the electron clouds of neighbouring molecules are forced to interpenetrate to an increasing extent. As a consequence, ligands between the atoms in any molecule may become more or less profoundly modified. Resultant effects depend, broadly, on the class of melt. For molecular liquids, it seems likely that the identity of individual molecules is maintained until a degree of interpenetration is attained which makes actual bond rupture followed by bond rearrangement thermodynamically favourable. Such rupture and rearrangement might not be expected to be reversible. However, this possibility has not been widely studied.

For classes of crystals and melts in which there is already appreciable electron cloud overlap between the molecules, more striking effects may be expected for considerably lower increases of pressure than with molecular liquids. Clearly this applies both to ionic melts, and molten metals. In general, electron delocalization may be expected to increase because of the increased overlap, and the changes are more likely to remain reversible than for molecular liquids.

### 2.6.2. *Interpretation of the Simon equation*

In view of its very general form, various attempts have been made to derive the Simon equation as part of a general theory of melting.

One line of development is to attempt the application of corresponding state theory to melting. Even for the inert gases, the melting curves cannot be made to agree with corresponding state theory using as adjustable variables the $\varepsilon$ attraction and $\sigma$ repulsion parameters of the conventional Lennard-Jones 12–6 potential to describe molecular interaction. Empirical reduction factors $\alpha'$ for temperature and $\beta'$ for pressure, i.e. $\alpha'T$ and $\beta'P$, can however be found which yield agreement for melting curves for the different gases, within the uncertainty of the temperature scale ($0.02°C$) (Michels and Prins, 1962). Even more extensive generalizations of corresponding state theory, when applied to melting (Domb, 1958), encounter the obstacle that only very few crystals such as those of the inert gases comprise units of structure sufficiently simple to satisfy the basic requirements about molecular interactions. Other discussions of the Simon equation have been given (Domb, 1951; De Boer, 1952).

Correlation of the Simon equation must be in principle possible with each of the various theories of melting (see Chapter 3). For example, various vibrational theories have been proposed (Simon, 1953; Salter, 1954) whose success and limitations correspond with those of the underlying theories of melting at 1 atm.

A phenomenological derivation (Voronel, 1958) starts with the Clausius–Clapeyron equation in the form

$$dP/d(\ln T) = H_f/\Delta V_f$$

and makes the empirical but plausible assumption that the ratio of the heat of fusion $H_f$ to the volume change $\Delta V_f$ is a linear function of the pressure i.e.

$$H_f/\Delta V_f = C(P+a) \tag{2.7}$$

where $C$ and $a$ are constants. Insertion into the Clausius–Clapeyron equation followed by integration gives, since

$$dP = d(P+a)$$

$$\frac{1}{c} \cdot \frac{d(P+a)}{(P+a)} = d(\ln T)$$

or

$$\left(\frac{a+P}{a+P_{\text{trip}}}\right) = (T/T_{\text{trip}})^c$$

and since usually $a \gg P_{\text{trip}}$ this becomes

$$(P/a) = (T/T_{\text{trip}})^c - 1$$

The physical interpretation of equation (2.7) involves the fraction $f$ of the latent heat which goes to expand the solid on melting. If the work of expansion $W_f$ on fusion is written ($U$ is the internal energy) as

$$W_f = f(H_f) = \left(P - \left(\frac{\partial U}{\partial V}\right)_T\right)\Delta V_f$$

comparison of coefficients with equation (2.6) shows that

$$c(P+a) = 1/f(P - \partial U/\partial V)_T)$$

indicating as corresponding terms $c = 1/f$ and $a = -(\partial U/\partial V)_T$ which is approximately the same as $a = -(\partial U/\partial V)_0$ for a solid at 0 K.

## 2.7 Melting at very high temperatures

Conditions for a very high melting point of any crystal are well known. Since $T_f = H_f/S_f$, high values of $T_f$ are favoured by high values of $H_f$ and low values of $S_f$. As will be apparent from other sections of this book, high values of $H_f$ are associated with large energy terms required to introduce defects into the crystals. This usually implies large bonding energies between the units of structure in a crystal lattice. Low values of $S_f$ imply a simple mechanism of melting. Refractory solids thus tend to have strong ionic or covalent network bonds between the atoms to keep $H_f$ high, and to have simple structures usually capable of increasing their positional disorder only, to keep $S_f$ low.

Table 2.5 lists some crystals with $T_f > 2500°C$ in various classes of compounds (Quill, 1950; see Nowotny, 1960; Brookes and Packer, 1968; Rudy and Progulski, 1967).

*Table 2.5.  Crystals with $T_f > 2500°C$*

| Compound | $T_f$ (°C) | | Compound | $T_f$ (°C) | |
|---|---|---|---|---|---|
| $Be_3N_2$ | 2470 | | $ThC_2$ | 3050 | |
| BN | 3270 | (sublimes) | VC | 3100 | |
| AlN | 2500 | | CbC | 3770 | |
| ScN | 2920 | | $Ta_2C$ | 3670 | |
| TiN | 3200 | | TaC | 4070 | |
| ZrN | 3255 | | $Mo_2C$ | 2960 | |
| HfN | 3580 | | $Mo_2C$ | 2965 | |
| TaN | 3360 | | $W_2C$ | 3000 | |
| Un | 2900 | | WC | 2900 | (decomposes) |
| $CaC_2$ | 2570 | | UC | 2700 | |
| SiC | 2970 | | $UC_2$ | 2770 | |
| TiC | 3450 | | BaS | 2460 | |
| ZrC | 3805 | | CeS | 2725 | |
| TaC | 3980 | | ThS | 2500 | |
| HfC | 4160 | | TaC | 3980 | |

## 2.8 Range of existence of condensed phases—sublimation

The observation that on heating some solids sublime but do not melt is no more than an accidental consequence of the fact that most investigations are carried out at one atmosphere pressure. Both the solid phase and the liquid phase independently can be in equilibrium* with a vapour phase. Equilibrium is characterized by the equality of free energy between any two phases. In the absence of air there is thus only one temperature, the triple point $T_{trip}$ at

---

* Equilibrium assumes the absence of disturbing factors such as hysteresis.

which each of the free energies

$$G_S = f_S(P, T)$$

$$G_L = f_L(P, T)$$

$$G_G = f_G(P, T)$$

can be the same. However, since the vapour pressures of most solids are low at $T_f$ (see p. 24) the partial pressure of atmospheric gases which makes the total pressure up to one atmosphere shifts the equilibrium temperature $T_f$ between solid and liquid only slightly from the triple point, $T_{trip}$. Special classes of solids are known, on the other hand, for which the vapour pressure of the solids exceeds one atmosphere even before the melting point is reached; under ordinary conditions such solids sublime without melting. Melting can however be observed by increasing the applied pressure till this becomes equal to or greater than the triple point pressure.

Mathematically, relationships between the solid, liquid and vapour phases of a substance can be completely expressed in terms of the free energy functions, provided each phase is in true statistical equilibrium. No knowledge about molecular structure is required in classical thermodynamics; this makes thermodynamic theory one of the most powerful generalizations of chemical physics. However, to obtain more physical insight it is interesting to know what kinds of molecular structures favour direct transition from solid to vapour, and what kinds of structures, in contrast, favour the liquid phase. A natural upper limit to the existence of a liquid phase is set by the critical temperature $T_{crit}$ for the change gas $\leftarrow$ liquid. The particular aspect of general relationships between melting and molecular structure under discussion here is sometimes expressed in terms of the *existence range* $(T_{crit} - T_f)$ for a liquid condensed state of matter. Substances for which the liquid state is favoured have exceptionally long liquid ranges. At the converse extreme others have abnormally short ranges.

Examination of the 'liquid existence range' can in principle only be really informative in terms of structural theories of melting, which are not discussed until later in this book. However, at the present time a great bulk of information about melting (particularly of organic substances) is still in a very empirical state. For this reason, in the present state of knowledge, certain empirical correlations can be most conveniently recorded here in connection with the general phenomenological phase relationships between solid, liquid and gaseous states of a substance.

In quite general terms, the molten state must be regarded as a statistical thermodynamic 'rival' to the crystalline state. Both are 'condensed' phases of comparatively high density, in which intermolecular attraction potentials for the majority of the molecules are near the upper limit set by molecular 'contact' between any pair of molecules. When the data are available, one

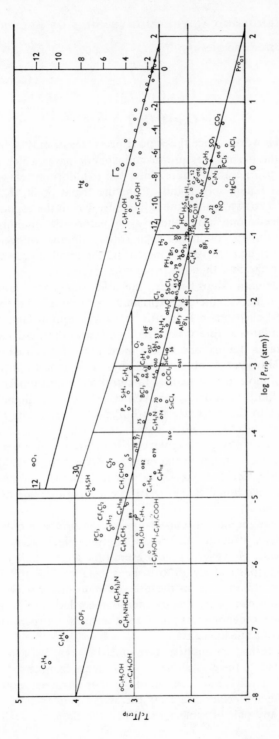

FIG. 2.2. General correlation between the liquid range $T_{crit}/T_{trip}$ and $\log P_{trip}$ (for points with a number but no formula, consult the original publication)

**Table 2.6. Existence ranges of liquids based on the ratio $T_b/T_f$**

*Inert gases*

| | $T_f$ (K) | $T_b/T_f$ (1 atm) | | $T_f$ (K) | $T_b/T_f$ (1 atm) |
|---|---|---|---|---|---|
| Ne | 24 | 1·11 | | | |
| A | 83 | 1·04 | | | |
| Kr | 116 | 1·03 | | | |
| Xe | 161 | 1·03 | | | |

| *Group IA metals* | | | *Group IIA metals* | | |
|---|---|---|---|---|---|
| Li | 459 | 3·5 | Be | 1551 | 2·1 |
| Na | 371 | 3·1 | Mg | 924 | 1·5 |
| K | 335 | 3·1 | Ca | 1019 | 1·5 |
| Rb | 312 | 3·1 | Sr | 1073 | 1·3 |
| Cs | 302 | 3·2 | Ba | 1123 | 1·3 |

| *Covalent halides* | | | | | |
|---|---|---|---|---|---|
| $CF_4$ | 90 | 1·6 | $TaF_5$ | 370 | 1·3 |
| $CCl_4$ | 249 | 1·4 | $TaCl_5$ | 480 | 1·1 |
| $CBr_4$ | 363 | 1·3 | $TaBr_5$ | 513 | 1·2 |
| $CI_4$ | 444 | 1·3 | $TaI_5$ | 640 | 1·1 |
| $SiF_4$ | 183 | 1·5 | $MoF_6$ | 290 | 1·1 |
| $SiCl_4$ | 206 | 1·6 | $WF_6$ | 273 | 1·1 |
| $SiBr_4$ | 278 | 1·5 | $WCl_6$ | 548 | 1·1 |
| $SiI_4$ | 394 | 1·4 | $HgF_2$ | 918 | 1·0 |
| | | | $HgCl_2$ | 550 | 1·0 |
| | | | $HgBr_2$ | 514 | 1·2 |
| | | | $HgI_2$ | 523 | 1·2 |

| *Alkali halides* | | | | | |
|---|---|---|---|---|---|
| NaF | 1268 | 1·6 | KF | 1130 | 1·6 |
| NaCl | 1073 | 1·6 | KCl | 1043 | 1·6 |
| NaBr | 1020 | 1·6 | KBr | 1015 | 1·6 |
| NaI | 935 | 1·7 | KI | 995 | 1·7 |
| RbF | 1048 | 1·6 | CsF | 955 | 1·6 |
| RbCl | 990 | 1·7 | CsCl | 915 | 1·7 |
| RbBr | 950 | 1·7 | CsBr | 905 | 1·7 |
| RbI | 911 | 1·7 | CsI | 894 | 1·7 |

| *Group IB halides* | | |
|---|---|---|
| AgCl | 728 | 2·5 |
| AgBr | 703 | 2·5 |
| CuCl | 703 | 2·7 |
| CuBr | 761 | 2·1 |

| *Alkaline earth halides* | | | | | |
|---|---|---|---|---|---|
| $MgF_2$ | 1536 | 1·6 | $CaF_2$ | 1691 | 1·6 |
| $MgCl_2$ | 987 | 1·7 | $CaCl_2$ | 1055 | 2·2 |
| $MgBr_2$ | 984 | 1·5 | $CaBr_2$ | 1033 | 2·0 |
| $MgI_2$ | 923 | 1·3 | $CaI_2$ | 1013 | 1·5 |
| $SrF_2$ | 1673 | 1·6 | $BaF_2$ | 1626 | 1·5 |
| $SrCl_2$ | 1145 | 2·0 | $BaCl_2$ | 1233 | 1·7 |
| $SrBr_2$ | 926 | 2·3 | $BaBr_2$ | 1120 | 1·9 |
| $SrI_2$ | 788 | 2·3 | $BaI_2$ | 984 | 2·0 |

fairly informative empirical phenomenological correlation (Zernicke, 1950) is to use the ratio $T_{crit}/T_{trip}$ as a measure of the existence range for a liquid. The correlation plot in Fig. 2.2 indicates in general that as the triple point pressure increases the liquid range decreases. The roughly linear correlation indicated in Fig. 2.2 can be valuable in suggesting some outstanding exceptions which can be pointers to unusual liquid structures. Such anomalies can however only be adequately discussed in relation to crystal and defect structures, as examined in later chapters of this book.

Unfortunately, despite the interest and importance of this particular parameter, for many liquids the critical temperature $T_{crit}$ is not yet known. In any case more sophisticated consideration of the existence range of liquids must take into account that a gradual change of behaviour is to be expected as the density decreases. As already emphasized, near the melting point most liquids in fact resemble the crystal state more closely, whereas near the critical point they resemble the gaseous state. It so happens that factors which particularly favour the molten state, regarded as a rival equilibrium assembly in thermodynamic competition with the crystal, lengthen the liquid range at the lower end of the temperature scale, i.e. far away from $T_{crit}$.

Accordingly instead of using the ratio $T_{crit}/T_{trip}$ for which the experimental information is in any case limited, there are theoretical as well as practical reasons for assessing the general comparison between different liquids by comparing the boiling point with the freezing point. Either the dimensionless ratio $T_b/T_f$ where $T_b$ is the boiling point at 1 atm pressure, or the absolute range $T_b - T_f$ may be conveniently compared for different substances. Table 2.6 illustrates some results for typical simple structures.

Some striking differences of liquid stability ranges have been identified on this basis for different types of structure (Weyl and Marboe, 1959). In the

Table 2.7. Values of $T_b/T_f$ and $T_{crit}$ for melts of some polyvalent metal atoms

| Metal | $T_f$ (K) | $T_b$ (K) (1 atm) | $T_b/T_f$ | $T_{crit}$ (K) (est.) |
|---|---|---|---|---|
| Cs | 301·6 | 958 | 3·2 | 2150 |
| Rb | 311·6 | 974 | 3·1 | 2190 |
| K | 335·4 | 1039 | 3·1 | 2440 |
| Na | 370·6 | 1163 | 3·1 | 2800 |
| Bi | 544·4 | 1832 | 3·4 | 4620 |
| Pb | 600·5 | 2024 | 3·4 | 5400 |
| Ga | 302·9 | 2510 | 8·3 | 7620 |
| Sn | 505·0 | 2960 | 5·9 | 8720 |
| Fe | 1808 | 3160 | 1·7 | 10000 |
| U | 1406 | 4200 | 3·0 | 12500 |
| Mo | 2893 | 5100 | 1·8 | 20500 |
| Re | 3440 | 5900 | 1·7 | 20500 |
| Ta | 3300 | 5700 | 1·7 | 22000 |
| W | 3643 | 5800 | 1·6 | 23000 |

same way, for exceptionally long existence ranges, large values of the ratio $T_b/T_f$ point to peculiarities (usually in the structure of the melt) which warrant special investigation. Some of these are discussed in subsequent chapters of this book.

The exceptionally long liquid range of some of the melts of polyvalent metal atoms (Grosse, 1962), such as Ga or Sn, point to thermodynamically favourable properties for the molten state of this kind of assembly. Table 2.7 indicates values of $T_b/T_f$ and $T_{crit}$ which may be compared with other types of melts, e.g. in Table 2.6. [For Ag see Kirshenbaum, Cahill, and Grosse, 1962.]

### 2.9 Thermodynamic criteria of continuous transition from crystal to melt

As ordinarily encountered, melting always involves a discontinuous jump in thermodynamic parameters such as the enthalpy and (usually) the volume. This is true even when there are precursor effects on either side of the discontinuity. However, a fundamental problem in general theories of melting is to determine whether any substances may attain 'critical' conditions for the transition between solid and liquid, at which there is NO discontinuity between these two condensed states of matter, i.e. in formal analogy with critical conditions for the transition between liquid and gas.

Since practically all substances increase in volume on melting, the melting temperature for practically all substances rises as the pressure increases, in accordance with the Clausius–Clapeyron equation (2.4). However the compressibility of the liquid state is normally greater than that of the solid state; the thermal expansion is also normally greater. According to which factor predominates, it could happen that the volume change $\Delta V_f$ on melting gradually decreases for some substances as the applied pressure and thus $T_f$ are increased. In the limit, a value of $T_f$ might be attained for which $\Delta V_f$ is zero. Examples of such behaviour are found, for example, with rubidium and caesium. A few crystalline substances are indeed known (see below) for which $\Delta V_f$ is actually negative at ordinary pressures, so that the condition $\Delta V_f \to 0$, though not often met experimentally, cannot be regarded as wholly exceptional. If the entropy change $S_f$ likewise were to decrease as $P$ and $T_f$ are increased, critical melting with *both* $\Delta V_f$ and $S_f =$ zero could in principle be attained, when the change of phase could be thermodynamically truly continuous.

Empirical tests of this possibility have been made in various ways. In the case of helium, considerable effort has been expended in searching for a critical melting point on the grounds that this substance shows an exceptionally large range of melting points over the range of pressures experimentally available. However, for helium no evidence of approach to critical melting is observed even up to 3000 atm (Dugdale and Simon, 1953). For the other substances various graphical tests have been applied to melting curves

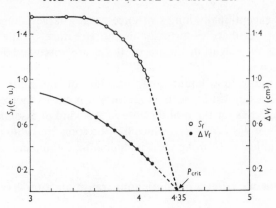

FIG. 2.3. Critical melting test for potassium. Reproduced with permission from Ebert, *Öst. Chem. Zeitung* **55,** 1 (1954)

determined at high pressures (Ebert, 1954). These could be particularly interesting with crystals containing molecules capable of more than one mechanism of melting, as discussed in subsequent chapters of this book. A fairly conventional procedure of searching for critical melting is to plot $S_f$ and $\Delta V_f$ as a function of $P$. In some cases the pairs of curves suggest a common zero, as would be required for critical melting. Figures 2.3–2.5 illustrate plots for metallic potassium, $CCl_4$ and chloroacetic acid, respectively. As will be seen from these Figs., the estimated critical pressure should be within the range of values now attainable experimentally. Unfortunately however,

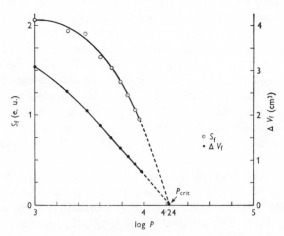

FIG. 2.4. Critical melting test for $CCl_4$. Reproduced with permission from Ebert, *Öst. Chem. Zeitung* **55,** 1 (1954)

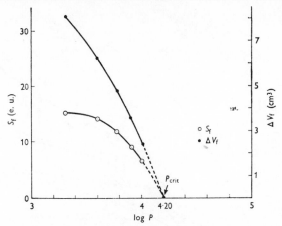

FIG. 2.5. Critical melting test for chloroacetic acid.
Reproduced with permission from Ebert, Öst. Chem.
Zeitung 55, 1 (1954)

experimental determinations of the pressure/melting point curve for potassium (Ponyatovskii, 1958) do not reveal any anomalies or phase transitions up to a pressure of $30\,000$ kg cm$^{-2}$ at which $T_f$ is 251°C. The value of $dT_f/dP$ decreases from $0.016$ K kg$^{-1}$ cm$^{-2}$ at $P = 1$ to $0.003$ K kg$^{-1}$ cm$^{-2}$ at $P = 30\,000$ kg·cm$^{-2}$.

Estimates for potassium at $22\,390$ kg cm$^{-2}$ and 225°C based on Fig. 2.3 lie well below these limits. Direct evaluation of either $\Delta V_f$ or $S_f$ might help to elucidate this discrepancy. The melting curves of rubidium and caesium show some remarkable anomalies; by analogy some information about changes of electronic properties of potassium with pressure might also give useful clues about why critical melting is not attained when expected (see p. 17).

A more discriminating plot, which also permits some estimates of the magnitude of the critical pressures, is to record $S_f$ as a function of $\Delta V_f$. It is necessary to use data obtained as close as possible to the estimated critical point, since the trend of the curve changes with increasing pressure owing to changes in the relative influence of different mechanisms of melting. Figure 2.6 illustrates some of the plots over a considerable range of values of these parameters. The general decrease of $S_f$ at higher densities points to the interesting conclusion that most mechanisms of melting become less favourable under conditions where repulsion potential energies are high. This can be readily understood in the light of structural interpretations of melting discussed in subsequent chapters of this book.

Experimental evidence thus shows that corresponding state principles can in fact seldom be applied to melting. Theoretical considerations about mechanisms of melting in relation to molecular structure and crystal

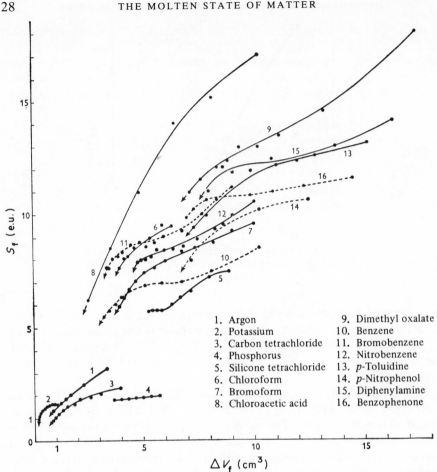

FIG. 2.6. Critical melting plot of $S_f$ against $\Delta V_f$ for molecules with several mechanisms of melting. Reproduced with permission from Ebert, *Öst. Chem. Zeitung* **55,** 1 (1954)

structure, which are examined later in this book, likewise do not point to any critical condition for the solid–liquid transition which can be regarded as really general. The reason is that various mechanisms which may all contribute to the melting process are likely to be diversely affected by increases of molecular density and by increases of temperature. For example, a different relationship would be expected for the inert gases, which exhibit only a single (positional) mechanism of melting, and polyatomic molecules which may have several ways of achieving increases of entropy for the melt. It is interesting to note here that the model for positional disordering leading to melting of crystals developed by Lennard-Jones and Devonshire (1939) points to a critical temperature $T_{\text{crit}} = 1 \cdot 1 \phi_0/k = 1 \cdot 57 T_{f(1 \text{ atm})}$ ($\phi_0$ is a potential energy term in this model) above which the isotherms lose the sigmoid character associated with the appearance of two phases; the pressures calculated to reach this limit should be readily accessible (for argon 2400 atm). These are well below the experimental values up to which no critical transformation has been in fact observed, so that this model obviously needs refinement in this respect (see Chapter 11).

Using revised critical constants for the gas → liquid transition, critical melting parameters for crystals with simple units of structure have been estimated as follows (Furukawa, 1960):

| Crystal | $P_{crit}$(katm) | $T_{crit}$(K) |
|---------|-----------------|---------------|
| He | 11–14 | 65–75 |
| $N_2$ | 22·8 | 302 |
| $CCl_4$ | 17·4 | 631 |
| K | 200–300 | 700–1000 |
| Fe | 3000–4000 | 3000–4000 |

Using as criterion the identity of internal energy $U$ and the product $P_{crit}/V_{crit}$ the following have been estimated (Alder and Jura, 1952) as critical parameters:

| Crystal | $P_{crit}$ (katm) | $T_{crit}$ (K) | $V_{crit}$ (cm$^3$ mole$^{-1}$) |
|---------|------------------|----------------|--------------------------------|
| He | 29 | 90 | 8·0 |
| Ne | 150 | 540 | 9·1 |
| A | 250 | 1840 | 17·4 |

### 2.9.1. Other criteria for critical melting

An alternative approach to critical melting, not specifying volume and entropy, is to calculate conditions under which various other parameters become the same for crystal and melt. A diversity of attempts might be made in principle. At present these all involve extensive extrapolation, but results are not without interest. None are free from theoretical as well as practical criticisms.

A general consideration is that at sufficiently high densities, such as result from the application of pressures of the order of $10^5$ atm or more, electron orbitals of individual molecules can overlap to such an extent that the most weakly held electrons become delocalized. Both crystal and melt may then exhibit metallic properties. Modifications in the trends of $\Delta V_f$ and $S_f$ with changes of density and pressure may be expected above thresholds where metallic behaviour sets in. These have not been at all fully explored, but are likely to affect the possible attainment of critical conditions for melting. This kind of change of bond structure is to some extent foreshadowed by changes of electrical properties of some semiconductors on melting (Chapter 9).

Electron delocalization could profoundly affect the trend of repulsion potential energies with increasing density. For crystalline helium, theoretical calculations applying perturbation theory to the Hartree–Fock equation indicate a transition to the metallic state at about $3 \times 10^7$ atm pressure (Trubitsyn and Ulinich, 1962). However, for other atoms and molecules considerably lower pressures can be foreseen in favourable cases (see Mott, 1961).

## 2.10  Melting of very small portions of solid

*Homomolecular effects.* When melting is discussed as an equilibrium between crystal and melt, the assumption is tacitly made that regions of solid and liquid are so large that contributions from their surfaces to either phase can be neglected. Under these conditions the freezing point $T_f$ is uniquely determined by the bulk properties of the two phases.

Strictly speaking, the thermodynamic equation of state of the surface layers in actual contact with the solid, and forming part of it, must differ from that of the bulk solid, owing to uncompensated molecular interactions at the surfaces. Very little is known about the surface energies of solids, even in contact with their own vapour. Special experiments, such as those on the motion of bodies under pressure through ice, reveal that the surface layers of 'solids' are more mobile under shear than the crystal lattice in bulk (Telford and Turner, 1963) (see p. 34). Differences between surface and bulk properties are much better understood for liquids than for solids, even for somewhat indirect quantities such as surface free energy and surface entropy. Various physical conditions arise, in which $T_f$ for the bulk material is displaced by what are known as surface effects. A broad separation is usefully made between the discussion of homomolecular and heteromolecular surface effects on melting.

Homomolecular effects are best known at the present time from their role in the spontaneous formation of crystal nuclei capable of growth in a supercooled melt, or in the occurrence of other kinds of (anticrystalline) cooperative fluctuations. These involve the concept of a free energy $\sigma_{SL}$ contributed to the system by the interface between crystal and melt and proportional to its area. Nucleation phenomena are further discussed in Chapter 15. When a solid is present as very small particles, suspended in the melt, or covered by a skin of the melt, the lowering of their freezing point is sometimes calculated from the Thomson equation.

$$T_f = -\frac{2\sigma_{SL}}{P_S} \cdot \frac{M}{S_f}$$

where    $M$ = molecular weight,
$S_f$ = entropy of fusion per mole,
$P_S$ = density of solid, and
$r$ = radius of particle.

However, this equation and various refinements proposed for it (Hanszen, 1960) can hardly be tested until direct experimental methods for evaluation of $\sigma_{SL}$ have been devised. For metals diverse semiempirical proposals have been made for calculating $\sigma_{SL}$ (Löhberg, 1972):

$$\sigma_{SL} = 0\cdot26\,Ya\ \mathrm{erg\ cm}^{-2}$$

where $Y$ is the shear modulus and $a$ is the separation of the atoms. Rumball and Kondic (1968) suggest

$$\sigma_{SL} = 0\cdot179 H_f / V^{\frac{2}{3}}\ \mathrm{erg\ cm}^{-2}$$

where the heat of fusion $H_f$ is in $\mathrm{erg\ cm}^{-3}$ and $V$ is the molar volume. This important but elusive parameter $\sigma_{SL}$ is not likely to follow any such simple generalizations for crystals giving rise to classes of melts other than metals, but no extensive discussion has been made (see also pp. 244, 315, 402).

Melting of small particles of lead and of bismuth (Peppiatt and Sambles, 1975) has been studied experimentally; the findings suggest that the effect of particle size is controlled by more than one dominant parameter such as $\sigma_{SL}$. It may be supposed that the structure of the melt if complex (Chapter 13) will depend on the dimensions of the region it occupies; use of a single interface parameter may not be realistic. Another awkward complication arises from the consideration that crystal faces with different indices will normally have different surface free energies, so that a crystal in free space must eventually take up an equilibrium shape with the areas $A_1$, $A_2 \ldots A_p$ of its different faces with respective surface free energies for unit area $G_1$, $G_2 \ldots G_p$ such that $\Sigma A_p G_p$ is a minimum (e.g. Shuttleworth, 1950). Basically, the atoms or molecules at an interface are less well ordered than at the interior of a crystal, but detailed calculations of resulting differences of free energy etc. need more information to be successful. Application of computer techniques of calculation offer some attractive prospects.

### 2.10.1. *Melting at grain boundaries*

Structural theories of cooperative defects such as dislocations in crystals show that the boundaries between grains in polycrystalline material frequently are loci of defects in high concentration. Since the melt is more disordered than the crystal (Chapter 6) transition to the liquid state might be expected to occur more readily by the more highly disordered material in the immediate proximity of such boundaries and at lower temperatures $T_f^*$ than the melting point $^0T_f$ for fully ordered crystalline materials (see Chapter 12).

By establishing a temperature gradient in polycrystalline specimens, one end of which is liquid and the other solid, lowered melting temperatures $T_f^*$ at grain boundaries have been verified for aluminium (Chaudron, Lacombe, and Yannacquis, 1948) and have also been reported for tin. These effects are interesting but like other influences of crystal disorder on melting (e.g. Chapter 12) they are subject to practical difficulties of experimentation. For example impurities not soluble in the crystals, but soluble in the melt, tend to be squeezed out and to be localized at grain boundaries, and would likewise give $T_f^* < {}^0T_f$. Though it would be of considerable theoretical interest, fully definitive evidence which proves that cooperative crystal defects (and not impurities) are responsible for this lowering of freezing point at grain boundaries is difficult to establish.

An exceptionally low activation energy for migration of atoms or molecules at grain boundaries provides further indirect evidence of lowered $T_f^*$ for such regions in crystals. To some extent, boundary thicknesses over which anomalous thermal and transport behaviour can become prominent in crystals will depend on the interatomic forces. Because of the ease with which they form well-compacted polycrystalline masses, and simple cooperative defects such as dislocations, crystals of metals provide a rich field for phenomena pointing to thermally activated material at grain boundaries (e.g. Wever, 1959). It may well be, however, that the 'plastic state' of molecular crystals of organic substances provide even more versatile examples of the phenomena to be studied (see p. 139). This warrants further study.

### 2.11 Heteromolecular effects—melting of sorbed layers on surfaces

Since crystal stability is determined by long-range order, normally extending over many unit cells of crystal structure in any direction, some modification of $T_f$ might be anticipated when the 'crystalline' and 'liquid' phases are present only as multilayers adsorbed on a substrate. Instead of constituting regions of condensed phases extending in three dimensions, the substance in such sorbed layers may extend in two directions and be only a few atoms or molecules thick normal to the adsorbing surface. In a truly 'floating' two-dimensional film, unlike a three-dimensional crystal, it becomes impossible to specify the positions to within one lattice spacing of particles distant from the chosen centre of reference, because of the amplitude of thermal vibrations. A 'floating' two-dimensional film may be expected to show extended short-range order, but melting will not be discontinuous (Dash, 1973). Experimentally studied examples always refer to melting of films 'anchored' on a substrate, whose interactions with the sorbed molecules make a distinctive contribution to the melting process; usually a sharp peak is then observed, with marked precursor effects on either side. Instances worked out in considerable detail include calorimetric studies of nitrogen sorbed titanium dioxide in layers 2–5 molecules thick (Morrison, Drain, and Dugdale, 1952). These suggest that such thin layers melt not far from the temperature $T_f$ for extended crystals. However as the layers get thinner the sharpness of the transition from crystalline to molten state decreases progressively, as is illustrated in Fig. 2.7. Detailed interpretations of such effects in terms of models of melting (see Chapter 11) have not yet been proposed. It is noteworthy that with different sorbed films anchored to the same substrate the sharpness of melting processes and the extent of precursor effects varies considerably, and in ways that often await further elucidation. For molecules

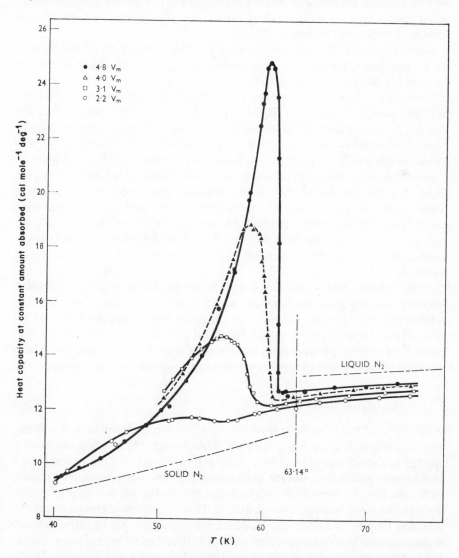

FIG. 2.7.  Heat capacity of $N_2$ sorbed on titanium dioxide. Reproduced by permission of the National Research Council of Canada from the *Canadian Journal of Chemisty* **30,** 890–903 (1952)

such as neon sorbed on graphon the melting peak lies close to that of neon crystallized in three dimensions (Huff and Dash, cited in Dash, 1973). For hydrocarbons of various kinds sorbed on graphon, progressive freezing is apparent (Findberg, 1971). Polar organic compounds sorbed on graphon show evidence of marked lateral interaction between the sorbed molecules, as well as with the substrate (Findenegg, 1972, 1973). Bromine and also xenon when sorbed on graphite show clearly marked phase transitions in addition to melting (Lander and Morrison, 1967).

Phase transitions have been reported for argon adsorbed on graphite at 65 K, but structural identification of the changes taking place is not yet satisfactory (Jura and Criddle, 1951).

Methods of studying changes in surface structure when a melt freezes have not been widely investigated. It is interesting to note, for example, that the change in over-potential for evolution of hydrogen at a mercury surface changes somewhat on freezing (Bockris, Parsons, and Rosenberg, 1951). However the structural implications have not been extensively developed.

In studies of the melting and freezing of sorbed molecules it must be kept in mind that the surface of the sorbing medium may exert strong orienting forces, sufficient even to modify lateral forces in the surface layers and even the polymorphic structure assumed by the sorbate. X-ray evidence for ice sorbed on silica gel (Brzhan, 1959), as well as theoretical consideration, indicate that the usual arrangement of $H_2O$ molecules in ice is disturbed on the walls.

Both benzene and water when sorbed on porous glass (Vycor) show a lowering of transition temperature and spread of the liquid→solid phase transformations. Details of the experimental findings are however by no means clear in all cases. It would be useful to correlate information about changes of several properties in the freezing region to elucidate the effects reported (Loisy, 1941; Hodgson and McIntosh, 1960).

### 2.11.1. *Regelation of ice*

Ice is one of the (very few) crystals that melt with a contraction of volume. This is attributed to its open network structure (p. 282). When pressure is applied to a small region in a block of ice at its melting point, for example by a loop of wire pulled by a weight, portions of the solid actually under pressure melt and flow to above the wire. Accordingly, the wire is slowly pulled through the block without ever cutting it. This penomenon, known as regelation, has long been known (p. 1). Accurate measurements of rates give rise to considerable problems of interpretation (Telford and Turner, 1963; Frank, 1967; Nunn and Rowell, 1967; Nye, 1967; Townsend and Vickery, 1967). An unresolved question is to what extent the layers of ice in the immediate vicinity of the free surface have the same thermodynamic and mechanical properties as for ice in bulk. As pointed out, little is known about the surface free energy $\sigma_{SL}$ of a crystal with respect to its own melt, and presumably the structure of the layers of the crystal at the free surface may differ considerably from the structure at the interior. In layers of water adjacent to biological materials, changes of its structure have often been suggested (see Drost Hansen, 1969) but for ice in bulk the evidence appears to be uncertain. A liquid-like layer several hundred molecules thick has been suggested on the basis of contact experiments between spheres (Jellinek, 1967). As an estimate for $\sigma_{SL}$, which has as yet not been measured independently (see Chapter 15), Kuhn suggests (disregarding the sign of the difference between densities of solid $\rho_S$ and of melt $\rho_L$)

$$\sigma_{SL} = \{(\rho_S - \rho_L)/\rho_L\}\sigma_{LG}$$

This would make

$$\sigma_{SL} \sim 10\% \, \sigma_{LG}$$

for water, which gives fair agreement with the mechanical properties of the gels when these indicate comparable crystallite sizes. (see p. 37).

## 2.12 Melting in capillaries

Somewhat similar investigations have dealt with the freezing of liquid retained in microcapillaries, for example in silica gel or ferric oxide gel. Making the simplifying assumption that all the capillaries have the same radius $r$, and that the contact angle between liquid and walls is zero, the lowering of vapour pressure of the liquid is given by the well-known Kelvin equation, which may be written in the form

$$\ln p/p_0 = -\frac{2\sigma_{LG}}{\rho_L r} \cdot \frac{M}{RT} \tag{2.8}$$

where

$\sigma_{LG}$ = the surface tension liquid/vapour,

$M$ = molecular weight

$\rho$ = density, and

$r$ = radius of capillary involved.

Use of this equation permits the parameter $r$ to be calibrated directly with the substance whose modified freezing is to be investigated.

If the material when still solid in the capillaries remains unaffected by any curvature or other influence of the walls, its vapour pressure presumably must be equated with that of the bulk solid phase. On this hypothesis, the freezing point depression for the sorbed material is related to the lowering of vapour pressure by the equation:

$$\ln p/p_0 = -\frac{S_f}{R} \cdot \frac{\Delta T}{(T_f - \Delta T)}$$

By making observations on various adsorbents saturated to varying extents with sorbed molecules, breaks in the vapour pressure curve and hence in $\Delta T$ have been reported in relation to the degree of saturation. Data for benzene on silica gel, alumina gel, bentonite, and charcoal support this interpretation (Puri, Singh, and Myer, 1957).

On the other hand, for water adsorbed on silica gel, and for dioxan adsorbed on ferric oxide gel, it has been claimed that the vapour pressure of the *solid* phase is likewise modified by the curvature of its free surface (Batchelor and Foster, 1944) leading to an equation analogous to (2.8), but replacing the term $\sigma_{LG}$ by the term $\sigma_{SG}$. Ethylenediamine sorbed by silica gel likewise shows a marked lowering of freezing point which is interpreted in a similar way (Brown and Foster, 1952). The apparent contradiction between the two possibilities for the frozen solids does not yet appear to have been definitely resolved. Possibly highly polar molecules which are strongly influenced by forces attributable to the walls show one type of behaviour, and less polar molecules, such as benzene, the other. However in that event, the effects of forces emanating from the capillary walls on the thermodynamic parameters of crystal and melt would have to be allowed for, in addition to the effects of the 'free' surfaces of solid and liquid introduced in the conventional equation (2.8).

In the region of space surrounded by the walls of a capillary, uncompensated forces whose magnitude may well be comparable with or even considerably greater than intermolecular forces in the bulk crystal may be acting on and orienting molecules in the capillary. In that case liquids are likely to have their structural arrangement of molecules profoundly altered whilst they are in the capillary. Freezing phenomena may be modified accordingly, and sharp freezing points may even be replaced by gradual immobilization of the molecules, thus resembling freezing in polymers (Chapter 17). Dielectric studies on water absorbed on alumina, for example, indicate that $H_2O$ molecules in the capillaries pass from liquid to the glassy state only gradually (Dransfield, Frisch, and Wood, 1961). Freezing phenomena for caged microregions of a substance can become more clearly focussed when a synthetic sorbent is used in which the cavities have uniform size. For example $H_2O$ molecules in a synthetic zeolite (Linde 5A) show no freezing peak at all with a water content less than $0·233$ g $H_2O$ g$^{-1}$ zeolite, whereas a clear peak is observed at about $-30°C$ for water contents between $0·233$ g and $0·28$ g per g of zeolite, with a calculated heat of fusion of about $36$ cal g$^{-1}$. A second peak observed between $-5°C$ and $0°C$ with water

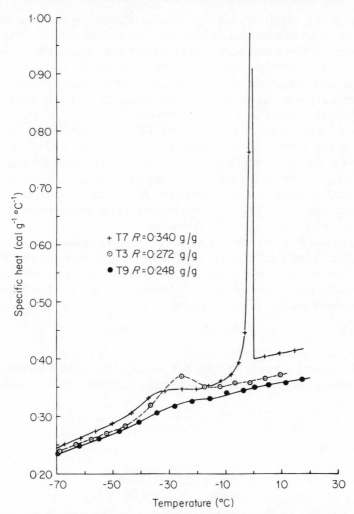

FIG. 2.8. Specific heat is plotted against temperature for
zeolite at the water contents given on the figure. Reproduced
with permission from Haly, *J. Phys. Chem. Solids*, **33**, 131
(1972)

contents between 0·28 and saturation at 0·306 g g$^{-1}$ is attributed to freezing
of films sorbed on the external surfaces of the zeolite particles (Haly, 1972).

## 2.13 Melting of microcrystals in gels

Various high polymers exhibit the phenomenon of swelling when immersed
in diverse solvents. One way of investigating the influence of size of crystal on
the freezing point is to follow the freezing curves of solvents in such swelled

gels, (e.g. Kanig, 1960). Particularly detailed information has been obtained in the case of hydrophilic gels such as polyvinyl alcohol and polyacrylic acid, or mixtures of these two components (Mayer and ·Kuhn, 1961). Swelling proceeds until it reaches from 75 to 96% by volume of water content, depending on the neutralization procedures used; all small ions are removed by dialysis, and the vapour pressure over such swelled gels is practically indistinguishable from that of the pure solvent.

The relative fraction $\theta$ of water frozen at different temperatures below 0°C leads to an excess heat capacity. Plots of $d\theta/dT$ are illustrated in Fig. 2.9 for the undissociated gel in which the acid has not been neutralized, and for the dissociated gel in which the network of polymer is about half the mesh size. Careful studies show that the depression of freezing temperatures is not wholly reversible owing to irreversible 'tearing' of the internal network structure resulting from the growth of ice crystals.

Standard calculations of the depression of the freezing point are related to a characteristic edge length for the crystallites, making their volume $V = a^3\phi$ and their surfaces $S = 6a^2\psi$ where $\phi$ and $\psi$ are geometrical parameters which depend on the crystal habit. For cubic crystallites $\phi = \psi = 1$. Changing the size of crystallites from $a$ to $(a + da)$ requires work

$$\sigma_{SL}\,dS = 12a\psi\sigma_{SL}\,da$$

Following the usual thermodynamic argument, the number of moles of material changing phase is $dn = 3a^2\phi(\rho/m)\,da$ where $\rho$ is the density of the crystals and $m$ the gaseous molecular weight and this involves a work term

$$RT\ln\!\left(\frac{p_{Smicro}}{p_{Smacro}}\right) \times dn$$

giving

$$\ln\!\left(\frac{p_{Smicro}}{p_{Smacro}}\right) \approx \frac{\Delta p}{p} = \frac{12\psi\sigma_{SL}}{3a\phi\rho}\frac{m}{RT}$$

To calculate the lowering of freezing point, application of the Clausius–Clapeyron equation to liquid and solid yields as final outcome, equating freezing point differences

$$\Delta T_f = -\frac{4\sigma_{SL}}{S_f\cdot\rho}\frac{m\psi}{\phi}$$

### 2.13.1. Effect of capillaries on nucleation

It may be noted that just as capillary sorption modifies $T_f$, the temperature of spontaneous nucleation $T_N$ (see Chapter 15) may also be affected (Takamura, 1958). This possibility at present remains speculative.

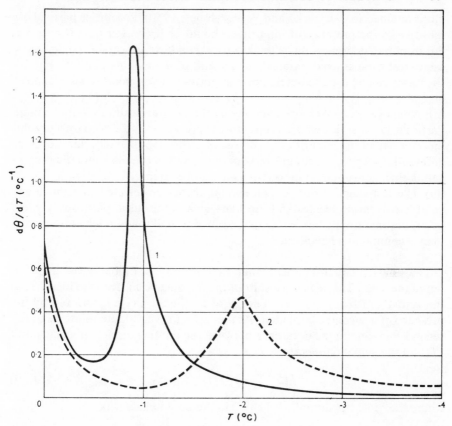

FIG. 2.9. Fractional change $d\theta/dT$ of water frozen in a gel of polyvinyl alcohol/poly-acrylic acid: curve 1, dissociated gel; curve 2, undissociated gel. Reproduced by permission of Akademische Verlagsgesellschaft, Wiesbaden

## 2.14 Effects of impurity on melting

### 2.14.1. *Impurities not soluble in the crystalline phase*

Studies of effects of impurities on the melting of solids have been of considerable historical importance for the general development of physical chemistry. Conventional criteria for purity are well known (e.g. Timmermans, 1958). This book is not primarily concerned with experimental methods, but it should be noted that, for stringent purification, using thermodynamically reversible situations, the following conditions should be fulfilled:

(i) Each operation of purification should be continued until the fractions rejected agree asymptotically in properties (within experimental error) with those retained (e.g. Rybicka and Wynne Jones, 1950). Properties particularly sensitive to the presence of impurities should be used as criteria if possible. For example, in organic melts the electrical conductance can be a much more useful test for dissolved water, than any shift of the melting point $T_f$. Dielectric losses can be used to detect polar impurities in non-polar solvents, and so on.

(ii) Whenever possible more than one method of purification should be used. Particularly for solids whose crystal forces are weak, and whose own crystals therefore tolerate impurities more readily, the use of molecular sieves or differential sorption on a solid sorbent may give a more discriminating segregation than conventional thermodynamic procedures.

(iii) The last stage of purification should always be selected to remove any volatile solvent molecules and the like—small molecules, particularly when polar, become trapped at crystal defects in 'purified' solids much more readily than is commonly recognized.

The simplest situation arises when the impurity is soluble in the melt but not in the solid. This leads to a shift of the freezing point from the ideal $^0T_f$ to the actual $T_f$ because the free energy of the liquid phase is lowered by the presence of a second component. At the limit, for very dilute solutions of the second component, the difference between the limiting or ideal freezing point $^0T_f$ and the actual temperature $T_f$ is given by

$$H_f(1/^0T_f - 1/T_f) = R \ln N_1 \qquad (2.9)$$

where the mole fraction of impurity 2 in the liquid phase 1 is

$$N_2 = 1 - N_1$$

Where $N_2$ is small, equation (2.9) approximates to the Raoult form $H_f(1/^0T_f - 1/T_f) = -RN_2$. As is well known, this expression is derived on the basis of ideal mixture laws for the two components in the liquid phase. The standard derivation is conveniently given here in view of its development for solid solutions discussed in Section 2.15.

At equilibrium, the partial molal free energy of either component must be the same in the solid and liquid phase, i.e.

$$\bar{G}_1 = [\bar{G}_1] \quad \text{and} \quad G_2 = [\bar{G}_2]$$

(square brackets refer to the solid phase).

When component 2 is practically insoluble in crystals of 1, $[N_2] = 0$; and when $N_2$ is small in the melt, approximately $\bar{H}_1 = H_1$ where $H_1$ is the molar heat content of the pure liquid and also

$$\bar{S}_1 = S_1 - R \ln N_1$$

Thus

$$\bar{G}_1 = \bar{H}_1 - T\bar{S}_1 = H_1 - T(S_1 - R \ln N_1)$$

and

$$[\bar{G}_1] = [H_1] - T[\bar{S}_1]$$

are equal at a temperature $T_f$ such that

$$-T_f\{[\bar{S}_1] - S_1 + R \ln N_1\} = H_1 - [\bar{H}_1]$$

For the pure solid

$$[\bar{H}_1] = [H_1] \quad \text{and} \quad [\bar{S}_1] = [S_1]$$

Also

$$H_f = H_1 - [H_1]$$
$$S_f = H_f/{}^0T_f = S_1 - [S_1]$$

where ${}^0T_f$ is the limiting ideal melting point in the absence of the second component. Substituting and rearranging gives equation (2.9).

Precision studies of freezing point depressions show that this simple derivation can become inadequate in a number of ways. For additives with long-range forces, as in mixed ionic melts, the assumptions made break down at surprisingly low mole fractions (Chapter 8). For molecular crystals with molecular impurities equation (2.9) can usually serve as a working approximation when $N_2 \ll 1$ especially when molecular weights are not too dissimilar. Network crystals (Chapter 10) probably do not obey the relations at all, and other principles must be applied to account for the effects of impurities on freezing. This is also the case for 'crystals' of polymers (Chapter 17).

Application of equation (2.9) has been used for many cases where evaluation of $H_f$ by direct calorimetry is inconvenient or impracticable. By using sensitive thermometry, as is now readily feasible, accurate measurements of freezing point depressions might thus yield a very general means for evaluating entropies of fusion. Unfortunately, some of the crystals of particular interest are unlikely to behave in a simple way when impurities are present. Since it is not fully informative, cryoscopy is never to be recommended when entropies of fusion can be determined by direct calorimetry. Some of the difficulties that arise in special cases include the following.

(1) The species undergoing dissolution may not be accurately known so that $N_2$ in equation (2.9) is not accurately specified. This difficulty arises particularly when the dissolving species can undergo association or dissociation, whose extent then needs to be evaluated by some other means.

As examples of associating species, molecules such as organic acids readily form dimers in solvents such as benzene. Freezing point depressions give entropies which tend to be too low for this reason. It may be anticipated that many molecules forming hydrogen bonds behave in this way, but systematic investigations are scanty.

Examples of dissociating species are provided by a large number of inorganic salts. When dissolved in water, these can in the majority of cases be treated as undergoing complete dissociation in very dilute solutions, so that equation (2.9) applies with a doubling of the number of particles acting as the impurity. However, electrostatic interaction between ions of opposite sign begins to be appreciable at quite low concentrations in aqueous solution. Correspondingly, in solutions of lower dielectric constant than water, appreciable departure from ideal behaviour, i.e. complete dissociation, begins at even lower concentrations, since electrostatic forces are stronger than in water at comparable separations of the ions. Thus ionic species generally give ideal depressions of freezing points only at extremely dilute mole fractions. When comparing different solvents, approximate quantitative threshold concentrations at which departures from ideal freezing point depressions become experimentally appreciable can be derived in principle, for example from the Debye–Hückel theory of strong electrolytes. Unfortunately, a well-known difficulty in this and other theories of electrolyte solutions is that the values of the local or microdielectric constants $D_1, D_2 \ldots$ etc. to be inserted in comparative calculations for different solvents are not necessarily the same, or even in the same sequence as the macroscopic dielectric constants as determined for fairly large volumes of the fluids. This discrepancy arises because the structure of the fluid in the immediate neighbourhood of an ion may differ very markedly from the structure of the pure liquid, generally in the sense that the dielectric constant is lowered in the neighbourhood microregion.

(2) A second obstacle in evaluating entropies of fusion from freezing point depressions arises for salts for which the structure of the melt may not correspond with the stoichiometric formula, when this assumes complete dissociation into positive and negative ions. For example, if a divalent salt such as lithium sulphate $Li_2SO_4$ is dissolved in molten lithium nitrate $LiNO_3$ it is not evident without experimental trial whether the 'impurity' behaves as $2Li^+$ and $SO_4^{2-}$ or as $Li^+$ and $(LiSO_4)^-$ and whether the solvent behaves as $LiNO_3$ or $[Li(NO_3)_2]^-$ and $Li^+$ or $Li^+$ and $NO_3^-$; various other ion associations may become prominent in other cases. Other aspects of ion association in melts are considered in relation to melting mechanisms of salts (Chapter 8).

When the amount of added impurity exceeds 1–2%, practically all two-component systems show considerable departures from the elementary equation (2.9). Complete phase diagrams can be calculated only in terms of thermodynamic activities. This book does not aim to give a complete discussion of phase-rule diagrams for even two-component systems. Special problems in relation to the melting of crystal compounds are however further examined in later chapters in relation to molecular structures which dissociate on melting.

## 2.15 Melting of Solid solutions

When phase diagrams are plotted over the whole range of compositions in a two-component system, at either extreme the most usual behaviour is for each pure substance to have its freezing point depressed by the other. Crystals of either form separate practically pure from the melt until the freezing point curves intersect at the eutectic temperatures, below which any composition of the two substances freezes to a conglomerate of crystals of the two pure substances.

As has long been known, the second component may however be appreciably soluble in the crystalline as well as in the molten state of the substance whose depression of freezing point is under examination. When this happens, free energies of both solid and liquid are affected by the dissolved impurity. Analytical representation of solid solution thermodynamics can start from the same basis as when component 2 is effectively insoluble in crystals of 1. The most obvious modification is to write for the solid lattice, by analogy with the liquid (see above)

$$[\bar{H}_1] = [H_1]$$

and

$$[\bar{S}_1] = [S_1] - R \ln [N_1]$$

The solubility equation now becomes

$$H_f(1/{}^0T_f - 1/T_f) = R \ln N_1/[N_1] \tag{2.9.1}$$

introducing as it were a distribution coefficient $k$ for the impurity between crustal and melt, where

$$k = (1 - [N_1])/(1 - N_1) = [N_2]/N_2$$

When $N_1$ and $[N_1]$ are both small, approximately

$$R \ln N_1/[N_1] = R\{(N_2] - N_2\} = RN_2(k - 1) \tag{2.9.2}$$

Although this formal analytical treatment is quite straightforward a number of instances are known for which it is clearly insufficient. Often, this arises because the assumption that the partial molal heat content $[\bar{H}_1]$ is practically the same as $[H_1]$ and remains effectively unmodified by the impurity in solid solution is far from adequate.

One group of crystals whose solid solutions have been studied with special references to changes in $[\bar{H}_1]$ is presented by the polymethylene hydrocarbons (n-paraffins). These show remarkable melting behaviour in mixtures

(Oldham and Ubbelohde, 1940; see also Chapter 7). When the chain length of a pair of these molecules does not differ by more than one or two carbon atoms, the partition coefficient $k$ is close to unity. In such a case, the solidus and liquidus practically coincide and the freezing point of one component is *raised* even by minimal additions of the other. As a specific example, solid solutions are formed between the pair of very closely related molecules n-hexadecane and n-hexacedecene-1 for which the freezing point obeys the linear equation $T_f = N_1{}^0 T_{f_1} + N_2{}^0 T_{f_2}$. Other similar systems have also been discovered.

Another class of crystals for which the assumption $[\bar{H}_1] = [H_1]$ proves to be inadequate arises where (for at least one of the pure substances) potential energy barriers oppose any orientational randomization at a theoretical temperature $T_c$ which for the pure crystals lies above $T_f$. (This is further discussed in Chapter 6). When a second molecule dissolves in the crystal, this may lower such barriers sufficiently to permit orientational disorder below $T_f$. Usually such disorder is then smeared over a range of temperatures, and the heat of fusion at $T_f$ is itself lowered.

Possibilities of this kind are only likely to arise when the partition coefficient $k$ of an impurity between melt and host crystal is near unity, which in turn would indicate that the impurity can pack into the host crystal lattice without introducing large repulsion terms. One example studied in some detail is presented by crystalline benzene. Various molecules that are analogues of benzene may lower $[\bar{H}_1]$ for this crystal appreciably, by promoting rotation in the crystal below $T_f$ (Thompson and Ubbelohde, 1950).

As is further discussed in later chapters, organic molecules offer rich possibilities for investigating the limiting differences in size and shape of repulsion envelopes which still just permit cocrystallization of two components in solid solution or in an ordered crystal compound. Pairs of globular molecules form solid solutions very readily (Timmermans, 1950), whereas, for long-chain molecules, increasing the size of side-chain substituents soon restricts the range of cocrystallization (Oldham and Ubbelohde, 1940).

Apart from some isolated instances, effects of solid solution on rotational barriers have as yet not been widely studied with particular reference to the solid–liquid transformation. The phenomena are somewhat easier of access in the case of solid–solid transformations: a number of instances are known where the presence of a solute shifts the peak temperature $T_c$ for such a transformation to a marked extent. As a classic example, effects of krypton on the lambda transformation in crystalline methane are further referred to in Chapter 4. Krypton atoms both lower the peak temperature of transformation and broaden its range (Eucken and Veith, 1938). Again, introducing $(NH_4)^+$ in solid solution in potassium nitrate causes the lattice to expand, and helps to stabilize a crystal structure which is only metastable for pure

$KNO_3$ (Coates and Crewe, 1961). Related consequences may be anticipated for $T_f$ but have not yet been reported.

### 2.15.1. Melting of crystals of isotopic molecules

In the course of experiments on effects of isotopic substitution on thermal transformations in crystals, and on other molecular parameters generally, a fairly substantial range of phenomena has been observed which concerns actual melting of isotopic atoms and molecules. Though complete interpretation can sometimes be quite complex, it is of considerable interest to note that change of isotopes, thus changing molecular masses and moments of inertia (but keeping the simpler kinds of crystal force fields constant), often hardly affects melting points. Some characteristic results are recorded Table 2.8. The melting point of $^6LiNO_3$ appears to lie about 0·03°C above that of $^7LiNO_3$ (Janson and Lunden, 1970) but no unique attribution of the cause has been made.

**Table 2.8. Melting parameters of isotopic atoms and molecules**

| Atom or molecule | $T_f (K)$ | $H_f (cal\,mol^{-1})$ | Ref. |
|---|---|---|---|
| $CH_4$ | 90·64 | 224·0 | Clusius and Popp, 1940 |
| $CH_3D$ | 90·42 | 217·5 | Chester and Dugdale, 1954 |
| $CD_4$ | 98·78 | 215·7 | Chester and Dugdale, 1954 |
| $H_2$ | 13·95 | — | Clusius and Weigand, 1940 |
| $D_2$ | 18·65 | — | Clusius and Weigand, 1940 |
| $^{20}Ne$ | 24·66 | 79·23 | Clusius, Flubacher, Piesbergen, Schleich and Sperandio, 1960 |
| $^{22}Ne$ | 24·84 | 79·74 | Clusius, Flubacher, Piesbergen, Schleich and Sperandio, 1960 |
| $^{14}N_2$ | 63·14 | 55·62 | Clusius, Sperandio, and Piesbergen, 1959 |
| $^{15}N_2$ | 63·19 | 56·61 | Clusius, Sperandio, and Piesbergen, 1959 |

In principle, even with the simplest structures, isotope replacement affects the structure by its influence on zero-point energies (Ubbelohde, 1936) and by its influence on vibrational energy and entropy (Chapter 3). Next to helium, probably one of the largest differences in the melting of isotopes is presented by hydrogen and deuterium, included in Table 2.8, for 1 atm pressure. At higher pressures, up to about 2800 kg cm$^{-2}$, plots of melting pressures as a function of melting temperatures remain parallel for the two isotopes, with a displacement of about 17 kg cm$^{-2}$ for the deuterium in the direction of the positive pressure axis (Chester and Dugdale, 1954). This parallelism implies that the Clausius–Clapeyron ratio $S_f/\Delta V_f$ is the same for the two isotopes.

Using a quasicrystalline model for the melts and a Debye model for vibrations of the crystals, attempts have been made to calculate the quantum effects in the melting of the isotopic molecules $^{20}$Ne and $^{22}$Ne (Toda, 1958). Theories of melting (Chapters 3 and 11) are not yet sufficiently precise to make full use of the very refined comparisons obtainable from experimental data on isotopic atoms and molecules. In principle, some of the most sensitive tests of melting theories should be based on isotope effects.

### 2.15.2. *Melting of helium*

Compared with other crystals, a very large amount of work has been carried out on the melting of helium. This crystal shows the peculiarity that the liquid phase is more stable than the solid right down to 0 K, until the pressure is raised above about 26 atm. The low mass of the atom introduces large quantum terms, which makes the melting of helium far from typical for simple solids. For the development of quantum mechanical theories of condensed phases, it is fortunate that comparisons can be made between isotopes $^3$He and $^4$He, which have quite different quantum mechanical

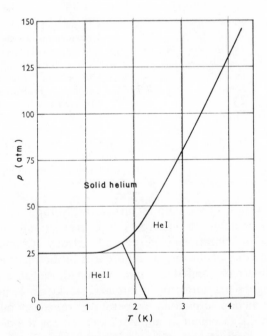

FIG. 2.10. Solid–liquid diagram for $^4$He. Reproduced by permission of Macmillan (Journals) Ltd

characteristics. Many of the peculiarities of liquid helium at very low temperatures involve mechanical or thermomechanical properties not normally included in theories of the equilibrium between crystalline and molten phases.

Discussions of many highly specialized aspects of liquid helium would be inappropriate in relation to the general survey of melting which determines the scope of this book. Melting of helium can however be reviewed briefly in thermodynamic terms. Figure 2.10 illustrates that $P$, $T$ curves for the melting of $^4$He. It will be noted that the solid–liquid curve cuts the pressure axis near 25 atm.

At the other extreme of pressure, hopes that helium would provide an example of 'critical' melting at sufficiently high pressures (see section 2.9) have not been realized. Owing to the predominant role of vibrational zero-point energy in the thermodynamic properties of solid and liquid helium around 1 K (Simon and Swenson, 1950 and see Chapter 3) for this particular melting transition $S_f \to 0$ as $T_f \to 0$ but $\Delta V_f$ remains finite. This stresses the need to satisfy *both* the criteria $S_f \to 0$ and $\Delta V_f \to 0$ to make sure that the liquid and solid show no discontinuities of properties and become indistinguishable at a critical melting point. Solid and liquid helium may both show anomalous behaviour even at pressures not too high above the melting pressure (Goldstein, 1961).

The way in which the entropy difference between solid and liquid $^4$He sinks to zero at 0 K, whereas the volume difference does not, is clearly apparent from Table 2.9.

**Table 2.9.  *Properties of $^4$He along the melting curve***

| $p$ ($kg\,cm^{-2}$) | $T_f$ ($K$) | $\Delta V_f$ ($cm^3\,mole^{-1}$) | $V_{liq}$ ($cm^3\,mole^{-1}$) | $S_f$ ($cal\,deg^{-1}\,mole^{-1}$) |
|---|---|---|---|---|
| 26·48 | 1·350 | 2·046 | 23·030 | 0·170 |
| 29·88 | 1·723 | 1·706 | 22·448 | 0·840 |
| 40·00 | 2·046 | 1·324 | 21·507 | 1·142 |
| 60·02 | 2·540 | 1·241 | 20·495 | 1·260 |
| 79·04 | 2·947 | 1·164 | 19·741 | 1·316 |
| 99·99 | 3·355 | 1·089 | 19·104 | 1·350 |
| 125·21 | 3·835 | 1·038 | 18·489 | 1·374 |
| 175·05 | 4·688 | 0·986 | 17·573 | 1·402 |
| 1120·3 | 14·746 | 0·6073 | 12·752 | 1·704 |
| 1422·8 | 17·158 | 0·5630 | 12·173 | 1·724 |
| 1778·6 | 19·774 | 0·5277 | 11·666 | 1·746 |
| 2134·2 | 22·210 | 0·4970 | 11·225 | 1·764 |
| 2347·7 | 23·602 | 0·4752 | 11·031 | 1·774 |
| 2417·8 | 24·050 | 0·4814 | 10·974 | 1·777 |
| 2930·1 | 27·190 | 0·4567 | 10·512 | 1·798 |
| 3555·6 | 30·770 | 0·4300 | 10·115 | 1·820 |

These data (see Grilly and Mills, 1959) also illustrate how at very high pressures $\Delta V_f$ decreases progressively but $S_f$ shows no sign of doing so, implying that critical melting is not in sight at high temperatures either.

In its melting mechanism $^3$He has a different behaviour from $^4$He. Its atomic nucleus, unlike $^4$He, carries a spin. This introduces additional degrees of freedom in the crystal and melt, with the possibility of a contribution to the entropy of fusion from changes in spin ordering on melting. Part ordering of these spins is already found at $0.4$ K in the liquid; it is not fully known how the process of crystallization affects this possible contribution to the overall entropy of fusion. Comparisons between $^3$He and $^4$He up to 3500 kg cm$^{-2}$ reveal a number of striking differences in the melting of these isotopic atoms (Grilly and Mills, 1959).

(i) At equal pressure, the volume change on melting $^3$He is consistently lower than that of $^4$He.

(ii) In the solid state, $^3$He shows a transformation from b.c.c. to h.c.p. as the pressure and temperature are increased. A corresponding transformation in $^4$He is much less prominent on the phase diagram (Schuch, Overton, and Brout, 1963).

(iii) At low pressures, the entropy of fusion is considerably greater for $^3$He than for $^4$He. This difference is gradually overtaken and reversed as is indicated in Table 10.

## 2.16  Melting of polymorphs: crystals yielding the same melt

Systematic discussion of specific aspects of the melting of molecular crystals is presented in later sections of this book, when groups of molecules showing similar behaviour are considered in relation to structures in which they crystallize. However, it may be noted here that the melting of polymorphs, particularly in organic chemistry, attracted the attention of physical chemists concerned with molecular structure long before thermodynamic arguments included any reference to the structure of the phases. Brief mention may be usefully made here of some of the historic conclusions arrived at when there was little or no reference to crystal structure.

When more than one crystal state of a substance can be found, classical thermodynamic considerations show that if they all melt to give the same liquid, the most stable polymorph (which therefore has a lower free energy) will have a higher melting point. This conclusion follows from the location of intersections of the free energy surfaces for the two solids with that for the liquid, to give two melting plots of $P$ against $T_f$. Examples of such phenomena are reported from time to time; verification is difficult unless the rate of transformation from one crystal form to the other is slow, compared with the rate at which either melts.

### Table 2.10. Properties of $^3He$ along the melting curve

| $p$ $(kg\,cm^{-2})$ | $T_f$ $(K)$ | $\Delta V_f$ $(cm^3\,mole^{-1})$ | $V_{liq}$ $(cm^3\,mole^{-1})$ | $S_f$ $(cal\,deg^{-1}\,mole^{-1})$ |
|---|---|---|---|---|
| 51·61 | 1·332 | 1·0366 | 23·700 | 0·842 |
| 53·17 | 1·375 | 1·0360 | 23·575 | 0·856 |
| 64·06 | 1·659 | 0·9700 | 22·767 | 0·930 |
| 69·28 | 1·783 | 0·9435 | 22·450 | 0·955 |
| 79·00 | 2·000 | 0·8880 | 21·917 | 0·988 |
| 91·27 | 2·255 | 0·8633 | 21·360 | 1·018 |
| 99·94 | 2·425 | 0·8523 | 21·015 | 1·033 |
| 100·00 | 2·427 | 0·8488 | 21·012 | 1·033 |
| 110·86 | 2·630 | 0·8153 | 20·625 | 1·048 |
| 118·85 | 2·775 | 0·8063 | 20·355 | 1·056 |
| 125·16 | 2·887 | 0·7850 | 20·152 | 1·060 |
| 128·40 | 2·943 | 0·7955 | 20·055 | 1·063 |
| | | $\beta$-Solid | | |
| 146·29 | 3·252 | 0·8868 | 19·543 | 1·186 |
| 160·13 | 3·490 | 0·8766 | 19·230 | 1·204 |
| 175·01 | 3·735 | 0·8583 | 18·935 | 1·223 |
| 204·57 | 4·205 | 0·8250 | 18·388 | 1·259 |
| 237·43 | 4·732 | 0·8066 | 17·848 | 1·299 |
| 1208·8 | 14·689 | 0·5617 | 12·991 | 1·617 |
| 1449·2 | 16·592 | 0·5394 | 12·511 | 1·648 |
| 1707·2 | 18·518 | 0·5231 | 12·123 | 1·669 |
| 2098·6 | 21·256 | 0·4880 | 11·595 | 1·695 |
| 2543·0 | 24·158 | 0·4664 | 11·140 | 1·714 |
| 2986·8 | 26·887 | 0·4373 | 10·800 | 1·727 |
| 3554·8 | 30·184 | 0·4179 | 10·398 | 1·738 |

For some crystals, transformation in the solid state between two polymorphs may be so slow that true thermodynamic equilibrium may never be reached with respect to one or even both of these forms, in their relationships with the melt. This appears to be the case for 1,2,4-trimethylbenzene for example, which shows evidence of melting points near 229.3 K (solid I) and 224.2 K (solid II), but for which the abnormalities are not completely resolved (Putnam and Kilpatrick, 1957). Structural differences in these polymorphs do not appear to have been elucidated.

With tetraethyl derivatives of tin and lead, an unusually large number of polymorphs has been observed; melting points are all comprised within a narrow range of temperatures (Staveley, Paget, Goalby, and Warren, 1950; Staveley, Warren, Paget, and Dowrick, 1954). For tin compounds, ten polymorphs have been claimed, melting from 137·4 K to 147·1 K. For the lead compounds at least six polymorphs with melting points ranging from 135·6 K to 141·5 K have been detected. Corresponding tetramethyl derivatives do not show polymorphism of this kind, which likewise appears to be absent in methyl and ethyl derivatives of carbon, silicon, or germanium.

This instance of polymorphism has been attributed to configurational isomerism (see also Chapter 7) of the tetraethyl tin and lead compounds. Owing to the increasing length of the heteroatomic bond in the sequence C—C (1·54), C—Si (1·93), C—Ge (1·98), C—Sn (2·18) and C—Pb (2·29 Å), when the carbon atom forms part of a methyl or ethyl group, rotation about this bond becomes progressively freer as the atomic number of the heteroatom increases. Thus in methyl derivatives of tin and lead such rotation does not lead to different configurational isomers which are stable. With the ethyl derivatives, the groups obstruct one another and probably different isomeric configurations of moderate stability crystallize in the different polymorphs. It appears to some extent to be a matter of chance which isomer is nucleated in spontaneous crystallization. However, in the melts configurational equilibrium occurs quite rapidly.

Molecules of soaps such as sodium palmitate exhibit at least five successive phase changes below the temperature of appearance of the isotropic melt ($T_f < 290°C$) (Vold and Vold, 1939). Molecular configurations, crystal packing and crystal defect formation may all contribute to the explanation for this exceptional variability of the most stable structure as the temperature is raised, and until crystal structural differences have been elucidated discussion remains rather superficial (see p. 222).

When the molecules in the crystal have simple structures it is sometimes possible to use these to characterize the relative stability of polymorphs with different melting points, and to combine it with accurate information about their thermodynamic parameters. Thus for $COCl_2$ (Giauque and Ott, 1960), there are three polymorphs whose respective melting points are I, 145·37 K, II, 142·09 K, and III, 139·19 K. At the melting point the first solid is the stable form. Solid III has a higher entropy than solid II by 1·08 e.u. Solid II, though less stable than I, appears to be less disordered than III.

## REFERENCES

B. J. Alder and G. Jura (1952) *J. Chem. Phys.* **20**, 1491.
S. E. Babb, Jr. (1963) *Rev. Mod. Phys.* **35**, 400.
S. E. Babb, Jnr. (1966) *Phys. Rev. Lett.* **17**, 1250.
J. Bassett (1939) *C.R. Acad. Sci., Paris* **208**, 169.
R. W. Batchelor and A. G. Foster (1944) *Trans. Faraday Soc.* **40**, 300.
J. O'M. Bockris, R. Parsons, and H. Rosenberg (1951) *Trans. Faraday Soc.* **47**, 766.
P. W. Bridgman (1941) *J. Chem. Phys.* **9**, 794.
P. W. Bridgman (1945) *Am. J. Sci.* **243A**, 90.
P. W. Bridgman (1946) *Rev. Mod. Phys.* **18**, 1.
C. A. Brookes and M. E. Packer (1968) in *Special Ceramics*, ed. P. Popper.
M. J. Brown and A. G. Foster (1952) *Nature, Lond.* **169**, 37.
V. S. Brzhan (1959) *Kristallografiya* **4**, 631.
G. Chaudron, P. Lacombe, and N. Yannacquis (1948) *C.R. Acad. Sci., Paris* **226**, 1372.

P. F. Chester and J. S. Dugdale (1954) *Phys. Rev.* **95**, 278.

J. B. Clark and C. W. F. T. Pistorius (1974) *High Temperatures—High Pressures* (*G.B.*) **5**, 319.

K. Clusius and L. Popp (1940) *Z. Phys. Chem.* **46B**, 63.

K. Clusius, A. Spreandio, and U. Piesbergen (1959) *Z. Naturforsch., Ergebn. Tieftempforschung* **14A**, 793.

K. Clusius, P. Flubacher, U. Piesbergen, K. Schleich, and A. Sperandio (1960) *Z. Naturforsch.* **15A**, 1.

K. Clusius and K. Weigand (1940) *Z. Phys. Chem.* **46B**, 1.

R. V. Coates and J. M. Crewe (1961) *Nature, Lond.* **190**, 1190.

J. G. Dash (1973) *Progr. Surface Membrane Sci.* **7**, 95.

J. De Boer (1952) *Proc. Roy. Soc. Ser. A* **215**, 4.

L. Deffet (1951) *Chim. Ind.* **64**, 290.

C. Domb (1958) *Suppl. Nuovo Cimento* **9**, Vol. 1, 9.

K. Dransfeld, H. L. Frisch and E. A. Wood (1961) *J. Chem. Phys.* **36**, 1574.

W. Drost Hansen (1969) *Ind. Eng. Chem.* **61**, 10.

J. S. Dugdale and F. E. Simon (1953) *Proc. Roy. Soc., Ser. A* **218**, 291.

L. E. Ebert (1954) *Ost. Chem. Ztg.* **55**, 1.

A. Eucken and H. Veith (1938) *Z. Phys. Chem.* **38B**, 393.

G. H. Findenegg (1971) *J. Colloid Interface Sci.* **35**, 249.

G. H. Findenegg (1972) *J. Chem. Soc. Faraday Trans. I* **68**, 1799.

G. H. Findenegg (1973) *J. Chem. Soc. Faraday Trans. I* **69**, 1069.

F. C. Frank (1967) *Phil. Mag.* **16**, 1267.

K. Furukawa (1960) *Nature, Lond.* **188**, 569.

W. F. Giauque and J. B. Ott (1960) *J. Am. Chem. Soc.* **82**, 2689.

L. Goldstein (1961) *Phys. Rev.* **122**, 726.

J. Goodkin, C. Solomons, and G. J. Janz (1957) *Rev. Sci. Instrum.* **29**, 105.

E. S. R. Gopal, S. N. Vaidya, R. D. Gambhir, and K. Govindarajan (1967) *J. Indian Inst. Sci.* **49**, 61.

E. R. Grilly and R. L. Mills (1959) *Ann. Phys. (New York and London)* **8**, 1.

A. V. Grosse (1962) *J. Inorg. Nucl. Chem.* **22**, 33.

A. R. Haly (1972) *J. Phys. Chem. Solids* **33**, 129.

K. J. Hanszen (1960) *Z. Phys.* **157**, 523.

W. H. Hardy, R. K. Crawford, and W. B. Daniels (1971) *J. Chem. Phys.* **54**, 1005.

C. Hodgson and R. McIntosh (1960) *Can. J. Chem.* **38**, 958.

Y. Horie (1967) *J. Phys. Chem. Solids* **28**, 1589.

V. A. Ivanov, I. N. Makarenko, and S. M. Stishov (1970) *Sov. Phys. Solid State* **12**, 7.

B. Janson and A. Lunden (1970) *Z. Naturforsch.* **25A**, 697.

A. Jayaraman (1965) *Phys. Rev.* **137A**, 179.

A. Jayaraman, W. Klement, and G. C. Kennedy (1963) *Phys. Rev. Lett.* **10**, 387.

H. G. Jellinek (1967) *J. Colloid Interface Sci.* **25**, 192.

G. Jura and D. Criddle (1951) *J. Phys. Colloid Chem.* **55**, 163.

G. Kanig (1960) *Kolloid Z.* **173**, 97.

M. Käss and S. Magun (1961) *Z. Kristallogr.* **116**, 354.

N. Kawai and Y. Inokuti (1968) *Japanese J. Appl. Phys.* **7**, 989.

N. Kawai and Y. Inokuti (1970) *Japanese J. Appl. Phys.* **9**, 31.

G. C. Kennedy, A. Jayaraman, and R. C. Newton (1962) *Phys. Rev.* **126**, 1363.

G. C. Kennedy and S. N. Vaidya (1970) *J. Phys. Chem. Solids* (*G.B.*) **31**, 2329.

A. D. Kirshenbaum, J. A. Cahill, and A. V. Grosse (1962) *J. Inorg. Nucl. Chem.* **24**, 333.

W. Klement, Jr., A. Jayaraman, and G. C. Kennedy (1963) *Phys. Rev.* **131**, 632.

I. A. Korsunskaya, D. S. Kamenetsky, and I. L. Aptokar (1972) *Fiz. mokel Metalloved* **34** (5), 942.
A. Kraut and G. C. Kennedy (1966a) *Phys. Rev. Lett.* **16**, 608.
A. Kraut and G. C. Kennedy (1966b) *Phys. Rev.* **151**, 668.
J. S. Lander and J. Morrison (1967) *Surface Sci.* **6**, 1.
J. E. Lennard-Jones and A. F. Devonshire (1939) *Proc. Roy. Soc., Ser. A* **170**, 464.
L. M. Libby and F. J. Thomas (1969) *J. Phys. Chem. Solids* **30**, 1237.
K. Löhberg (1972) *Z. Metall.* **63**, 1.
M. Loisy (1941) *Bull. Soc. Chim. Fr.* **8**, 587.
H. D. Luedemann and G. C. Kennedy (1968) *J. Geophys. Res.* **73**, 2795.
Y. Maeda and H. Kanetsuna (1975) *J. Polym. Sci. Polym. Phys. Ed.* **13**, 637.
H. Mayer and W. Kuhn (1961) *Z. Phys. Chem. N.F.*, **30**, 289.
A. Michels and C. Prins (1962) *Physica (Utrecht)* **28**, 101.
J. A. Morrison, I. E. Drain, and J. S. Dugdale (1952) *Can. J. Chem.* **30**, 890.
N. F. Mott (1961) *Phil. Mag.* **6**, 62.
H. Nowotny (1960) *Bull Soc. Chim. Fr.* 1881.
K. R. Nunn and D. M. Rowell (1967) *Phil. Mag.* **16**, 1281.
J. F. Nye (1967) *Phil. Mag.* **16**, 1249.
J. W. H. Oldham and A. R. Ubbelohde (1940) *Proc. Roy. Soc., Ser. A* **176**, 50.
S. R. Peppiatt and J. R. Sambles (1975) *Proc. Roy. Soc., Ser. A* **345**, 387, 401.
D. W. Plester, S. E. Rogers, and A. R. Ubbelohde (1956) *J. Sci. Instrum.* **33**, 211.
E. G. Ponyatovskii (1958) *Kristallografiya* **3**, 508.
P. Pruzan, L. Ter. Minassian, P. Figuiere, and M. Swarc (1976) *Rev. Sci. Instrum.* **47**, 66.
B. R. Puri, D. D. Singh, and Y. P. Myer (1957) *Trans. Faraday Soc.* **58**, 530.
W. E. Putnam and J. R. Kilpatrick (1957) *J. Chem. Phys.* **27**, 1075.
L. L. Quill (1950) *Chemistry and Metallurgy of Miscellaneous Materials (Thermodynamics)*, McGraw-Hill, N.Y., London, Toronto, p. 40.
E. Rudy and J. Progulski (1967) *Planseeber. Pulvermet.* **15**, 13.
H. Rumball and V. Kondic (1968) in *The Solidification of Metals*, ISI Publication **110**, 149.
S. M. Rybicka and W. F. K. Wynne Jones (1950) *J. Chem. Soc., London* 3671.
H. Sackmann and F. Sauerwald (1950) *Z. Phys. Chem. (Leipzig)* **195**, 4.
H. Sackmann and P. Venker (1952) *Z. Phys. Chem. (Leipzig)* **199**, 100.
L. Salter (1954) *Phil. Mag.* **45**, 369.
H. Schinke and F. Sauerwald (1961) *Z. Phys. Chem. (Leipzig)* **216**, 26.
A. F. Schuch, W. C. Overton, and R. Brout (1963) *Phys. Rev. Lett.* **10**, 10.
R. Shuttleworth (1950) *Proc. Phys. Soc., Lond.* **63**, 444.
F. E. Simon and C. A. Swenson (1950) *Proc. Am. Acad. Arts Sci.* **82**, 319.
F. E. Simon and C. A. Swenson (1950) *Nature, Lond.* **165**, 829.
L. A. K. Staveley, H. P. Paget, B. B. Goalby, and J. B. Warren (1950) *J. Chem. Soc. London* 2290.
L. A. K. Staveley, J. B. Warren, H. P. Paget, and D. J. Dowrick (1954) *J. Chem. Soc., London* 1992.
S. M. Stishov (1967) *J.E.T.P. Lett.* **6**, 97.
S. M. Stishov (1969) *Sov. Phys. Usp.* **11**, 816.
S. M. Stishov, I. N. Makarenko, V. A. Ivanov, and A. M. Nikolaenko (1973) *Phys. Lett.* **45A**, 18.
G. M. Stocks and W. H. Young (1969) *J. Phys. Chem.* **2**, 680.
G. C. Straty and E. D. Adams (1966) *Phys. Rev.* **150**, 123.
T. Takamura (1958) *Sci. Rep. Tokyo Univ.* **42**, 16.

J. W. Telford and J. S. Turner (1963) *Phil. Mag.* **8,** 527.

L. Ter Minassian and P. Pruzan (1977) *J. Chem. Thermodynamics* **9,** 375.

F. W. Thompson and A. R. Ubbelohde (1950) *Trans. Faraday Soc.* **46,** 349.

J. Timmermans (1958) *Bull. Soc. Chim. Fr.* Centenary No. 135.

J. Timmermans (1950) *3me Congrès National des Sciences Brussels Chimie Physique* **1.**

M. Toda (1958) *Suppl. Nuovo Cimento* **9,** 39.

D. W. Townsend and R. P. Vickery (1967) *Phil. Mag.* **16,** 1275.

V. P. Trubitsyn and F. R. Ulinich (1962) *Sov. Phys. Dokl.* **7,** 45.

A. R. Ubbelohde (1936) *Trans. Faraday Soc.* **32,** 525.

S. N. Vaidya, J. Akella, and G. C. Kennedy (1969) *J. Phys. Chem. Solids* **30,** 1411.

R. D. Vold and M. J. Vold (1939) *J. Am. Chem. Soc.* **61,** 808.

A. V. Voronel (1958) *Sov. Phys. Tech. Phys.* **3,** 2408.

C. E. Weir (1971) *Japanese J. Appl. Phys.* **10,** 714.

H. Wever (1959) *Z. Elektrochem.* **63,** 931.

W. A. Weyl and E. C. Marboe (1959) *J. Soc. Glass Technol.* **43,** 417.

T. Yoshida and S. Kamakura (1972) *Progr. Theoret. Phys.* **47,** 1801; **48,** 2110.

J. Zernicke (1950) *Rec. Trav. Chim. Pays-Bas* **69,** 116.

# 3. PHENOMENOLOGICAL THEORIES OF FUSION

## 3.1 Some general aspects of phenomenological theories of fusion

One aim of this book is to discuss fusion in terms of the structures of crystals and their melts. This approach lays stress on the fact that different structures may have different mechanisms of melting For any particular crystal structure several mechanisms may operate simultaneously, whereas in related crystals they may be spread over a range of temperatures.

Before discussing detailed mechanisms of melting of different types of crystals, it is useful to survey some general lines of investigation that have guided the development of modern theories.

One extremely general consideration stems from surveys of the thermodynamic characteristics of the solid and liquid states of a great variety of substances. For all known substances the entropy of the molten state is greater than that of the solid state at $T_f$. Consideration of this fact combined with application of the simple Boltzmann expression for entropy changes

$$S_f = R \ln(W_L / W_S)$$

where $W_L$ is the number of independent ways of realizing the molten state and $W_S$ the solid state, suggests that the transition from solid to melt involves some kind of increase in 'randomness'. Thermodynamics alone does not suffice to indicate what its nature may be. In fact, with different types of crystal structures, various ways may arise of increasing randomness on melting. This diversification forms a convenient basis of the classification followed in subsequent chapters of this book.

A second rather less general thermodynamic conclusion is that melting practically always involves an increase in volume. The number of crystalline substances that shrink on melting is surprisingly small. Some of the longest known examples from inorganic chemistry are elements given in Table 3.1. Recently, the same peculiarity has been discovered for some ionic crystals (Chapter 8; see also pp. 203).

*Table 3.1. Crystals melting with a shrinkage in volume*

| Atom or molecule | No. of nearest neighbours in crystal | Fractional volume change on melting $\Delta V_f / V_s$ |
|---|---|---|
| Sb | 3 + 3 | −0·0095 |
| Bi | 3 + 3 | −0·0335 |
| Ga | 1 + 6 | −0·032 |
| $H_2O$ | 4 | −0·083 |

Detailed consideration of specific types of crystal structures must be deferred till later chapters, but a common characteristic that tends to accompany volume shrinkings on melting can conveniently be mentioned here. Crystals that show this very unusual behaviour often have a small coordination number or a small number of nearest neighbours of atoms or molecules that are the units of structure. This imparts an 'open' structure to the crystal; various evidence of the kinds discussed below shows that such open network structures in part collapse on melting, thus aiding shrinkage.

Further to these very general thermodynamic facts about melting, consideration of the actual magnitudes of changes of entropy and volume, and of their derivatives, can give highly significant information which may be discussed in relation to various mechanisms of melting. One conclusion of wide importance is that for most substances the heat of fusion $H_f$ is only a small fraction of the heat of vaporisation $H_{vap}$ measured at the melting point. Examples are quoted in later chapters in connexion with specific crystal structures. This general observation suggests that the molten state at the freezing point is likely to be nearer to the crystal than to the gas in structure and properties. As already stated, a convenient way of recording this consideration is to refer to liquids near their freezing points as 'melts'. By contrast, well above the melting point, as the critical temperature is approached, it is well known that liquids can be treated with fair success in statistical thermodynamics, by attributing to them a quasigaseous behaviour. One of the reasons for some of the current confusion in theories of 'the liquid state' is that each of these models may be most appropriate for the same substance, but over different temperature ranges.

### 3.1.1. *Corresponding state calculations*

On a rather naïve view, solids that are similar might be expected to show similarities in melting, provided these are measured in appropriate units. Because of the diversity and also the complexity of some melting mechanisms, no approach of this kind has however had wide general success. Expectations of similitude are often contrary to the facts, but if and when they are realized they throw a useful light on the mechanisms of melting that operate for the substances being compared.

Following a well-known procedure for other properties of molecules, derivation of any general law of melting in terms of the forces between the atoms or molecules that constitute the units of structure might (for example) be examined in relation to the Principle of Corresponding States. Dimensionless quantities such as $\Delta V_f / V_{crit}$ and $T_{trip} / T_{crit}$ or $P_{trip} / P_{crit}$ where $\Delta V_f$ is the change of volume on fusion, $T_{trip}$ the triple point temperature, $P_{trip}$ the triple point pressure, should be the same for all substances to which the principle can be applied ($V_{crit}$, $T_{crit}$, $P_{crit}$ are the corresponding critical parameters).

Empirically, as is well known, the principle works quite well for the *liquid–gas* transition over a fairly wide range of substances. This is because for such liquids a quasigaseous model gives a reasonable approximation for both the gaseous state, and also for the liquid state not too far from the critical temperature $T_{crit}$. By contrast, for most solids the principle unfortunately proves to be not applicable (Guggenheim, 1945). For crystals of the inert gases the constancy of some of these ratios is however quite notable. A full discussion has been given by Crawford (1976). Table 3.2. illustrates some of the leading similarities for the inert gases. Constancy is also observed in many cases for values of $\Delta V_f / V_S$ and of $S_f$ at the melting point (Chapters 6 and 8); (these parameters are in many cases practically the same at $T_f$ as at the triple point temperatures to which the Principle strictly refers).

**Table 3.2. Corresponding state parameters in the melting of inert gas crystals**

|  | $T_{trip}/T_{crit}$ | $100 P_{trip}/P_{crit}$ | $(V_s)_{trip}/V_{crit}$ $(cm^3)$ | $(V_L)_{trip}/V_{crit}$ |
|---|---|---|---|---|
| Ne | 0·553 | 1·63 | 0·336 | 0·387 |
| Ar | 0·556 | 1·42 | 0·329 | 0·377 |
| Kr | 0·553 | 1·32 | 0·329 | 0·376 |
| Xe | 0·557 | 1·40 | 0·327 | 0·376 |

See also Street and Staveley (1971).

Limitations of the principle of corresponding states applied to melting are readily apparent. For example, the dimensionless ratio of the entropy of the crystal measured from 0 K as reference, divided by the entropy of fusion, $S_f$, proves to be in no way comparable for the different inert gas atoms at corresponding temperatures (Guggenheim and McGlashan, 1960). However, when comparisons are restricted to crystals of a given type, evidence for similitude appears to have some significance, even though agreement is not perfect. For example, values of the ratio $T_f/T_{crit}$ for alkali halide crystals show only a moderate trend with increasing ionic size (Gopal, 1965). For crystals of metals, this is likewise the case, but the ratio has a different value (Table 3.3).

Search for empirical constancy (or approximate constancy) of ratios such as $U^*/T_f$ for 'similar' solids where $U^*$ is the lattice energy of the crystals (corrected for zero-point energy in the case of the lighter atoms) reduces quite large variations in $T_f$ to much more limited ranges (Gopal, 1955a,b,c; 1965). When they exist (see Table 3.4) such correlations point to broad relationships between the interatomic or intermolecular forces in the crystal, and $T_f$. Without much more detailed models of mechanisms of melting, not much further progress can be made by this kind of survey of properties. Only glimpses of similitude can be seen, since there are disturbing factors in both the crystals and the vapour, which introduce irregularities.

**Table 3.3. Similitude ratios for $T_f/T_{crit}$ of crystals**

Alkali halides

| Cation | Anion F | Cl | Br | I |
|--------|---------|------|------|------|
| Li | 0·28 | 0·26 | 0·27 | 0·22 |
| Na | 0·29 | 0·32 | 0·32 | 0·32 |
| K | 0·34 | 0·33 | 0·32 | 0·33 |
| Rb | 0·32 | 0·32 | 0·31 | 0·30 |
| Cs | 0·33 | 0·31 | 0·31 | 0·30 |

metals

| | | | | |
|-----|------|-----|------|
| Li | 0·13 | Ca | 0·15 |
| Na | 0·13 | Sr | 0·15 |
| K | 0·14 | Ba | 0·15 |
| Rb | 0·14 | | |
| Cs | 0·14 | | |

**Table 3.4. Correlation of $U^*/T_f$ ($cal\ deg^{-1}\ mole^{-1}$) with lattice type**

| Crystal | | $T_f(K)$ | $U^*/T_f$ |
|---------|-----|----------|-----------|
| Ne | | 24·50 | 24·12 |
| A | | 83·80 | 24·22 |
| Xe | | 115·9 | 24·32 |

Alkali halides

| Cation | | $T_f(K)$ | $U^*/T_f$ |
|--------|-----|----------|-----------|
| Li | F | 1143 | 58 |
| | Cl | 886 | 64 |
| | Br | 820 | 63 |
| | I | 719 | 54 |
| Na | F | 1253 | 50 |
| | Cl | 1077 | 54 |
| | Br | 1028 | 55 |
| | I | 924 | 52 |
| K | F | 1153 | 55 |
| | Cl | 1048 | 52 |
| | Br | 1003 | 51 |
| | I | 996 | 53 |
| Rb | F | 1033 | 60 |
| | Cl | 988 | 50 |
| | Br | 957 | 48 |
| | I | 913 | 46 |

Group 1B halides

| Cation | | $T_f(K)$ | $U^*/T_f$ |
|--------|-----|----------|-----------|
| Tl | Cl | 703 | 22·8 |
| | Br | 733 | 22·4 |
| | I | 713 | 22·3 |
| Ag | Cl | 708 | 30·9 |
| | Br | 728 | 27·9 |

| Crystal | | $T_f(K)$ | $U^*/T_f$ |
|---------|-----|----------|-----------|
| Divalent cation halides | | | |
| Mg | F | 1669 | 41·7 |
| | Cl | 981 | 60·8 |
| | Br | 973 | 58·3 |
| Ca | F | 1633 | 37·8 |
| | Cl | 1045 | 49·1 |
| | Br | 1033 | 46·9 |
| | I | 845 | 54·1 |
| Sr | F | 1673 | 38·0 |
| | Cl | 1146 | 42·7 |
| | Br | 916 | 50·4 |
| | I | 675 | 64·6 |
| Ba | F | 1553 | 37·8 |
| | Cl | 1198 | 37·8 |
| | Br | 1120 | 38·4 |
| | I | 1013 | 40·5 |

Even for 'globular' molecules (Chapter 6) whose repulsion envelope is roughly spherical, correspondence rules do not apply at all generally for $T_f$ (Trappeniers, 1950).

## 3.2 Mechanical theories of melting

Instead of expressing the equilibrium between a solid and its melt in classical thermodynamic terms, which always involves reference to both phases, various attempts have been made to characterize melting on the basis that a crystal becomes mechanically 'unstable' or 'unrealizable' above $T_f$. Melting is then regarded as a kind of thermal catastrophe for the crystal lattice. For example, a characteristic 'one-phase' asymptotic interpretation of melting develops the notion that a liquid differs from a crystal in having zero resistance to shear. If the shear modulus $\mu$ of a crystal decreases with rising temperature, as will normally be expected for simple solids (because their thermal expansion increases the distance between the atoms and lessens restoring forces), this could lead to a temperature-dependence of the form suggested by Sutherland $\mu = \mu_0(1 - T/T_f)^2$. On this model for melting, $\mu$ sinks to zero at the melting point $T_f$. This or any similar approach to a theory of a 'mechanical melting point', $T_{mech}$, suffers from the same theoretical drawback as other 'one-phase' theories of melting such as the well-known Lindemann theory of a vibrational melting point $T_{vib}$ discussed below (section 3.3). One-phase theories predict a limit to the stability of the solid regardless of the nature of the phase into which the solid transforms on fusion. On the other hand, thermodynamics imposes a two-phase theory of melting. Depending on the molecular characteristics and on the crystal type, the molten state benefits from diverse ways of increasing its entropy that may have only very indirect relation to the crystal lattice energy. Both the crystalline and molten states of matter are condensed phases, and delicate questions concerning their *relative* range of stability on the temperature scale cannot be resolved without thermodynamic considerations.

A further (experimental) objection to mechanical theories of melting is that in those cases where the difficulty of measuring true elastic properties of crystals at high temperatures has in fact been overcome, the trend of $\mu$ with $T$ does not give support to a shear instability limit to the crystalline phase, but suggests that usually

$$T_f \ll T_{mech}$$

This kind of criticism about the inadequacy of mechanical theories of melting may, perhaps, lose its force in special cases. One of the experimental difficulties in testing 'mechanical' limits to the crystalline state is to obtain single crystals sufficiently large to yield valid measurements of elastic properties up to $T_f$, and to avoid the complications arising from crystal defects that lead to mechanical breakdown near $T_f$ for trivial reasons. For example, 'globular' crystals show behaviour somewhat akin to mechanical melting when they become 'plastic' at temperatures below $T_f$ (Ubbelohde, 1961) (and see p. 139). Mechanical instability of such crystals under shear must, however, be attributed to crystal defects; it has been estimated for example in $C(SCH_3)_4$ that the proportion of vacancy defects rises up to above 1% at $T_f$, and accounts for plastic flow of this globular molecule (Dunning, 1961).

Again, it may not be at all easy to distinguish experimentally between a solid for which the elastic/plastic shear threshold is finite but very low, and a true liquid for which this threshold is zero. A clear example is presented by the mesophases of a group of molten alkali carboxylates (Duruz, Michels, and Ubbelohde, 1972) which have very high viscosities ($>3$ poise) but most of which behave as authentic liquids when a sensitive 'horizon' test is applied (Michels and Ubbelohde, 1974).

These difficulties which refer to static experiments limit the theoretical usefulness of theories about the shear instability of solids. Other objections to mechanical theories of melting arise from basic theoretical considerations about actual viscous flow in liquids. These require the specification of relaxation times for *each* of the structurally different possible processes in the liquid which can promote momentum transfer and thus lead to viscosity. For stresses of sufficiently short duration, less than the shortest relaxation time, liquids would in fact behave like crystals anyhow. As the time of application of shearing stress is prolonged, processes with sufficiently small relaxation periods can operate to relieve the stress, and thus to permit 'flow' of the liquid. Non-Newtonian viscous flow in liquids can also complicate mechanical concepts about melting in other ways; for example, some fluids show much lower resistance to flow once they have been set in motion. Again, when a fluid → glass transition is approached and traversed (Chapter 16) distinctive mechanical behaviour is observed. Such transport properties can throw useful light on mechanisms of fusion, particularly with certain classes of fluids, such as network melts. Examples relevant to melting are given below; general discussion of non-Newtonian flow as a guide to the structure of liquids is not however an objective of this book.

Analytically, a theory of a mechanical melting point defines $T_{mech}$ as the temperature of breakdown of resistance to shear (see Ubbelohde, 1940; 1941) and aims to calculate the condition for the elastic shear coefficient $C_{44}$ *vanishing* in terms of the interatomic forces and the temperature. It seems unlikely, however, that the change from solid→liquid can normally be continuous at $T_f$. A jump seems plausible on account of the discontinuous change in volume on melting at constant pressure. Conditions for shear instability of crystals as a result of a sufficiently high concentration of dislocations have been worked out for simple close-packed metallic structures with some success (Scol, 1961) (see also section 11.5). If melting could be studied at constant volume, the mechanical criterion for a limit to the crystalline phase might be more promising as a basis for a general theory of melting; but in most cases the temperatures of fusion calculated would be much higher than those actually found experimentally.

A further difficulty about mechanical theories of melting arises from considerations about specific heats of crystals and their melts. In the Debye model for a vibrating crystal there is one set of longitudinal waves and another set, twice as numerous, of transverse waves. It is generally assumed that $C_{44}$ has sunk to zero in the liquid; transverse waves might perhaps be regarded as having been replaced by rotational motion. This hardly corresponds with the experimental observation that near $T_f$ $(C_P)_L > (C_P)_{SOLID}$ in many cases. For metals the two specific heats are nearly equal (p. 239).

As already pointed out above, with regard to structural relaxation of all kinds in liquids another objection to evaluating $T_f$ in terms of vanishing contributions particularly from shear waves to the enthalpy is that values of $C_{44}$ must be frequency-dependent for processes with time intervals near the critical relaxation time. Even for the liquid, $C_{44} \neq 0$ for shear waves of sufficiently high frequency. For all these reasons, attempts to develop phenomenological theories of mechanical melting, disregarding details of molecular structure, seem too naive to throw much light on mechanisms of melting in general.

Despite the weaknesses of any one-phase approach to the concept of a limit to the stability of the solid state under ordinary conditions, mechanical (and vibrational) theories do present some suggestive possibilities, and have been repeatedly studied for this reason. Even in the absence of any fully satisfactory theory, present evidence suggests that the thermodynamic melting point $T_f$ will normally lie below any mechanical melting point $T_{mech}$ or vibrational melting point $T_{vib}$ which can be calculated as a stability limit to the existence of a given crystal lattice. The liquid state appears to become thermodynamically preferred discontinuously, before either of these limiting temperatures has been reached. It is a pity that thermodynamic melting is a very rapid process, and that no serious kinetic obstructions to the transformation solid → liquid have been recorded (see Chapter 2). At present it does not therefore seem probable that $T_{mech}$ or $T_{vib}$ will be observable even in extremely rapid superheating of solids, since they lie above $T_f$ at ordinary pressures. On increasing the pressure, $T_f$ may however rise more rapidly than $T_{mech}$ or $T_{vib}$ and may eventually become higher than either. In that event, the solid state, even though of lower entropy than the liquid state, and of lower specific volume, may yield to minimal external shearing stresses above a temperature $T_{mech}$. This would correspond with mechanical melting at a lower temperature than thermodynamic melting. Completely plastic ordered states of matter have not yet been identified (see above); but there are some theoretical grounds for anticipating them if conditions can be found for heating the solid to sufficiently high temperatures whilst avoiding thermodynamic transformation to the liquid state. Though of considerable interest for theories of transport in condensed states of matter, non-thermodynamic melting cannot as yet be discussed profitably at greater length. To investigate mechanical melting, more information is required about changes in creep and cooperative migration of molecules in solids as the temperature rises. A further obstacle in any case may be that, at the high densities and temperatures which appear to be required, collapse of the outer electrons of the atoms or molecules into a 'metallic' assembly may have occurred (Chapter 8), thereby making any theories of critical melting meaningless with regard to crystal lattices with conventional atomic or molecular structures.

### 3.3 Vibrational theories of melting: one-phase models

Preceding sections have discussed the instability of crystal lattices as their temperature is raised, without any detailed inquiry into the structural changes brought about by the temperature increase. Within recent years many authors have sought to consider such structural changes, but only in general terms, which lead to melting. Theories can be broadly grouped into those which stress consequences of the volume expansion, and those which concentrate on the vibrational behaviour of crystal lattices. Volume expansion theories of melting play a prominent role in 'hard sphere' models of solids, as well as in statistical thermodynamic models of the creation of lattice defects. These are considered in Chapter 11 following chapters descriptive of melting of specific crystal types. By way of an example of volume expansion theories, when attempts are made to correlate the entropy of fusion of molecular crystals with the structure of the *molecules* (because the structure of the crystals is not known), one fairly general approach is to attribute part to the change of entropy at constant volume, and part to the effects of volume increase which involve changes of molecular configurational energy and the like (Bondi, 1968), i.e.

$$S_f = (\Delta S_f)_V + (\Delta S_f)_{\text{config}} \qquad (3.1)$$

$$(\Delta S_f)_{\text{config}} = \left(\frac{\partial S}{\partial V}\right)_P \Delta V_f = \alpha \beta_0 \Delta V$$

where $\alpha$ is the thermal expansion coefficient and $\beta_0$ is the bulk modulus at zero pressure. Both components have been calculated for molecules of fairly simple structure (see Table 3.5).

Vibrational theories of melting are conveniently discussed here in relation to the general phenomenology of melting. The general idea that melting involves some kind of increased heat motion seems to be a fairly obvious outcome of any atomic theory of matter. Concrete mathematical form was

**Table 3.5. Notional contribution of different entropy changes to fusion**

| Molecule | $(\Delta S_f)_{config}$ (e.u.) | $(\Delta S_f)_v$ (e.u.) |
|---|---|---|
| Ar | 0·90 | 0·79 |
| $N_2$ | 0·57 | 0·79 |
| $O_2$ | 0·36 | 0·64 |
| $C_6H_5F$ | 2·10 | 3·77 |
| $C_6H_5Cl$ | 1·42 | 3·68 |
| $C_6H_5Br$ | 1·60 | 3·69 |
| $C_6H_5I$ | 1·83 | 3·56 |
| $p$-$ClC_6H_4Cl$ | 3·26 | 3·54 |
| $C_6H_6$ | 1·76 | 2·49 |
| $C_6H_5CH_3$ | 2·0 | 2·48 |
| $p$-xylene | 3·25 | 3·95 |
| $m$-xylene | 2·06 | 4·12 |
| $o$-xylene | 1·63 | 4·98 |

The original publication should be consulted for these and other calculations that attempt to rationalize equation (3.1).

first given to this general idea in the Lindemann theory of vibrational instability of crystal lattices leading to fusion (Lindemann, 1910). Using the simplest model for a vibrating solid, with a single characteristic frequency $\nu_E$ as first proposed by Einstein, Lindemann developed the conclusion that as the temperature rose the vibrations of the atoms about their equilibrium positions in the lattice must increase progressively in amplitude. He introduced the concept that when this amplitude of movement between two extreme positions attains a critical fraction of the distance between the equilibrium positions of neighbours, their vibrations would 'interfere' mutually to such an extent that the crystal would become mechanically unstable. Simple algebraical approximations lead to the Lindemann melting point formula,

$$T_{vib} = C\nu_E^2 V_S^{\frac{2}{3}} M \qquad (3.1.1)$$

which relates the characteristic frequency of the Einstein solid with the vibrational melting point $T_{vib}$ and with other physical magnitudes that are known for the particular solid, such as the molar volume $V_S$ and the atomic mass $M$. The general constant $C$ was assumed to be the same for crystals of similar structure and was therefore evaluated from the melting temperature of one particular crystal.

Table 3.6 records the ratio of $(\nu_E)_f$ calculated from the melting point, to $(\nu_E)_{C_p}$ calculated by fitting an Einstein equation to the specific heat at low temperatures. This table assumes as norm for platinum $(\nu_E)_f = (\nu_E)_{C_p} = 3\cdot1 \times 10^{-12}\,s^{-1}$. As will be seen, even with metal crystals whose structure is comparatively simple, the ratio departs from unity considerably and the

**Table 3.6. Ratio of characteristic frequencies calculated from melting points and specific heats of simple solids (see Roberts and Miller, 1960)**

| Metal | $(\nu_E)_k/(\nu_E)_{C_p}$ |
|-------|---------------------------|
| Sn | 0·67 |
| Sb | 0·74 |
| Cd | 0·78 |
| Al | 0·82 |
| Zn | 0·84 |
| Cu | 0·89 |
| Ag | 0·97 |
| Pt | (1·00) |
| Mg | 1·06 |
| Pb | 1·17 |

correlation is only fair. No obvious relation can be seen with either the crystal structures, or $T_f$.

More sophisticated treatments of the notion that for sufficiently high temperatures a vibrating lattice must cease to be stable have taken various forms. Semiempirically, and without making any *a priori* assumptions about the critical amplitude of vibration expressed as a fraction of the inter-molecular lattice spacing, one can seek for uniformities in a 'Lindemann parameter'

$$L_f = \theta_D V^{\frac{1}{3}}(M/T_f)^{\frac{1}{2}} \tag{3.2}$$

which is based on experimental parameters that are generally accessible, such as the Debye characteristic temperature $\theta_D$. Even for metal crystals expectation of uniformity does not seem very hopeful (Ishizaki, Spain, and Bolsaitis, 1975) (Table 3.7).

If the Debye temperature $\theta_D$ of the crystal could be evaluated near $T_f$ this would probably reduce the spread in $L_f$, but the limited range of crystals for which this improvement can be tested is not very rewarding. Perturbations resulting from the formation of melts with unusual structural make up are likely to upset uniformity; thus the agreement between melting point and characteristic Debye vibrational temperatures is poor for semiconductor compounds (Aggarwal, Raju, and Verma, 1973). Much more refined lattice dynamical calculations (Shapiro, 1970) of the critical amplitude of vibration as a fraction of the interatomic spacing can bring out similitudes, e.g. for the crystals of the alkali metals, more clearly (Table 3.8).

As a further refinement, the selection of specific vibrational modes in the overall phonon spectrum, whose increasing amplitude should lead to lattice instability, has received considerable attention. One plausible choice is to stress the role of shear vibrational modes; with only limited success however (Dickey and Paskin, 1969; Shimada and Yokota, 1974). A somewhat less *a priori* theory calculates the mechanical melting point $T_{mech}$ by extrapolation

**Table 3.7. Lindemann parameters $L_f$ (crystals grouped by types). $T_f$ in K units**

*Metal crystal*

| f.c.c. | $L_f$ | $T_f$ | b.c.c. | $L_f$ | $T_f$ | r.c.p. | $L_f$ | $T_f$ |
|---|---|---|---|---|---|---|---|---|
| Ag | 148 | 1235·0 | Ba | 147 | 1983·1 | Be | 150 | 1558·1 |
| Al | 138 | 933·5 | Cr | 125 | 2133·1 | Cd | 168 | 594·2 |
| Au | 137 | 1337.5 | Cs | 118 | 301·7 | Co | 132 | 1767·1 |
| Ca | 124 | 1113.1 | Fe | 121 | 1813.1 | Gd | 129 | 1583.1 |
| Cu | 143 | 1357.6 | K | 122 | 336.3 | Hf | 143 | 2503·1 |
| Ir | 155 | 2720·1 | Li | 124 | 453.1 | Ir | 155 | 2720·1 |
| Ni | 143 | 1728·1 | Mo | 138 | 2893·1 | Mg | 134 | 923·1 |
| Pb | 149 | 600·6 | Nb | 104 | 2698·1 | Re | 134 | 3453·1 |
| Pd | 138 | 1827·1 | Na | 114 | 370·9 | Ti | 140 | 1943·1 |
| Pt | 151 | 2045·1 | Rb | 118 | 311·9 | Tl | 138 | 577·1 |
| Rh | 167 | 2236·1 | Ta | 110 | 3273·1 | Zn | 151 | 692·7 |
| Sr | 140 | 1043·1 | | *Rhomb.* | | Zr | 133 | 2123·1 |
| Th | 159 | 1973·1 | Bi | 201 | 472·4 | Se | 91 | 493·1 |
| V | 123 | 2193·1 | Hg | 171 | 234·2 | Te | 178 | 723·1 |
| W | 135 | 2640·1 | Sb | 138 | 903·8 | | | |
| | | | As | 170 | 886·1 | | | |

*Orthorhomb*

| | $L_f$ | $T_f$ |
|---|---|---|
| Ga | 261 | 302·9 |
| V | 192 | 2193·1 |

*Tetragonal*

| | $L_f$ | $T_f$ |
|---|---|---|
| In | 142 | 429·7 |
| Sn | 259 | 505·1 |

*Salts*

| | $L_f$ | $T_f$ | | $L_f$ | $T_f$ |
|---|---|---|---|---|---|
| NaCl | 210 | 1081·1 | RbBr | 206 | 955·1 |
| KCl | 195 | 1049·1 | AgBr | 229 | 705·1 |
| AgCl | 239 | 728·1 | | | |
| KI | 180 | 959·1 | LiF | 200 | 1115·1 |
| RbI | 189 | 915·1 | $CaF_2$ | 211 | 1633·1 |

*Network crystals*

| | $L_f$ | $T_f$ |
|---|---|---|
| Ge | 209 | 1232·1 |
| Si | 195 | 1683·1 |
| C | 154–198 | ~4000 |
| $H_2O$ | 217 | 273 |

from experimental values for the elastic constants (Jackson and Liebermann, 1974). For alkali halides $T_{mech}$ always lies above $T_f$ at ordinary pressures, but the values run parallel (Table 3.9).

It may be added that actual measurements of ultrasonic velocities on indium show that longitudinal acoustic modes stiffen near $T_f$. Longitudinal modes show no measurable premelting (Chung, Gunton, and Saunders, 1975). Clearly for this crystal $T_{mech} > T_f$. This appears to be contrary to the hypothesis of Ida discussed below.

**Table 3.8.** *Critical fractional amplitude for melting*

Metal crystal
b.c.c.

| | |
|---|---|
| Li | 0·116 |
| Na | 0·111 |
| K | 0·112 |
| Rb | 0·115 |
| Cs | 0·111 |

For *f.c.c.* metals a different critical fraction is calculated

| | |
|---|---|
| Al | 0·072 |
| Cu | 0·069 |
| Ag | 0·071 |
| Au | 0·075 |
| Pb | 0·065 |
| Ni | 0·077 |

**Table 3.9.** *Mechanical (shear) ($T_{mech}$) and thermodynamic ($T_f$) melting points of alkali halides in °C*

| Cation | F | | Cl | | Br | | I | |
|---|---|---|---|---|---|---|---|---|
| | $T_{mech}$ | $T_f$ | $T_{mech}$ | $T_f$ | $T_{mech}$ | $T_f$ | $T_{mech}$ | $T_f$ |
| Li | 914 | 842 | 655 | 614 | 572 | 547 | 499 | 450 |
| Na | 1122 | 988 | 932 | 801 | 861 | 755 | 772 | 651 |
| K | 1022 | 846 | 931 | 776 | 890 | 730 | 822 | 686 |
| Rb | 959 | 775 | 890 | 715 | 886 | 682 | 838 | 642 |

Other variants of vibration melting as a one-phase phenomenon have been proposed from time to time. A phonon assembly which instead of asymptoting to zero elasticity at $T_{vib}$ avalanches into collapse of the lattice has been discussed (Khomskii, 1969). In a different approach, in the Percus–Yevick formulation where a liquid is described in terms of $3N$ collective coordinates, the entropy of fusion $S_f/R$ can be correlated with the long wavelength limit of the liquid structure factor (Omini, 1972). The original papers should be consulted for details.

Much subsequent discussion has made it clear that as a 'one-phase' theory of vibrational melting the Lindemann approach suffers from a basic weakness with respect to fundamental thermodynamic principles. However, as in the case of mechanical melting, this objection may perhaps lapse for calculation of limits at very high temperatures and pressures, as indicated in the previous section. Even under ordinary conditions, Lindemann's formula does provide a convenient means for correlating various physical properties of a crystal related to $T_f$ (Lennard-Jones and Devonshire, 1939). One reason appears to be that the energy of creating positional disorder in a crystal is closely related to the restoring forces operative in crystal vibrations; either quantity may be used to characterize other physical properties of the crystal.

It should further be stressed that the original Lindemann model for vibrational melting, like many of its more sophisticated successors, in a strict sense refers only to crystals with the simplest possible structures, i.e. that are assemblies of close-packed atoms. Crystals containing more complex molecules as units of structure exhibit a vibrational complexity which rules out any simple limit to lattice stability, determined merely by vibrational amplitudes of the molecular centres of mass. In such more complex structures, specific vibrational modes nevertheless may become unstable above critical amplitudes. This may influence thermal transformations from one condensed state to another, in certain cases. It is unfortunate that up to the present vibrational theories have been restricted to the simplest possible crystal lattices; complexities of melting discussed in subsequent chapters in this book are simply overlooked in conventional theories of vibrational melting.

Since even the simplest crystals in principle could exhibit various mechanisms of melting, and since it is a matter of contingency whether $T_f \gtrless T_{vib}$ or $T_f \gtrless T_{mech}$, thus determining melting behaviour, a realistic model of the vibrating crystal with adequately characterized anharmonic vibrations is essential in order to determine whether *any* vibrational mechanism could give rise to a melting process competitive at *any* pressure with various thermodynamic mechanisms of disordering. To reduce mathematical complexities with anharmonic vibrations, this problem has sometimes been tackled theoretically for the so-called linear chain of atoms representing a crystalline solid in one dimension.

Before discussing this it may be added that the basic concept that certain types of vibrations of a crystal lattice must become unstable when their amplitude exceeds a critical magnitude is probably highly relevant to specific *solid–solid* transformations in which certain types of vibration undergo 'phonon fading' on approaching the transition point (see p. 73). Phonon fading is however a much more specialized situation than is generally assumed for any widely applicable mechanisms of melting.

### 3.4 Vibrational melting of a line solid

For a linear chain in which the interaction between neighbours can be represented by a Morse potential function, it has been shown (Dugdale and McDonald, 1954) that its (anharmonic) vibrations give rise to a number of features analogous to those often shown by actual simple three-dimensional solids. These include the gradual rise in $C_p$ above $3R$ and the phenomenon of thermal expansion. It does not appear however that vibratrional anharmonicity can account for the premelting phenomena observed experimentally for some crystals as $T_f$ is approached, e.g. for the alkali metals (Chapter 12).

Another interesting feature of the vibrational behaviour of the one-dimensional solid in the theory of Dugdale and McDonald is a transition from 'quasicrystalline' to 'quasigaseous' behaviour in a narrow range of temperatures. This lends some support to the view discussed above, that, as the degree of compression of a solid is progressively increased and $T_f$ is thereby raised, vibrational or mechanical melting to a 'quasigaseous' fluid may catch up with and even overtake thermodynamic or defect melting to a 'quasicrystalline' liquid. Any discussion of critical melting clearly needs to keep the possibility of various competitive mechanisms in view, as different temperatures and pressures lead to different liquid structures predominating.

### 3.5 Vibrational properties in relation to thermodynamic behaviour of solids

Correlations between $T_f$ and various manifestations of the vibrational content of crystals have been often proposed, and are sometimes fairly close. As one example, the product of the thermal expansion and $T_f$ is approximately constant, i.e. $\alpha T_f = $ constant. Since $\alpha$ itself varies with temperature, the values at room temperature plotted in Fig. 3.1 are to some extent chancy and conceal the true relationship (Nowotny, 1958).

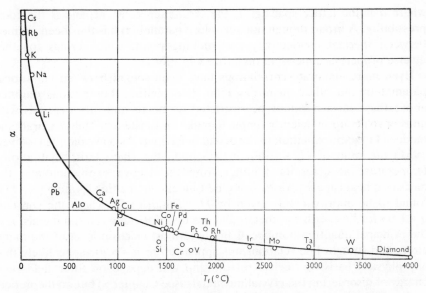

FIG. 3.1. Relation between coefficient of thermal expansion $\alpha$ (ordinate) and $T_f$ (abscissa). Reproduced by permission of Planseeberichte fur Pulvermetallurgie

A 'corresponding state' Plot of $\alpha_T/\alpha_{T_{f/2}}$ against $T/T_f$ gives a single curve for the ionic crystals NaCl, KCl and CsBr, which remains a straight line to $T/T_f \sim 0{\cdot}75$ and thereafter curves steeply upwards. The thermal expansion $\alpha_T$

is determined by X-ray diffraction and (unlike the bulk thermal expansion) is only indirectly affected by premelting vacancy formation, so that the origin of the steep rise warrants further investigation (Pathak and Vasavada, 1970).

A somewhat less direct correlation attempts to link the melting point with the entropy of the solid at $T_f$, i.e. $[S_S]_{T_f}$ (Turkdogan and Pearson, 1953). If the solid obeys the Nernst–Planck third law of thermodynamics $[S_S] = 0$ at 0 K. At $T_f$ the entropy of the solid is determined by the specific heat integral:

$$[S_S]_{T_f} = \int_0^{T_f} C_p \, d(\ln T)$$

$C_p$ is normally determined by the crystal vibrations so that $T_f$ in turn would be determined by the vibrational energy of the solid. Examination of the rather scanty data for which this criterion has been tested suggests that some correlation may indeed exist, but that reasons for it must be indirect.

Again, the absolute melting temperature may be used to eliminate the elastic constants from a semiempirical equation for thermal conductivity $\lambda$ of solids (Keyes, 1959), assuming the Lindemann relationship

$$\lambda = a^3 Y^2 T_{vib} \beta^{\frac{3}{2}} P^{\frac{1}{2}} \tag{3.3}$$

where $a$ is the lattice spacing, $Y$ the Gruneisen constant, and $\beta$ the compressibility. A broad though not very close parallelism has thus been claimed between thermal conductivity and melting point, which brings together diverse properties dependent on lattice vibrations.

Even more elaborate correlations have been sought between vibrational parameters and other quantities related to melting. For reasons outlined above these must normally be restricted to metals and other crystals with units of structure sufficiently simple to make an Einstein or Debye vibrational model a fair approximation to the actual behaviour. As one example (Jollivet, 1951) the ratio of $H_f$ to $R\theta_D$ in which $\theta_D$ the characteristic vibrational temperature as determined either from the Debye expression and the measured heat capacity, or $\theta_L$ from the Lindemann melting formula (3.1.1) is found to be approximately given by $H_f = R\theta_D q(n-1)$, where the constant $q = 1 \cdot 63$ for f.c.c. or h.c.p. metals and $q = 0 \cdot 90$ for cubic centred metals; $n$ is the principal quantum number of the outermost electron level of the metal atom, and straight line relationships appear up to $n = 5$. It seems likely that any physical basis for such a relationship will depend on force fields and energy of disordering the crystalline metals (see Chapter 8) but no theoretical interpretations have yet been developed.

## 3.6 Tests of the Lindemann melting formula

Conventional tests of the Lindemann formula for different crystals of simple structure (Table 3.6 above) are inevitably somewhat insensitive owing to the occurrence of the term $(T_{vib})^{\frac{1}{2}}$. More accurate and more refined examination of the correlations implied by the formula may be made as follows.

(i) By comparing the melting points of isotopic molecules. Some of these are tabulated in Chapter 2; in general they do not confirm the vibrational theory of melting as incorporated in equation (3.1.1). For metallic lithium, it has been estimated that the difference in melting points for $^{6}$Li and $^{7}$Li is about 0·23°C (Crawford and Montgomery, 1957).

In relation to vibrational models for melting it is interesting to note that for many metals the velocity of sound in the melt at $T_f$ is about one-half that in the crystal just below $T_f$. Simple considerations about the vibrational frequency of atoms in a rather crude model (Rao, 1944) indicate that approximately this ratio would be expected, owing to the greater amplitude of vibration of atoms in the melt. For bismuth the velocity of sound in the melt is however actually higher than in the crystal.

(ii) By testing the formula

$$\theta_L = C'\{T_f/MV^{\frac{2}{3}}\}^{\frac{1}{2}} \tag{3.4}$$

for solids for which $\theta_L$ and $T_f$ have been determined for a range of densities and pressures from accurate calorimetric experiments.

Although none of the criteria discussed give complete support to a 'vibrational catastrophe' as the origin of a transition from a crystalline to a fluid state, there can be no doubt that precise measurements of the velocity and of the damping of amplitude of sound over the widest possible range of frequencies can give most valuable information about melting, as well as about other transitions between condensed phases. Reference to current work is made in various places in what follows (pp. 73 seq.); even more general use of phonon studies promises to be particularly rewarding in the investigation of highly defective crystals, and of the molten state generally.

(iii) The empirical success of Lindemann's melting formula and the fact that it can be correlated with the Lennard-Jones and Devonshire theory of melting (Chapter 11) has stimulated attempts to develop somewhat less crude models of melting of solids of simple structure, in relation to the role of crystal vibrations. One example of such a model has been discussed in detail by Gilvarry (1956a,b,c). The basic assumption is again made, as in Lindemann's crude model, that fusion occurs when the root mean square amplitude of thermal vibration reaches a critical fraction $\rho$ of the nearest neighbour separation $\gamma_a$ of the atoms of the crystal at $T_f$; $\rho$ is assumed to be the same for all isotropic monatomic solids, i.e.

$$\rho^2 \gamma_a{}^2 = \bar{u}_a{}^2$$

Amplitudes of vibration are calculated from the Debye–Waller formula for intensity of X-ray scattering

$$I_T = I_0 \exp(-2M)$$

where

$$M = 8\pi^2(\sin \theta/\lambda)^2 \bar{u}_a^2$$

$\lambda$ is the wave length of X-rays; $\theta$ is the Bragg angle, and $\bar{u}_a^2$ the mean square amplitude. In principle the vibrational frequency $\nu_f$ is calculated at $T_f$ from the elastic constants. Unfortunately, these are known right up to $T_f$ for a few solids only, such as ionic crystals (Quimby and Siegel, 1938; Hunter and Siegel, 1942). For this and for various other reasons, those refinements of one-phase Lindemann models have not greatly advanced the vital question of their correlation with experiment.

Other changes can also be proposed for the way in which a simple crystal lattice may become mechanically unstable, when the amplitude of thermal vibration exceeds a critical value. For example increasing anharmonicity of oscillations must result from thermal expansion of the crystal. Ultimately the increasing separation of the time-average position of the atoms leads to a state where *lattice* vibrations can no longer be defined (Ida, 1969; Ishizaki, Bolsaitis, and Spain, 1973). As for other vibrational instability theories, $T_{vib} > T_f$ when calculated on this basis. If expansion is prevented, by increasing the pressure sufficiently, this particular route to instability would seem to be less effective than the alternatives discussed above.

### 3.6.1. *Vibrational melting of helium and other inert gases*

The melting of crystalline helium provides a rich source of data for studying the role of vibrational energy in fusion. The amount of melting data for $^3$He and $^4$He much exceeds that for all the rest of the inert gases combined. Quantum effects for helium greatly outweigh effects from thermal vibrations, so that these isotopic substances are not really characteristic for the phenomenology of melting throughout the crystalline state of matter. Specialist discussions of the helium isotopes have been provided in a number of publications (Wilks, 1967; Keller, 1969; Grilly, 1971; Trickey, Kirk, and Adams, 1972; Crawford, 1976). Only relationships of melting of helium to thermal melting of crystals generally will be discussed here. Melting parameters for helium (Dugdale and Simon, 1953) show only a surprisingly small drift of the empirical constant $C$ in (3.1), as is illustrated in Table 3.10.

Analysis of the above findings shows that the constant $C$ has only about 0·6-times the usual value for metals. If this mathematical model is useful at all, those frequencies in the phonon spectrum that are most important for melting clearly differ from the Einstein frequency put forward as of predominant importance in the original Lindemann model for melting. In fact

*Table 3.10. Melting of helium under pressure—*
*computation of nominal vibrational temperatures*
*(Lindemann model)*

| $\theta$ (K) | $T_f$ (K) | $V$ (cm³) | $C$ |
|---|---|---|---|
| 32 | 3·1 | 18·3 | 96 |
| 55 | 7·9 | 14·4 | 95 |
| 72 | 11·3 | 13·1 | 102 |
| 92 | 17·3 | 11·6 | 100 |
| 110 | 23·3 | 10·6 | 101 |

the vibrations which are considered to lead to the melting of helium must be very different (much longer) in wavelength than those whose predominant contribution to the zero-point energy determines the existence of helium at one atmosphere pressure as a liquid down to the lowest temperatures. This could be a pointer to a dislocation mechanism for the melting of helium (see below).

The solid–liquid transformation of helium is indeed unique, owing to the predominant role of zero-point energy for the thermodynamics of the process. As pointed out previously (Chapter 2), for helium $S_f \to 0$ as $T_f \to 0$ but the experimental evidence is quite contrary to the simultaneous possibility $\Delta V_f \to 0$ as $T_f \to 0$. Thus 0 K is not a *critical* melting temperature for helium, in the sense discussed in Chapter 2 since the solid and liquid phases do not become indistinguishable.

Since $S_f \to 0$, $H_f \to 0$ *a fortiori*. Because the specific volumes remain different, the equation $H_f = \Delta E + P \Delta V$ indicates that $\Delta E = -P \Delta V$, i.e. on compressing the liquid to form the solid at these low temperatures the free energy put into the system must increase the internal energy of the solid above that of the liquid. The vibrational thermal energy of either phase is of the order $0·01$ cal mole$^{-1}$ at 1 K, but zero-point energies are about 50 cal mole$^{-1}$ for the solid and 39 cal mole$^{-1}$ for the liquid, both under an external pressure of 25 atm (Simon and Swenson, 1950).

Although the Lindemann expression leads apparently to quite a reasonable correlation between $T_f$ and $\theta$ in Table 3.10, consideration of the relative content of zero-point energy and thermal energy at $T_f$ as pressure (and hence $T_f$) are increased would lead on the Lindemann model to the conclusion that 'thermal' vibrational energy is many more times effective than zero-point vibrational energy, in leading to fusion. Actually, melting of helium occurs because the expansion to form the liquid permits a lowering of the characteristic frequency, owing to the fact that the atoms no longer vibrate in such strong force fields as in the crystals. At 25 atm pressure $V_L = 23·25$ cm³ and $V_S = 21·2$ cm³. Melting of helium thus takes place because repulsions between neighbouring atoms interact with the zero-point vibrations. In a very sophisticated sense, this could be described as a Lindemann mechanism of melting through vibrational instability. Primarily however this particular process is volume and not temperature-dependent.

Isotope effects with $^3$He are extremely prominent (e.g. Johnson and Wheatley, 1970; Johnson, Lounasmaa, Rosenbaum, Symko, and Wheatley, 1970), but will not be discussed in any detail in this book.

Results for this unique element provide further evidence that the attribution of melting to the attainment of a critical amplitude of vibration, as in the theory put forward originally by Lindemann, is not ordinarily physically meaningful. For argon, detailed experimental tests of the Lindemann expression have been made over a range of temperatures from about 120 K to 205 K, and a range of high pressures (Ishizaki, Spain, and Bolsaitis, 1975). Both longitudinal and transverse velocities of sound were measured close up to the melting point at each pressure. A calculated 'Lindemann parameter' $L_f = 118 \cdot 0 \pm 0 \cdot 6$ cm g$^{\frac{1}{2}}$ mole$^{-\frac{5}{6}}$ was found to apply over the whole of this high-pressure range, to within 2% accuracy. On this basis the root mean square amplitude as a fraction of the crystal spacing is $0 \cdot 133$. There is evidence that below about 100 K anomalous values are found for the Lindemann parameter, because at these lower pressures vacancy formation in the crystals contributes to the melting mechanism (Ross, 1969). General phenomenological aspects of melting of inert gas crystals are discussed in Chapter 2. Crystals of inert gases are still often used as models to test mechanical theories of melting. Despite the inadequacy of a mechanical approach to provide a complete theory of melting, force field considerations must determine the equation of state of these crystals (Dobbs and Jones, 1957). Interatomic forces directly affect the heat of fusion $H_f$ and indirectly the entropy of fusion $S_f$, and are thus intimately bound up with the melting point $T_f = H_f / S_f$.

### 3.7 Vibrational solid–solid transformations—two-phase theories

Many structural aspects of the transformation from a crystal to its melt are likewise found in transformations from one crystal state to another. At present structural aspects of solid–solid transformations can be characterized with considerably greater precision in favourable cases, being less subject to the experimental and theoretical difficulties found in dealing with the statistical thermodynamics of the molten state, and also because the structure of many crystals is much better known that that of their melts. These considerations make it useful to examine solid–solid transformations in some detail as an introduction to various mechanisms of melting. Furthermore, certain transformations which take place separately in one kind of crystal well below $T_f$ may in other types coincide with the actual melting process, and must then be treated additively with other melting mechanisms in the overall change taking place on fusion.

Classical thermodynamics describes the transformation from one solid phase to another at a transition temperature $T_c$ in the same general terms as for the transformation from solid to melt at $T_f$. The free energies $G_1 = H_1 - TS_1$ and $G_2 = H_2 - TS_2$ become equal at $T_c$. Except as discussed in relation to precursor phenomena (Chapters 12 and 13), free energy surfaces intersect sharply at $T_c$. Differences between the tangent slopes determine the entropy and volume changes on passing from one phase to another. This is represented by the conventional equations

$$S_1 - S_2 = \left(\frac{\partial G_2}{\partial T}\right)_P - \left(\frac{\partial G_1}{\partial T}\right)_P$$

$$V_2 - V_1 = \left(\frac{\partial G_2}{\partial P}\right)_T - \left(\frac{\partial G_1}{\partial P}\right)_T$$

The form 2 stable at the higher temperature must have higher enthalpy, but the volume relationships are not restricted in any way.

Structural thermodynamics aims to give reasons why the entropy of structure 2 catches up with that of structure 1 as the temperature rises, and thereby compensates for its higher enthalpy. The present chapter is particularly concerned with vibrational causes for a phase transformation.

### 3.8 Vibrational softening leading to transformations in crystals

Probably the simplest origin of a thermal transformation between one polymorph and another arises from differences in the vibrational energy and entropy that accompany differences of structure. An example which has been studied in very considerable detail is the transformation from grey (subscript g) to white (subscript w) tin. At atmospheric pressure $T_c = 292$ K and $H_w - H_g = 522$ cal g atom$^{-1}$. Crystal X-ray diffraction studies show that grey tin has cubic symmetry with each atom surrounded by 4 nearest neighbours coordination number 4 at $2 \cdot 80$ Å. The looser structure of white tin is accompanied by a lower vibrational characteristic temperature and its specific heat is uniformly greater than that of grey tin. Since the transformation is quite sluggish, both forms can be followed calorimetrically down to the lowest temperatures where specific heats sink towards zero. Figure 3.2 illustrates plots of $\Delta G$, $\Delta H$ and $\Delta S$ against $T$ for the two forms. At $T_c$, $\Delta G$ sinks to zero since $T \Delta S$ compensates for the greater vibrational enthalpy of white tin. Figure 3.3 illustrates plots of $C_p$ against $\ln T$ from which the vibrational entropies can be directly obtained by graphical integration.

This particular solid–solid transformation can thus be directly attributed to the respective contributions of vibrational enthalpy and vibrational entropy to the free energy of two different crystal polymorphs. Many other instances may probably exist of similar vibration controlled polymorphism. A number

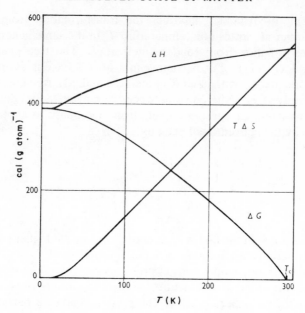

FIG. 3.2. Vibrational transformation grey tin to white tin

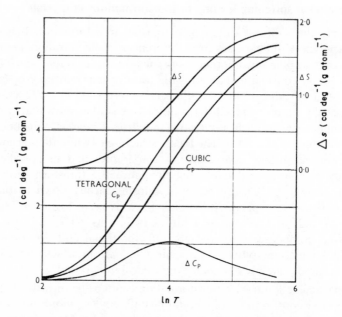

FIG. 3.3. Heat capacity plots for the vibrational transformation
in tin

of metals transform from a low-temperature close-packed structure to a high-temperature b.c.c. phase (Grimvall, 1975). It seems likely that the principal operative factor is the change in vibrational free energy on passing to a crystal with somewhat lower Debye temperature $\theta_D$ as in the example discussed above.

Any anharmonicity of vibrations that may be found is to a large extent incidental in this kind of transformation, in which the entire phonon spectrum contributes. This should be contrasted with phenomena of 'phonon fading' of specific frequencies, which accompanies some other solid–solid transformations and which is discussed in Chapter 4.

As a general rule, if equilibrium can be established at all between different solid states of a substance, as a result of vibrational effects, this implies that enthalpy differences must be quite small, since the transition must be occasioned by the build-up of a sufficient difference of vibrational entropy between the two crystal forms to compensate for enthalpy differences before the melting point is reached. The likelihood of establishing such transformations unequivocally seems most promising for the case of fairly simple crystal structures. Possibilities for constructing various crystal structures with about the same lattice energy, from a given molecule, are multiplied when more than one type of interaction force can be set up between them. In this way enthalpies of different structures can be matched fairly closely by compensating changes of interaction that are both positive and negative, on passing from one crystal polymorph to another. As one example, molecules capable of forming hydrogen bonds are remarkable in respect of the number of polymorphs of approximately equal stability that can be achieved. Ice is well known for this peculiarity. Polymorphs of ice are too numerous to discuss in detail in this book.

## 3.9 Changes of vibrational parameters on melting

As discussed in preceding sections, one-phase theories of melting cannot give a satisfactory account of what is, except perhaps in the limit, a two-phase process. Normally there must be some discontinuity between the phases. Nevertheless, in elucidating the role of vibrational enthalpy and entropy in the stability of crystals, it is interesting to consider to what extent these quantities change on passing from crystal to melt. For thermodynamic computations elementary theories of the vibrational energy of solids often yield reasonably accurate values. When the units of structure are atoms, usually a Debye model with a single characteristic temperature $\theta_D$ yields fair approximations for the quantities

$$[H_S]_{T_f} = \int_0^{T_f} C_p \, dT; \; S_S = \int_0^T C_p \, d(\ln T)$$

The total vibrational contribution to $H_S$ and $S_S$ up to $T_f$ is usually assessed by graphical means. It sometimes happens in practice that specific heats appear in marked excess above the Dulong and Petit value $C_V = 3R$ which elementary theory indicates as the limiting vibrational specific heat, for crystals whose units of structure are atoms. A decision has to be made in each instance whether the rise observed is due to marked anharmonicity of the vibrations, or to contributions from other modes of energy increase in the crystals, such as the formation of crystal defects in thermal equilibrium (see Chapter 9).

For crystals in which the units of structure are identifiable molecules, the total vibrational energy content shows a more complex temperature-dependence than when the units are atoms. Often the vibrational energy can be approximately represented by a Debye function to account for the vibrations of the molecules as whole entities, together with a number of Einstein functions for the internal molecular vibrations i.e.

$$H_{vib} = D(\theta_D) + E(\theta_{E_1}) + E(\theta_{E_2}) + \cdots$$

Such a vibrational model is appropriate only if the crystal forces are weak compared with internal bond forces of the molecules, so that the characteristic temperatures $\theta_{E_1}$, $\theta_{E_2}$ etc. are not greatly modified from those for isolated molecules, and can be approximately represented by independent Einstein vibrator functions.

When the melt can be treated as quasicrystalline, a large fraction of its thermal energy might conceivably be attributed to vibrations of the molecules in local potential energy wells constituted by their neighbours. So far as such an approximation applies, the characteristic temperature for these vibrations must be considerably lower than that appropriate to the environment in the crystal, so that the vibration enthalpy is higher. It should be noted that if an important part of the heat of fusion $H_f$ is due to the change in intramolecular vibrational energy of the assembly, this could modify simple rules of similitude based on positional disordering only. It is also interesting to note that in a dislocation theory of positional melting referred to on p. 296 there are considerable vibrational contributions to the overall entropy of fusion.

In general terms, the vibrational transition from one solid polymorph to another, discussed above, indicates how the higher vibrational entropy compared with the solid can compensate for the higher heat content of the melt, thus making it the stable phase at temperatures above $T_f$.

Various two-phase models have been examined from time to time to account for melting of crystals of simple structure entirely in terms of the vibrational change on passing from solid to melt. However in principle it seems unlikely that other basic mechanisms, in particular positional disordering, will make no appreciable contribution to melting so that they may be justifiably neglected. Brief reference may nevertheless be made to some calculations referring wholly to vibrational changes.

An equation of state for a simple vibrating solid can be constructed for which the available energy $A$ contains only three dominant terms (Herzfeld and Goeppert-Mayer, 1934):

$$A = \Phi + \tfrac{3}{2}Nh\nu + F^*$$

where $\Phi$ is the potential energy of the atoms under their mutual forces, and the second term is the zero-point energy of $3N$ vibrations. $F^*$, the contribution of the vibrational motions to $A$, is the only term directly dependent on temperature. The pressure can be expressed as

$$P = -\left(\frac{\partial A}{\partial V}\right)_T = P_1 + P_2 + P_3$$

in which the first two terms depend on the changes in equilibrium positions when the volume is changed homogeneously, and give the equation of state, and

$$P_3 = \left(\frac{E}{V}\right)\left(\frac{\partial(\log \nu)}{\partial(\log V)}\right)_T$$

$E$ is the vibrational energy at temperature $T$, given by

$$E = \int_0^T C_v \, dT$$

where $C_v$ is the specific heat at constant volume. $P_3$ is always positive for simple interaction forces between the atoms.

Empirically, for many solids the parameter

$$Y = \left(\frac{\partial(\log \nu)}{\partial(\log V)}\right)_T$$

known as the Gruneisen parameter, is practically constant. $P_1 + P_2$ together give a minimum at the equilibrium lattice spacing at 0 K; at higher temperatures the contribution of $P_3$ leads to thermal expansion and to a new series of minima. These have been tentatively identified with $T_f$ at the appropriate pressures, in a kind of mechanical theory of melting, based on the grounds that at still higher temperatures $P$ and $V$ increase together and the crystal is therefore mechanically unstable.

Calculations for argon show values of the melting temperature that are numerically too low on this mechanical theory of melting. On the other hand, using the criterion $\partial A_{vib}/\partial v = 0$ to define the melting temperature $T_{vib}$ in terms of the vibrational free energy $A_{vib}$ of an Einstein solid with characteristic frequency $\nu$ leads to values of $T_f$ for the inert gases, too high by a factor of about 3 (Bazarov, 1961). This emphasizes in another way that vibrational melting mechanisms are not thermodynamically competitive, taken alone. It is a fact that in addition to any changes of vibrational parameters, the liquid state is favoured by other factors; one such factor of quite general importance arises from the increase in positional entropy on melting (Chapter 5).

### 3.9.1. *Communal entropy*

A positional contribution can probably be introduced in the simplest possible way, as communal entropy. Though it has attracted some attention as being very general, this concept is unsatisfactory in specific applications. The usual argument is that if a melt is regarded as a quasicrystalline assembly, relative to the crystal the positional disorder makes a contribution to the entropy of fusion which can be expressed by the communal entropy $R$. In all, this leads to the expression

$$S_f = 3R \ln(\nu_S/\nu_L) + R \qquad (3.5)$$

where $\nu_S$ is the Einstein frequency of the crystal, and $\nu_L$ that of the atoms of the quasicrystalline melt (see Fowler and Guggenheim, 1939). However, application of this expression even to simple ionic melts yields values of $S_f$ that are too low; presumably this arises from the neglect of (more complex) positional defect contributions such as those discussed in Chapters 8 and 11. In any case the above models are too crude to justify simple additive treatment of the various entropy contributions as suggested in (3.5) (McQuarrie, 1962; see also Chapter 7). Communal entropy is further discussed in Chapter 11; see also Brusset, Kaiser, and Perrin (1968).

### 3.10 Vibrational properties of melts in relation to the crystalline state

Other evidence reviewed in subsequent chapters indicates that vibrational energy and entropy seldom constitute the only mechanism of melting  Vibrations of atoms or molecules in both crystal and melt are nevertheless of controlling importance for various physical properties related to melting. In the crystal, where long-range order permits the transmission of interactions between remote neighbours, a whole spectrum of vibrations is to be expected, with both longitudinal and transverse modes. In a melt, each atom or molecule vibrates subject to restoring forces which are more predominantly contributed by its nearest neighbours. The extent to which major changes in the phonon spectrum arise on melting can to some extent be predicted from the crystal structures. Many metals and ionic crystals with charged atoms provide the simplest cases, for which the melts have vibrational properties quite close to those of the crystals, allowing for the change in mean separation of units resulting from the change in volume on melting.

For other types of crystals, distinctive vibrational relationships between the solid and its melt are found, according to the general class of melt under discussion. Thus for molecular crystals (Chapter 6) interesting problems can arise concerning torsional oscillations and the onset of rotation either below $T_f$, or even not until well above it. When the molecules are flexible (see Chapter 7) changes in the vibrational spectrum on melting can be even more far-reaching. For network crystals the progressive growth of linkage fractures below and above the melting point (Chapter 10) must likewise affect the phonon spectrum in a special way, both with respect to their frequency distribution and their damping. This has been comparatively little studied up to the present.

### 3.10.1. *Specific heat*

In relation to the preceding paragraph it is noteworthy that the mean specific heats of melts of many metals lies close to that of the crystals (Chapter 9). When the specific heat of the crystals is predominantly due to the phonon spectrum, it seems likely that vibrational contributions also account for most of the thermal expansion of the melt near $T_f$. However, as the melt expands above $T_f$ non-vibrational contributions to the specific heat become progressively more important. One illustration is provided by simple models for positional defect melting mechanisms (e.g. Lennard-Jones and Devonshire, 1939) (and Chapter 11). If the rate of increase of defects (such as vacancies in the quasicrystalline melt) is $dn/dT$ per mole, their contribution to the specific heat is $\delta C = \varepsilon(dn/dT)$. The energy $\varepsilon$ of creating a defect such as a vacancy or 'hole' in the melt is of the same order as the heat of vaporization. Tentatively, near the freezing point the specific heat of the melt, on this basis, might be written

$$C_L \sim C_{vib} + H_{vap}\, dn/dT$$

Other factors leading to specific heats much above $3R$ per g atom are discussed on pp. 246 seq.

### 3.10.2. *Thermal expansion*

Defect creation also increases the volume. If the volume increase per defect is $\delta v$, approximately

$$\frac{dV}{dT} = \left(\frac{dV}{dT}\right)_{vib} + \delta v \cdot \frac{dn}{dT}$$

The parallelism between thermal expansion and specific heat may be fairly close for simple crystals. However, it should be stressed that no simple model for the structure of a melt proves to be universally applicable, as is also discussed in later chapters.

Like the specific heats, thermal expansions of melts of crystals of simple structure such as the metals are practically the same, usually slightly larger, than for the crystals (Chapter 9). Unfortunately and rather surprisingly, for melts of crystals with complex units of structure, no systematic investigation has yet been made of thermal expansions. The fraction of the total thermal expansion which may be attributed to vibrational contributions is an important quantity, whose value is not generally known for different classes of crystals.

## 3.11 Transport coefficients of melts as a diagnostic guide to their structure.

Theoretical interpretation of transport processes in liquids has often concentrated on those with particularly simple units of structure, such as molten metals. However, in melts near $T_f$ transport processes may show various kinds of anomalies; a systematic study of these for certain classes of melt could provide valuable diagnostic information about structure. Unfortunately, such information is as yet far from systematic. Examples are cited in various chapters of this book. With regard to transport properties in relation to vibrations the following paragraphs give some examples of simple theories. However, these theories do not allow for the fact that, even in fairly simple melts containing molecules, motions are coupled. Coupling of translation and rotation can be inferred for example from isotope effects on various transport coefficients, such as $NH_3/ND_3$ (Alei and Litchman, 1972).

### 3.11.1. *Viscosity*

The transmission of momentum from one layer to the next in a liquid in which there is a velocity gradient involves collisions between the atoms or molecules in these layers. Such 'collisions' in the melt can also be regarded (in a rather elementary way) as 'vibrations' of these molecules, whose amplitude is limited by repulsion barriers attributable to nearest neighbours. This consideration suggests that a close relationship might be found between the characteristic frequency of vibration $\nu_L$ and the viscosity of the melt, provided this has a simple structure. If a quasicrystalline structure is assumed for simple melts, the characteristic frequency $\nu_L$ differs from that of the crystal $\nu_S$ primarily because of the different specific volume. For an Einstein or a Debye model of the vibrating assembly a characteristic frequency $\nu_E$ or $\nu_D$ can be readily estimated from experimental specific heats for the crystals.

As already stated the crystal structure of many metals is apparently particularly simple and thus appears to lend itself to this kind of correlation between vibrational frequencies and viscosity of the melts. In fact structures may however be more complex (Chapter 9). In order to account for the viscosity of molten metals not too much above $T_f$ simple vibrational models have been proposed with a characteristic temperature of vibration $\theta_{\text{vis}} = h\nu_{\text{vis}}/k$ given by

$$\nu_{\text{vis}} = \tfrac{3}{4}(N^2/\rho M^2)\eta_{\text{f}} \tag{3.6}$$

where $\eta_{\text{f}}$ is the viscosity of the molten metal extrapolated to $T_{\text{f}}$. $N$ is Avogadro's number $6{\cdot}06 \times 10^{23}$, $\rho$ is the density of the melt in g cm$^{-3}$ and $M$ is the atomic weight.

An alternative formulation applicable to many metals (Andrade, 1934) is to write

$$\eta_{\text{f}} = CV^{-\frac{2}{3}}(MT_{\text{f}})^{\frac{1}{2}} \tag{3.7}$$

where $C$ is a constant of the order $5{\cdot}7 \times 10^{-4}$. Four molten metals studied

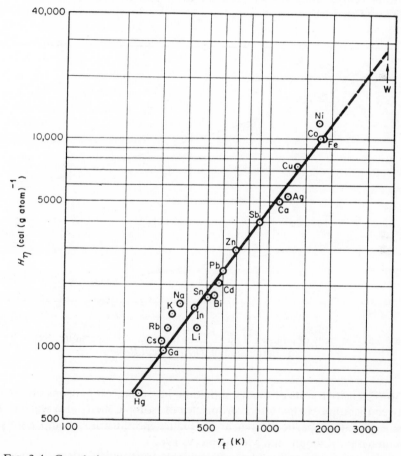

FIG. 3.4. Correlation between heat of activation for viscous flow $H_\eta$ and $T_{\text{f}}$ for molten metals. Reproduced with permission from Grosse, *J. Inorg. Nucl. Chem.*, **25**, 317 (1963)

with some accuracy (Rothwell, 1962) give values ($\times 10^4$) Sn 5·33; Pb 5·42; Al 4·44; Bi 3·76. This provides interesting verification of the model proposed for a molten metal as a vibrating close-packed assembly with $\nu_L$ not far from $\nu_S$ assuming (according to this model of viscosity) that at the extreme of every vibration the atoms take up the translational velocity corresponding with the plane of motion parallel to the directions of flow.

It seems likewise significant that $\theta_{vis}$ calculated from equation (3.6) agrees quite closely with characteristic Debye $\theta_D$ temperatures for the solid alkali metals, Li, Na, and K for which the ratios $\theta_{vis}/\theta_D$ are 1·03, 1·11, and 1·00 respectively (Andrade and Dobbs, 1952). An empirical correlation has been reported between the activation energy $H_\eta$ for viscous flow in the simple equation representing its temperature-dependence, and $T_f$ (Grosse, 1961, 1963 and Fig. 3.4).

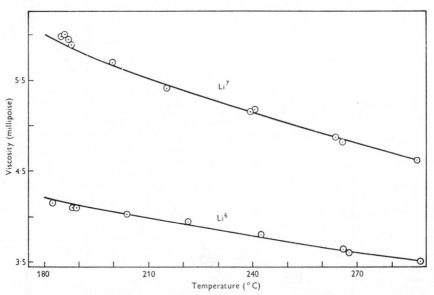

FIG. 3.5. Viscosity of melts of $^6$Li and $^7$Li as function of temperature. Reproduced by permission of *The Physical Review*

More sensitive tests of theory should be based on comparisons of viscosity between isotopic melts. Other properties of isotopic melts also warrant further exploration, particularly near $T_f$. Figure 3.5 illustrates comparisons for lithium (Ban, Randall, and Montgomery, 1962).

Until more experimental data are available from which comparisons can be made between $C_p$(solid) and $C_p$(liquid) at the melting point, other interpretations of melting in relation to the viscosity of the melts must be left with many queries unanswered. Considerable additional information is available for liquids that can be supercooled to form glasses. General problems are further discussed in Chapter 16.

### 3.11.2. Thermal conductivity

Insofar as heat transfer in melts can be described in terms of vibrational energy of the molecules, a change in thermal conductivity $\lambda$ on passing from crystal to melt should be directly related with the change in $\nu$ discussed in preceding sections.

On the basis of a vibrational theory of melting, it has been proposed that the product $\lambda \sqrt{M}$ should be constant at $T_f$ for melts whose crystals have the same Lindemann constant (see section 3.3). $M$ is the molecular weight. For a number of simple aromatic and aliphatic molecules this product is found to have the value $2 \cdot 4 \times 10^{-3}$ (Visvanathan and Sanghi, 1959).

As has repeatedly been emphasized, and as is further discussed below for melting of crystals with more complex molecular structures, the difficulty in applying any simple empirical rule such as the preceding is that melting may lead to other degrees of freedom becoming operative, such as torsional vibrations, and various kinds of molecular rotations. The extent to which any observed changes in thermal conductivity on melting are to be attributed to the role of such internal energy modes becoming unfrozen has hardly been examined. Theories of scattering of vibrational waves, or of transfer of internal energy by collision, have not been correlated to any marked extent with specific molecular structures. This is unfortunate, since thermal conductivity is difficult to interpret in detail just because it depends in a sensitive way on cooperative disordering in the solicd and melt near $T_f$. More sophisticated theories could make it a valuable research tool in studying the topology of melts.

As one example of what might be achieved, measurement of isotope effects in the thermal conductance of melts suggests that internal degrees of freedom of molecules may not contribute to thermal conductances to the same extent as their contribution to the specific heat (Horrocks, McLaughlin, and Ubbelohde, 1963). This could be explained in terms of relaxation effects in melts (see p. 149); though important in principle these relaxation effects have not yet been sufficiently explored to warrant detailed discussion here.

In addition to making comparisons between thermal conductivities of crystal and melt, comparisons can also be usefully made between crystalline and glassy forms of the same substance (see Chapter 16). For examples at 0°C for crystalline quartz $\lambda$ in cal cm$^{-1}$ s$^{-1}$ ranges between 0·016 and 0·017 per °C according to the different determinations; $\lambda_{cryst}$ decreases with rising $T$. In striking contrast with this, glassy quartz at 0°C shows values ranging between 0·00023 and 0·0036 at 0°C; $\lambda_{glass}$ increases with rising temperature. This notable difference illustrates the general comment that development of methods of measurement of the scattering of vibrations of various kinds promise to yield much information about melting problems.

### 3.11.3. *Temperature coefficients of electrical resistivity of metals*

One factor which accounts for the rise in the electrical resistivity of metals (such as is normally observed with increasing temperature) stems from the scattering of electrons by the thermal vibrations of the atoms. Changes in electrical resistivity of metals on passing from crystal to melt have been attributed to changes in the atomic vibrations. Since other more complex factors may also intervene, discussion of this aspect of the melting of metals is postponed to Chapter 9.

## REFERENCES

M. D. Aggarwal, V. Raju, and J. K. D. Verma (1973) *Rev. Roum. Phys.* **18,** 89.

M. Alei and W. M. Litchman (1972) *J. Chem. Phys.* **57,** 4106.

E. N. Da C. Andrade (1934) *Phil. Mag.,* **17,** 497. Supplement.

E. N. Da C. Andrade and E. R. Dobbs (1952) *Proc. Roy. Soc., Ser. A* **211,** 12.

N. T. Ban, C. M. Randall, and D. J. Montgomery (1962) *Phys. Rev.* **128,** 6.

L. P. Bazarov (1961) *Sov. Phys. Dokl.* **5,** 1293. cf. A. Bondi (1968) *Properties of Molecules, Crystals and Liquids,* Wiley, New York.

H. Brusset, L. Kaiser, and F. Perrin (1968) *J. Chim. Phys.* France, **65,** 260.

D. Y. Chung, D. J. Gunton, and G. A. Saunders (1976) *Phys. Rev. B,* **13,** 3239.

R. K. Crawford in *Rare Gas Solids* (1976) ed. M. L. Klein and J. A. Venables, Academic Press, New York.

R. K. Crawford and D. J. Montgomery (1957) *Bull. Am. Phys. Soc.* **2,** 299.

J. M. Dickey and A. Paskin (1969) *Phys. Lett.* **30A,** 209.

E. R. Dobbs and G. O. Jones (1957) *Rep. Progr. Phys.* **20,** 516.

J. S. Dugdale and D. K. C. McDonald (1954) *Phys. Rev.* **96,** 57, 1905.

J. S. Dugdale and F. E. Simon (1953) *Proc. Roy. Soc., Ser. A* **218,** 291.

W. J. Dunning (1961) *J. Phys. Chem. Solids* **18,** 21.

R. Fowler and E. A. Guggenheim (1939) *Statistical Thermodynamics,* Cambridge University Press.

J. J. Gilvarry (1956a) *Phys. Rev.* **102,** 3, 308, 317.

J. J. Gilvarry (1956b) *Phys. Rev.* **103,** 1700.

J. J. Gilvarry (1956c) *Phys. Rev.* **104,** 909.

R. Gopal (1955a) *Z. Anorg. Chem.* **278,** 42.

R. Gopal (1955b) *Z. Anorg. Allg. Chem.* **279,** 229.

R. Gopal (1955c) *Z. Anorg. Allg. Chem.* **281,** 217.

R. Gopal (1965) *J. Indian Chem. Soc.* **42,** 297.

E. R. Grilly (1971) *J. Low Temp. Phys.* **4,** 16.

G. Grimvall (1975) *Physica Seripta* **12,** 168.

A. V. Grosse (1961) *J. Inorg. Nucl. Chem.* **23,** 333.

A. V. Grosse (1963) *J. Inorg. Nucl. Chem.* **25,** 317.

E. A. Guggenheim (1945) *J. Chem. Phys.* **13**, 253.
E. A. Guggenheim and M. L. McGlashan (1960) *Mol. Phys.* **3**, 563.
K. F. Herzfeld and M. Goeppert-Meyer (1934) *Phys. Rev.* **46**, 995.
J. Horrocks, E. McLaughlin, and A. R. Ubbelohde (1963) *Trans. Faraday Soc.* **59**, 1110.
L. Hunter and S. Siegel (1942) *Phys. Rev.* **61**, 84.
Y. Ida (1969) *Phys. Rev.* **187**, 951.
K. Ishizaki, P. Bolsaitis, and I. L. Spain (1973) *Phys. Rev.* **7B**, 5412.
K. Ishizaki, I. L. Spain, and P. Bolsaitis (1975) *J. Chem. Phys.* **63**, 1401.
J. N. S. Jackson and R. C. Liebermann (1974) *J. Phys. Chem. Solids* **35**, 1115.
R. T. Johnson and J. C. Wheatley (1970) *J. Low Temp. Phys.* **2**, 423.
R. T. Johnson, O. V. Lounasmaa, R. Rosenbaum, O. G. Symko, and J. C. Wheatley (1970) *J. Low Temp. Phys.* **2**, 403.
L. Jollivet (1951) *C.R. Acad. Sci., Paris* **232**, 966.
W. E. Keller (1969) *Helium³ and Helium⁴*, Plenum Press, New York.
R. W. Keyes (1959) *Phys. Rev.* **115**, 564.
D. I. Khomski (1969) *Sov. Phys. Solid State* **11**, 163.
J. E. Lennard-Jones and A. F. Devonshire (1939) *Proc. Roy. Soc., Ser. A* **169**, 317; **170**, 464.
F. A. Lindemann (1910) *Phys. Z.* **11**, 609.
D. A. McQuarrie (1962) *J. Phys. Chem.* **66**, 1508.
H. Michels and A. R. Ubbelohde (1974) *Proc. Roy. Soc., Ser. A* **338**, 447.
H. Nowotny (1958) *Hochschmelzende Metalle 3*, Plansee Seminar.
M. Omini (1972) *Phil. Mag.* **26**, 287.
D. D. Pathak and N. G. Vasavada (1970) *Acta Cryst.* **26A**, 655.
S. L. Quimby and S. Siegel (1938) *Phys. Rev.* **54**, 293.
M. R. Rao (1944) *Current Sci.* **13**, 225.
Roberts and Miller (1960) *Heat and Thermodynamics*, Blackie and Son London.
M. Ross (1969) *Phys. Rev.* **184**, 233.
E. Rothwell (1962) *J. Inst. Metals* **90**, 389.
M. Scol (1961) *Z. Phys.* **164**, 93.
J. N. Shapiro (1970) *Phys. Rev.* **1B**, 3982.
K. Shimada and M. Yokota (1974) *J. Phys. Soc. Japan* **36**, 1356.
F. E. Simon and C. A. Swenson (1950) *Nature, Lond.* **165**, 829.
W. B. Street and L. A. K. Staveley (1971) *J. Chem. Phys.* **55**, 2495.
N. Trappeniers (1950) *3me Congrès National des Sciences*, Brussels, Chimie Physique **1**.
S. B. Trickey, W. P. Kirk, and E. D. Adams (1972) *Rev. Mod. Phys.* **44**, 668.
E. T. Turkdogan and J. Pearson (1953) *J. Appl. Chem.* **3**, 495.
A. R. Ubbelohde (1940) *Ann. Rep. Chem. Soc.* **36**, 167.
A. R. Ubbelohde (1941) *Ann. Rep. Chem. Soc.* **37**, 150.
A. R. Ubbelohde (1961) *J. Phys. Chem. Solids* **18**, 90.
A. R. Ubbelohde, H. J. Michels, and J. J. Duruz (1972) *J. Physics E* **5**, 283.
S. Visvanathan and I. Sanghi (1959) *Naturwiss.* **46**, 647.
J. Wilks (1967) *Liquid and Solid Helium*, Oxford University Press.

# 4. SOLID–SOLID TRANSFORMATIONS RELATED TO FUSION

## 4.1 Structural interpretations of phase transformations in crystals

From the standpoint of classical thermodynamics, the change from one solid phase to another can be discussed in exactly the same way as the change from a solid to a melt. Considered in all its aspects, the general theme of solid–solid transformations has a range extending far beyond the scope of the present book. However, when regard is paid to the structures of the phases involved, certain aspects of solid–solid transformations also prove to be of particular importance for the investigation and interpretation of fusion. Generally, this is so when the solid polymorphs have fairly closely related crystal structures. In such cases, the precision with which crystal structures can often now be determined makes it possible to identify with considerable certainty the *mechanisms of entropy increase* on passing from the form stable below to the form stable above the transition temperature. The importance of such studies for melting lies in the fact that some of these mechanisms of entropy increase in crystals are also found on passing from the crystal to the melt.

A second way in which study of polymorphic transformations in solids has proved particularly valuable for understanding melting originates from those changes which prove to be thermodynamically more or less continuous. Precursor phenomena over a range of temperatures on either side of a jump in properties at a transition point are quite common when two solid polymorphic forms are closely related. In the limit, when this relationship is extremely close, so called 'lambda points' or 'transformations of higher order' are found in crystals. These may be associated with hybrid crystals containing microregions, in a way which points to analogous conglomerate structures in certain melts.

In what follows, only those aspects of solid–solid transformation particularly relevant to melting are discussed at all fully. Even within this selective restriction, detailed knowledge about phase transformations has grown enormously since the first edition of this book. Refined structural studies have revealed many clear examples of the *coexistence* of closely allied structures in *hybrid* crystals (see p. 111). In many other examples where such structural evidence is less clear cut, more or less pronounced hysteresis has been observed which very probably can be attributed to the mechanisms discussed below. Essentially, these involve nucleation and growth of domains of the transform starting at the surface or at defect regions within single crystals.

Acoustic studies in the neighbourhood of phase transformations have also led to the detection of precursor break-up of single crystals into hybrids, whose absorption of ultrasonic waves can show notable increases.

### 4.1.1. *Some mechanisms of entropy increase in crystal transformations*

Increases of the overall vibrational entropy in crystals, such as can lead to polymorphic transformations in tin (Chapter 3) arise from momentary displacements of the structural units of the crystal lattice, from definable positions of lowest potential energy. These displacements, which lead to the well-known vibrations in crystal lattices, have relaxation times of the same order as the vibrational frequencies associated with them. Vibrations can thus be expected to be in thermal equilibrium for all ordinary properties of condensed states of matter. Comment has already been made on conditions in a crystal, under which phonon anomalies arise, particularly near a crystal transformation.

Improvements in acoustic measurements have favoured the rapid growth of studies on specific 'soft vibrational modes' whose amplitude becomes unstable; for these the phonon frequency tends to zero, as the temperature of a crystal is raised to the transition at $T_c$ from one crystal lattice to another. Even when the transition involves a discontinuous jump of thermodynamic properties, the increasing anharmonicity as $T_c$ is approached usually influences 'average' properties such as the specific heat and thermal expansion, as well as certain optical properties (see Fleury, 1971, 1972; Worlock, 1971).

'Soft phonon modes' can thus make an important contribution to precursor anomalies of these physical parameters. By way of examples, this appears to be the case for the transitions in $KNO_2$ (Ema, Hamano, Hatta 1975), $RbAg_4I_5$ (Graham and Chang, 1975), $C_sPbCl_3$ (Yamada, 1973), $Ti_2O_3$ (Chi and Sladek, 1973), $P_2Cl_{10}$ (Chihara, Nakamura, and Tachiki 1973), tetra-chloro-$p$-benzoquinone (Chihara and Nakamura, 1973), $TlN_3$ (Iqbal and Christoe, 1974), $AuCuZn_2$ (Mori and Yamada, 1975), $Gd_2(MoO_4)_3$ (Courdille and Dumas, 1974), $K_2ReCl_6$ (Van Driel, Armstrong, and McEnnan 1975), and $RbCaF_3$ (Rushworth and Ryan, 1976).

In these fairly complex crystal structures it is often possible to locate regions of particular ligand instability, for example because of a steep dependence of force fields on interatomic distances. Such effects are reminiscent of the overall vibrational instability postulated in the primitive Lindemann theory of melting, but they can be much more localized in the crystal lattice.

Whilst phonon studies near a solid–solid transformation temperature quite often establish precursor anomalies, their attribution to specific structural peculiarities may remain ambiguous. For example, in $K_2SO_4$ (Kolontsova, Kulago, Byakhova and Mikhailenko 1973) and in AgI (Shaw, 1974) an accumulation of point defects has been cited as a likely cause of pretransition anomalies.

Structural information about thermal transformations in condensed phases is still too patchy to permit any systematic survey of all the consequences of vibrational instability. A possible example may be found in the transformation in caesium chloride at 469°C (Menary, Ubbelohde, and Woodward, 1951). The *gradual* precursor changes preceding certain transformations are also probably due to vibrational effects (Ubbelohde, 1963). For specific vibrational modes, if not for the whole spectrum, extreme vibrational anharmonicities are likely to precede jumps to new minima of potential energy, made by the units of structure in a solid–solid transformation. Build-up of anharmonicity arises from the form of the restoring potential for displacements in a cooperative assembly, which in many cases is far from parabolic, and which is often quite sensitive to repulsion energies and thus to the interunit distance or crystal volume as this undergoes thermal expansion.

When isolated units make the jump to a new minimum of potential energy, this would constitute a crystal lattice defect. Generally, for isolated defects the vibrational frequencies and other parameters such as electronic excitation will differ from those in undisturbed parts of the parent lattice. As more and more units jump to assume a defect condition, they will exert a cooperative influence on the parent lattice also. Usually at a critical concentration of defects their cooperative influence favours an alternative lattice structure, for thermodynamic reasons. The interesting question arises, whether cooperative disordering of this kind shows nucleation and growth of domains within the more perfect lattice, or whether it is more or less smoothly diffused. For some transformations experimental evidence suggests a clustering of defects into regions or domains (Thomas, Giray and Parks 1973; Tong and Wayman, 1974).

### 4.2 Positional randomization in solid–solid transformations

Amidst the great diversity of cooperative changes that can occur in crystals, and that are accompanied by changes of crystal structure (see Jaffray, 1948; Ubbelohde, 1952, 1957) positional disordering and orientational disordering are most general. Their study has been of major importance in developing structural interpretations of melting.

Solid–solid transformations in which there is an increase of entropy for the high-temperature form, when the units of structure merely take up alternative positions at random, with different minima of potential energy, appear to be comparatively rare. Such cases are nevertheless highly significant for melting, since positional randomization is a universal mechanism of fusion.

One class of crystal showing positional randomization above $T_c$ includes certain salts of silver, such as AgI (Helmholz, 1935) and $AgHgI_2$ (Ketelaar, 1934). X-ray studies (see Wells, 1962) show that below $T_c$ the small silver cations ($r_+ = 1.26$ Å) in these crystals occupy a regular array of lattice points between the large iodine anions ($r_- = 2 \cdot 16$ Å). Above $T_c$ these cations become redistributed at random between a group of sites, for some of which the potential energy of occupation is somewhat higher. Hopping from one type of site to the other under external influence, such as an electric field, occurs with great readiness particularly in AgI, so that the electrical conductivity of its high-temperature form ($1 \cdot 3$ ohm$^{-1}$ cm$^{-1}$) is comparable with that of many ionic melts. One way of describing this transformation is that the cation lattice has actually 'melted' above $T_c$ whereas the anion lattice retains its crystalline long range order up to the thermodynamic melting point $T_f$. (see chapter 8).

Marked precursor phenomena are found in thermodynamic properties such as the specific heat of silver halides. Electrical conductivities of the solids likewise show precursor phenomena (see p. 219). Related defect structures are found for sulphides, selenides, and tellurides of $Cu^+$ and $Ag^+$, but these have been less fully characterized (see Wells, 1962).

The combination of a rigid anion lattice with a 'quasimolten' cation lattice has become technically useful in providing solids with high ionic conductivities, sometimes termed 'solid electrolytes'. This fact has intensified but also channelled scientific study of such crystals. Modes of transformation from the low-temperature form, in which binding may be ionic or even covalent, and the high-temperature form, in which one of the ionic species has high mobility, involve features which can only be briefly summarized with special reference to matters discussed in the present chapter. Some examples are listed in Table 4.1.

**Table 4.1 Positional ionic randomization in crystals**

| Crystal | $T_c$ (°C) | Authors | Feature |
|---|---|---|---|
| $RbAg_4I_5$ | | Pardee and Mahan (1974) | Presence of ordered micro-domains seems likely |
| $KAg_4I_5$ $RbAg_4I_5$ | | Bradley and Munro (1969) | Both these salts show high conductivities owing to positional disordering of $Ag^+$ but no lambda peak was found. Others (Owen and Argue, 1967) find a lambda peak for both—possibly owing to stoichiometric differences in the precise composition |
| $Ag_2HgI_4$ | 50 | Browall and Kasper (1974) | Transition is from single crystals of the high-temperature form to multi-option domains of the low-temperature form |
| AgI | | Perrott and Fletcher (1968a,b) | The peak $C_p$ anomaly is found only for the stoichiometric compound and the peak is suppressed when excess silver is taken up by the lattice |

In a similar kind of way, study of the polymorphic form II of $AgNO_3$ which is stable above 159·4°C, at 1 atm pressure has verified that its electrical resistance sinks to very low values at pressures above about 5 kbar, so that there is practically no break on melting around 345°C under this pressure (Pistorius, 1961). It is not yet clear whether this convergence of properties between crystal and melt originates from high lattice mobility arising from positional defects, particularly for the $Ag^+$ cation, or whether some kind of electronic conductance may be responsible. Analogy with AgI suggests that cation melting may have occurred. Phenomena of this kind in which specific physical properties of solid and melt become indistinguishable illustrate certain kinds of 'critical' melting. However, unless the thermodynamic properties of solid and melt likewise become indistinguishable at the same time, this kind of asymptotic melting temperature is separated and distinct from a possible thermodynamic critical point of fusion. It has obvious analogies with $T_{mech}$ or $T_{vib}$ as discussed in Chapter 3.

A second group of crystals showing marked positional randomization above $T_c$ includes a number of ordered alloys, such as $Fe_3Al$, CuZn, $Cu_3Pt$, etc. (see Ubbelohde, 1952, for an elementary discussion). In these metallic solids, the free energy below $T_c$ is lowest if the dissimilar atoms occupy sites on two interpenetrating lattices. Above $T_c$, the entropy increase on permitting redistribution of the atoms on either lattice at random gives lower free energy to the disordered crystal, despite the increase in enthalpy involved.

Order–disorder transformations in alloys at one time proved extremely valuable in pointing to workable statistical treatments of positional melting. In view of the subsequent comparatively successful development of theories directly applied to fusion (as is discussed later, Chapter 11), the statistical thermodynamics of the solid–solid positional disordering transformation need not be discussed here. A special feature refers to the formation of cooperative defects. Unfortunately, studies of positional disordering transformations in ordered alloys or other solids have seldom or never been investigated with single crystals. Significant information thus still remains to be derived about this kind of change, with regard to domain formation (e.g. Thomas, Giray and Parks 1973). Domain formation is made likely by various indirect evidence, such as the reported occurrence of hysteresis in certain order–disorder transformations. This important aspect of phase transformations is at present better discussed in relation to other kinds of cooperative changes in crystals, for which structural information about crystal hybridization is available (section 4.5).

## 4.3 Orientational randomization in crystals with quasispherical molecules

In studies of the chemical physics of melting, transitions in solids that involve orientational disordering have played an important part in drawing attention to the fact that crystals can comprehend several mechanisms of melting. In certain cases, positional and orientational disordering may operate simultaneously at the melting point $T_f$ whereas, in others, mechanisms of disordering other than positional may be strung out stepwise on the scale of temperatures, which may considerably facilitate their investigation and elucidation.

Examples of orientational transitions in solids appear to have been first discovered in specific heat studies of ammonium salts and of crystalline methane (see Ubbelohde, 1940; Extermann and Weigle, 1942, for references). As illustrated in Figs. 4.1 and 4.2, the specific heat in such

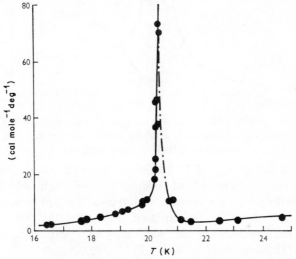

FIG. 4.1. Molar heat capacity of methane $CH_4$ around
its lambda point

transitions rises to a peak value at a characteristic temperature $T_c$ and falls more or less gradually back to values characteristic of vibrations in the crystal lattice. From the shape of the specific heat curves, which are rather fancifully compared with the Greek capital lambda, such effects are generally known as 'lambda type' phenomena. By plotting the specific heat against $\ln T$, the integral under the curve $\int C_p \, d(\ln T)$ can be evaluated. Subtraction of the estimated increase of vibrational entropy yields the excess entropy $S_c$ taken up by the solid on passing through the transition region. Illustrative values are recorded in Table 4.5 below. Accompanying this specific heat and entropy increase, various other changes can frequently be detected, provided sufficiently refined methods of observation are used. Dilatometry measures changes in the specific volume, which are frequently only small and which may even be negative for some substances. X-ray diffraction in favourable cases has revealed small changes of spacing, as illustrated for the ammonium

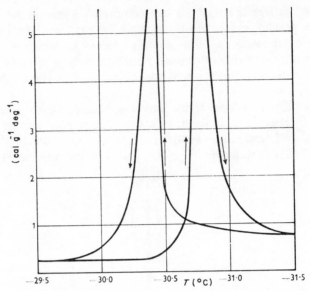

FIG. 4.2. Specific heat of ammonium chloride (showing
hysteresis) around its lambda point. Reproduced by
permission of Birkhauser Verlag, Basel

chloride transition in Fig. 4.3. Extensive research has revealed lambda
phenomena for a great diversity of systems, in all of which some parameter in
the crystal passes from a cooperative ordered state at low temperatures, to a
disordered state above $T_c$. For melting, only positional and orientational
disordering are mechanisms of wide general importance as occurring
separately, or simultaneously at $T_f$.

In the most general sense, the increase of vibrational entropy of a crystal is
only one version of its general increase of disorder as the temperature rises,
since on average the increasing amplitude of vibrations implies less time spent
in the ideal lattice positions. Restoring forces in vibrations are similar to those
operative for positional and orientational disorder in crystals. However unlike
positional disorder, for example, vibrational disorder has a very short relaxation time to attain equilibrium. When a phonon pulse is passed through the
solid, its relaxation time is probably of the order of a few lattice spacings
divided by the velocity of sound. As already stated, in nearly all physical
measurements on condensed states of matter, vibrations may therefore be
assumed to be in statistical thermal equilibrium unless a temperature gradient
is maintained. Other significant modes of entropy increase available to crystals as their temperature rises have considerably longer relaxation times, since
the units of structure have to move to new minima of potential energy.
'Frozen-in' distributions are quite often found with respect to such modes of
entropy increase.

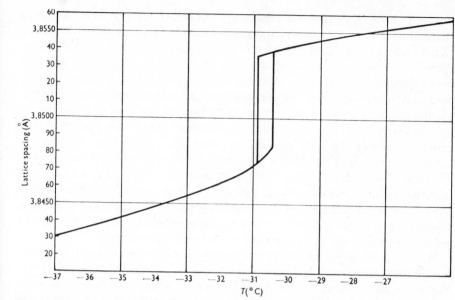

FIG. 4.3. Lattice spacing of ammonium chloride single crystal around its lambda point. Reproduced by permission of Birkhauser Verlag, Basel

A model which has proved to be of wide importance for crystals whose units of structure are polyatomic molecules refers to the mutual *orientations* of such units in the crystal lattice. According to classical theory of crystallographic space groups and of the underlying molecular point-group symmetry, whenever the unit of structure is non-spherical the directions of the molecular axes undergo oscillations about equilibrium directions in the crystal. Though they may exhibit unusual anharmonicities as mentioned in section 4.1.1, such oscillations must be regarded as forming part of the ordinary vibrational spectrum of the crystal. In addition, jumps of orientation over a potential barrier to new equilibrium orientations of higher energy can also arise. In the limit, non-spherical molecules may actually rotate more or less freely through all possible orientations, even if certain directions of the molecular axes present energy minima and are thus preferred.

At one time, it was thought that thermal transformations involving orientational randomization were always due to the onset of actual 'free rotation' (or spinning about specific molecular axes) above the peak transition temperature $T_c$. On this basis, the autocatalytic shape of a lambda curve was explained by the fact that the first few molecules to rotate in a crystal would have to surmount the maximum barrier $V_{rot}$ opposing random orientation of their axes with respect to their neighbours. However, as more and more neighbours pass into the rotating conditions, with rise in temperature, the barrier opposing the rotation of still more neighbours would be on average reduced. Thus $V_{rot}$ may be expected to decrease steeply with increasing fraction $n/N$ of rotating units. Statistical mechanical theory shows, however, that similar effects would already be found in crystals containing 'hindered' rotators. In these the units of structure all have the same orientations of lowest potential energy with respect to the crystal lattice at low temperatures: torsional oscillations about these energy minima occur in the normal manner. As the temperature rises, an increasing fraction of the units of structure flip over to orientations in selected alternative directions in the crystal lattice with potential energy $U_{\theta_1}$, $U_{\theta_2}$ etc. higher than $U_{\theta}$. These new minima are separated from one another by potential barriers opposing free rotation. Whenever the values of the minima of potential energy $U_{\theta_1}$, $U_{\theta_2}$ etc. are affected by the fraction $n/N$, a lambda shape is likely for the additional specific heat introduced by the randomization. The general qualitative argument given above does not make it clear whether or not there will also be a discontinuous jump at $T_c$ from one crystal state to another.

Although thermodynamic measurements *per se* cannot yield any very detailed information, it has been shown in certain cases that even above $T_c$ the molecules behave as 'hindered' rotators. Since vibrations about the minima $U_{\theta_1}$, $U_{\theta_2}$ etc. involve both kinetic and potential energy these make an orientational contribution $C_{or} = 3R$ per mole. 'Free' rotators would make a lower specific heat contribution $C_{or} = 3R/2$.

Studies particularly of the heat capacity of crystals containing non-spherical molecules have revealed a wide variety of thermal transitions which can be attributed to increasing randomization of molecular axes. In the description followed up to the present, the low-temperature form 1 would contain non-spherical molecules all with axes oriented in specific ways, and the high-temperature form would contain the same molecules but with their axes in alternative directions, with a greater degree of randomization. X-ray diffraction studies of such transformations often show that the crystal structures of forms 1 and 2 have practically the same lattice spacings, as is to be expected from the comparatively small change in the interactions between the molecules, owing to randomization of their molecular orientations to a greater or lesser degree. Randomization of molecular axes above $T_c$ has been verified by various refined measurements not all of which can be detailed here. By way of example, nuclear magnetic resonance studies have been carried out on the alkali salts $M^+[PF_6]^-$ where M is Na, K, Rb, or Cs (Miller and Gutowsky, 1963), which show lambda transformations between 100 K and 160 K with pronounced hysteresis whose dependence on crystal size and previous history indicates that domains of the transform nucleate and grow within the alternative structure.

Figures 4.2 and 4.4 illustrate some general features of specific heat data and volume changes accompanying lambda transformations for ammonium salts. As is discussed in following sections, the marked precursor rise in specific heat

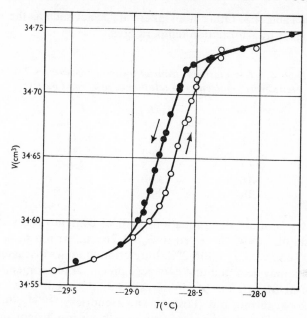

FIG. 4.4. Molar volume of $NH_3DCl$ around its lambda point

and apparent absence of latent heat of transformation have sometimes been considered to be characteristic of all lambda-type transformations. In conformity with the attribution of the lambda transformation to the randomization of orientation of the $NH_4^+$ cation, the peak transformation temperature $T_c$ is practically the same for this cation in salts with a diversity of anions. This is illustrated in Table 4.2. 'Tetrahedral' molecules analogous in structure to $NH_4^+$ show corresponding transformations at temperatures which are lower in the case of weaker crystal fields, for example in crystalline methane, but on a rising trend as the force fields become stronger, for example $SiH_4$ as

**Table 4.2** *Width of hysteresis loop* $\Delta T_c$ *in relation to* $\Delta V_c$

| Salt | $T_c\,(K)$ | $\Delta T_c\,(K)$ | $\Delta V_c$ ($cm^3\,mole^{-1}$) |
|------|------------|-------------------|----------------------------------|
| $NH_4Br$ | 234·4 | 0·06 | 0·03 |
| $ND_3HCl$ | 247·95 | 0·07 | 0·07 |
| $ND_4Br$ | 214·9 | 0·11 | 0·08 |
| $NH_3DCl$ | 244·6 | 0·12 | 0·11 |
| $NH_4Cl$ | 242·8 | 0·35 | 0·15 |
| $(NH_4)_2SO_4$ | 223·4 | 1·2 | 0·35 |
| $(ND_4)Br$ | 168·1 | 9·0 | 0·6 |

See Ubbelohde (1957).

illustrated in Table 4.3. Other non-spherical species such as the nitrate ions have even higher transformation temperatures.

**Table 4.3. Orientational transformation temperatures $T_c$ of tetrahedral molecules in relation to rotational barriers**

| Molecule | $T_c(K)$ | $T_f(K)$ |
|---|---|---|
| $CH_4$ | 20·5 | 90·6 |
| $SiH_4$ | 63·5 | 88·5 |
| $CF_4$ | 76·3 | 84·5 |
| $CCl_4$ | 222·5 | 250·3 |
| $NH_4Cl$ | 242·8 | — |
| $CBr_4$ | 320 | 365·5 |

For $NaNO_3$ accurate determinations of heat content observed by calorimetry (Mustajoki, 1957) have led to $S_c = 1·26$ e.u., with a peak $T_c = 276°C$ and $S_f = 6·03$ e.u. with $T_f = 306·2°C$. In relation to topics discussed in following sections, it may be noted that extensive precursor phenomena accompany the transformation at $T_c$. No hysteresis was detected. The crystal form remains trigonal throughout (Posnjak and Hendricks, 1931); the degree of refinement investigated may of course have been insufficient to reveal the differences involved.

An important thermodynamic means of information about orientational transformations in crystals is to compare the behaviour of isotopic molecules. In the majority of cases, these comparisons have been made for the molecules $AH_n$ and $AD_n$ where A is a central atom, such as C, S, Se, etc., bonded to either hydrogen or deuterium.

Probably the most striking differences in orientational transitions of isotopic molecules are found in the series of carbon derivatives $CH_4$, $CH_3D$, $CD_4$. Specific heat curves at ordinary pressures are illustrated in Figs. 4.5 and 4.6 (Clusius and Popp, 1940). In ordinary determinations, whereas $CH_4$ has only a single clearly defined transition at 20·42 K (Fig. 4.1) a double peak is found for the deuterated molecules. Experimental evidence shows that the lower transition is much less sharp in $CH_4$ but can be detected below 10 K in refined observations.

No complete theoretical interpretation has been put forward, but a plausible view is that mutual orientation of these tetrahedral molecules can exhibit several minima in the crystal. Although no precise X-ray measurements have been made, there is reason to anticipate that at the same temperatures the crystal lattice of $CD_4$ will have slightly smaller unit cell dimensions than $CH_4$, from the fact that the C—D bond length in deuteromethane must be slightly smaller than for C—H in methane, owing to zero-point energy differences (Ubbelohde, 1936). The repulsion envelope of the deuterated molecules should thus be slightly smaller. Since the intermolecular attractions in the

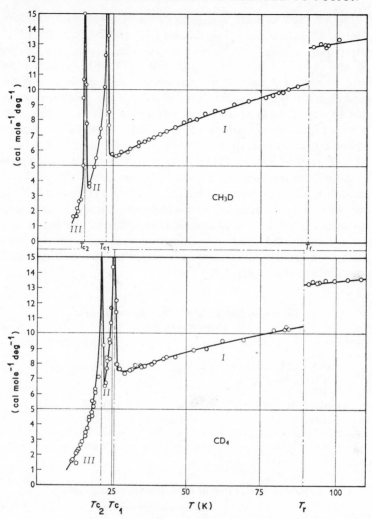

FIG. 4.5. Molar heat capacity of $CH_3D$
FIG. 4.6. Molar heat capacity of $CD_4$

crystal are due to the peripheral electrons, they are practically identical in these isotopic molecules. Moments of inertia are very different, so that the zero-point energy and rotational quanta of these molecules show large divergences. (For detailed discussion of alternative interpretations of the differences observed see Colwell, Gill, and Morrison, 1962.)

It is interesting to note, in relation to later chapters in this book, that despite often quite marked differences in transition points, melting points of these isotopic crystals are practically the same (see p. 45). Thus for $CH_4$,

$T_{trip} = 90 \cdot 675$ K, and for $CD_4$, $T_{trip} = 89 \cdot 784$ K (Colwell, Gill, and Morrison 1962). Entropies of fusion are $2 \cdot 48$ e.u. for $CH_4$ and $2 \cdot 42$ e.u. for $CD_4$.

Corresponding studies have been made on other deuterium compounds. For $H_2Se$ and $D_2Se$ (Kruis, 1941) substitution of D for H leads to a lowering of melting point and a rise in two transition points below $T_f$. Within the experimental error the transformation processes for mixtures of isotopic molecules appear to be as sharp as for any single species. It does not seem possible to attribute these isotope effects to any single factor, but melting points show a linear trend from $207 \cdot 4_8$ K for $H_2Se$ to $206 \cdot 2_4$ K for $D_2Se$ (99·6% D). This gives a ratio $T_f(H)/T_1(D) = 1 \cdot 005_8$ whereas the vibrational melting theory would require a ratio {see equation (3.11)}$(V_1/V_2)^{2/3}$. Unfortunately the isotopic volumes $V_1$ and $V_2$ of solid $H_2Se$ and $D_2Se$ are not yet known with sufficient accuracy to test this comparison. Other examples of isotopic effects on thermal transformations in solids will be found in Tables 4.2 and 4.4. Usually, the solid–solid transformation peaks for isotopic molecules differ by a greater temperature interval, than the melting points. An example is illustrated in Table 4.5.

**Table 4.4. Isotope effects in some solid–solid transformations**

| Salt | $T_c(H)(K)$ | $T_c(D)(K)$ |
|------|-------------|-------------|
| $KNaC_4H_4O_6 4H_2O$ (Rochelle salt) | 255–258 | 250–251 |
| $KH_2PO_4$ | 122 | 213 |
| $(NH_4)H_2AsO_4$ | 215·8 | 298·6 |
| $KH_2AsO_4$ | 95·6 | 162·0 |
| $RbH_2AsO_4$ | 109·9 | 177·8 |
| $CsH_2AsO_4$ | 143·3 | 212·4 |

See refs. in Ubbelohde (1957).

**Table 4.5. Orientational transformations and melting**

| Molecule | $T_f(K)$ | $S_t$ (e.u.) | Orientl. transformations $T_{c2}$ | $S_{c2}$ | $T_c(K)$ $T_{c1}$ | $S_c$ (e.u.) $S_{c1}$ |
|----------|----------|--------------|-----------------|----------|-------------------|-----------------------|
| KCN | 907 | — | 167 | — | — | — |
| $KNO_3$ | 606 | 4·0 | 401 | — | — | — |
| $O_2$ | 54·3 | 2·0 | 43·7 | 4·0 | 23·7 | 0·9 |
| $N_2$ | 63·1 | 2·7 | 35·4 | 1·5 | — | — |
| CO | 68·1 | 2·9 | 61·5 | 2·5 | — | — |
| HCl | 158·9 | 3·0 | 98·4 | 2·9 | — | — |
| HBr | 186·2 | 3·1 | 114 | 0·7 | 89·0 | 0·7 |
| HI | 222·3 | 3·1 | 125 | 1·5 | 70·0 | 0·3 |
| $H_2S$ | 187·6 | 3·0 | 126·2 | 0·9 | 103·6 | 3·5 |
| $D_2S$ | 187·1 | 3·0 | 132·8 | 0·9 | 107·8 | 3·7 |

Refs in Ubbelohde (1940); for KCN in Cimino, Parry, and Ubbelohde (1959), and for $KNO_3$ in Kennedy, Ubbelohde, and Woodward (1953).

### 4.3.1. *Orientational transformations in crystals with quasicylindrical molecules*

The entropy of specific crystal structures can in principle be increased by a variety of randomizations of the mutual orientation of molecules. Whenever the molecular interactions are small, even molecules whose repulsion envelope is far from 'roughly spherical' may present diverse packing arrangements in the crystal lattice for which the energy difference $\varepsilon$ leads to a characteristic temperature $\varepsilon/R$ less than $T_f$; these are then likely to appear as orientational transformations in the crystals below $T_f$. Preceding examples have referred to groups with quasispherical symmetry. Quasicylindrical symmetry can lead to allied orientational transformations. A group of molecules which offers many examples of such transition have a polymethylene chain, $R_1(CH_2)_n R_2$. The end groups $R_1$, $R_2$ may be methyl (for paraffins) or various polar groups such as OH, COOH, I, etc. A typical transition of this kind is illustrated in Fig. 4.7 for n-pentadecane (Ubbelohde, 1938). Transitions in solids of this kind may be highly sensitive to impurities in solid solution in the crystals, which can have a profound effect on repulsion barriers. This modifies the energy difference $\varepsilon$: the kinetics of onset of the transformation can also be modified, leading to varying degrees of hysteresis. Such influences appear to account for the shift in $T_c$ for hexadecyl alcohol with changing impurity content (Kachiuthi and Sakurai, 1949).

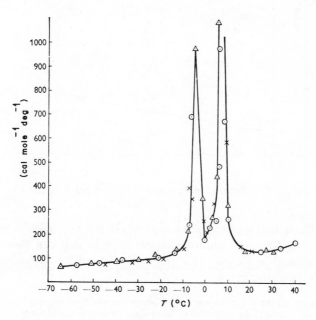

FIG. 4.7. Molar heat capacity of n-pentadecane

4.3.2. *Sensitiveness of orientational transformations to crystal lattice inter-actions*

Various general sources of information indicate that the onset and the peak temperature $T_c$ of orientational randomization in a crystal can be particularly sensitive to influences from neighbours in the crystal. Presumably this is because the potential barriers opposing randomization are the outcome of cooperative interactions involving a number of molecules and because repulsion energies change steeply with changes in the distance of nearest approach*.

(i) Systematic introduction of impurities in solid solution has been effected to lessen these barriers. A well-known example refers to the introduction of krypton in solid solution in crystalline methane (Fig. 4.8). Progressive addition of these 'lubricating' atoms lowers and in this instance broadens the transformation.

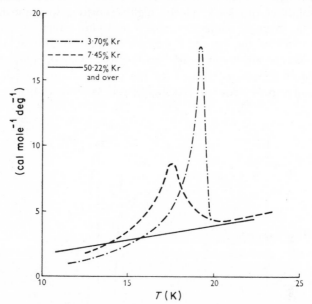

FIG. 4.8. Effects of progressive additions of krypton to solid methane

It seems likely that unsuspected impurities in solid solution in many other crystals may modify orientational barriers in ways that can also affect premelting, discussed in Chapter 12. Impurities in solid solution also modify hysteresis as discussed below.

* It has even been suggested that in some crystals, oscillations of large amplitude, such as those of benzene molecules at $-3°C$ about their hexad axes (Cox, Cruikshank, and Smith, 1958) as well as rotations, must occur 'in phase', to minimize these repulsions.

(ii) Application of external pressure forces the molecules in a crystal closer together, and thereby increases barriers opposing orientational randomization. In the limit, it would be most informative to investigate solid–solid orientational transformations at constant volume. When orientational randomization contributes in one way or another to mechanisms of melting, effects of increasing the pressure can have characteristic consequences (see Ubbelohde, 1961). Up to the present, only isolated studies are available of effects of pressure on orientational transformations generally. One important investigation (Lawson, 1940) showed that the sharp peak in the ammonium chloride transformation is broadened and somewhat displaced, at constant volume, as illustrated in Fig. 4.9. Enhanced pressures reduce the entropy of transition markedly (Trappeniers and Van der Molen, 1966).

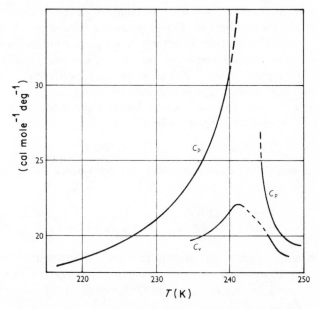

FIG. 4.9. Broadening of lambda transition in $NH_4Cl$ at constant volume. Reproduced by permission of *The Physical Review*

## 4.4 Thermodynamic transformations of 'higher order'

Any lambda transformation appears to differ in a rather profound way in its thermodynamic features from the conventional conditions for the transformation from one solid polymorph to another. According to classical thermodynamics, polymorphs have wholly independent crystal structures, and, in consequence, free energy surfaces for the two solid forms $G_1 = f_1(p, T)$ and $G_2 = f_2(p, T)$ are likewise wholly independent. Changes of the substance from one state to the other involve *discontinuous* jumps in the tangent slopes to these surfaces at their intersection, as specified in Chapter 2, equations (2.2) and (2.3). On the other hand, for lambda transformations there are no discontinuous jumps in $S$ or $V$. Such changes are therefore appropriately termed *continuous* thermodynamic transformations (Ubbelohde and Woodward, 1945).

Studies of thermal expansion $\alpha$ on passing from one crystal state to the other show anomalies similar to those found in the heat capacity. Plots of either $C_p$ or $\alpha$ against $T$ frequently show lambda form. Figure 4.1 illustrates a well-known example for crystalline methane (refs. in Ubbelohde, 1952).

### 4.4.1. *Structural interpretation of transformations of 'higher order'*

Following a traditional line of mathematical development of classical thermodynamics, thermodynamic functions can be treated purely geometrically, disregarding any molecular structure. Anomalous transformations of the lambda kind have been interpreted on this geometrical basis, as transformations of 'higher order'. These show the peculiarity that at the transformation temperature $T_c$ the free energy curves for species 1 and 2 are postulated not only to intersect but to touch. Conventional mathematical theory shows that surfaces intersect when

$$G_1 = f_1(p_c, T_c) = G_2 = f_2(p_c, T_c)$$

If they also touch at the intersection the first derivatives are likewise equal, i.e.

$$\left(\frac{\partial G_1}{\partial T}\right)_p = \left(\frac{\partial G_2}{\partial T}\right)_p; \qquad \left(\frac{\partial G_1}{\partial p}\right)_T = \left(\frac{\partial G_2}{\partial p}\right)_T$$

An immediate consequence of this geometrical situation would be the absence of any sharp discontinuous changes of entropy or of volume on passing from 1 to 2, in view of equations (2.2) and (2.3). Of course the higher derivatives such as

$$\frac{1}{V}\left(\frac{\partial V}{\partial T}\right)_p = \alpha \quad \text{or} \quad -\frac{1}{V}\left(\frac{\partial V}{\partial p}\right)_T = \beta$$

need not be equal. In fact, as will be evident from the consideration of lambda curves such as Figs. 4.2 and 4.4 quite steep changes are found in $C_p$ and $\alpha$; these may imply actual discontinuities in the next order of derivatives. (Information about the compressibility $\beta$ is unfortunately much more restricted (e.g. Lawson, 1940) than about specific heats or thermal expansions, because of the difficulty of measuring this property particularly for single crystals.)

From a purely geometrical standpoint, this extension of classical thermodynamics to the case where free energy curves touch as well as intersect seems very plausible. Two additional considerations are as follows.

(1) To describe a real transformation, state 1 must have its free energy surface above state 2 at the high temperatures and below it at low temperatures. The surfaces must therefore intersect as well as touch (Ehrenfest, 1933; Justi and von Laue, 1934).

(2) Since $\Delta V = 0$ and $\Delta S = 0$ at $T_c$ the Clausius–Clapeyron equation becomes indeterminate. Lambda points are however sensitive to pressure. Conventional mathematical theory (e.g. Hurst, 1955) shows that the equation which becomes indeterminate must be replaced by equations involving first derivatives of those quantities whose difference vanishes at $T_c$. One convenient pair of equations is

$$\frac{\mathrm{d}p}{\mathrm{d}T} = \frac{\Delta C_p}{VT\,\Delta\alpha} = \frac{\Delta\alpha}{\Delta\beta}$$

(see Ubbelohde, 1962).

Much additional information about lambda transformations can be gathered by including structural considerations. From a structural standpoint the purely geometrical analysis above is in conformity with the general finding that lambda-type transitions in fact arise *only* when the differences in structure between 1 and 2 are comparatively minute; often they are only detectable by very special means. Whenever there are large differences in structure, phase 1 and phase 2 are sharply differentiated, and a discontinuous transition is found from one to the other. This suggests some kind of intuitive explanation of why the free energies of 1 and 2 are so closely similar that the surfaces actually touch. However, classical geometrical theory completely fails to account for the important observation that lambda transitions are very often, indeed almost unavoidably, accompanied by marked hysteresis. Figures 4.2, 4.3 and 4.4 illustrate some of the instances of hysteresis that are found in connection with orientational randomization. Even the simplest transformations, as in $CH_4$ and $CD_4$, show hysteresis (Colwell, Gill, and Morrison, 1962). This finding shows that despite its acceptance in physical chemistry, the classical Phase Rule is in fact inadequate for structural reasons, which purely geometrical analysis fails to reveal (Ubbelohde, 1956).

## 4.5 Hybrid crystals in a solid–solid transformation

A clue to the structural interpretation of continuous thermodynamic transformations is obtained when instead of making measurements with a polycrystalline mixture, as if often done for convenience, a single crystal is taken through a cycle of temperatures including $T_c$ (see Ubbelohde and Woodward, 1945, 1946, 1947). This may serve to establish formation of hybrid single crystals.

X-ray measurements of high precision on single crystals can be used to give quite detailed information about the course of a lambda transformation, since the averaging of diverse local changes in a polycrystalline assembly is avoided. As has been stated above, lambda transitions are *only* observed if the structures of 1 and 2 are very close to one another, e.g. both 1 and 2 may have cubic symmetry, but may differ slightly in lattice spacing. As one example, in the case of $NH_4Cl$ (Dinichert, 1942, 1944). Fig. 4.3 illustrates the course of the transformation for this substance. In other cases, as for transformations in the alkali nitrates, the high-temperature form may have a different crystal symmetry but with a unit cell practically the same in forms 1 and 2 (Kennedy, Ubbelohde, and Woodward, 1953).

### 4.5.1. *Persistence of crystal axes in a cyclic lambda transformation*

X-ray studies on lambda transformations using single-crystal techniques have laid stress on the persistence of crystal axes when the single crystal is taken round a cycle of temperatures comprising the transition point $T_c$. In some characteristic instances, when methods of observation with sufficiently high resolving power are used, it is verified that as the single crystal is taken from the low-temperature form* 2 towards $T_c$, regions or domains of structure 1 begin to appear. In the close neighbourhood of $T_c$ a *hybrid* single crystal is formed in which forms 1 and 2 coexist, both being present in significant proportions. As the temperature rises still more above $T_c$, the hybrid crystal passes into a single crystal of form 1.

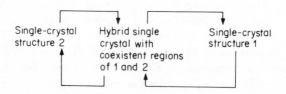

The general scheme implies that nucleation of structure 1 occurs *within* the substance of structure 2. Since 1 and 2 are not completely identical, their coexistence in the hybrid state involves mechanical strains, and internal surface energy at the boundaries of separation of 1 and 2. Nucleation occurs with predetermined orientation of the new form, since this can only grow within the structure of 1, without disrupting the single crystal, if the unit cells

---

* With different authors, the numbering sequence 1, 2 etc. is often arbitrary, and also depends on whether $T_c$ lies above or below room temperature.

of the initial and final states are practically identical. If this orientation is perfect, there is complete persistence of crystal axes throughout the cycle.

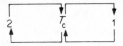

Study of some actual examples shows that although persistence of crystal axes can be practically perfect, some residual strain can usually be detected after each cycle.

### 4.5.2. *Appearance of new polymorphic forms*

In the hybrid region, entirely new polymorphic forms sometimes appear as intermediates, presumably because their nucleation within the matrix of the parent single crystal is easier than direct conversion to the normal alternative form. Such freak intermediates in the transformation from one single crystal to another *ex hypothesi* have crystalline arrangements even closer to the generating structure than that into which it normally transforms. Typical examples of this remarkable behaviour are found in $KNO_3$ and in $KCN$ (Kennedy, Ubbelohde, and Woodward, 1953; Cleaver, Rhodes and Ubbelohde, 1963; Cimino, Parry, and Ubbelohde, 1959).

Although occurrence of such intermediates can be made to appear as loop lines in the thermal cycle, further change of temperature in the same direction eventually leads to the normal end-form. Except in special circumstances, for example when a new structure appears, on heating above $T_c$, the crystal passes through the hybrid back to a single crystal of form 1, whether it starts from the single or the multiple option version of 2.

The discovery of hybrid single crystals has thrown essential light on the structural interpretation of phase transitions of higher order and also, by implication, on the solid–liquid transition.

### 4.5.3. *Independent nucleation*

Clearly the basic assumption in the classical Phase Rule that two phases are wholly independent is only valid when the nucleation of the transformation

can take place in a manner wholly independent of the starting point. For a change such as vapour → liquid this independence can usually be guaranteed, since structural rearrangements in both phases normally have relaxation times short compared with that of experimental observations. Even the nucleation liquid → solid normally satisfies the basic requirement of 'no-memory' in the transformation, though certain phenomena of nucleation call for rather careful discussion (see Chapter 15). However, when both structures 1 and 2 are crystalline, unless the transformation Solid 2 → Solid 1 occurs by way of the liquid or vapour phase (so that the nuclei of 1 have to be formed from a wholly disordered state and can thus be chosen with arbitrary orientations) there is usually at least partial structural correlation between the orientations of domains in the two solids. Even if the transformation occurs by migration of atoms or molecules within the solids, when the differences between the crystallographic unit cells are sufficiently great, the appearance of the new form leads to disruption of the crystals, and only very incomplete correlation of the new crystallites is found relative to the axes of the single crystal from which they originated. A typical example of incomplete correlation is found in the spontaneous formation of α-resorcinol from β-resorcinol (Robertson and Ubbelohde, 1938). Shattering of crystals occurs as a result of intracrystalline nucleation in the transformation on cooling

$$\beta(N_2) \rightleftharpoons \alpha(N_2)$$
$$\text{cubic} \qquad \text{h.c.p.}$$

Gannon and Morrison (1973). The interesting question as to what happens in cycles repeated to exhaustion has not yet been elucidated.

### 4.5.4. *Transformation by way of hybrid crystals*

One thermodynamic consequence of transformations through internal nucleation by way of hybrid crystals is that additional terms must be included in the free energy expression for each state, to represent the contributions from the internal surface energy $\sigma_{12}$ at the boundaries between domains of the two structures in the hybrid crystals, and from the compressive or tensile energy $\chi_{12}$ when a domain of one structure has to be produced within the matrix of the other, which normally has slightly different size and shape of unit cell. One may write

$$G_1 = f_1(p, T, \sigma_{12}, \chi_{12}) \quad \text{and} \quad G_2 = f_2(p, T, \sigma_{12}, \chi_{12}) \qquad (4.1)$$

These additional 'memory' energy terms are to some extent arbitrary, since the location and size of domains of (say) form 1 to be nucleated within a single crystal of form 2 will depend on its past history and thus on accidental factors such as the incorporation of dislocations and other defect sites in the crystal. Incorporation of impurities may also be expected to influence the pattern of domains in the hybrid. For example, molecules of $H_2O$ held extremely strongly at defects in $KNO_3$ have produced a pronounced influence on the course of the transformation (Kennedy, Ubbelohde, and Woodward, 1953).

In the transformation of KCN foreign anions (Cimino, Parry, and Ubbelohde, 1959) may likewise be important in deciding the detailed pattern of domains of the newly appearing structure.

In such transformations the single crystals are only markedly hybridized in a fairly narrow range of temperatures around $T_c$. Well above or well below this range, only form 1 or form 2 is found. It may happen, of course, that the unit cell of 1 has several equivalent options in undergoing the transformation to form 2. For example, the tetragonal cell of $KH_2PO_4$ has 4 equivalent options in passing to the monoclinic form stable below about 120 K. In such cases, when there is no external constraint, single crystals of 1 eventually pass into crystals of 2 with all options equally represented (see Fig. 4.10). Quite mild constraints may however favour one of these options at the expense of its rivals. For example, if the low temperature form 2 is ferroelectric, imposition of an external field may favour one option (see refs. in Ubbelohde, 1962). Application of external mechanical constraints may likewise favour certain options.

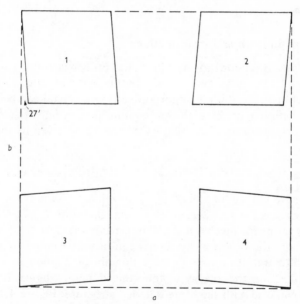

FIG. 4.10. Four options in the hybrid crystal formed from tetragonal $KH_2PO_4$ in the transition range. Section of crystal hybrid normal to $c$ axis

Nucleation conditions affecting solid–solid transformations may likewise depend on the method of preparation of the crystals. Thus for sodium sulphate the transformation

$$Na_2SO_4(V) \rightarrow (I)$$

$$(T_c 230-265°C)$$

is affected by fine grinding, and is subject to hysteresis, which can be suppressed by added moisture (Kracek, 1929). The lowest temperature for onset (231·8–231·9°C) is found to differ notably for $Na_2SO_4$ prepared by recrystallization (thenardite) and by efflorescence of the decahydrate at room tempeature in the presence of drying agents (232·4–235·8°C). The effloresced material is finely divided and has about one-half the bulk density of thenardite.

A different modification $Na_2SO_4$(III) nucleates on cooling $Na_2SO_4$(I) and transforms into it on reheating at 248·5°C (Schmidt and Sokolov, 1961).

A general conclusion can be derived from these and similar studies in a diversity of transformations for which the nature of the structural change can be quite diverse. Nucleation within single crystals, and the persistence of crystal axes in hybrid crystals in a thermal cycle, can occur provided that the unit cells of the two forms do not differ too greatly. This is necessary both for kinetic reasons, and to ensure that neither the mechanical breaking strength nor stress limits for plastic flow are exceeded.

### 4.6 Hysteresis in lambda transformations

Quantitative expressions for hybrid crystal transformations can be derived as follows (see Ubbelohde, 1962, 1963). In the region of temperatures around $T_c$ over which hybridization is important, a hybrid crystal can be regarded as containing fraction $x$ of the high-temperature form 1 and fraction $(1-x)$ of the low-temperature form 2, so that its free energy is given by the composite expression

$$G_{\text{hybrid}} = x f_1(p, T, \sigma_{12}, \chi_{12}) + (1-x)f_2(p, T, \sigma_{12}, \chi_{12}) \qquad (4.2)$$

It is important to realize that this expression will not normally follow the same path in the process $1 \rightarrow \underline{T_c} \rightarrow 2$ as in the reverse direction $2 \rightarrow \underline{T_c} \rightarrow 1$ because of the arbitrary terms for mechanical and internal energy.

Algebraically, the origin of hysteresis can be represented by attributing a somewhat different strain energy for a domain of 2 when it is wholly surrounded by 1, from the energy of the same domain when it is largely surrounded by material with the same detailed structure. Figure 4.11 illustrates how an increase in volume for the process $2 \rightarrow \underline{T_c} \rightarrow 1$ necessitates that domains of 1 first formed in 2 are under compression, whereas domains of 2 formed in 1 are under tension. It is convenient to symbolize crystal substance originating in either direction in a box, and for residual substance to omit this box. At equal values of $x$ the two expressions

$$\begin{matrix} G_{\text{hybrid}} \\ x:(1-x) \end{matrix} = x f_1(p, T, \underline{\sigma_{12}}, \underline{\chi_{12}}) + (1-x)f_2(p, T, \sigma_{12}, \chi_{12}) \qquad (4.2.1)$$

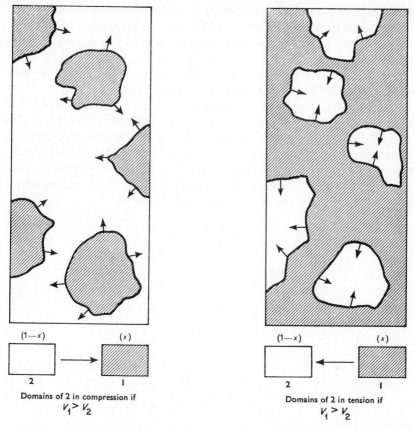

FIG. 4.11. Hybrid single crystal around the lambda point, illustrating domains in tension or compression. Similar fractions of 1 and 2 but quite different strain parameters

and

$$\frac{G_{\text{hybrid}}}{x:(1-x)} = xf_1(p, T, \sigma_{12}, \chi_{12}) + (1-x)f_2(p, T, \underline{\sigma_{12}}, \underline{\chi_{12}}) \qquad (4.2.2)$$

will not be the same because of the difference between the terms with and without the box. If the molar volume $V_1 > V_2$ it seems likely that $\underline{\chi_{12}} > \chi_{12}$. Differences between $\underline{\sigma_{12}}$ and $\sigma_{12}$ are less easy to generalize, since these refer to internal energy attributable to misfits at the domain boundaries. (See also pp. 247 seq.).

This differentiation, according to the direction of transformation of single crystals, leads to a quantitative expression for the well-known phenomenon of hysteresis, whose magnitude will generally be largest for the largest differences between unit cells of 1 and 2, provided these differences are not so great as to lead to actual disruption and thus loss of persistence of crystal axes. One such difference which has been explored in the ammonium halide transformation that occurs around 240 K can be expressed quantitatively in terms of the change in overall specific volume. Hysteresis is found to be largest when the volume difference is largest, as is illustrated in Table 4.2.

A structural description of the 'contact' between free energy surfaces in a lambda transformation, which takes into account the considerations discussed in this section, is that around $T_c$ where they intersect these surfaces show a certain arbitrariness or indeterminacy, owing to the role of the additional parameters in the free energy functions. This indeterminacy permits a kind of smearing overlap of the curves that simulates true contact, as illustrated in Fig. 4.12 (Ubbelohde, 1951).

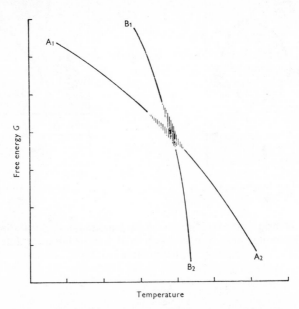

FIG. 4.12. Overlap of free energy curves owing to arbitrary contributions from strain energy, simulating contact at $T_c$

## 4.7 Coexistence of related transforms in hybrid crystals as origin of hysteresis

Theories of solid–solid transformations by way of hybrid crystals have been further developed by diverse lines of investigation.

Evidence for the coexistence within hybrid crystals of transforms over a range of temperatures has been reported for quite a variety of structures. In favourable cases sufficient resolving power is available in the methods used to permit their direct observation. When a transformation involves coexistence this implies that nucleation and growth of the alternative form occur preferentially within the single crystals, or at favourable regions of the crystal surface. One consequence is that small amounts of impurities in solid solution, and various built-in lattice strains, can play a major role in such 'cannibal transformations' of any crystal. This consideration can be important when comparing results obtained by different methods of observation, or by workers using a common method of observation but on different material. Homogeneous nucleation within a defect free crystal is considered to be impossible (Mnyukh and Panfilova, 1976). Examples of coexistence include coexistence over the range 50–70°C for $Cu_2HgI_4$ (Chivian, Clayton, and Eden, 1969) and from 300–700 in $V_2O_3$ (Chandrasekhar, Sinha, and Honig 1974). Hybrid phase transformations in cyclohexane have been detected at 186·5 K (Zhizkin, Terpugov, Moskaleva, Bagdanskis, Balabunov and Vasil'ev, 1973; see Zhizkin and Usmanov, 1971). Stress-induced hybrid crystal transformations in 1, 8-dichloro-10-methylanthracene (Jones Thomas and Williams 1975) have been followed by infrared absorption spectra. Pressure-induced hybrid crystal transformation has been observed in paratellurite $TeO_2$ (Skelton, Feldman, Liu, and Spain, 1975). For $ND_4Cl$ neutron diffraction has established a hysteresis loop of breadth 0·035 K at 249 K.

### 4.7.1 Acoustic detection of coexistence

When single crystals pass into the hybrid state, numerous domain interfaces are developed, which affect the attenuation of elastic waves passing through the material. The transition in $NaNO_2$ at 163·8°C has been used to investigate coexistence, using a diversity of physical methods (Hatta, 1970; Hatta, Ishiguro, and Mikoshiba, 1970, and earlier papers). This transformation shows the steep increase of attenuation of longitudinal (but not transverse) acoustic waves near the peak transformation temperature. Probably this may be attributed to the break-up into hybrid crystals.

Analogous acoustic effects have been observed in the lambda transformation in $NH_4Cl$ (Garland and Snyder, 1970, Garland and Pollina, 1973).

When acoustic attenuation is observed with polycrystalline material, as in the lambda transformations in $CH_4$ (Wolf, Stahl and Watrous 1974) or in $CD_4$ (Wolf and Stahl, 1973) attribution to hybrid crystal formation is less sure; hysteresis data in cyclic measurements are however a good test.

Acoustic methods have also been applied to the transformation in $KMnF_3$ to study attenuation (Matsuda, Hatta, and Sawada, 1973). Acoustic attenuation in $Ti_2O_3$ likewise may be attributed to phonon viscosity losses (Chi and Sladek, 1973). For the transformation in AgI direct measurements of the elastic constants have shown hysteresis (Fjeldy and Hanson, 1974).

## 4.8  Generalized thermodynamics of hysteresis

On transforming one polymorph into another, various kinds of thermodynamic 'hold up' may be observed. Even when the transformation is wholly discontinuous and without precursor effects, some overpotential must usually be built up to permit the alternative form to nucleate and grow at the expense of the less stable form. In the case of discontinuous transitions, geometrical plots against temperature of experimental superheating and supercooling boundary lines may inscribe a region or loop in the phase diagram. This region probably has geometrical analogies with other hysteresis loops. However, in a discontinuous transformation no state within the loop can be attained experimentally; the loop depends on external factors influencing nucleation and growth from the vapour or disordered melt. Hysteresis loops arising from coexistence of domains of either form in a continuous transformation are basically different, since local microregions in the mass can range over all intermediate values of parameters such as the specific volume. Hysteresis curves whose origin stems from coexistence may be regarded as envelopes to thermodynamic parameters defined by equations (4.2.1) and (4.2.2).

Rubidium nitrate shows a number of polymorphic transformations that illustrate various features of hysteresis (Kennedy, 1970)

$$V \underset{164°C}{\rightleftharpoons} III \underset{219°C}{\rightleftharpoons} II \underset{284°C}{\rightleftharpoons} I \underset{310°C}{\rightleftharpoons} Melt$$

The transformation II → I is to an intermediate NaCl type f.c.c. structure not shown by other alkali nitrates. It has 4 molecules per unit cell, but there is not quite sufficient room for the $NO_3^-$ to randomize orientation; this probably explains the very unusual feature that the crystal contracts on melting, which permits non-lattice coordination but still has coordination number 4 (Cleaver and Williams, 1968). On passing from I → II complex lamellar twinning is observed (i.e. a form of coexistence), and in temperature cycles II ⇌ I rapid healing of scars is observed.

The transformation III ⇌ IV is rapid, with $\Delta V = 0·903$ cm$^3$ mole$^{-1}$. It is thus particularly suitable for dilatometric study of the influence of impurities in solid solution on the hysteresis (Pöyhönen, Sivonen, and Hilpela, 1964).

Figures 4.13 and 4.14 illustrate the dilatometric hysteresis loops for two different samples:

$$(1) \text{ with } 0{\cdot}21 \text{ weight\% } KNO_3,$$

$$1{\cdot}15. \text{ weight\% } CsNO_3;$$

$$(2) \text{ with } 0{\cdot}05 \text{ weight\% } KNO_3.$$

It would appear that the greater impurity content 'lubricates' this particular transformation and gives narrower hysteresis loops (Fig. 4.13).

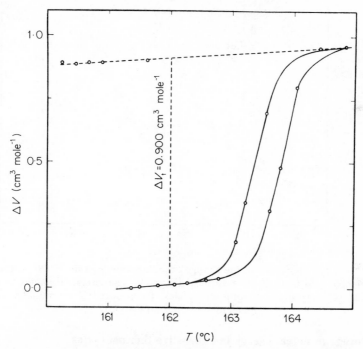

FIG. 4.13. Hysteresis of the change of molar volume in the $RbNO_3$ impure sample (see text) at the transition III ⇌ IV. Reproduced by permission of Academia Scientarum Fennica

Nitrates have been studied in several other ways, in relation to hysteresis in their transformations, and coexistence of transforms in hybrid crystals. X-ray diffractometry can be used to show coexistence in $Rb(NO_3)$ and $Cs(NO_3)$ (Kennedy, Taylor, and Patterson, 1966). Alternative nucleation options in the transformation of single crystals of $TlNO_3$ have been studied microscopically (Kennedy and Patterson 1965). Electrical measurements on changes of resistivity and dielectric constants in the transformations below 0°C show hysteresis loops for nitrates of Li, Na, K Rb, Cs, Ag, Tl and $NH_4$

(Fermor and Kjekshus, 1968). The order–disorder transformation in graphite nitrates around −20°C is of special interest because of the layer structure. Hysteresis and phase-boundary scattering are found for this transformation also (Ubbelohde, 1966).

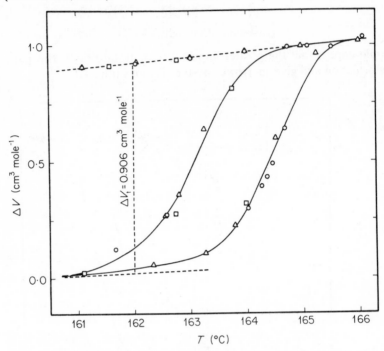

FIG. 4.14. Hysteresis of the change of molar volume in the $RbNO_3$ pure sample (see text) at the transition III ⇌ IV. The circles, triangles and squares correspond to different cycles. Reproduced by permission of Academia Scientarum Fennïca

## 4.9  Storage of defect energy in exhaustive thermal cycles

Specific heat measurements show a 15° broad loop in the transformation in metallic samarium at 835 K (Polovov, 1973). When hysteresis is due to internal nucleation and growth effects, repeat cycles can yield information about the location and movement of strain centres and impurities, as in the transformation in RbCl under pressure (Lacam and Peyronneau, 1973). Even when no closed cycle can be observed, strain energy is stored in the crystals during a transformation, e.g. in $NH_4I$ and $NH_4Br$ (Kennedy, Patterson, Chaplin, and McKay, 1974) and this modifies the 'ideal' thermodynamic equations as indicated above. When the crystals shatter on cooling through a transformation as in octafluoronaphthalene (Pawley and Dietrich, 1975), it would be interesting to observe the effects on crystallite size of repeated temperature cycles traversing $T_c$. Presumably a limiting storage of strain energy would be attained after repeat cycles, and hence a limiting reduction in crystallite size. This kind of exhaustive cycling does not appear yet to have been followed through (see also p. 105).

## 4.10 Polymorphic transitions in condensed microphases

The theory outlined in section 4.5 refers to hybrid single crystals in which the transformations of neighbouring microregions have orientational and positional correlations imposed by their genesis within a single crystal. Other aspects of the general fact that in condensed phases transformations often first begin (i.e. 'nucleate') as microregions are discussed in relation to the cluster theory of prefreezing viscosity enhancement (Chapter 13), the theory of glass formation by the aggregation of clusters (Chapter 13), and the theory of crystal nucleus formation in a melt (Chapter 15).

At the boundaries between microregions, the units of structure are 'activitated' states of higher energy, since the environment which normally ensures lowest minima of potential energy cannot be realized owing to mismatching between neighbouring regions. Various instances of enhanced kinetic activity in the neighbourhood of $T_c$ for solid–solid transformations appear to originate from the presence of such excited states at the boundaries between microregions (see Hedvall, 1938). Enhanced kinetic activity has been reported for radical recombination and radical polymerization (Magat, 1962, see Chapter 15). Again enhanced magnetocatalytic effects have been reported at ferromagnetic–paramagnetic transitions (Schwab, 1962). The significance of such states of higher energy at domain walls, which are in effect cooperative defect sites in a crystal, is discussed in relation to premelting in Chapter 12.

### 4.10.1. *Effects of impurities on transformation temperatures*

When impurity atoms or molecules are present in solid solution in any single crystal it is unlikely that they will be diffused uniformly throughout it. For thermodynamic reasons, as discussed, for example by Willard Gibbs impurities will tend to segregate preferentially either towards the boundaries of intracrystalline domains, or to their interior (a less likely situation). As a consequence of this non-uniformity of distribution, classical thermodynamic equations describing the shift of a transition temperature by added impurity, as a function of its concentration, are not at all accurately applicable. Furthermore, by their segregation to intracrystalline domains impurities can have magnified influence on the $\sigma_{12}$ terms discussed in the preceding paragraphs, even at quite low nominal concentrations. For these reasons, the breadth and general shape of a hysteresis loop in solid–solid transformations can give much more sensitive indications about impurities present than many other thermodynamic parameters.

# REFERENCES

R. S. Bradley, D. C. Munro, and S. I. Ali (1969) *High Temp. Pressures* **1**, 103.

K. W. Browall and J. S. Kasper (1974) *J. Solid State Chem.* **10**, 20.

G. V. Chandrasekhar, A. B. P. Sinha, and J. M. Honig (1974) *Phys. Lett. A* **47A**, 185.

T. C. Chi and R. J. Sladek (1973) *Proc. 5th Int. Conf. Internal Friction*, Springer, p. 127.

H. Chihara and N. Nakamura (1973), *J. Chem. Phys.*, **59**, 5392.

H. Chihara, N. Nakamura, and M. Tachiki (1973) *J. Chem. Phys.* **59**, 5387.

J. S. Chivian, R. N. Claytor, and D. D. Eden (1969) *Appl. Phys. Lett.* **15**, 123.

A. Cimino, G. S. Parry, and A. R. Ubbelohde (1959) *Proc. Roy. Soc., Ser. A* **252**, 445.

B. Cleaver and J. F. Williams (1968) *J. Phys. Chem. Solids* **29**, 877.

B. Cleaver, E. Rhodes, and A. R. Ubbelohde (1963) *Proc. Roy. Soc., Ser. A* **276**, 437.

K. Clusius and L. Popp (1940) *Z. Phys. Chem.* **46B**, 63.

J. H. Colwell, E. K. Gill, and J. A. Morrison (1962) *J. Chem. Phys.* **36**, 2223.

J. M. Courdille and J. Dumas (1974) *Satellite Symposium 8th Int. Congr. Acoustics*, Lancaster University, England, p. 130.

E. G. Cox, D. W. J. Cruikshank, and J. A. S. Smith (1958) *Proc. Roy. Soc. Ser. A* **247**, 1.

P. Dinichert (1942) *Helv. Phys. Acta* **15**, 462.

P. Disichert (1944) *Helv. Phys. Acta* **17**, 389.

P. Ehrenfest (1933) *Con. Amsterd. Proc.*, **362**, 153, *Leyden Comm. Suppl.* 75b.

K. Emma, K. Hamano, and I. Hatta (1975) *J. Phys. Soc. Japan* **39**, 726.

R. Extermann and J. Weigle (1942) *Helv. Phys. Acta* **15**, 455.

J. H. Fermor and A. Kjekshus (1968) *Acta Chem. Scand.* **22**, 2054.

T. A. Fjedy and R. C. Hanson (1974) *Phys. Rev. B* **10**, 3569.

P. A. Fleury (1971) *J. Acoust. Soc. Am.* **49**, 1041.

P. A. Fleury (1972) *Comments on Solid State Physics*, Bell Laboratory, New Jersey, U.S.A., **4**, 149, 157.

D. J. Gannon and J. A. Morrison (1973), *Can. J. Phys.* **51**, 1590.

C. W. Garland and R. J. Pollina (1973) *J. Chem. Phys.* **58**, 5002.

C. W. Garland and D. D. Snyder (1970), *J. Phys. Chem. Solids* **31**, 1759.

L. J. Graham and R. Chang (1975) *J. Appl. Phys.* **46**, 2433.

I. Hatta (1970) *J. Phys. Soc. Japan* **28**, 1266.

I. Hatta, T. Ishiguro, and N. Mikoshiba (1970) *J. Phys. Soc. Japan* **28**, 211.

J. A. Hedvall (1938) *Reaktionsfähigkeit Fester Stoffe*, Barth.

L. Helmholz (1935) *J. Chem. Phys.* **3**, 740.

C. Hurst (1955) *Proc. Phys. Soc., Lond.* **68B**, 521.

Z. Iqbal and C. W. Christoe (1974) *Chem. Phys. Lett. (Netherlands)* **29**, 623.

J. Jaffray (1948) *Ann. Phys. Paris (Ser 12)* **3**, 5.

W. Jones, J. M. Thomas, and J. O. Williams (1975) *Phil. Mag.* **32**, 1.

E. Justi and M. von Laue (1934) *Phys. Z.* **35**, 945.

Y. Kachiuthi and T. Sakurai (1949) *J. Phys. Soc. Japan* **4**, 365.

S. W. Kennedy (1970) *Phys. Status. Solidi A* **2**, 415.

S. W. Kennedy and J. H. Patterson (1965) *Proc. Roy. Soc. Ser A* **283**, 498.

S. W. Kennedy, G. F. Taylor, and J. H. Patterson (1966) *Phys. Status. Solidi* **16**, K175.

S. W. Kennedy, A. R. Ubbelohde, and I. Woodward (1953) *Proc. Roy. Soc., Ser. A* **219**, 303.

S. W. Kennedy, J. H. Patterson, R. P. Chaplin, and A. I. Mackay (1974) *J. Solid State Chem.* **10**, 102.

J. A. A. Ketelaar (1934) *Z. Kristallogr.* **87**, 436.

E. V. Kolontsova, E. E. Kulago, N. I. Byakhova, and I. E. Mikhaelenko (1973) *Sov. Phys. Dokl. (U.S.A.)* **18**, 332.
F. C. Kracek (1929) *J. Phys. Chem.* **33**, 1281; see F. C. Kracek and C. J. Ksanda (1930), *J. Phys. Chem.* **34**, 1741.
A. Kruis (1941) *Z. Phys. B* **48**, 321.
A. Lacam and J. Peyronneau (1973) *J. Phys. (Paris)* **34**, 1047.
A. W. Lawson (1940) *Phys. Rev.* **57**, 417.
M. Magat (1962) *Pure Appl. Chem.* **5**, 487.
M. Matsuda, I. Hatta, and S. Sawada (1973) *Ferroelectrics (G.B.)* **8**, 595.
J. W. Menary, A. R. Ubbelohde, and I. Woodward (1951) *Proc. Roy. Soc., Ser. A* **208**, 158.
G. R. Miller and H. S. Gutowsky (1963) *J. Chem. Phys.* **39**, 1983.
Yu V. Mnyukh and N. A. Panfilova (1976) *Sov. Phys. Dokl. (U.S.A.)* **20**, 346.
M. Mori and Y. Yamada (1975) *Solid State Commun. (U.S.A.)* **17**, 127.
A. Mustajoki (1957) *Ann. Acad. Sci. Fenn.* A 6. Paper 5.
B. B. Owens and G. R. Argue (1967) *Science* **157**, 308.
W. J. Pardee and G. D. Mahan (1974) *J. Chem. Phys.* **61**, 2173.
G. S. Pawley and O. W. Dietrich (1975) *J. Phys. C.* **8**, 2549.
C. M. Perrott and N. H. Fletcher (1968a) *J. Chem. Phys.* **48**, 2143.
C. M. Perrott and N. H. Fletcher (1968b) *J. Chem. Phys.* **48**, 2681.
C. W. F. T. Pistorius (1961) *Z. Kristallogr.* **115**, 291.
V. M. Polovov (1973) *Zh. Eksp. Teor. Fiz.* **65**, 1557.
E. Posnjak and S. B. Hendricks (1931) *J. Am. Chem. Soc.* **53**, 3339.
J. Pöyhönen, T. Sivonen, and M. Hilpela (1964) *Ann. Acad. Sci. Fenn. IV Physica* **170**, 3.
J. M. Robertson and A. R. Ubbelohde (1938) *Proc. Roy. Soc., Ser. A* **167**, 122, 137.
A. J. Rushworth and J. F. Ryan (1976) *Solid State Commun* **18**, 1239.
N. F. Schmidt and V. A. Sokolov (1961) *Russ. J. Inorg. Chem.* **12**, 1321.
G. M. Schwab (1962) *Pure Appl. Chem.* **5**, 655.
G. H. Shaw (1974) *J. Phys. Chem. Solids (G.B.)* **35**, 911.
E. F. Skelton, J. L. Feldman, C. Y. Liu, and I. L. Spain (1976) *Phys. Rev. B* **13**, 2605.
G. A. Thomas, A. B. Giray, and R. D. Parks (1973) *Phys. Rev. Lett. (U.S.A.)* **31**, 241.
H. C. Tong and C. M. Wayman (1974) *Phys. Rev. Lett. (U.S.A.)* **32**, 1185.
N. J. Trappeniers and T. J. van der Molen (1966) *Physica* **32**, 1161.
A. R. Ubbelohde (1936) *Trans. Faraday Soc.* **32**, 525.
A. R. Ubbelohde (1938) *Trans. Faraday Soc.* **34**, 289.
A. R. Ubbelohde (1940) *Ann. Rev. Chem. Soc.* **36**, 150.
A. R. Ubbelohde (1951) *Nature, Lond.* **169**, 832.
A. R. Ubbelohde (1952) *Introduction to Modern Thermodynamical Principles*, 2nd edn. Oxford University Press.
A. R. Ubbelohde (1956) *Brit. J. Appl. Phys.* **7**, 313.
A. R. Ubbelohde (1957) *Quart. Rev. Chem. Soc.* **9**, 246.
A. R. Ubbelohde (1961) *J. Phys. Chem. Solids* **18**, 90.
A. R. Ubbelohde (1962) *Proc. Kon. Ned. Akad. Wetens. Proc.* **65B**, 459.
A. R. Ubbelohde (1963) *Z. Phys. Chem.*, n.f. **37**, 183.
A. R. Ubbelohde (1966) *Nature, Lond.* **212**, 70.
A. R. Ubbelohde and I. Woodward (1945) *Nature, Lond.* **155**, 170.
A. R. Ubbelohde and I. Woodward (1946) *Proc. Roy. Soc., Ser. A* **185**, 448.
A. R. Ubbelohde and I. Woodward (1947) *Proc. Roy. Soc., Ser. A* **188**, 358.
H. M. Van Driel, R. L. Armstrong, and J. M. M. McEnnan (1975) *Phys. Rev. B* **12**, 488.

A. F. Wells (1962) *Structural Inorganic Chemistry*, 3rd edn., Oxford University Press.

R. P. Wolf and F. A. Stahl (1973) *J. Chem. Phys.* **59,** 115; *Low Temp. Phys. Proc. 13th Int. Conf. Low Temp. Phys.* **2,** page 210.

R. P. Wolf, F. A. Stahl and J. A. Watrous (1974) *J. Phys. Chem. Solids* **35,** 1047.

J. M. Worlock (1971) *Structural Phase Transitions and Soft Modes*, ed. E. J. Samuelson *et al.*, Universitets Foleget, p. 329.

Y. Yamada (1973) *Ferroelectrics (G.B.)* **8,** 591.

G. N. Zhizkin and A. Usmanov (1971) *Sov. Phys. Solid State* **13,** 1989.

G. N. Zhizkin, E. L. Terpurgov, M. A. Moskaleva, N. I. Bagdanskis, E. I. Balabanov, and A. I. Vasil'ov (1973) *Sov. Phys. Solid State* **14,** 3028.

# 5. STRUCTURAL MELTING OF CRYSTALS. INTRODUCTION OF POSITIONAL DISORDER

## 5.1 Melts as disordered versions of their crystalline counterparts: X-ray and other physical evidence

The earliest applications of X-ray diffraction showed that practically all solids are crystalline. This means that their units of structure (atoms or molecules) recur in regular succession, so as to fill space completely by an ordered lattice arrangement of these units which extends in three dimensions. Generally, this regular filling of space is referred to as the 'long-range order' of solids, though the term 'order' is too naive to be adequate except for the very simplest crystals. The term 'long-range correlation' is preferable. Progressively, refinements and modifications of complete regularity in three dimensions have been identified for specific solids; where these affect melting they will be referred to again in what follows.

Applications of X-ray diffraction to the study of liquids has shown that on average the distance between the units of structure and their average arrangement are much the same as in the crystal. This could already have been guessed from the similarity of density and other thermodynamic parameters between most crystals and their melts. A major difference revealed even by fairly simple analysis of X-ray diffraction data is that in liquids long-range lattice correlations of position are absent. However, it would be most desirable to extract somewhat more specific information about the structure of liquids from X-ray diffraction data, since these can now be readily obtained with precision.

The angular distribution of X-ray diffraction intensities from a liquid is generally represented in the form of a radial distribution function which describes probabilities of positional location in a melt. Following well-known procedures originated by Zernike and Prins and extended by Debye and Menke (see refs. in Furukawa, 1962), the density of atoms is calculated from X-ray diffraction intensities in relation to the distance from the centre of any one atom taken as origin. Since the sizes and repulsion envelopes of the atoms are usually well known, the number of nearest neighbours to any one atom in the melt can be calculated with considerable certainty from such radial distribution functions (RDF). This is usually termed the coordination number (CN). Frequently the number of second nearest neighbours can be determined but only with considerably less accuracy. Various reviews with some critical assessment of the data have been published (e.g. Furukawa, 1962; Kruh, 1962).

Up to now, X-ray studies of liquid structure have been somewhat disappointing in relation to problems of melting. The general indications they give about the absence of long-range lattice correlations of position lead to highly important conclusions about positional melting, but any more detailed information provided by X-rays is blurred by a number of limiting factors. Some of these may be avoided by careful planning of the measurements, which makes it desirable to summarize current limitations briefly.

(1) In many cases, the liquid structure has been studied too far from $T_f$ to be fully characteristic of the melt. For example, for liquid argon X-ray diffraction measurements near the critical temperature $T_{crit}$ ( 165·7 K) show that the mean distance between nearest neighbours is 3·78 Å with 10·5 atoms on average as nearest neighbours (Gingrich and Thompson, 1961). At present this can be compared only with 12 atoms as nearest neighbours in the crystal, at the melting point $T_f$ which is 71° lower! Even distribution curves obtained by improved methods for cyclohexane, benzene and silica, have not been correlated (Mendel, 1962) as closely with melting as might be wished.

Clearly to permit really close correlation with theories of positional melting (Chapter 6), information about the positional location of units in the liquid is needed as near as possible to $T_f$.

(2) Even for crystals whose units of structures are simple atoms, the precision of many radial distribution functions evaluated experimentally is not very high. The nature of the calculations tends to give mean values. Local differences of packing, for example in anticrystalline clusters (Chapter 13) would often not be brought out by X-ray methods as used hitherto. Attempts to improve the quality of RDFs, for example for molten metals (Furukawa, 1960), and to correlate these with other melting parameters, point to particularly prominent departures from smooth random packing in melts such as those of the B metals Zn, Ge, Sn, Bi, and Sb. These anomalies are of interest in relation to problems of prefreezing (Chapter 13), but are not yet sufficiently well characterized to be fully informative (see Ladyanov, Arkbarov and Velyukhanov, 1973 and also Chapter 9).

(3) A difficulty of principle is that RDFs reduce the description of spatial distribution to a smoothed one-dimensional function. This averaging process in itself tends to obscure local differences of structure in a melt.

One way of testing the degree of significance of RDFs for describing nearest neighbour arrangements in melts is to evaluate mean fluctuations from the average smoothed value. There is evidence (Prokhorenko and Fisher, 1959) that the microstructure of a number of melts of simple molecules shows large fluctuations from the mean, of the order of 25% for the first coordination sphere and 50% for the second coordination sphere. As is discussed in Chapter 11, such findings might be qualitatively accounted for by attributing a conglomerate structure to such melts, with a proportion of *anticrystalline* clusters whose density is somewhat higher or lower than the overall average (see Lemm, 1970).

With liquids containing molecules instead of atoms, the well known Zernike and Prins method as extended by Debye for obtaining RDFs is no longer strictly appropriate, owing to the intramolecular diffraction contributions to the overall intensities of X-ray diffraction. For molecules with effectively rigid skeletons such as mesitylene, cyclohexane, and benzene (see Chapter 6) molecular contributions can be subtracted (Otvos and Mendel, 1962) but the information obtained about the structure of the melt still remains far from precise, and does not encourage further developments.

When melts give rise to several diffraction maxima, as for molten naphthalene which shows at least three, attempts have even been made to suggest that a quasicrystalline *unit cell* can be attributed to the structure of the melt, not very different in dimensions to those for the crystal. Thus for the solid

$$a:b:c = 8\cdot34:6\cdot05:8\cdot69$$

$$\alpha = \gamma = 90° \qquad \beta = 122° \, 49'$$

whereas for the melt

$$a:b:c = 7\cdot78:8\cdot44:8\cdot92$$

Insofar as this suggestion is applicable, it may be noteworthy that for such a model

$$V_S = 368\cdot7 \text{ Å}^3 \quad \text{and} \quad V_L = 492\cdot7 \text{ Å}^3$$

with a ratio $V_L/V_S = 1\cdot34$, considerably greater than the experimentally determined macroscopic ratio $V_L/V_S = 1\cdot20$ (Cennamo, 1952). If the proposed model for the melt is realistic, to compensate for the greater specific volume required by the quasicrystalline regions, regions of more closely packed (e.g. roughly parallelized molecules would also need to be present in order to reconcile these two values. This would in fact be consistent with a conglomerate structure for the melt, further discussed in Chapters 11 and 13.

At present these findings can be regarded as suggestive but only very tentative. Possibilities of a mesophase when the molecules do not rotate unhindered in the melt above $T_f$ (see p. 149) cannot be disregarded (Kurik and Shayuk, 1975).

Observations on physical properties other than RDFs can be more revealing to indicate any 'cooperative fluctuations' of local packing of molecules in a melt, which are sometimes also called 'modulations' or 'domains'. Studies of viscous flow and of other cooperative movements of molecules in melts sometimes prove to be particularly sensitive to topological arrangements of the molecules. A general review of their importance for melting is given in

Chapter 13. In other cases direct spectroscopic observations have been revealing. By way of illustration, the spectral line of Hg = 2537 Å when mercury atoms are dissolved in molten or crystalline argon shows marked variations. In argon crystals, Hg atoms in solid solution give rise to a narrow triplet shifted by 1200 cm$^{-1}$ to higher energy with respect to the free Hg atom. The median line of this triplet agrees with the high-energy component of a doublet which is observed from liquid argon. The formation of this doublet is considered to point to Hg atoms with two different kinds of environment, possibly quasicrystalline in clusters and quasigaseous outside them (Robinson, 1960). Liquid argon does not offer many options for closer packing domains, but in neopentane an even more complex spectral line shape-dependence on density is observed, supporting the view that more research seems warranted to utilize this particular method for probing local nearest neighbour environments in melts consisting of non-spherical molecules. Such developments seem likely to be rewarding particularly if carried out systematically for liquids with related molecular structures.

As is discussed for some specific examples below (pp. 275 seq.), a topological description of the texture of a liquid is often more informative, and is in any case an essential complement to structural studies using X-rays of the range of wavelengths ordinarily available (1·5–4 Å). X-rays of longer wavelength, and the use of glancing angle diffraction can sometimes give valuable topological information. In other cases, light scattering or sonic vibrations have been used with success as probes. To resolve problems of the molten state of matter, much more use of vibrations of all kinds with wavelengths between 20 Å and 6000 Å seems desirable.

### 5.1.1. *Melts as quasicrystalline lattices: the role of positional defects.*

An approach to the structure of melts which is quite as valid but conceptually quite different from that of the radial distribution function starts with the ideal crystal and leads on to the melt by progressive incorporation of crystal defects. Such a development forms a basis for *quasicrystalline* models for the structure of melts. As a warning, this term has been used rather diversely by different authors. Its usage in the present text warrants rather careful definition in special relation to problems of melting. Early determinations of 'crystal structures' were dominated by the concepts of classical crystallography. It was often tacitly assumed that most crystals (except at their free surfaces) assume ideal perfect structures as represented by the conventional crystallographic lattices. As is now well known, however, many physical properties indicate that real crystals incorporate various types of lattice defect. When these arise from accidents of growth, annealing the crystal under conditions that permit migration of atoms or molecules in the lattice should remove these defects and would in time bring the crystal into an equilibrium state. However, even for the fully annealed equilibrium states of many crystals, it is found that at any given temperature a finite proportion of crystal defects persists permanently, indicating that such defective structures are thermodynamically more stable than ideal perfect crystals.

A general explanation of this finding can be derived by considering the make-up of the free energy $^0G_S = {}^0H_S - T^0S_S$ of the ideal crystal, with respect to the value $G_S$ for a more stable defective crystal containing $n$ defect sites in $N$ total sites. To focus discussion, these defects could be vacant sites, i.e. absences of atoms or molecules in regular close-packed structures. When the concentration of defects is small, the work $\varepsilon$ of creating any vacancy can usually be regarded as independent of any neighbouring defects, i.e. $\varepsilon$ is independent of $n/N$. The defect solid then has a heat content

$$H_S = {}^0H_S + n\varepsilon$$

and entropy

$$S_S = {}^0S + f(n/N)$$

Simple statistical arguments (e.g. Ubbelohde, 1952) show that, provided the presence of defects does not disturb the vibrational energy appreciably, the free energy is a minimum not for the ideal but for the defect solids; i.e.

$$G_S = {}^0H_S - T^0S_S + \{n\varepsilon - Tf(n/N)\}$$
$$= {}^0G_S + \{n\varepsilon - Tf(n/N)\} \tag{5.1}$$

is a minimum when $n/N \sim \exp(-\varepsilon/kT)$. Under the conditions $n/N \ll 1$ the implied assumption that presence of the equilibrium concentration of defects does not seriously affect thermodynamic properties of a solid such as its specific heat or other vibrational properties is justifiable. Furthermore, at very low concentrations, various types $1, 2, 3 \ldots$ of structural defects can be harboured in a crystal with energies of formation $\varepsilon_1$, $\varepsilon_2$, $\varepsilon_3$ practically independent of one another, i.e. such defects are not directly cooperative.

$\varepsilon$ for most types of defects includes repulsion terms for nearest neighbours. For this reason, any factor causing the crystal lattice to expand generally reduces any value of $\varepsilon$ progressively. If $n/N$ approaches $10^{-3}$ or over, interaction between neighbouring defects usually begins to introduce cooperative action between defect sites to a significant extent. This tends to reduce the related value of $\varepsilon$ which then becomes appreciably dependent on $n/N$. Both this general influence of expansion of volume, and the eventual cooperation between neighbouring defect sites give an autocatalytic character to the formation of any type of defect in a crystal in thermal equilibrium, as the temperature increases. Nevertheless, for most but not all crystals, though it increases steeply even at the melting point, the absolute value of $n/N$ remains too small to affect the volume or heat content in any easily observed way. Relatively few crystal structures are exceptions for which really marked thermodynamic premelting can be observed (see Chapter 12). On passing from crystal to melt, the jump in volume $\Delta V_f / V_S$ lowers $\varepsilon$ enormously in all simple cases; in consequence $n/N$ becomes a fraction comparable with unity in the melt at the freezing point.

## 5.2 Positional disordering on melting and quasicrystalline melts

### 5.2.1. *Positional melting*

Transitions involving positional disordering in changing from one solid structure to another were referred to in Chapter 4. One statistical thermodynamic view of melting is to regard it as an analogous process of positional disordering, with the difference that the polymorph stable at higher temperatures (i.e. the melt) also happens to have the mechanical and physical properties of a fluid. On this model of melting, melts are naturally regarded as quasicrystalline. In fact on this version of the quasicrystalline model for liquids, melts are to be considered as polymorphic forms having the (markedly defective) *lattice* of a crystal, and therefore showing fluid flow.

Simultaneous influences from increasing vibration amplitudes and increasing concentration of defects in the crystal are not excluded by this approach; these two modes of energy uptake are in fact likely to interact in leading to melting. This behaviour is illustrated for example by the two-dimensional melting of bubble rafts subjected to vibration (Fukushima and Ookawa, 1955).

Full positional thermal disorder in a crystal thus leads to an idealized structural model for the melt, generally described as the quasicrystalline model. Melting can be achieved with reference to the crystal lattice by a sequence of notional operations.

(a) The crystal lattice is expanded to give the average separation of units of structure observed in the melt.

(b) Positions of the units are disordered with respect to the crystal lattices, to give the average positional disorder observed in structural studies on the melt.

(c) Other types of disorder also found in the melt are introduced in this disordered lattice.

However, a quasicrystalline melt formed by expanding the lattice and by introducing various types of disorder up to the equilibrium concentration is a workable model only for simple liquids. Even present knowledge of liquid structures indicates that it has serious limitations for describing many real liquids. Nevertheless, this specific and deliberately somewhat restrictive sequence of operations to arrive at a quasicrystalline melt from a crystal can be a convenient reference basis since it is clearly defined.

5.2.1.1. *Cell models for liquids.* The fact that various properties of liquids indicate considerable non-random distribution of nearest neighbours to any molecule has been losely described by some authors also by the general term 'quasicrystalline structure'. When this refers merely to the microenvironment of any molecule this term should not be used. For example, thermal, optical and dielectric comparisons support certain general analogies between liquids and crystals (Debye, 1939). But in view of the fact that each molecule vibrates in a characteristic cell formed by its nearest neighbours, these less specific analogies between liquids and crystals which merely refer to the *nearest neighbour* environment are more accurately and better described as 'cell' models. Indiscriminate use of the term 'quasicrystalline' merely to denote general nearest neighbour resemblances between crystals and their melts, can be very misleading (see Stuart, 1941). In particular, when discussing certain characteristic differences between crystals and their melts the important consideration emerges that some types of liquid cannot be represented at all as quasicrystalline melts, according to the specific definition adopted in this book. Such melts may nevertheless contain high concentrations of *anticrystalline* structures with non-random packing of nearest neighbours, and may thus show well marked 'cell' character. Vague use of the term 'quasicrystalline' can lead to numerous other inconsistencies (Hildebrand and Archer, 1961) which are avoided by consistent and precise definition of the term. Reasons are further developed in subsequent chapters for the important conclusion that not every kind of crystal structure may be expected to give rise to a quasicrystalline melt.

### 5.2.2. *Classification of types of liquids*

An important advantage of setting out on the basis of a narrowly defined quasicrystalline model for a melt is that this naturally suggests a fairly simple and straightforward structural classification of types of disorder that may make major contributions to the melting process. Such a classification immediately brings out the fact that there are several mechanisms of melting, or ways of increasing the disorder of a crystal at the melting point, $T_f$, to produce a melt of the same free energy and thus in equilibrium with the solid. This important conclusion is illustrated by diverse examples in subsequent Chapters of this book.

### 5.2.3. *Similitude theories of positional melting*

It is convenient to examine positional melting from a phenomenological standpoint before attempting to formulate detailed models so as to calculate statistical thermodynamic functions for the solid and melt. For this purpose, crystals of the inert gases are obviously the most suitable. Atoms of the inert gases all crystallize as cubic close-packed structures. In changing to quasi-crystalline close packed melts, only positional disorder can be introduced, in the form of crystal defects.

Obvious kinds of defects to be introduced to the quasicrystalline model include the following.

(1) Creation of vacant lattice sites. The energy required to create vacancies is related to the latent heat of sublimation $H_{vap}$ per atom. Lattice vacant sites are sometimes referred to as Schottky defects (Fig. 5.1a).

(2) Insertion of additional atoms at interstitial sites between the normal lattice points. For cubic close-packed structures, each atom has 12 nearest neighbours. Interstitial sites with a higher energy, but still presenting minima of potential energy, have 6 nearest neighbours. A complication may need to be considered according to whether an interstitial atom is to be regarded as 'tied' to a specific vacancy, or as completely independent (Fig. 5.1b).

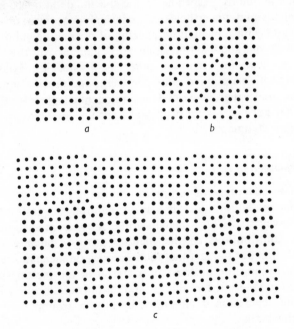

FIG. 5.1. Types of isolated and cooperative positional defects

Both types (1) and (2) of positional disorder normally decrease the density of atomic packing. This is obvious for vacancy formation. For inserting interstitial atoms, unless the neighbours are permitted to bulge outwards, very high repulsion energies of formation would be required (Fig. 5.1b).

A third type of positional disorder which may be of importance for melting involves the introduction of *cooperative* defects (Oldham and Ubbelohde, 1940) as illustrated in Fig. 5.1c. Cooperative defects can involve much smaller energies of formation and smaller increases of volume per unit of crystal structure (atom or molecule) than isolated defects such as vacancies or interstitial insertions. However, the fact that they require some degree of correlation through juxtaposition of defects in neighbours implies that their location is far from completely random. This reduces the increase of positional entropy achieved by their introduction in the crystal. Cooperative defects are discussed below in relation to 'dislocation' theories of melting. These involve a specific type of cooperative defect.

By analogy with other examples of physical similitude, a quasicrystalline model suggests a theorem of *structural similitude* in the case of crystal whose units can undergo only positional disorder on melting.

According to this theorem, crystals of similar structure should reach equilibrium with melts of similar (quasicrystalline) structure. This would imply the following:

(1) The entropy increase of positional disorder on melting would be the same for all crystals of the same type.

(2) The fractional increase in volume brought about by disordering on melting, $\Delta V_f / V_s$ would be the same, insofar as a single type of positional disorder applies.

(3) The melting point $T_f = H_f / S_f$ would show a constant relationship with the energy $\varepsilon$ of creating isolated positional defects. The ratio $H_f / H_{vap}$ should be constant for crystals of similar structure. If entropies of fusion are the same, as postulated under (1), this implies that $H_f / H_{vap}$ is constant, or $T_f / H_{vap}$ is constant.

Nature provides too few inert gases to test such correspondence relationships at all thoroughly; in one of them, helium, comparisons are further perturbed by major quantum effects. (For the other inert gases, small quantum corrections are made quite readily, e.g. Gopal, 1955). This shortage of suitable atomic crystals has stimulated a search for other crystals whose units of structure, though not single atoms, are linked by centrosymmetric forces, like the inert gases. In the range of molecular crystals, for the purpose of testing mathematical models, interactions between neighbours can frequently be represented in terms of forces of attraction, and in terms of repulsive forces described by reference to a repulsion envelope. Diatomic molecules of the permanent gases thus behave approximately as ellipsoids; simple polyatomic molecules likewise can be treated as deformed spheres. When mutual orientations of such molecules in crystal or melt can be disregarded (see below) approximate similitudes can again be looked for in the melting parameters.

Similitude effects have likewise been looked for in at least two other types of crystals:

(i) when the units of structure are ionic—particularly for salts the electronic structure of whose ions corresponds with that of the inert gases;
(ii) when the units are metal atoms.

Special factors affect the melting of each of these types of crystals; discussion is deferred to chapters of this book dealing with special crystal types. The chapter following deals more specifically with positional and other features of the melting of molecular crystals.

## REFERENCES

F. Cennamo (1952) *Rend. Acc. Naz. Lincei*, 8, **12**, 3, 294.
P. Debye (1939) *Z. Elektrochem.* **45**, 174.
E. Fukushima and A. Ookawa (1955) *J. Phys. Soc. Japan* **10**, 970.
K. Furukawa (1960) *Sci. Rep. Res. Inst. Tohoku Univ.* **12**, 368.
K. Furukawa (1962) *Rep. Progr. Phys.* **25**, 396.
N. S. Gingrich and C. W. Thompson (1961) 'Physics' Ginn, Boston.
R. Gopal (1955) *Z. Anorg. Allg. Chem.* **281**, 217.
J. H. Hildebrand and G. Archer (1961) *Proc. Nat. Acad. Sci., Wash.* **47**, 1881.
R. F. Kruh (1962) *Chem. Rev.* **62**, 319.
M. V. Kurik and V. A. Shayuk (1975) *Fiz. Tverd. Tela* **17**, 2320.
V. I. Ladyanov, V. I. Arkbarov, and V. P. Velyukhanov (1973) *Sov. Phys. Dokl. (U.S.A.)* **18**, 531.
K. Lemm (1970) *Mol. Cryst. Liquid Cryst.* **10**, 259.
H. Mendel (1962) *Acta Cryst.* **15**, 113.
J. W. H. Oldham and A. R. Ubbelohde (1940) *Proc. Roy. Soc. Ser. A* **176**, 50.
J. W. Otvos and H. Mendel (1962) *Acta Cryst.* **15**, 657.
V. K. Prokhorenko and I. A. Fisher (1959) *Z. Fiz. Khim.* **33**, 1852.
G. W. Robinson (1960) *Mol. Phys.* **3**, 301.
H. A. Stuart (1941) *Kolloid Z.* **96**, 140.
A. R. Ubbelohde (1952) *Introduction to Modern Thermodynamical Principles*, Oxford University Press.

# 6. STRUCTURAL MELTING.
# MOLECULAR CRYSTALS OF RIGID MOLECULES

## 6.1 Positional melting of inert gases and related crystals

Table 6.1 records values for the entropy increase and the volume increase on melting the inert gases. It will be seen that each property is closely similar throughout the series, (Helium is not included since the large contribution of zero-point energy of lattice vibrations to the equations of state of the solid and liquid (Chapter 3) swamp ordinary positional entropy and energy terms.) When the inert gases are compared with small polyatomic molecules, an

**Table 6.1. Volume and entropy changes on melting molecular crystals $S_f < 4$ e.u.**

| Ref. | Molecule | $T_f (K)$ | $\Delta V_f / V_s$ (%) | $S_f$ (e.u.) |
|------|----------|-----------|------------------------|--------------|
| 1 | Ne | 24·57 | 15·3 | 3·26 |
| | Ar | 83·78 | 14·4 | 3·35 |
| | Kr | 115·95 | 15·1 | 3·36 |
| | X | 161·36 | 15·1 | 3·40 |
| | $CH_4$ | 90·67 | 8·7 | 2·47 |
| | $CH_3D$ | 90·42 | — | — |
| | $CD_4$ | 89·78 | — | 2·42 |
| | $PH_2$ | 139·41 | 5·5 | 1·92 |
| | $H_2$ | 13·95 | 12·2 | |
| | HD | 16·60 | 12·2 | |
| | $D_2$ | 18·65 | 13·0 | |
| 2 | $SiI_4$ | 395·5 | — | 3·0 |
| 3 | $O_2$ | 54·3 | — | 2·0 |
| | $N_2$ | 63·1 | 7·5 | 2·7 |
| | CO | 68·1 | — | 2·9 |
| | HCl | 158·9 | — | 3·0 |
| | HBr | 186·2 | — | 3·1 |
| | HI | 222·3 | — | 3·1 |
| | $H_2S$ | 187·6 | — | 3·0 |
| | $D_2S$ | 187·1 | — | 3·0 |
| | $H_2Se$ | 207·4 | — | 2·9 |
| | $D_2Se$ | 206·2 | — | 2·9 |
| | $SiH_4$ | 88·5 | — | 1·8 |
| | $CF_4$ | 84·5 | — | 2·0 |
| | $C(CH_3)_4$ | 256·5 | — | 3·0 |

*References*
1. Clusius and Weigand (1940).
2. McCarty, Landauer, and Binkowski (1960).
3. Ubbelohde (1950).

unexpected degree of similitude is in fact also found for the melting of quite a large number of molecular crytals, despite the fact that their units of structure have distorted spherical repulsion envelopes (see Table 6.1). Heat capacity curves for these solids down to the lowest temperatures, and in some cases other physical properties, reveal the explanation: whenever crystals of poly-atomic molecules exhibit particularly small entropies of fusion, about 3 e.u. or even lower, this may often be attributed to disordering only of their positions in a lattice occurring on melting.

### 6.1.1. *Orientational disordering of molecules in crystals.*

In such cases, lambda transitions are quite generally found in the crystals at temperatures $T_{C_1}$, $T_{C_2}$, etc. considerably below the melting point, $T_f$. It seems plausible to suppose that such lambda transitions involve disordering of mutual orientations of axes of these non-spherical molecules. After such randomization of axes has been achieved, little further increase of orien-tational disorder can be expected at the melting point itself. At one time, it was even postulated that such molecules were indeed 'rotating freely' in the crystal lattice above $T_c$. However, 'free rotation' proves to be rare in crystals. Indeed even many melts do not provide the physical space for their molecules to rotate 'freely' near $T_f$ (section 6.6.3). Though of great importance for other physical properties, these kinetic aspects of orientational randomization are of only subsidiary significance for the thermodynamic properties of crystal and melt. Provided the relaxation time for the mutual orientation of molecu-lar axes in a crystal is small compared with the time required for making thermodynamic measurements, entropy of randomization of orientation $S_{or}$ can already be introduced above $T_c$ into the crystal. The question of its uptake at $T_f$ does not then arise. According to this interpretation, above $T_c$ molecules in these crystals behave in a quasispherical manner. Melting parameters for the positional disordering of molecules with a small number of atoms that have achieved quasispherical behaviour in the crystals are recorded in Table 6.1.

It is not even necessary that the molecules should be very small for them to show such behaviour. A class of crystals which exhibits predominantly posi-tional melting even though the number of atoms in each molecule is quite large comprises so called 'globular' molecules. Discussion of these is con-veniently deferred until section 6.5.2.

## 6.2 Orientational disorder on melting

Examples of orientational randomization in solid–solid transformations as discussed in Chapter 4 have been progressively identified. For a range of crystals containing even fairly small molecules it is found that the onset or absence of an orientational transition in the solid state below $T_f$ is chancy.

This sensitiveness of $T_c$ to diverse influences on the crystal lattice has been investigated in several ways. It can determine whether orientational disordering is included at $T_f$ in the melting process, so that

$$S_f = \{S_{pos} + S_{or}\}$$

or whether $T_f$ lies considerably higher so that the two types of entropy increase are clearly separated on the temperature scale.

Table 6.1 illustrates molecules with $S_f \sim S_{pos}$ only, and by contrast Table 6.2 illustrates molecules with $S_f = S_{pos} + S_{or}$ (Eucken, 1942; Ubbelohde, 1950).

**Table 6.2. Non-spherical molecules with two or more mechanisms of melting $S_t > 4$ e.u.**

| Molecule | Ref. | $T_f$ (K) | $S_f$ (e.u.) | $\Delta V_f / V_s$ |
|---|---|---|---|---|
| $HCF_3$ | 1 | 117·97 | 8·22 | — |
| $CO_2$ | 2 | 216·5 | 9·25 | 0·285 |
| $N_2O$ | 2 | 182·3 | 8·58 | — |
| COS | 2 | 134·3 | 8·41 | 0·116 |
| $CS_2$ | 2 | 161·1 | 6·51 | 0·067 |
| $Br_2$ | 2 | 267 | 9·66 | — |
| $Cl_2$ | 2 | 172·1 | 8·89 | — |
| $(CN)_2$ | 2 | 245·3 | 7·90 | — |
| HCN | 2 | 259·8 | 7·73 | — |
| $C_2H_4$ | 2 | 103·97 | 7·70 | 0·116 |
| $C_2H_6$ | 2 | 89·9 | 7·60 | 0·12 |
| $SO_2$ | 2 | 197·6 | 8·95 | — |
| $CHCl_3$ | 2 | 210·0 | 10·8 | — |
| $SiCl_4$ | 2 | 203·4 | 9·06 | 0·14 |
| $C_6H_6$ | 2 | 278·5 | 8·44 | 0·134 |
| $SF_6$ | 2 | 218 | 5·22 | — |
| $H_2O$ | 2 | 273·15 | 5·25 | — |
| $NH_3$ | 2 | 195·4 | 6·91 | — |
| NO | 2 | 109·4 | 5·03 | — |
| $CH_3CF_3$ | 3 | — | 10·15 | — |

*References*
1. Valentine, Brodale, and Giauque (1962).
2. Ubbelohde (1950).
3. Guthrie and McCullough (1961).

Molecules which do not show orientational randomization in the crystals and which can be readily purified can constitute useful model substances for sophisticated researches on multiple mechanisms of melting. Convenient examples are becoming identified progressively.

For molecular crystals which combine two mechanisms of fusion, the entropy increase for orientational disorder $S_{or}$, may be considerably greater than $S_{pos}$. Thus in 1,1,1-trichloroethane, the entropy of randomization is nearly four-times as large as the entropy of fusion. $S_{or}$ corresponds to the randomization of orientation of the dipoles in the crystals in this case.

## 6.3. Repulsion envelopes in the melting of molecular crystals with rigid molecules

In discussing the orientational randomization of molecules in crystals, various considerations indicate that repulsion effects play predominant roles. A convenient description of molecular shape in this regard is provided by its 'repulsion envelope', as obtained by taking each of the peripheral or surface atoms in turn, and constructing the 'repulsion sphere' for this atom. According to Pauling (1960) (who appears to have first introduced crystallographic uses of this concept) the repulsion radius of an atom in directions where it is not bonded to neighbours is approximately constant. This geometrical representation is useful in discussing diverse problems concerning the melting of molecular crystals (see refs. in Al Mahdi and Ubbelohde, 1953). Figure 6.1 illustrates some repulsion envelopes constructed on this basis for a number of aromatic hydrocarbons.

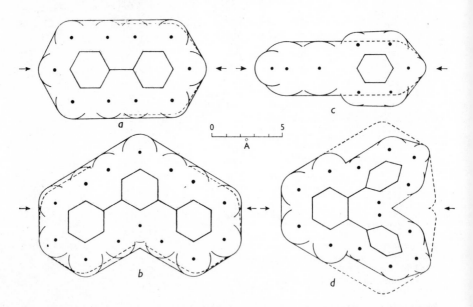

FIG. 6.1. Repulsion envelopes for some aromatic hydrocarbons. (a) Diphenyl; (b) *m*-terphenyl; (c, d) *o*-terphenyl viewed in the plane of the middle ring and from above it

Except in ionic crystals, repulsion forces are very short range. As a result of intermolecular attractions, due to dipole, dispersion, and other interactions between non-bonded electrons in neighbouring molecules, molecules in molecular crystals shrink together until the repulsion envelopes are in contact, in such a way as to minimize the volume, and thus maximize the potential energy of attraction. Comparison between the crystal volume found experimentally and that calculated from considerations of closest packing of repulsion envelopes give good agreement in a number of cases, provided that these molecular skeletons can be regarded as rigid.

In this connection 'rigidity' implies that the atoms vibrate about a single equilibrium configuration of the molecule, and that new configurations can only be produced through energy inputs $\gg kT$. For example, many aromatic molecules can be treated as rigid in this sense.

As an extension to principles of similitude in a series of structurally related molecules, similar melting mechanisms may be looked for whenever similar repulsion envelopes lead to similar crystal structures.

### 6.3.1. *Effects of pressure on rotational disordering*

One reason such comparatively small changes in the shape of the repulsion envelopes change the character of the melting process of many molecular crystals from merely positional with $S_{pos} \sim 3$ e.u. to a more complex summation of mechanisms is that the energy $\varepsilon_{or}$ required to achieve randomization of orientation depends in a very sensitive way on the peak repulsions, for the most unfavourable arrangement of neighbours. Thus $\varepsilon_{or}$ is not solely dependent on average values. These peak repulsions act as barriers to orientational randomization without necessarily affecting the total lattice energy to a comparable extent. Application of external pressure by forcing the molecules closely together should increase these peak repulsions quite strongly, thereby raising the temperature $T_c$ for orientational radomization in the crystals. Until the pressure attains sufficiently high values for any given crystal to bring $T_c$ close to $T_f$ (see Chapter 11), melting mechanisms could thus be affected to an abnormal extent by applied pressure. Classical thermodynamics throws little light on the shift of transformation points in molecular crystals by pressure, nor does this interesting problem appear to have been much investigated experimentally. For a typical globular crystal, adamantane, the solid–solid transition has been measured under pressure at 20°C (Pistorius and Reising, 1969).

Addition of more nearly spherically shaped molecules in solid solution in the crystals may help to 'lubricate' randomization of orientation and to depress $T_c$ well below $T_f$, thus changing a crystal with a complex mechanism of melting to one showing merely positional melting at $T_f$. This aspect of the melting of solid solutions was referred to previously (section 2.15). It necessitates particular care in the purification of crystals which show an orientational transformation near $T_f$. For example, it seems likely that effects of impurities in solid solution on premelting of KSCN (Rhodes and Ubbelohde, 1959) can be attributed to lubrication of randomization.

**6.4 Coexistence of related structures in rotator transitions in crystals**

In molecular crystals the change from specific to random orientation of molecular axes in a solid–solid transition often does not require major structural changes. Accordingly, transformation routes by way of hybrid single crystals seems fairly common, with the accompanying phenomena of hysteresis, as a consequence of the need to nucleate domains internally. Bypassing intermediate structures when a transformation cycle is traversed in a given sense has also been observed in a number of cases, for molecular crystals. These include groups of molecules in which the high temperature solid is plastic, e.g.

$$CCl_4, \quad CH_3CCl_3, \quad CHCl_2CH_3$$
$$(CH_3)_n CCl_{(4-n)}$$

(Rudman, 1970 quoted in Silver and Rudman, 1970) and a number of closely related derivatives of pentaerythritol $R-C(CH_2OH)_3$ in which the existence range of the plastic phase depends in a systematic way upon the substituent (Doshi, Furman and Rudman, 1973) (see Table 6.3).

*Table 6.3. Rotator transitions in pentaerythritol analogues*

| Molecule (RC(CH$_2$OH)$_3$ | $T_c (K)$ | $T_f (K)$ |
|---|---|---|
| R = —NO$_2$ | 347·4 | 438 |
| —CH$_3$ | 351·3 | 497 |
| —COOH | 394·7 | 492 |
| —NH$_2$ | 407·3 | 446 |

In pentaerythritol itself, proton magnetic resonance and calorimetry have established the following thermal parameters (Smith, 1969)

$$(CH_2OH)_4C \quad T_c\,453 \quad T_f\,534$$

with a heat of transition $H_c$ from the tetragonal to the (plastic) f.c.c. crystal of 10 kcal mole$^{-1}$. This heat of transition is too small to disrupt all the hydrogen bonds in the tetragonal crystal so that some kind of disordered network must persist in the plastic state. The entropy of fusion is 3.2 e.u.

**6.5 Trends in melting parameters showing similitude for structurally related molecules**

One example of similitude is shown by tetrahalides and similar molecules based on elements of Group 4 of the Periodic System, when their repulsion envelopes can be regarded as quasispherical (Sackmann and Cloos, 1958; see also Guthrie and McCullough, 1961).

In Table 6.4 the lower values for $\Delta V_f/V_s$ and of $S_f$ for carbon tatrachloride may be related to a transformation monoclinic $\rightarrow$ cubic at $-47 \cdot 5°C$, which is likely to involve orientational randomization. Two crystal forms can coexist very near $T_f$ (Koga and Morrison, 1976). As can be seen, all the other molecules show quite close similitude in their melting parameters (Table 6.4). As is further discussed later in this book, for none of these molecules does the volume of the melt permit 'free' rotation.

*Table 6.4. Similarities in melting parameters of tetrahalides*

| Molecule | $T_f(°C)$ | $\Delta V_f/V_s (\%)$ | $V_L (ml)$ | $S_f (e.u.)$ |
|----------|-----------|----------------------|------------|--------------|
| $CCl_4$  | $-22 \cdot 7$ | $5 \cdot 30$  | $91 \cdot 73$  | $2 \cdot 31$ |
| $SiCl_4$ | $-67 \cdot 7$ | $12 \cdot 42$ | $102 \cdot 59$ | $9 \cdot 08$ |
| $GeCl_4$ | $-49 \cdot 5$ | —             | $104–106$      | —            |
| $TiCl_4$ | $-23 \cdot 0$ | $12 \cdot 40$ | $105 \cdot 22$ | $9 \cdot 07$ |
| $SnCl_4$ | $-33 \cdot 0$ | $11 \cdot 88$ | $110 \cdot 07$ | $9 \cdot 11$ |

### 6.5.1. *Trends in melting points with increasing molecular size*

For rigid molecules whose crystals have similar entropies of fusion, heats of fusion are likely to show a trend which is more or less systematic with increasing molecular size, simply owing to the increase in energy required to create similar positional or orientational crystal defects. Freezing temperatures $T_f = H_f/S_f$ are then likely to show systematic trends, further discussed on p. 136. Broad phenomenological correlations of melting point with molecular shape of organic molecules are sometimes attempted (e.g. Bunn, 1955). For rigid molecules with similar repulsion envelopes, for example, quasispherical, or disc-like, or rod-like, compounds of any type with the higher cohesive energies will generally have the higher melting points. A so-called 'universal' plot of melting point against cohesive energy does accordingly segregate molecules onto the members of a fan distribution. Rigid quasispherical molecules lie at the left and flexible molecules lie at the right of the fan (Fig. 6.2).

On the basis of the general model for structural melting discussed above, this fan distribution would be expected simply because molecules to the left tend to have the fewest mechanisms of disorder and lowest entropies of melting, whereas flexible molecules on the right have the largest number of such mechanisms for a given cohesive energy. (A different interpretation (Bunn, 1955) is based on the transmission of vibrations, in a qualitative extension of the Lindemann model for melting as a catastrophic breakdown of lattice vibrations; see p. 68 for a discussion of inadequacies of that model.)

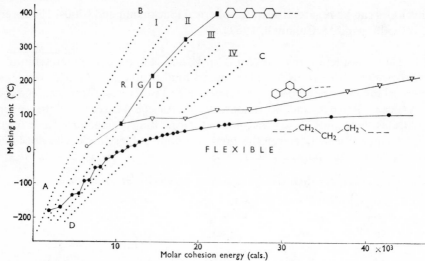

FIG. 6.2. Melting points of monomeric substances as function of their molecular cohesive energy. Reproduced, with permission, from Bunn, *J. Polym. Sci.* **16,** 323 (1955)

### 6.5.2. *Melting of globular molecules*

Particularly striking melting behaviour is observed for rigid molecules of large molecular weight whose repulsion envelope is so nearly spherical that solid–solid transitions involving orientational transformations have $T_c \ll T_f$. Despite the large molecular weights, above $T_c$ these solids behave in important respects similarly to monatomic crystals. Entropies of fusion involve practically only positional disordering, and in some cases are even below $S_f = 3$ e.u. Table 6.5 contrasts globular molecules which have this characteristic, with others for whose crystals $T_c$ and $T_f$ coincide. For these, $S_f$ is much larger since more than one mechanism of disordering is involved at the melting point.

*Table 6.5.* **Melting parameters of globular molecules**

| Group $C_5$ | $T_c$ | $T_f$ | $S_c$ | $S_f$ | $\Delta S_{(total)}$ |
|---|---|---|---|---|---|
| Cyclopentane | −186·1 | −133·7 | 1·3 | 6·0 | 7·3 |
| Cyclopentanone | −51·3 | — | — | 8·3 | 8·3 |
| Cyclopentanol, $C_5H_9OH$ | $\left.\begin{array}{c}-37\cdot4\\-71\cdot3\end{array}\right\}$ | −19·5 | $\begin{array}{c}0\cdot4\\1\cdot1\end{array}$ | 1·5 | 3·0 |
| $C_5H_9Cl$ | −105·1 | −94·3 | 4·6 | 1·6 | 6·2 |
| $C_5H_9CN$ | — | −75·2 | — | — | — |
| $C_5H_9NH_2$ | — | −85·7 | — | 7·4 | 7·4 |
| **Group $C_6$** | | | | | |
| $C_6H_{11}CN$ | −13·0 | 12·05 | 0·86 | 3·1 | 3·9 |

*Reference*
Labruyere Verhavert (1951).

Best known amongst globular molecules is a group of terpene derivatives. Other alicyclic systems with bridged rings can show similar behaviour. For example, additions to the range of large 'globular' molecules now known to have low entropies of fusion of the order 2–3 e.u. have been made in a systematic study of derivatives of 'bent' polycyclic molecules (Pirsch, 1955). Figure 6.3 illustrates some of the molecular skeletons involved. Table 6.6 records some actual molecular species with low entropies of fusion. Because

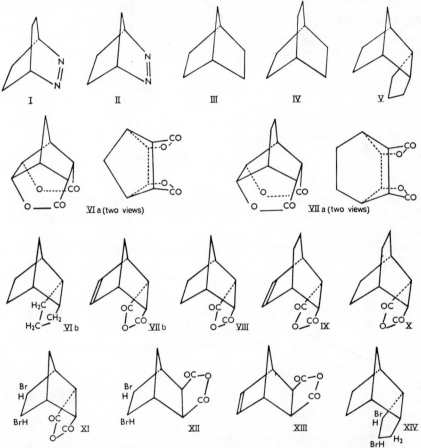

—FIG. 6.3. Skeleton structures of some globular molecules with low entropies of fusion. Atomic skeletons correspond to the following molecules: (I) 1,4-endo-azocyclohexane; (II) endomethylenedehydropiperidazine; (III) bicyclo-2,1,2-heptane; (IV) bicyclo-2,2,2-octane; (V) α-dicyclopentadiene; (VIa) see Table 6.6; (VIb) hydrogenated form of V; (VIIa) (VIIb) (VIII) (IX) and (X) see Table 6.6; (XI) dibromo-α-cis-3,6-endomethylenehexahydrophthalic acid anhydride; (XII) dibromo-β-cis-3,6-endomethylenehexahydrophthalic acid anhydride; (XIII) β-cis-3,6-endomethylene-Δ⁴-tetrahydrophthalic acid anhydride; (XIV) dihydro-α-cyclopentadiene dibromide. Reproduced by permission of Springer-Verlag of the high molecular weights these compounds can constitute useful solvents with very large cryoscopic constants.

Molecules of types XI, XII, XIII, and XIV on the other hand apparently fail to randomize their orientations in the crytals and have entropies of fusion in the range 10–17 e.u. as a consequence.

Large three bar cage hydrocarbons have been investigated, though it is not known whether the bars behave rigidly or show configurational isomers.

Bicyclo-2,2,2-octane also has analogues where tervalent N replaces either or both CH groups numbered (1) and (2).

$$\begin{array}{ccc} & \mathrm{CH} & \\ H_2C & CH_2 & CH_2 \\ | & | & | \\ H_2C & CH_2 & CH_2 \\ & CH_2 & \end{array}$$

3-Oxobicyclo-3,2,2-nonane has also a related molecule with the oxo group replaced by NH.

$$\begin{array}{ccc} & \mathrm{CH} & \\ H_2C & CH_2 & CH_2 \\ | & O & | \\ H_2C & CH_2 & CH_2 \\ & CH & \end{array}$$

Thermodynamic parameters have been reported by Amzel and Becka (1969), Westrum, Wong, and Morawetz (1970), and Amzel, Cucarella, and Becka (1971), who give earlier references.

Points of interest in connection with such crystals of globular molecules are as follows.

(i) By suitable synthetic procedures comparatively small modifications of external shape can be introduced which nevertheless suffice to suppress the

**Table 6.6.** *Some large polyatomic molecules with low entropies of fusion*

| | Molecule skeletal type (Fig. 6.4) | $T_f$ (K) | $S_f$ (e.u.) |
|---|---|---|---|
| $\alpha$-cis-3,6-Endomethylene-$\Delta^4$-tetrahydrophthalic acid anhydride | VIIb | 438 | 2·67 |
| $\alpha$-cis-3,6-Endomethylenehexahydrophthalic acid anhydride | VIII | 441 | 2·70 |
| $\alpha$-cis-Endoethylene-$\Delta^4$-tetrahydrophthalic acid anhydride | IX | 420 | 2·54 |
| $\alpha$-cis-3,6-Endoethylenehexahydrophthalic acid anydride | X | 460 | 3·21 |
| Dilactone of 4,5-dioxy-3,6-endomethylene-hexahydrophthalic acid | VIa | 539 | 2·89 |
| Dilactone of 4,5-dioxyendoethylene-hexahydrophthalic acid | VIIa | 578 | 3·10 |

quasispherical behaviour. Some examples are recorded in Table 6.7, which should be examined in conjunction with the other tables and with Fig. 6.3 Semi-quantitative comparisons based on general behaviour near $T_f$ (Timmermans, 1961) include as characteristics of globular molecules a centre or axis of symmetry, high $T_f$, high vapour pressure at $T_f$ and $S_f \leqslant 3$ e.u. Most globular molecules crystallize in the cubic system at $T_f$ and show marked plastic flow.

**Table 6.7.  Related molecules forming globular and non-globular crystals**

Fatty series

| Globular | Non-globular |
|---|---|
| $CH_4$, $CD_4$, $CH_3D$ | $C_2H_6$ |
| $C_2H_2$ | $CH_2{=}CH_2$ |
| $C(CH_3)_4$ | $C(C_2H_5)_4$ |
| $CF_4$, $CCl_4$, $CBr_4$ | $CH_2Cl_2$ |
| $CCl(CH_3)_3$, $CCl_2(CH_3)_2$ | $CHCl_3$ |
| $CCl_3CH_3$, $CClF_2CH_3$ | $CClH_3$ |
| $CCl(CH_3)_2NO_2$ | $CCl_2HCH_2Cl$ |
| $C(NO_2)_4$ | $CH(NO_2)_3$ |
| $C(SCH_3)_4$ | $C(SC_2H_5)_4$ |
| $C_2Cl_6$, $C_2F_6$, $C_2F_5Cl$ | $C_2HCl_5$ |
| $C_2(CH_3)_6$, $C_2Cl(CH_3)_5$ | — |
| $C(CH_3)_3C_2H_5$, $CH(CH_3)_2CH(CH_3)_2$ | — |
| $CH(CH_3)_2C(CH_3)_2C_2H_5$ | — |
| $C(CH_3)_3C(CH_3)_2C_2H_5$ | $C(CH_3)_3CH(CH_3)C_2H_5$ |
| $C(CH_3)_3CH(CH_3)CH(CH_3)_2$ | — |
| $C(CH_3)_3CH{=}CH_2$ | $CH_3CH_2CH{=}CHCH_2CH_3$ |
| $CH_3CH{=}CHCH{=}CHCH_3$ | — |
| $CH_3OH$, $CH_3OD$ | $C_2H_5OH$ |
| $(CH_3)_3COH$, $(CH_3)_3CSH$ | — |
| $CH_3CH(OH)_2$ | — |
| $(CH_3)_2C_2H_5COH$ | $CH(CH_3)_2CH_2CO_2H$ |
| $C(CH_3)_3CO_2H$, $CCl_3CO_2H$ | $CH_2ClCO_2H$ |
| $CCl_3C_6H_5$ | $CH_2ClC_6H_5$ |
| $(CNCH_2)_2$ | malonitrile, glutaronitrile. |

Cyclic series
Polymethylenes

| | |
|---|---|
| In $C_4$: perfluorcyclobutane | cyclobutanone |
| In $C_5$: 1,1 and 1,2-*cis*- dimethylcyclopentane | methylcyclopentane, 1,3 and 1,2-*trans*- |
| 1,2,2-*trans*-trimethylcyclopentane | dimethylcyclopentane |
| cyclopentanol, cyclopentadienol | cyclopentene, cyclopentanone, |
| cyclopentyl chloride, perfluorcyclopentane | cyclopentylamine, cyclopentylnitrile, |
| | cyclopentanone, methylcyclopentanol, |
| In $C_6$: cyclohexane, cyclohexylchloride | methylcyclohexane, 1,3-, 1,4- and 1,2- |
| cyclohexanol, cyclohexanone, | *trans*-dimethylcyclohexane |

α-hexachlorcyclohexane,     β-hexachlorcyclohexane,
cyclohexene     cyclohexylamine
1,3-cyclohexadiene,     cyclohexanoxime
1,4-cyclohexadione
cyclohexylfluoroform, 1,1- and     menthane
1,2-*cis*
dimethylcyclohexane

## Heterocyclics

| *Globular* | *Non-globular* |
|---|---|
| Azocyclohexane | piperidine |
| Endomethylene piperazine | piperazine |
| Tropinone | thiophene |
| Camphor anhydride | cineol |
| Tetrahydrofurane | furane |
| Thiocyclohexane | dioxane |

## Condensed cyclics

| | |
|---|---|
| Decahydronaphthalene | tricyclopentadiene |
| Dicyclopentadiene | |
| Dihydro and tetrahydrodi- | |
| cyclopentadiene-α: alcohol | |
| Ketone, oxide and bioxide | |
| Tetrahydrohydrindane | |

## Bridged cyclics

| | |
|---|---|
| Endomethylenecyclohexane, | Fenchane: alcohol and ketone |
| Camphene | |
| Camphenilane, isocamphane, | |
| Camphenilone | |
| Norcamphor, norpinane, | |
| Pinane, tricyclene | |
| Endoethylenecyclohexane | |

## Camphor group

| | |
|---|---|
| Camphane, isobornylane, | 3-Camphor: bromide, sulphone, |
| Bornylene, bornyl and | sulphonamide, sulphoxide, and |
| Isobornyl chloride, 2,6- | sulphochloride |
| and 2,10 | |
| Dichlorocamphane, borneol and | |
| Isoborneol, bornylamine, | |
| Camphor, thiocamphor, 2,3- | |
| camphanedione, 3-cyano- | |
| camphor, camphoroxime and | |
| camphorquinone | |

Potential barriers opposing randomization of orientation of molecules in this kind of crystal originate almost entirely from interlocking of repulsion envelopes of neighbouring molecules. When these molecules are non-spherical, maxima and minima of repulsion are encountered if any molecule is rotated with respect to its neighbours in the crystal lattice. Such repulsive forces are very sensitive to the intermolecular distance, and can be affected by apparently comparatively minor changes in the crystal. Expansion of the crystal lattice by thermal or other influences normally reduces potential barriers to randomization of orientation.

Detailed thermodynamic and related studies have been made for various bridged cyclooctanes and cyclononanes, whose skeletal structures are indicated in Fig. 6.3. These skeletal structures are not wholly 'rigid' and critical attribution must be made, for example, of different vibrational modes in crystal and in melt (see also Chang and Westrum, 1960; Guthrie and McCullough, 1961; Barber and Westrum, 1963; Trowbridge and Westrum, 1963; Wolff and Westrum, 1964). Unfortunately changes of volume at transition and melting points often have not been determined for many other globular compounds of interest. It is expected for such compounds that increased pressure would have a particularly steep effect in raising the temperature at which orientational randomization can occur, even raising this to above $T_f$, i.e. into the melt. If this occurs a mesophase may intervene before a completely isotropic melt is attained (see 301 seq.).

One sensitive means of investigating consequences of detailed differences of repulsive forces for orientational transitions is to measure dielectric properties as a function of temperature. For globular molecules whose repulsion envelope is practically spherical (Timmermans, 1950) the transition temperature $T_c$ for rotation in the crystals can be determined with particular simplicity when such molecules carry a fixed dipole. $T_c$ frequently lies well below $T_f$. In the temperature interval between $T_c$ and $T_f$ the activation energy for 'rotation' is often even lower than in the melt. Presumably in such melts the lowered symmetry of interaction forces upsets some of the electrostatic compensation of forces in the crystal (see Chapter 8).

In Table 6.8 (see Clement and Davies, 1960) heats of activation are derived by applying the Eyring equation to the relaxation time $\tau$ observed from dielectric measurements at different frequencies and different temperatures

$$\frac{1}{\tau} = \frac{kT}{h} \exp\left[ -\frac{\Delta H^{\neq}}{RT} + \frac{\Delta S^{\neq}}{R} \right]$$

Insofar as molecules in the globular range of crystal behaviour act as if they were approximately spherical, similitude (or at least close analogy) may be expected with melting parameters for inert gas crystals. Although this still needs verification from case to case, generally on melting $\Delta V_f / V_s$ tends to be

**Table 6.8.** *Heats of activation for rotation of dipoles*
*carried by globular molecules* $\Delta H^{\neq}$ *kcal/mole$^{-1}$*

| Molecule | $\Delta H^{\neq}$ crystal (in range $T_c$–$T_f$) | $T_c$ (°C) | $T_f$ (°C) | $\Delta H^{\neq}$ melt |
|---|---|---|---|---|
| 2-Chloro-2-nitropropane $(CH_3)_2CCl(NO_2)$ Succino- | 1·31 | −60 | −22 | 1·46 |
| nitrile $CH_2CN$ | 2·0 | −44 | — | 3·9 |
| $\mid$ $CH_2CN$ | | | | |
| Methyl chloroform $CH_3CCl_3$ | 1·16 | −50 | −33 | 1·16 |
| t-Nitrobutane $(CH_3)_3CNO_2$ | 0·5 | — | — | 1·0 |

in the higher range of values for globular molecules (see Timmermans, 1950). This strengthens the analogy with melting of the crystals of inert gases for which $\Delta V_f/V_s$ is likewise remarkably high (see p. 129).

### 6.5.3. *Plastic crystals*

Crystals with globular molecules frequently show a remarkable degree of plasticity near the melting point (Michils, 1948). Presumably this is because it is very difficult to 'pin' dislocations in the crystals. When external shearing forces are applied, these unpinned dislocations permit extensive slipping. The general theme of plastic flow in crystals is too extensive for discussion in detail here. Physical investigations on the plasticity of crystals of globular molecules do however have special relevance to general theories about melting, since the changes in various physical properties at the transition temperature $T_c$ in the sequence

$$\text{rigid crystal} \underset{T_c}{\rightleftharpoons} \text{plastic crystal} \rightleftharpoons \text{melt}$$

may make this temperature analogous to a mechanical melting point $T_{mech}$ (Clement and Davies, 1962, 1964). For such crystals, in a sense $T_c \leqslant T_{mech} < T_f$ whereas for brittle crystals the thermodynamic melting point $T_f < T_{mech}$ i.e. $T_f$ lies considerably below any hypothetical mechanical melting point $T_{mech}$ (see Chapter 3). Plasticity in globular molecular crystals probably differs from the plasticity in metallic crystals (Campbell, 1975) since 'coldworking' does not appear to have been reported for the former. In metallic cubic crystals plastic flow can perhaps be attributed to 'high temperature creep' (Dunning, 1961). Effects of pressure on the plasticity of molecular crystals could help to illuminate possible differences in their flow mechanisms from that in metals (Ubbelohde, 1961), which in turn might throw light on the defect structures of the crystals. Although macroscopic measurements of 'plasticity' have been most widely used, ultrasonic data may help to detect features of molecular motion on the microscale (Green and Scheie, 1967), Valuable advances are being made by such correlations between more than one property of plastic crystals (McKay and Sherwood, 1975). The extent of the change in response to applied shear, at the transition to the plastic state depends greatly on the molecule. Though plastic near $T_f$ solid biphenyl is about ten times less so than globular rotator molecules (Corke and Sherwood, 1971). For $CCl_4$ a plot of shear strength against the reduced temperature $T/T_f$ shows that this crystal behaves like argon or krypton and shows only about one-tenth the resistance to applied shear, to that shown by a salt such as AgCl (Towle, 1968, where earlier references are quoted). It can be very informative to contrast physical properties of plastic and brittle crystals at reduced temperatures $T/T_f > 0.75$ and to examine more than one physical property (Sherwood, 1972; Smith, 1975). Thus to achieve the same strain rate of $10^{-6} s^{-1}$ the applied stress (in $kN\ m^{-2}$) is $\sim 180$ for white phosphorus, compared with only 20 for cyclohexane (Hampton, McKay, and Sherwood, 1974). Dynamic factors in the response of plastic crystals to applied shear can help to probe changes in the premelting temperature range of molecular crystals, though analysis of the acoustic data raises problems still unresolved (Green and Scheie, 1967; Dobromyslov and Koshkin, 1970).

Use of PMR with molecular crystals which show a 'rotator' transition below the melting point generally confirms the expectation that below $T_c$ hindrances to orientational freedom of the molecules control their kinetic behaviour. Between $T_c$ and $T_f$ molecular self-diffusion in the crystals becomes the dominant kinetic process. However, certain crystal types including white phosphorus and pentaerythritol do not follow the general pattern. Probably this is because their structure is of the network type (Chapter 10) so that progressive defect formation leading to melting follows a different sequence (Smith, 1969).

## 6.6 Melting of rigid linear molecules

In molecular crystals with quasispherical molecules, intermolecular forces increase only slowly with increasing molecular weight. This is because dispersion and dipole forces have a comparatively short range of action so that the atomic nuclei and electrons at the interior of a globular molecule make only minor contributions to the intermolecular potential. As indicated above, crystals containing globular molecules of quite high molecular weights still have accessible melting points, well below the temperature of any thermal decomposition of the compound. Programmes for systematic syntheses to prepare and study the melting of progressively larger and larger globular molecules, retaining the same general shape, would perhaps be difficult to implement; in any event no extensive ranges of trends of $T_f$ with molecular size have yet been established. Globular proteins are not without interest in this connexion.

With molecules that are stretched out in their structure, homologous series are more easily prepared. The melting behaviour of flexible molecules shows special characteristics and the behaviour of their homologous series is discussed in Chapter 7. Rigid stretched out molecules are unfortunately much less common so that the effect of increasing molecular size is difficult to generalize. One group of rigid molecules whose repulsion envelopes are roughly cylindrical is constituted by polyenes $R_1(CH=CH)_n R_2$ and another by polyines $R_1(C\equiv C)_n R_2$.

For the larger values of $n$ these are stable usually only when the groups $R_1$ and $R_2$ have suitable electronic properties. Data for some polyenes are recorded in Table 6.9.

*Table 6.9. Melting points of some rigid molecules*
*with quasicylindrical repulsion envelopes*

| Molecule | $T_f\,(°C)$ | Ref. |
|---|---|---|
| *Diphenylacetylenes* | | |
| Diphenylacetylene | | |
| $\quad C_6H_5-C\equiv C-C_6H_5$ | 62·5 | 1 |
| 1,4-Diphenylbutadiyne | | |
| $\quad C_6H_5-C\equiv C-C\equiv C-C_6H_5$ | 88 | 3 |
| 1,6-Diphenylhexatriyne | | |
| $\quad C_6H_5-C\equiv C-C\equiv C-C\equiv C-C_6H_5$ | 96 | 4 |
| 1,8-Diphenyloctatetrayne | | |
| $\quad C_6H_5-C\equiv C-C\equiv C-C\equiv C-C\equiv C-C_6H_5$ | 110 | 4 |

*Tetraphenylpolyenes*

$$\begin{array}{c} C_6H_5 \\ \diagdown \\ \quad\quad C=CH-(CH=CH)_n-CH=C \\ \diagup \\ C_6H_5 \end{array} \begin{array}{c} C_6H_5 \\ \diagup \\ \\ \diagdown \\ C_6H_5 \end{array}$$

| | $T_f\,(°C)$ | Ref. |
|---|---|---|
| $n=0$ | 202 | 5 |
| $n=1$ | 202–204 | 6 |
| $n=2$ | 198–199 | 6 |
| $n=3$ | 227–228 | 6 |
| $n=4$ | 213–214·5 | 6 |

*References*
1. Smith and Falkof (1942).
2. Jones (1950).
3. Armitage, Entwistle, Jones, and Whiting (1954).
4. Wittig and von Lupin (1928).
5. Wittig and Klein (1936).

A particularly important sequence of rigid quasicylindrical molecules is constituted by the *para*-phenylhydrocarbons and their substituents. Phenyl groups in such molecules can undergo oscillation or even rotation about a molecular axis formed by the join of all the *para* atoms, but the repulsion envelopes are practically rigid with respect to this axis. It is disappointing that despite their great interest for solid state organic chemistry melting parameters of such homologous series have hardly been explored systematically. Table 6.10 illustrates the fact that $T_f$ rises steeply, with the probable consequence that the interval between melting point and boiling point diminishes progressively as the molecular weight increases. Indeed, ultimately practically no liquid phase may be stable at all at 1 atm pressure, for this class of rigid molecules.

**Table 6.10.** The melting points of some stretched polyphenyls

| Meta and para polyphenyls Name | | | $T_f (°C)$ | |
|---|---|---|---|---|
| | Meta | Ref. | Para | Ref. |
| Benzene ($n = 1$) | 6 | | 6 | |
| Diphenyl ($n = 2$) | 70 | | 70 | |
| Terphenyl ($n = 3$) | 87 | 1,2 | 210 | 3,4,5 |
| Quaterphenyl ($n = 4$) | 86 | 6 | 318 | 7, 8 |
| Quinquephenyl ($n = 5$) | 112 | 9 | 395;380 | 6, 10 resp. |
| Sexiphenyl ($n = 6$) | — | | 475 | 6, 9, 11 |
| Septiphenyl ($n = 7$) | — | | 544 | 6, 9 |
| Noniphenyl ($n = 9$) | 166 | 9 | — | |
| Deciphenyl ($n = 10$) | 184 | 9 | — | |
| Undeciphenyl ($n = 11$) | 202 | 9 | — | |
| Duodeciphenyl ($n = 12$) | 223 | 9 | — | |
| Tredeciphenyl ($n = 13$) | 245 | 9 | — | |
| Quatuordeciphenyl ($n = 14$) | 270 | 9 | — | |
| Quindeciphenyl ($n = 15$) | 292 | 9 | — | |
| Sedeciphenyl ($n = 16$) | 321 | 9 | — | |

*References*

1. Chattaway and Evans (1896).
2. France, Heilbron, and Hey (1939).
3. France, Heilbron and Hey (1938).
4. von Braun (1927).
5. Basford (1936).
6. Florin and Mears (1955).
7. Ullmann and Meyer (1904).
8. Gelissen and Hermans (1925).
9. Busch and Weber (1936).
10. Gerngross and von Dunkel (1924).
11. Pummerer and Bittner (1924).

For allenes (Table 6.11) $T_f$ likewise rises steeply as the molecular size increases. It is satisfactory confirmation of the general principles under discussion concerning the melting of molecular crystals that the *meta*-phenyls

**Table 6.11. Melting points of some rigid molecules
with quasicylindrical repulsion envelopes**

| Molecule | $T_f (°C)$ | Ref. |
|---|---|---|
| Tetraphenylallene | | |

Tetraphenylallene

$$C_6H_5 \diagdown \diagup C_6H_5$$
$$C{=}C{=}C$$
$$C_6H_5 \diagup \diagdown C_6H_5$$

164–166    1

1,1,4,4-Tetraphenylbutatriene

$$C_6H_5 \diagdown \qquad\qquad \diagup C_6H_5$$
$$C{=}C{=}C{=}C$$
$$C_6H_5 \diagup \qquad\qquad \diagdown C_6H_5$$

237    2

1,1,6,6-Tetraphenylhexapentaene

$$C_6H_5 \diagdown \qquad\qquad\qquad \diagup C_6H_5$$
$$C{=}C{=}C{=}C{=}C{=}C$$
$$C_6H_5 \diagup \qquad\qquad\qquad \diagdown C_6H_5$$

302    2

1,1,8,8-Tetraphenyloctaheptaene

$$C_6H_5 \diagdown \qquad\qquad\qquad\qquad \diagup C_6H_5$$
$$C{=}C{=}C{=}C{=}C{=}C{=}C{=}C$$
$$C_6H_5 \diagup \qquad\qquad\qquad\qquad \diagdown C_5H_5$$

decomposes
before melting    2

*References*
1. Vorlander and Weinstein (1923).
2. Kuhn and Wallenfels (1938); Kuhn and Platzer (1940); Kuhn and Zahn (1951).

(included in Table 6.10 for comparison) because of their considerable configurational flexibility show a much less steep rise in $T_f$ with increasing molecular weight. This effect of molecular flexibility is referred to again in Chapter 7.

Rigid quasicylindrical molecules are of considerable interest in regard to their melting behaviour and should lead to molten states with unusual properties. Their melts would be expected to show marked local anisotropy of molecular packing, and probably even mesophases with liquid crystal behaviour (Chapter 14) intervening on the temperature scale between crystals and isotropic melts. The work of creating positional defects, such as vacancies, 'top to toe' molecular inversions, or interstitial packings would be expected to grow with increasing length of the molecule.

Tentatively, for rigid molecules, the relationship to be anticipated in a homologous series might be of the form $H_f = H_0 + nH_s$ where $H_0$ and $H_s$ are parameters which are constant in a given homologous series. By contrast with flexible molecules (see Chapter 7) since the molecules are rigid, mechanisms of melting (including positional and orientational disorder) are not likely to lead to entropies of fusion rapidly increasing with $n$. It may be anticipated that

$$S_f = \Delta S_{pos} + \Delta S_{or} + \text{other terms}$$

will not vary much with $n$. As a consequence, in such series the melting point $T_f = H_f/S_f$ rises steeply as the molecular size increases. This steep rise can in fact be seen from Tables 6.8, 6.9, and 6.10; more examples could be wished for. It must be remembered that these high melting points unfortunately result in very low solubilities of the molecules in ordinary solvents, a fact which has in turn hindered synthetic manipulation. Low solubilities have also restricted the attention paid by physical organic chemists to this interesting group of molecules, particularly with regard to melting phenomena.

### 6.6.1 Quantitative calculations of entropy increases on orientational melting

In certain cases, the values of $S_f = \Delta S_{pos} + \Delta S_{or}$ can be calculated with reasonable accuracy. To give one example as a model, optical evidence from Raman spectra indicates that the mercuric halides consist of rigid linear molecules. The melts also contain a small proportion of ions, probably resulting from autocomplexing processes such as (see Chapter 8)

$$2HgX_2 \rightleftharpoons HgX^+ + HgX_3^-$$

but the proportion of species other than $HgX_2$ is too small to detect by optical means. A somewhat tentative calculation (Janz and McIntyre, 1962) assumes that $\Delta S_{pos} = 2$ e.u. (see Chapter 11).

$\Delta S_{or}$ is the difference between entropy of libration of the linear molecules about mean directions in the crystal with characteristic frequency $\nu$ and the entropy of rotation i.e.

$$\Delta S_{or} = (S^{T_f}_{rot} - S_{lib}^{T_f}]$$

where

$$S_{lib}^{T_f} = 2R \,(\ln kT/h\nu + 1)$$

and the entropy of 'free rotation' in the melt

$$S_{rot}^{T_f} = R \,(\ln\left[\frac{8\pi^2 IkT}{\sigma h^2}\right] + 1)$$

where $I$ is the moment of inertia of the molecules, and $\sigma$ is a symmetry factor.

On the basis, using $\nu = 40 \text{ cm}^{-1}$ for each of the salts, observed and calculated entropies of fusion may be compared from Table 6.12.

The fact that calculated changes are larger than those observed may plausibly be attributed to some coordinated rotation or 'clustering' in the melts (see below).

**Table 6.12. Entropy of fusion
of rotating mercury halides**

| Salt | $S_{f_{calc.}}$ | $S_{f_{obs.}}$ (e.u.) |
|------|------|------|
| $HgCl_2$ | 7·6 | 7·5 |
| $HgBr_2$ | 9·7 | 8·4 |
| $HgI_2$ | 10·7 | 8·6 |

### 6.6.2. *Melting of rigid planar molecules—vibrational entropy contributions to fusion*

As a result of progress in synthetic organic chemistry, as well as improvements in petroleum refining, a considerable diversity of molecules has become available of the important classes with fused aromatic rings, and which can be regarded as having an approximately planar repulsion envelope. Such disc-like molecules provide a useful range of models in terms of which to investigate structures of melts. Some of the characteristics of polyphenyl molecule melts are discussed in Chapter 13. One of the problems in this group is to account for the quite large entropies of fusion that are found (Al Mahdi and Ubbelohde, 1953) (see also data recorded in Table 6.13). In these melts, the molecules have insufficient room to rotate, and must lie roughly parallel to their neighbours (as discussed in section 6.6.3). A suggestion that warrants further investigation is that the vibrational entropy change on fusion makes a substantial contribution to the total entropy of melting of these crystals. In the simplest theory for the vibrational entropy of a crystal containing $3N$ vibrators all with the same characteristic frequency $\nu$ when $h\nu \gg kT$, the change on melting is $\Delta S_{vib} = 3R \ln(\nu_S/\nu_L)$ if the melt can be treated as quasicrystalline.

The suggestion has been made that planar rigid molecules in these crystals may have considerably greater vibrational freedom with values of $\nu_L$ for the melt considerably lower than for the crystals. Part of the total entropy of fusion would thus have a vibrational origin; considerable refinements of theory are however needed to calculate this term for molecular solids, at all realistically.

**Table 6.13. Entropy and volume changes on melting
crystals of aromatic molecules (Al Mahdi
and Ubbelohde, 1953)**

| Molecule | $\Delta V_f/V_s$ (%) | $S_f$ (e.u.) |
|----------|------|------|
| Benzene | 13·4 | 8·5 |
| Diphenyl | 12·4 | 13·0 |
| Naphthalene | 16·7 | 13·1 |
| Acenaphthane | 12·4 | 13·6 |
| Anthracene | 16·8 | 14·1 |
| Phenanthrene | 8·7 | 12·1 |
| Chrysene | 12·4 | 14·9 |

### 6.6.3. *Room to rotate—orientational correlations in melts*

In 6.2 reference has been made to orientational disordering as a chancy but important contribution to the total entropy of fusion of crystals. Generally, the expansion in volume on melting lessens potential barriers opposing rotation. Unless orientational disordering has already occurred in the crystals below $T_f$, at the melting point itself, a significant contribution to the entropy of fusion is highly likely from this mechanism of producing disorder.

Calculation (Al Mahdi and Ubbelohde, 1953, 1955; Andrews and Ubbelohde, 1954; Magill and Ubbelohde, 1958) shows that near $T_f$ many melts of molecules with non-spherical repulsion envelopes (see Joliet, 1951) simply do not have room for the constituent molecules to rotate freely. In these melts the nearest neighbours of any molecule must lie therefore roughly parallel with it, so as to permit economical close-packing without requiring long range order. The kind of arrangement taken up can be inferred from studied with assemblies in so-called 'model experiments' in which a large number of non-spherical units (e.g. cylinders) are shaken together in a limited space, and the packing assumed by them is investigated.

Various refinements can be usefully made to these simple criteria about 'room to rotate' in the melt of non-spherical molecules. Departures from precise relationships of similitude for example in the molecules Ar, $CH_4$ and $N_2$ have been attributed to an orientational correlation in the two latter which is however only local (Crawford, Daniels, and Cheng, 1975). Optical data for example for liquid $CClF_3$ can indicate almost free rotation in this (and presumably in similar melts) (Davies and Evans, 1975).

Figure 6.4 (from Stuart, 1941) illustrates in a pictorial way the orienting influence of non-spherical neighbours for solids shaped similarly to the rod-like repulsion envelope of carbon disulphide. With disc-shaped molecules, there is evidence that orientational randomization of simple molecules such as benzene and a number of its analogues such as thiophene and pyridine does occur at a temperature $T_{c_1}$ in the crystals below $T_f$, about the axis perpendicular to the plane of the aromatic rings, but not about the other two principal axes in the molecules. Rotation about these two axes is not achieved since it would frequently involve very large repulsion forces in the crystals. Relevant transformation temperatures $T_{c_2}$ and $T_{c_3}$ must be much greater than $T_f$ and thus are not apparent in the melt near its freezing point. Gradual increases of freedom may be expected as such 'oriented melts' expand on heating above $T_f$.

Table 6.14 illustrates for some larger rigid disc-shaped molecules to what extend the volume required for 'free rotation' about certain of the molecular axes is not available even in the melt. Such rigid planar molecules must lie roughly parallel in the melt to an extent even more marked than with carbon disulphide. Room for free independent rotation in the melt is available (even

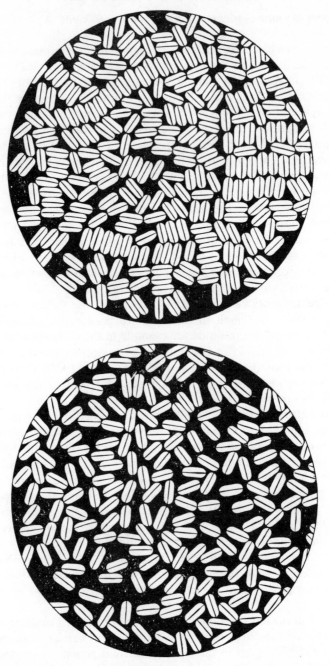

FIG. 6.4.  Increased orientational disorder of local packing in
melts of ellipsoidal molecules as the density decreases

in principle) only with respect to those principal axes marked with an asterisk in Table 6.14.

**Table 6.14.**

| Molecule | Molar volume of melt at $T_t$ | Volume required for free rotation about principal axes $(cm^2)$ | | |
|---|---|---|---|---|
| | | L | M | N |
| Benzene | 133·5 | 183·8 | 198·5 | 125·4* |
| Diphenyl | 225·4 | 225·9* | 355·4 | 200·2 |
| Naphthalene | 211·1 | 269·5 | 358·8 | 201·8* |
| Acenaphthene | 242·1 | 406·6 | 389·6 | 223·5* |
| Anthracene | 297·1 | 381·0 | 597·0 | 393·0* |
| Phenanthrene | 281·0 | 457·0 | 540·6 | 382·3 |
| Chrysene | 348·5 | 515·8 | 741·9 | 452·4 |

This approach to the structure of melts of molecules with non-spherical repulsion envelopes brings out a feature that also emerges in other ways in subsequent chapters. Although 'long-range positional lattice correlation', to the extent required to produce X-ray diffraction peaks of high order, is absent in the liquid, starting from any one molecule microregions are likely to be arranged in a highly characteristic manner to ensure economical packing of the molecules in space in the melt. These regions are conveniently described as 'clusters' and are not necessarily easy to detect by X-ray diffraction since they are usually *anticrystalline* in structure. In the past over-emphasis of 'long-range (lattice) correlation of positional order' has led to inadequate attention to other kinds of cooperative interaction in the molten state. Development of various scattering techniques has led to more precise information about short-range correlation of orientation of molecular axes, in rotator crystals in which there is a long range orientational disorder above $T_c$ but below $T_f$. This remark applies for the melts in which absence of such short-range correlation was wrongly assumed hitherto.

Quantitative evaluation of the freedom of orientational motion in plastic crystals using a variety of methods of measurement has been reviewed by Courtens (1974). Light scattering measurements point to considerable dynamical coupling between orientational and translational motion. Cold neutron scattering leads to the (at first surprising) conclusion that rotational relaxation times for cyclopentane are much the same in the plastic crystals as in the melts (about $1·3 \times 10^{-12}$ s for cyclopentane, De Graf, 1969). Proton magnetic resonance has been used to follow changes of molecular freedom on passing from brittle crystal to plastic crystal and to the melt for a variety of molecules including hexamethylethane $(CH_3)_3C_2$ (Smith, 1971), triethylene-diamine (Smith, 1965a) and a group of tetramethyl derivatives of metals $M(CH_3)_4$ where M = Sc, Ge, Sn or Pb (Smith, 1965b). Observed differences in behaviour again illustrate the important fact of how sensitive relative molecular motions and orientations can be to intermolecular force fields, as discussed above (see also Chapter 11). Molecular reorientation and self-diffusion in the crystals have also been studied in hexamethylsilane (Chadwick, Chezeau, Folland, Forrest, and Stranger, 1975) and trimethylsilyl, germyl and stannyl hydrazines (Veith, 1975).

Brillouin spectroscopy on sonic parameters for frequencies around $10^9$ Hz has been used to follow changes of molecular mobility (positional and orientational) for pivalic acid $(CH_3)_3CCOOH$ from the onset of the rotator crystal phase at $T_c = 6 \cdot 9°C$ to the melting point $T_f = 36 \cdot 5°C$ (Bird, Jackson, and Powles, 1973). The correlation time for reorientation is about $0 \cdot 6 \times 10^{-8}$ s even just below $T_f$; this is much longer than in plastic crystalline succinonitrile or molten nitrobenzene, both of which have reorientation times about $10^{-11}$ s near $T_f$ (Fontaine and Moriamez, 1968) indicating much greater freedom of the molecules.

## REFERENCES

A. A. K. Al Mahdi and A. R. Ubbelohde (1953) *Proc. Roy. Soc., Ser. A* **220**, 143.
A. A. K. Al Mahdi and A. R. Ubbelohde (1955) *Trans. Faraday Soc.* **51**, 3.
L. M. Amzel and L. Becka (1969) *J. Phys. Chem. Solids* **30**, 529.
L. M. Amzel, M. C. M. Cucarella, and L. Becka (1971) *J. Phys. Chem.* **75**, 1073.
J. N. Andrews and A. R. Ubbelohde (1955) *Proc. Roy. Soc., Ser. A* **228**, 435.
J. B. Armitage, N. E. Entwistle, E. R. H. Jones, and M. C. Whiting (1954) *J. Chem. Soc., London* 147.
C. M. Barber and E. F. Westrum (1963) *J. Phys. Chem.* **67**, 2372.
F. R. Basford (1936) *J. Chem. Soc. London* 1593.
M. J. Bird, D. A. Jackson, and J. G. Powles (1973) *Mol. Phys.* **25**, 1051.
J. von Braun (1927) *Ber. Dt. Chem. Ges.* **60**, 1180.
C. W. Bunn (1955) *J. Polym. Sci.* **16**, 323.
M. Busch and W. Weber (1936) *J. Prakt. Chem.* **146**, 1.
J. D. Campbell (1975) *Arch. Mech. Poland* **27**, 407.
A. V. Chadwick, J. M. Chezeau, R. Folland, J. W. Forrest, and J. H. Strange (1975) *J. Chem. Soc. Faraday Trans. I* **71**, 1610.
S. S. Chang and E. F. Westrum (1960) *J. Phys. Chem.* **64**, 1551.
F. D. Chattaway and R. C. T. Evans (1896) *Trans. Chem. Soc.* **62**, 980.
C. Clement and M. Davies (1960) *Arch. Sci., Genève* **13**, 77.
C. Clement and M. Davies (1962) *Trans. Faraday Soc.* **58**, 1719.
C. Clement and M. Davies (1964) *Trans. Faraday Soc.* **60**, 10.
G. Corfield and M. Davies (1964) *Trans. Faraday Soc.* **60**, 10.
N. T. Corke and J. N. Sherwood (1971) *J. Materials Sci.* **6**, 68.
E. Courtens (1974) *Phys. Rev.* **10A**, 967.
R. K. Crawford, W. B. Daniels, and V. M. Cheng (1975) *Phys. Rev.* **12A**, 1690.
G. J. Davies and M. Evans (1975) *J. Chem. Soc. Faraday II* **71**, 1275.
L. A. De Graaf (1969) *Physica (Utrecht)* **40**, 497.
N. A. Dobromyslov and N. I. Koshkin (1970) *Sov. Phys. Acoustics* **15**, 386.
N. Doshi, M. Furman, and R. Rudman (1973) *Acta Cryst.* **29B**, 143.
W. J. Dunning (1961) *J. Phys. Chem. Solids* **18**, 21.

A. Eucken (1942) *Angew. Chem., Neue Folge* **55**, 163.

R. E. Florin and T. W. Mears (1955) *U.S.A. Energy Commission Patent BNL 2446*, pp. 89–102.

H. France, I. M. Heilbron, and D. H. Hey (1938) *J. Chem. Soc., London* 1364.

H. France, I. M. Heilbron and D. H. Hey (1939) *J. Chem. Soc., London* 1288.

H. Gelissen and P. H. Hermans (1925) *Ber. Dt. Chem. Ges.* **58**, 290.

O. Gerngross and M. V. Dunkel (1924) *Ber. Dt. Chem. Ges.* **57**, 739.

J. R. Green and C. E. Scheie (1967) *J. Phys. Chem. Solids* **28**, 383.

G. B. Guthrie and J. P. McCullough (1961) *J. Phys. Chem. Solids* **18**, 53.

E. M. Hampton, P. McKay, and J. N. Sherwood (1974) *Phil. Mag.* **30**, 85.

G. J. Janz and J. D. E. McIntyre (1962) *J. Electrochem. Soc.* **109**, 842.

J. F. Joliet (1951) *C.R. Acad. Sci., Paris* **232**, 509.

E. R. Jones (1950) *J. Chem. Soc. London* 754.

Y. Koga and J. A. Morrison (1976) *J. Chem. Phys.* **62**, 3360.

R. Kuhn and K. Wallenfels (1938) *Ber. Dt. Chem. Ges.* **71**, 785, 1510.

R. Kuhn and G. Platzer (1940) *Ber. Dt. Chem. Ges.* **73**, 1410.

R. Kuhn and H. Zahn (1951) *Ber. Dt. Chem. Ges.* **84**, 566.

M. L. Labruyere Verhavert (1951) *Bull Soc. Chim. Belg.* **60**, 270.

J. H. Magill and A. R. Ubbelohde (1958) *Trans. Faraday Soc.* **54**, 1811.

L. V. McCarty, L. C. Landauer, and J. M. Binkowski (1960) *J. Chem. Eng. Data* **5**, 365.

P. McKay and J. N. Sherwood (1975) *J. Chem. Soc. Faraday Trans. I* **71**, 2331.

A. Michels (1948) *Bull. Soc. Chim. Belg.* **57**, 575.

L. Pauling (1960) *The Nature of the Chemical Bond*, Cornell University Press.

J. Pirsch (1955) *Monatsh. Chem.* **86**, 216, 226.

C. W. F. Pistorius and H. A. Reising (1969) *Mol. Cryst. Liquid Cryst.* **5**, 353.

R. Pummerer and K. Bittner (1924) *Ber. Dt. Chem. Ges.* **57**, 87.

E. Rhodes and A. R. Ubbelohde (1959) *Trans. Faraday Soc.* **55**, 1705.

H. Sackmann and G. Kloos (1958) *Z. Phys. Chem. (Leipzig)* **209**, 319.

J. N. Sherwood (1972) *Bull. Soc. Fr. Mineral. Cristallog.* **95**, 253.

L. Silver and R. Rudman (1970) *J. Phys. Chem.* **74**, 3134.

G. W. Smith (1965a) *J. Chem. Phys.* **43**, 4325.

G. W. Smith (1965b) *J. Chem. Phys.* **42**, 4229.

G. W. Smith (1969) *J. Chem. Phys.* **50**, 3595.

G. W. Smith (1975) *Adv. Liquid Cryst.* **1**, 189.

L. I. Smith and M. M. Falkof (1942) *Org. Syn.* **22**, 50.

H. A. Stuart (1941) *Kolloid Z.* **96**, 149.

J. Timmermans (1950) *3me Congrès National des Sciences, Brussels, Chimie Physique*, **1**.

J. Timmermans (1961) *J. Phys. Chem. Solids* **18**, 1.

L. C. Towle (1968) *Science* **159**, 629.

J. C. Trowbridge and E. F. Westrum (1963) *J. Phys. Chem.* **67**, 281.

A. R. Ubbelohde (1950) *Quart Rev. Chem. Soc.* **4**, 356.

A. R. Ubbelohde (1961) *J. Phys. Chem. Solids* **18**, 90.

F. Ullman and G. M. Meyer (1904) *Ann. Chem.* **332**, 52.

R. H. Valentine, G. E. Brodale, and W. F. Giauque (1962) *J. Phys. Chem.* **66**, 392.

M. Veith (1975) *Acta Cryst.* **31A**, S191.

D. Vorlander and P. Weinstein (1923) *Ber. Dt. Chem. Ges.* **56**, 1136.

E. F. Westrum, W. K. Wong, and E. Morawetz (1970) *J. Phys. Chem.* **74**, 2542.

G. Wittig and Freih von Lupin (1928) *J. Phys. Chem.* **61**, 1627.

G. Wittig and A. Klein (1936) *J. Phys. Chem.* **69**, 2087.

C. A. Wolff and E. F. Westrum (1964) *J. Phys. Chem.* **68**, 430.

# 7. MELTING OF FLEXIBLE MOLECULES

## 7.1 Increases of molecular configurational entropy on melting

In the preceding chapter various aspects of positional and orientational disordering of essentially rigid units have been discussed in relation to mechanisms of melting of molecular crystals. Fluids produced by these two types of disorder can in principle have their molecular arrangements referred to the crystal lattice as norm, and may thus be treated as 'quasi-crystalline' in the narrow precise sense of the term used in this book and discussed in Chapter 5.

By contrast, a large number of molecules are known which are flexible in the sense that they can assume more than one configuration of about the same energy. Most flexible molecules belong to the class of molecular crystals and are organic compounds. The widespread importance of this configurational factor is conveniently discussed following other aspects of melting of molecular crystals examined in preceding chapters. For the present purpose, a flexible molecule can be specified as one which can assume various arrangements of the atoms, with different minima of potential energy, involving changes of free energy not more than $2 \times$ or $3 \times kT$. This definition implies that in a dilute gaseous assembly, appreciable concentrations of each of the various configurations are still present in the proportions

$$n_1:n_2:n_3 \ldots = 1:\exp(-\Delta G_{12}/RT):\exp(-\Delta G_{13}/RT) \ldots \quad (7.1)$$

### 7.1.1. *Melting of n-alkane hydrocarbons*

A class of molecules whose configurational isomerism has been extensively studied is based on the normal paraffins. Starting with normal butane, which can assume configurations *syn* and *anti* (see Fig. 7.1) of practically equal free energy, the number of configurational isomers rapidly increases as the number of carbon atoms in the chain increases.

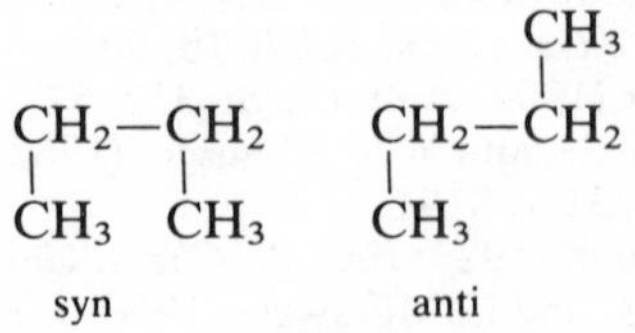

$$\begin{array}{cc} & \mathrm{CH_3} \\ & | \\ \mathrm{CH_2-CH_2} & \mathrm{CH_2-CH_2} \\ |\quad\quad | & | \\ \mathrm{CH_3}\ \ \mathrm{CH_3} & \mathrm{CH_3} \\ \text{syn} & \text{anti} \end{array}$$

FIG. 7.1. *Syn* and *anti* form of n-paraffins

The free energy of most of these isomers is practically the same when the molecules are separated far from one another in the gas phase; some of the most highly crumpled isomers have their free energies reduced by repulsions between the $CH_2$ groups and increased by the dispersion forces between the same groups (see ref. in Ubbelohde and Woodward, 1952). Experimental evidence suggests that crumpled forms appear to be present in quite high concentrations in the gas. On condensing the gas to a liquid, some gain in attractive potential energy can be achieved by increasing the proportion of stretched forms. On passing from the liquid to the crystal, an even more restrictive selection of isomers takes place. In most known cases only the fully stretched form is actually picked out for packing into the crystal lattice, since only by this means can the potential energy be minimized so as to fill space with a crystal lattice extending in three dimensions.

Figure 7.2 illustrates with dimensional models the change from a single (fully stretched) configuration in the crystal to a number of stretched and crumpled configurations in the melt. In configurational randomization, formation of a proportion of even the most crumpled isomers by rotation about C—C bonds increases the entropy of any fluid phase by increasing the diversity of species present. However, in the melt (though to a somewhat lesser extent than in the crystals) considerations of minimum potential energy favour the more stretched forms, which can lie roughly parallel, thus maximizing intermolecular attraction potentials in microregions of the liquid. Structural studies suggest that when the chains are fairly long, as in n-hexatriacontane $C_{36}H_{74}$ the melt near $T_f$ contains predominantly stretched forms, rotating about the chain axis (Vand, 1953). At higher temperatures, as the melt expands this bias towards stretched configurations may be expected to decrease. In the dilute gas, even for much shorter n-alkanes, crumpled forms are known to predominate (McCoubrey and Ubbelohde, 1951; McCoubrey, McCrea, and Ubbelohde, 1951; Cummings, McCoubrey, and Ubbelohde, 1952).

Infrared vibrational studies on n-alkanes from $C_4H_{10}$ to $C_{17}H_{36}$ illustrate the great variety of configurations assumed as the density of the melts changes with temperature (Snyder, 1967). Lattice vibrations in the crystal also change in relative intensity (Brunel and Dows, 1974).

This process of randomization of molecular *configuration* on melting constitutes a mode of disordering for flexible molecules distinct from and additional to the positional and orientational disorder previously discussed. For crystals containing flexible molecules

$$S_f = \Delta S_{pos} + \Delta S_{or} + \Delta S_{config} + \dots \qquad (7.2)$$

This formulation can be developed in various ways in the attempt to arrive at more quantitative expressions. A phenomenological approach makes

Fig. 7.2(a)

FIG. 7.2(b)

FIG. 7.2. Dimensional repulsion envelopes of n-decane, *a* crystals, *b* melt. Only a few of the configurations are illustrated.

plausible assumptions about the trend in heats of fusion $H_f$, and entropies of fusion $S_f$ as the number $n$ of flexible links in the chain molecule increases.

There are various consequences of this additional mechanism of configurational melting. The most important of these is that as the molecular weight of the molecules is increased progressively by increasing the number of units of structure in homologous series of molecules, the term $\Delta S_{config}$ likewise increases progressively without any obvious limit. Tentatively, one may write $\Delta S_{config} = S_0 + nS$ where $S_0$ and $S$ are constants characteristic of the homologous series under discussion.

On the other hand positional disordering referred for example to the centres of gravity of the molecules can make only a limiting contribution to the entropy of fusion. For any rotational isomer the term $\Delta S_{or}$ is likewise limited to a few entropy units at most, depending on the orientational options available in a given crystal structure. However in a homologous series of flexible molecules the total entropy of fusion can grow indefinitely owing to the configurational term $\Delta S_{config}$ in the total, i.e.

$$S_f = (\Delta S_{pos} + \Delta S_{or} + S_0) + nS \qquad (7.3)$$

Heats of fusion are a measure of the work of creating crystal defects. Defects mostly involve a separation of the molecules, as can be seen from the increase in volume $\Delta V_f$ which is normal on melting. Melting can be treated as taking place in two stages for crystals of flexible molecules. Most of the heat of fusion is absorbed in a first stage in separating the molecules on melting to the ultimate volume of the melt $V_L$ at $T_f$. In general, the second stage which involves a change from one configuration in the crystal to multiple configurations in the melt, to produce the final mixture of isomers, will not be expected to require a large heat change, particularly if at equilibrium the forms are present in comparable proportions. On this basis $H_f$ grows as the absolute change in molar volume on fusion $\Delta V_f/V_s$ grows, and may plausibly be assumed to be roughly proportional to the size of a molecule and to the number of units of structure in it.

As a first tentative approximation one may write $H_f = H_0 + nH$ where $H_0$ and $H$ are constants in any given homologous series. So far as these approximations are adequate, the melting points in a homologous series will be expected to take the form

$$T_f = H_f/S_f = \frac{H_0 + nH}{(\Delta S_{pos} + \Delta S_{or} + S_0) + nS} \qquad (7.4)$$

As $n$ becomes larger, the factors multiplied by it in the numerator and denominator swamp the other terms. In the limit, where $n$ is sufficiently large

$$(T_f)_{lim} \to \frac{nH}{nS} = H/S \qquad (7.5)$$

This conclusion implies that however large the molecular weight becomes, (e.g. in high polymers) melting points in a homologous series of flexible molecules converge to a limit $H/S$ known as the 'convergence temperature'. Actual examples of convergence behaviour have been followed in detail chiefly for the normal paraffins and their derivatives. Various means of study show that the potential barriers separating one minimum of configurational energy from the next are of the order $2\times$ or $3\times kT$ (McCoubrey and Ubbelohde, 1951). Isomerization proceeds quite rapidly provided packing considerations permit it (as is the case in the melt, and especially in the vapour) so that the equilibrium mixture of configurational isomers at any given temperature and volume may be assumed. Empirical convergence equations for melting points of these homologous series can be written in various forms. For example, for normal paraffins, the melting equation

$$(T_f)_n = \frac{T_c(n+a)}{(n+b)} \tag{7.6}$$

is in good agreement taking $T_c = 414\cdot3$ K, $a = -1\cdot5$, $b = 5\cdot0$. For polyethylene, regarded as a series limit, small modifications to equation (7.6) have been suggested. (Broadhurst, 1966). As a model series, these molecules present minor anomalies in behaviour, which disturb the broad general trend indicated above. Detailed analysis of the effects of chain length and of chain tilt in various polymorphic crystal structures of the same molecule has clarified some of these anomalies in the homologous series of paraffins (Broadhurst 1962b) (Fig. 7.3) and also throws light on effects of impurities in solid solution. More careful consideration of the elementary treatment just given shows (Broadhurst, 1962a) the following:

(i) The melting rule (7.6) should only be applied to homologues crystallizing in the same structure. Taking polyethylene (see Chapter 17) as the model for an infinite normal paraffin, its structure corresponds with an orthorhombic subcell which is also assumed by paraffins once $n > 41$. Polyethylene as ordinarily crystallized also contains microamorphous regions which complicate the behaviour somewhat, see Chapter 17). Necessarily, these lower its limiting melting point to about 135–138°C which is nevertheless still quite close to the calculated convergence temperature for normal praraffins of 141·1°C. Shorter chains, with $n < 40$, melt from a triclinic or hexagonal crystal. This difference of crystal structure may somewhat modify $(T_f)_n$. An alternation which is observed in $(T_f)_n$ for the shorter chains can be attributed to modifications of *crystal* packing of the molecules according to whether $n$ is odd or even (see Larsson, 1966).

(ii) Linear equations attributed empirically to $H_f$ and $S_f$ do not allow for the normal temperature-dependence of these quantities in both crystal and melt. For similitude to apply accurately, comparisons should be made at

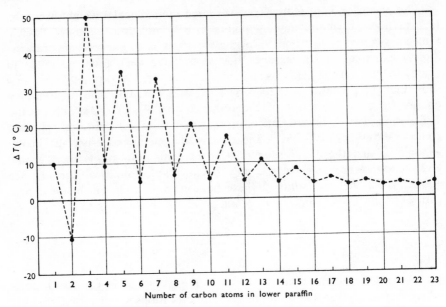

FIG 7.3. Differences $\Delta T$ between freezing points of successive n-alkyl paraffins

suitably adjusted temperatures (analogous to corresponding temperatures) and not at $T_f$ itself.

(iii) More refined considerations point to the presence of isomeric configurations of certain flexible alkane molecules in the crystals at different temperatures. Specific. examples have been described for n-tritriacontane (Piesczek, Strobl, and Malzahn, 1974) in relation to its three solid–solid phase transitions below $T_f = 71 \cdot 8°C$ and for the formation of 'kink faults' in planar chains of n-paraffins (Blasenberg and Pechold, 1967). A general review of solid–solid transitions in flexible alkanes gives other examples (Reinisch, 1968).

### 7.1.2. Melting of other flexible molecules

With any aliphatic molecule of the type $R(CH_2)_n CH_3$ if R is a group such as $-OH, -NH_2, -COOH$ etc., a whole collection of homologous series can be formed by preparing diverse derivatives. For example, about forty different types of derivatives have been investigated with the aim of characterizing the n-aliphatic alcohols (Salmon-Legagneur and Neveu, 1962).

Some of these homologous series show marked alternation of freezing points according to whether $n$ is odd or even, whereas others, including the n-primary alcohols themselves, show a uniform trend without alternation beyond about $n = 5$. Usually, alternation is associated with minor differences in crystal packing according to whether $n$ is odd or even, but plausible interpretations of this difference can only be substantiated when the crystal structures of the homologous compounds with $n$ even and $n$ odd have been determined.

Many instances of flexible molecules may be anticipated, but as yet few have been studied sufficiently to establish the parameters in full detail in relation to crystal structure. As one example, for molecules such as the ethers of dibenzoyl-methane, various configurations can be formulated whose energies of isomerization will not be very large (R is an n-alkyl group), for example; see Fig. 7.4.

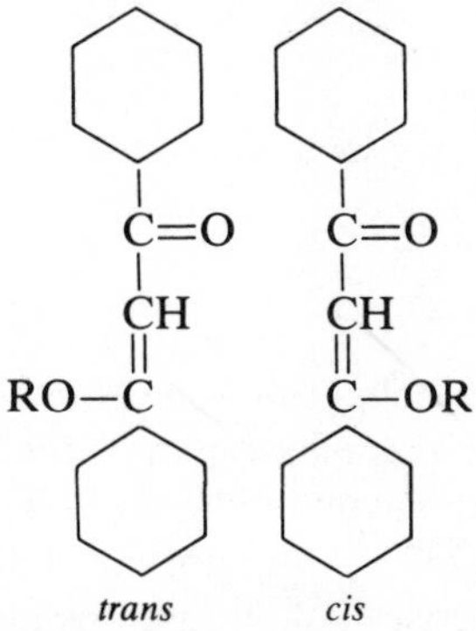

FIG. 7.4. Configurations of the ethers of dibenzoyl methane with similar energies of formation

The process of crystallization appears to select the *cis*-form whereas in melt or solution the *trans*-form appears to be favoured (Urushibara, Imura, and Ikeda, 1955). Presumably if *trans*-crystals could be suitably nucleated a different melting point would be observed.

*Meta*-polyphenyls can assume a variety of configurations by rotation of the planes of the phenyl groups about the C–C bonds joining them. Presumably this explains the trend of melting points in a homologous series, illustrated in Chapter 2, Table 2.1 (p. 12), though more experimental evidence about these melts is desirable.

A process akin to configurational melting may take place for certain polypeptide chains such as deoxyribonucleic acid (DNA). Two complementary chains of equal length in this molecule can exist in helical and random forms. Essentially, randomization is the outcome of unpairing of hydrogen bonds which, in the ordered form of the helix, link matched portions of the two chains. Assuming that the transformation is thermally reversible, a statistical mechanical model for melting can be constructed (Zimm, 1960). As might be

**Table 7.1.** *Melting of n-alkyl chains with*
*napthalene loading*

| Length of normal alkyl side chain | $T_f °C$ 1-alkyl | 2-alkyl |
|---|---|---|
| 1 | −31 | 34·2 |
| 2 | −13·5 | −7 |
| 3 | −12·9 | −3 |
| 4 | −22 | −5 |
| 5 | −26 | −4 |
| 6 | −18 | −5·5 |
| 7 | −9 | 0·6 |
| 8 | −2 | 13·0 |
| 9 | 8 | 11 |
| 10 | 15 | 19·6 |
| 11 | 22·4 | 20 |
| 12 | 27·5 | 27·0 |
| 13 | 35 | 33 |
| 14 | 38·5 | 42 |
| 15 | 43 | 43 |
| 16 | 46·5 | 49·5 |

expected, melting and the helix–coil transition in DNA depends to some extent on the mix of component base pairs adenine–thymine and guanine–cytosine, and the transition may extend over several degrees of temperature (Lehman and McTague, 1968). An important difference from the examples previously discussed may arise if there are strong *attractive forces* between neighbouring chains such as hydrogen bonds.

## 7.2 Vibrational melting of flexible molecules

In order to proceed with mathematical refinements of the concepts of vibrational melting discussed in Chapter 3, it is necessary to introduce somewhat drastic simplifications, in order to describe the behaviour of a model molecule. One such model (Longuet Higgins, 1958) discusses the behaviour of a flexible chain with $n$ links each of length $l$ that can be accommodated when fully stretched in a one-dimensional lattice. Forces of interaction between atoms on neighbouring chains are extremely short range and can thus be equated to attraction energies $\varepsilon$ between nearest neighbours.

In the model considered, vibrations of the chains whilst in the crystal are neglected, since their frequencies are assumed to lie too high to contribute appreciably to the heat content at $T_f$. In the melt, the 'free' chain has a spectrum of torsional vibrational frequencies $\nu_0, \nu_1, \nu_2 \ldots \nu_n$ with the fundamental frequency $\nu_0$ where $\kappa$ is a force constant and

$$\nu_0 = \frac{l}{2\pi} \sqrt{\frac{\kappa}{m}}$$

It is convenient to replace each of the above spectrum of frequencies by a common mean value $\nu$. With these simplifications, a critical temperature $T_{vib}$ is found which is given by

$$\left(\frac{8\pi^2 m l^2}{h^2}\right) T_{vib} = \left(\frac{\nu_0}{\nu}\right) \exp\left(\frac{\varepsilon}{\kappa T_c}\right) \qquad (7.7)$$

Above this temperature, the crystal has melted to give a mixture of random configurations in place of the crystal with fully stretched molecules only.

Though highly simplified, this model is interesting in presenting in a sophisticated way a mechanism of melting that is primarily due to an increase of vibrational entropy on passing from crystal to melt. It thus belongs to the group of two-phase theories of vibrational melting discussed in Chapter 3, with the special feature that molecular flexibility greatly increases the range of thermally excited vibrations in the melt.

## 7.3 Melting of 'loaded' polymethylene chains

In homologous series (referred to in section 7.1), end groups $CH_3$, OH, COOH, etc. constitute disturbing influences in the crystal packing, whose effect is large on short chains but tends to be smoothed out as the alkane chain attached to them is lengthened. When these end groups are very

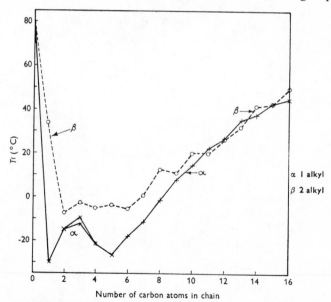

FIG. 7.5. Freezing points of n-alkylnaphthalenes. Reproduced with permission from Anderson, Smith, and Rallings, *J. Chem. Soc.* 443 (1953)

large, the disturbance created by them in the adlineation of flexible chains in the crystal may however not be smoothed out until quite long alkane chains are attached to them. As examples, Fig. 7.5 illustrates how, for the 1-alkylnaphthalenes, regular series behaviour does not begin until $n = 5$ and for the 2-alkylnaphthalenes until $n = 6$ (Anderson, Smith, and Rallings, 1953).

Until the crystal structures have been determined, the trend towards regular series behaviour exemplified in Table 7.1. is suggestive but cannot be definitely attributed to regular changes of configurational entropy and heat of fusion as the alkane chain is lengthened.

For n-aliphatic amides, with $CH_3$ and $-CONH_2$ as end groups (see Table 7.2), amide groups constitute heavy loading of the chain, since these exert strong dipole and hydrogen bond interactions with similar groups on neighbouring molecules. In the crystals, the amide groups of two molecules are coupled by two $-N-H-O-$ hydrogen bonds of length 2·99 Å to form a dimer (Turner and Lingafelter, 1955a,b). In these compounds strong attractions between these groups appear to overwhelm any marked series effect in

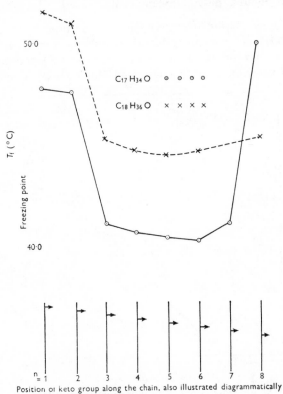

FIG. 7.6. Freezing points of n-alkane molecules loaded with one keto group, $C_{17}H_{34}O$; $C_{18}H_{36}O$

**Table 7.2. Melting points of n-alkane amides (Turner and Lingafelter, 1955a)**

| No. of C atoms in chain | $T_f$ (°C) |
|---|---|
| 3 | 78 |
| 4 | 113·5 |
| 5 | 105 |
| 6 | 100 |
| 7 | 95 |
| 8 | 105 |
| 9 | 99 |
| 10 | 98·5 |
| 11 | 99·0 |
| 12 | 102 |
| 13 | 100 |
| 14 | 104 |
| 16 | 104 |

the configurational contribution to the entropy of fusion, even for long alkane chains. $T_f$ thus shows no decided trend. To verify this interpretation more fully, it would be desirable to examine the melts of homologues of this series for any differences of actual crystal structure and to determine actual entropies and heats of fusion.

Derivatives of n-aliphatic aldehydes have also been investigated for series effects. The 2, 4-dinitrophenylhydrazones load the chain in a similar way as in the amides discussed above with the result that after a steep initial drop the melting point remains practically independent of chain length (Malkin and Tranter, 1951).

Again, mono-n-alkylaniline derivatives (see Table 7.3) show regularities of behaviour only for alkyl groups with more than 5 carbon atoms, beyond

**Table 7.3. Melting of 2–4 dinitrophenylhydrazones of n-alkane aldehydes**

| No. of C atoms in alkane chain | $T_f$ (°C) |
|---|---|
| 1 | 166 |
| 2 | 167·5 |
| 3 | 155 |
| 4 | 119 |
| 5 | 106 |
| 6 | 105 |
| 7 | 106 |
| 8 | 105·5 |
| 9 | 106·5 |
| 10 | 106 |
| 11 | 106·5 |
| 12 | 106 |
| 13 | 106·5 |

which a gradual rise is observed in $T_f$, with slight alternation for $n$ odd and $n$ even. With alkyl anilines containing smaller groups the aromatic part of the molecule appears to control melting behaviour, which then shows an erratic sequence as the molecule grows in size (Foster and Hammick, 1955).

### 7.3.1. *Effects of point of loading of chain molecules on $T_f$*

When the n-alkane chain contains a single polar group somewhere in the middle of the molecules, it is possible (by appropriate synthetic procedure) to obtain a chain of constant length and mass, and thus to compare melting behaviour as the polar group is moved along the chain.

A particularly simple attachment arises from the insertion of a keto group, in the molecule

$$R_1-CO-R_2$$

The sizes of the repulsion envelopes are such that little or no bulge of the paraffin crystal structure results from introduction of the keto group (Oldham and Ubbelohde, 1939). n-ketones of constant chain length thus generally all have similar crystal structure, except when the group is near one end of the chain.

On inserting the keto group, the magnitude of the additional loading of the chains has been energetically well defined (Ubbelohde, 1938). Figure 7.6 illustrates how the freezing point of these crystals is highest when the CO group is near one end, and also when it is at the middle of a chain, i.e. when $R_1 = R_2$. (This is only possible for chains when $n$ is odd.) Such series behaviour may probably attribute to effects arising from enhanced attractions between CO groups in neighbouring molecules, competing with attractions between neighbouring $CH_2$ groups. This comparatively mild loading of the chains modifies the spectrum of torsional vibrations in a way that depends on where the loading occurs along their length. The effect is reminiscent of the consequences of plucking a stretched wire at various points along its length on the vibrational overtones excited.

Further consequences of configurational isomerization of flexible molecules are discussed later in relation to the melting of high polymers (Chapter 17) and to nucleation processes (Chapter 15), as well as to network melts (Chapter 10).

### 7.4 Configurational melting—non-thermodynamic evidence

Particularly clear evidence for the increase of configurational disorder on melting flexible molecules is obtained from Raman spectra. For example for normal paraffins capable of configurational isomerism, Raman spectra are much simpler for the solid than for the melt. The reason is that (as stated earlier) only a single rotational isomer—usually the fully stretched zigzag configuration—is selected from the melt for packing into the crystal lattice. On the other hand, in the melt all possible configurations are present in concentrations dependent on the energy differences and entropy differences between them. According to one estimate, the difference between the configurations

$$
\begin{array}{cc}
& CH_3 \\
& | \\
CH_2-CH_2 & CH_2-CH_2 \\
| \quad\quad | & | \\
CH_3 \quad CH_3 & CH_3 \\
syn & anti
\end{array}
$$

for n-butane is about $800\ cal\ mole^{-1}$. For longer chains the difference increases at a rate not quite proportional to the number of flexible links in the chain (Sheppard and Szasz, 1949). In the same way, infrared studies confirm the change from a single rotational isomer in the crystal, to a mixture of isomers in the melt, for various n-alkyl bromides (Brown and Sheppard, 1954) and for $\omega, \omega'$-dicarboxylic acids (Corish and Davison, 1955). In these melts hydrogen-bonding further complicates the range of configurational options. Crystalline polyesters (Corish and Davison, 1955) show analogous increases in configurational complexity on melting. Organic molecules which can assume alternative configurations in the melt probably quite frequently show selective crystallization on freezing, though no comprehensive review relating melting parameters to molecular structure, or to the rate of isomerization of molecular structure on melting appears to have yet been made. Recent examples of some generality include tertiary butyl fumarates and maleates (Kingsland and Spedding, 1971), acryclic acid (Umemura and Hayashi, 1975), simple aldehydes and ketones (Caspary, 1968) and chlorinated hydrocarbons (Mizushima and Ichisima, 1971). These examples indicate how refined optical measurements can throw light on melting parameters of flexible molecules.

Again, using Raman spectroscopy, iso-paraffins, even when they contain some flexible links reveal much less difference between the melt and the crystal. For substituted butanes, it has not yet been possible to discriminate between alternative possibilities. (Szasz and Sheppard, 1949). In these cases the absence of any selectivity on crystallization (which is less marked than with the preceding examples), might be due to the fact that different configurations of these molecules (which in any case are more bunched than n-alkanes) can be packed together in the crystal in ordered or perhaps even in random solid solution. Alternatively, (and this has not yet been disproved) the difference in energy between different rotational isomers may be so large that one configuration predominates over the whole range of temperatures studied, not only in the crystal but even in the melt and in the gas phase.

Ultraviolet spectroscopy suggests (Banerjee, 1957) that molten $o$-chloro-anisole contains isomeric molecules with the methoxy group *cis* and *trans* with respect to the chlorine atom, whereas only one isomeric form is packed into the crystal. Again 4, 4'-alkoxyazoxybenzenes, with end groups from methyl to heptyl, can exist in the form of various rotational isomers. Yet as shown by infrared spectroscopy, the crystals contain only a single form. Even the liquid crystal stage contains bundles of parallel molecules (Maier and Englert, 1959; see Chapter 14). In the isotropic melt packing is always more random, owing to configurational isomers having much greater room to form.

### 7.5 Melting of homologous flexible molecules loaded at several points

When more than one factor contributes to $H_f$ and to $S_f$ in a homologous series of chain molecules, anomalous sequences of melting points may be found. For example, for the diamide chains (Cannepin and Parisot, 1954) $R_1.CO.NH.CH_2.NH.CO.R_1$, as $R_1$ is progressively lengthened, the melting point actually falls gradually. A tentative explanation may well be that $H_f$ increases much less rapidly than $S_f$ as the n-alkyl chain $R_1$ is lengthened.

Such behaviour can be readily understood in view of the strong hydrogen bonding at the amido groups between neighbouring chains in the crystal. This would contribute the predominant part of $H_f$ at least until $n$ is quite large, so that approximately $H_f = H_0$ (const.). On the other hand, $S_f$ probably grows with $n$ in the simplest case according to the equation $S_f = a + nb$ owing to the flexibility of the n-alkyl chain which permits many different configurations to be assumed in the melts. On this basis

$$T_f = H_f/S_f = H_0/(a + nb)$$

which falls as $n$ increases, as actually found. Direct determinations of entropies of fusion would help to verify this explanation, but do not appear to be yet available.

### 7.6 Crystals with mixed flexible tautomers in the melt

Examples discussed in the previous section refer to flexible molecules where configurational equilibrium is rapidly established in the melt, but in which crystallization selects one out of a mixture of possible configurational isomers, which is then packed economically into the crystal lattice. In principle, analogous processes of selection by crystallization could arise from a mixture of tautomers, i.e. which tautomerize only slowly, if crystal lattice forces demand it. For example, when molecules can assume tautomeric structures as in the keto–enol tautomerism

$$R.CO.CH_2.CO.CH_3 \rightleftharpoons R-\underset{\underset{OH}{|}}{C}=CH.CO.CH_3$$

or the geometrical isomerism of oximes (see Campbell and Smith, 1951)

$$C_6H_5.CH \qquad C_6H_5CH$$
$$\| \qquad\qquad \|$$
$$HON \qquad\qquad NOH$$

$anti\ T_f\ 34.5°C \qquad syn\ T_f\ 130°C$

the greater freedom in the melt generally permits all the tautomers to appear, if sufficient time is allowed. On crystallization, one or other tautomer may be preferred according to nucleation and growth conditions.

Similar suggestions have been put forward from time to time about molecules in which valency isomerism might be expected. Tests are not often available, however, to establish this kind of melting behaviour. In the case of $H_2O_2$ careful calorimetric measurements do not support any extensive tautomeric change on melting, as might be suggested by the possible bond rearrangement

$$HO-OH \rightleftharpoons \overset{H}{\underset{H}{\diagdown\!\!\diagup}} O \rightarrow O$$

(Giguere, Liu, Dugdale, and Morrison, 1954).

Melting of crystals in which the lattice forces impose a single structure, whereas the melt permits the coexistence of various tautomers, seems plausible in quite a number of cases but this effect has generally been detected by somewhat indirect chemical evidence. As a possible instance, some coordination complexes of transitional metals undergo a marked change of magnetic susceptibility, accompanying changes of configuration. For example, bis($N$-alkylsalicylaldimine) complexes of nickel(II) are diamagnetic in the solid state, whereas the melts are paramagnetic (Sacconi, Cini, and Maggio, 1958). Details of the change in bonding leading to this change in magnetic susceptibility are still unresolved. Possibilities such as a change from planar configuration of coordination bonds in the crystals, to a tetrahedral (paramagnetic) configuration for at least some of the molecules in the melt, are considered to be unlikely on the basis of dipole moment measurements on solutions of the crystals in benzene. These solutions are likewise paramagnetic. Formation of free $Ni^{2+}$ ions by dissociation is rejected since the melts do not conduct electricity sufficiently. Some kind of autocomplexing (see Chapter 8) may perhaps be reponsible for the change diamagnetic crystals $\rightleftharpoons$ paramagnetic melt which is observed.

The magnetic susceptibility of the melts changes quite rapidly with temperature, pointing to a shift of molecular equilibrium between tautomers of different configurations above $T_f$.

Molecular tautomerism in liquid sulphur has long been known (see Campbell and Smith, 1951). Ring molecules of sulphur $S_8$ can reform into chain structures in the melt. In this instance the change is accompanied by absorption of heat, so that the proportion of chain molecules increases as the temperature rises above $T_f$. At $T_f$ the concentration of chain polymer molecules can be determined only somewhat indirectly, since the average chain length varies with temperature (Paulis and Derbyshire, 1963). (Poulis, Massen, and v.d. Leiden, 1962). Thus the contribution of this tautomerism to the overall entropy of fusion of sulphur has not been accurately determined (see also p. 360).

### 7.7 Configurational disordering giving anticrystalline melts

As illustrated in the preceding sections of this chapter, melting of crystals of flexible molecules permits a diversity of configurational isomers to appear, all of which have practically the same energy in the melt. On the other hand, in the crystal, most of these isomers could be packed in solid solution only by accepting exceptionally large crystal defect energies. Such defect sites are normally highly improbable in the crystal because of the repulsive forces from stretched neighbouring molecules.

With the definition specified in this book, melts of flexible molecules can never be regarded as quasicrystalline, since structure of the melt cannot be derived from the crystal by simple operations on the lattice; additional local isomerization operations are called for which have a molecular but not a lattice specification (see Ubbelohde, 1964). Subsequent chapters of this book will also deal with other types of melts which are *anticrystalline* in this sense, but for other reasons.

REFERENCES

D. G. Anderson, J. C. Smith, and R. J. Rallings (1953) *J. Chem. Soc. London* 443.
S. B. Banerjee (1957) *Indian J. Phys.* **31,** 135.
S. Blasenberg and W. Pechold (1967) *Rheologica Acta* **6,** 171.
M. G. Broadhurst (1962a) *J. Chem. Phys.* **36,** 2578.
M. G. Broadhurst (1962b) *J. Res. Nat. Bur. Stand.* **66A,** 241.
M. G. Broadhurst (1966) *J. Res. Nat. Bur. Stand.* **70A,** 481.
L. C. Brunel and D. A. Dows (1974) *Spectrochim. Acta* **30A,** 929.
A. N. Campbell and N. O. Smith (1951) *The Phase Rule and its Applications,* Dover Publications, London.
A. Cannepin and A. Parisot (1954) *C.R. Acad. Sci., Paris* **239,** 180.
R. Caspary (1968) *Appl. Spectrosc.* **22,** 689.
P. J. Cornish and W. H. T. Davison (1955) *J. Chem. Soc. London* 2428.
G. A. Cummings, J. C. McCoubrey and A. R. Ubbelohde (1952) *J. Chem. Soc., London* 2725.
R. Foster and D. L. Hammick (1955) *Nature, Lond.* **175,** 255.
P. A. Giguere, I. D. Lin, J. S. Dugdale, and J. A. Morrison (1954) *Can. J. Chem.* **32,** 117.

M. Kingsland and H. Spedding (1971) *Chem. Ind.* p. 507.

K. Larsson (1966) *J. Am. Oil Chem. Soc.* **43**, 559.

G. W. Lehman and J. P. McTague (1968) *J. Chem. Phys.* **49**, 3170.

H. C. Longuet Higgins (1958) *Discuss. Faraday Soc.* **25**, 86.

W. Maier and G. Englert (1959) *Z. Phys. Chem. (Leipzig)* **19**, 168.

T. Malkin and T. C. Tranter (1951) *J. Chem. Soc.* 1178.

J. C. McCoubrey and A. R. Ubbelohde (1951) *Quart. Rev. Chem. Soc.* **4**, 364.

J. C. McCoubrey, J. N. McCrea, and A. R. Ubbelohde (1951) *J. Chem. Soc. London* 1961.

S. Mizushima and I. Ichishima (1971) *Essays in Structural Chemistry*, ed. D. A. Long, MacMillan, London, pp. 229.

J. W. H. Oldham and A. R. Ubbelohde (1939) *Trans. Faraday Soc.* **35**, 328.

W. Piesczek, G. R. Strobl, and K. Malzahn (1974) *Acta Cryst.* **30B**, 1278.

J. A. Poulis and W. Derbyshire (1963) *Trans. Faraday Soc.* **59**, 559.

J. A. Poulis, C. H. Massen, and P. v.d. Leiden (1962) *Trans. Faraday Soc.* **58**, 474.

L. Reinisch (1968) *J. Chim. Phys. Physicochim. Biol.* **65**, 1903 (*France*).

L. Sacconi, R. Cini, and F. Maggio (1958) *J. Inorg. Nucl. Chem.* **8**, 489.

F. Salmon-Legagneur and Mme L. C. Neveu (1962) *Bull. Soc. Chim. Fr.* 2130.

N. Sheppard and G. J. Szasz (1949) *J. Chem. Phys.* **17**, 86.

R. G. Snyder (1967) *J. Chem. Phys.* **47**, 1316.

G. J. Szasz and N. Sheppard (1949) *J Chem. Phys.* **17**, 93.

J. D. Turner and E. C. Lingafelter (1955a) *Acta Cryst.* **8**, 549.

J. D. Turner and E. C. Lingafelter (1955b) *Acta Cryst.* **8**, 551.

A. R. Ubbelohde (1938) *Trans. Faraday Soc.* **34**, 282.

A. R. Ubbelohde (1964) *J. Chimie. Phys.* **5**, 8.

A. R. Ubbelohde and I. Woodward (1952) *Trans. Faraday Soc.* **48**, 113.

J. Umemura and S. Hayashi (1975) *Bull. Inst. Chem. Res. Kyoto* **52**, 585.

Y. Urushibara, F. Imura, and K. Ideda (1955) *J. Chem. Phys.* **23**, 1724.

V. Vand (1953) Acta Cryst. **6**, 797.

B. H. Zimm (1960) *J. Chem. Phys.* **33**, 1349.

# 8. MELTING OF IONIC CRYSTALS

## 8.1 Introduction

One of the most rapid developments of the physical chemistry of the molten state during the past two decades, even since an earlier book on melting published by the author, (Ubbelohde, 1965) has been concerned with ionic melts. Precise physicochemical study of ionic melts has until quite recently been almost exclusively concerned with inorganic salts. Indeed, modern inorganic chemistry can hardly dispense with a detailed discussion of the molten state. However, amongst the newer ventures, successful methods of some generality for investigating organic molten salts have now been achieved. In view of the degree of innovation still attached to molten organic salts, it is convenient to discuss these in separate sections of this chapter from p. 220 onwards.

In crystal chemistry considered as a whole, ionic crystals constitute a highly distinctive class. Individual atoms or simple molecules can still be treated as centres from which intermolecular forces emanate, as in the case of molecular crystals. However, the range of action of primary electrostatic forces is much longer than in molecular crystals. This means that in condensed phases it never is sufficient to calculate attractions or repulsions in terms of simple pairing between neighbouring centres. In this respect, ionic crystals and ionic melts both involve many body summations. Fortunately, these take a simplified form in some characteristic ionic crystals.

## 8.2 Positional melting in ionic crystals—ions of inert gas type

In general, descriptions of mechanisms of melting and types of disorder in ionic crystals are broadly similar to those in molecular crystals and need not be recapitulated (see chapter 5) (Ubbelohde, 1960). The importance of the longer-range electrostatic interactions which arise when the units of structure are charged is however apparent in various ways.

At one time it seemed likely that crystals with ions whose electronic structure resembles the inert gases would provide particularly simple instances of positional melting, resembling the inert gases themselves. Positional defects in their crystals are well characterised by various physical measurements. For ions formed by the alkali metals and by the halogens, alternative types of linkages such as covalent bonding in the melt seem unlikely. Only positional mechanisms of melting need thus be considered for these simple salts. Rules of similitude and principles of corresponding states might therefore be expected to apply for this group of solids, with considerable accuracy. Accumulated evidence has shown, however, that even with ionic crystals of such ideally simple electronic structure, no completely regular behaviour is found, chiefly because polarizabilities of atoms vary so much throughout the Periodic System.

The absence of 'corresponding state' similitudes sometimes makes it inconvenient to choose the best temperature $T^*$ at which to compare a specific melting property in any series of ionic compounds. A value $T^* = 1·05T_f$ is chosen by some authors. Alternatively, if the thermal expansion coefficient $\alpha$ is known, a temperature $T^*$ calculated from the empirical equation $\alpha T^* = 0·4$ has been proposed (Ketelaar, 1975, 1976), but not all salts depend on simple vibrational thermal expansion as the sole contributor to $\alpha$ so that use of $T^*$ defined in this way may bring other difficulties.

Many of the general features of melting of ionic crystals are nevertheless seen most clearly for these simple salts, which thus constitute useful models as an introduction to mechanisms of melting of ionic crystals generally. In the absence of information to the contrary, these ionic melts may be assumed to have a quasicrystalline structure. The structure may be considered as formed by first expanding the lattice and then introducing various kinds of positional disorder. X-ray studies give some support to this model for halides of Groups I and II (Zarzycki, 1957; Levy and Danford, 1964). Various features of special relevance for ions of inert gas type ions are noted below. The general similarity between crystals and melts in the case of alkali halides is shown in various properties. For example, from 1120 K to 1300 K (i.e. to above $T_f$) there is relatively little change in the lower frequency end of the infrared spectrum, and at the upper frequency and the change does not exceed 10% for the typical salts KBr, NaCl, and LiF (Barker, 1972).

## 8.3 Ion defect formation in simple ionic crystals

Reliable information about the energy required to create defect sites in ionic lattices can in principle help to illuminate a number of related problems concerning their melting, as well as any premelting that may be observable (Chapter 12). Evidence is still rather contradictory, however, depending on the methods of observation and on their precision.

An empirical relationship has been proposed which states that

$$\varepsilon_{\text{defect}} \text{ in eV} \sim 2·14 \times 10^{-3} T_f$$

with fair agreement for all the halides of Li, Na, K, Gs, as well as for TlCl, $PbCl_2$, and $PbBr_2$. The correlation with $\varepsilon_{\text{defect}}$ for Schottky cation holes would suggest that it is this parameter which is predominant in the positional disordering; the original paper should however be consulted (Barr and Dawson, 1971).

The 'holes' in fluid condensed states which are plausibly considered to control properties such as the activation energy $E_\eta$ for viscous flow, and for self-diffusion, often conform to an empirical rule (Bockris and Emi, 1970)

$$E_\eta \sim E_{\text{diff}_+} \sim E_{\text{diff}_-} \sim 3 \cdot 7 R T_f$$

and its validity might be regarded as a consequence of positional melting being a dominant factor in ionic crystals. Probably the chief use of such empirical rules is to signalize those ionic melts departing from them, particularly those which constitute major exceptions and which demand appropriate special studies to elucidate reasons for the anomaly.

### 8.3.1. Interpenetrating sublattices

In relation to some of the problems discussed below, it may be assumed for a quasicrystalline ionic melt that the anions and cations are each preferentially surrounded by a shell of ions of opposite charge. On this basis, for example in a mixture of two salts, cations $1^+$ and $2^+$ form a random statistical mixture, as do anions $1^-$ and $2^-$; on this model ions of opposite charges constitute two interpenetrating quasicrystalline lattices. An alternative model in which anions and cations occupy all positions in the melt indifferently would for example lead to a different entropy of mixing, but no experimental justification seems available for it at normal temperatures (Doucet, 1959).

### 8.3.2. Hydrogen in ionic lattices

Because of its small size the polarizing effect of the proton on electron clouds of anions is so great that covalent bonding proves to be the state of lower energy for compounds of hydrogen with halogen atoms, even in their crystalline state. Thus there is no known analogy of behaviour between alkali metal cations and protons as cations in simple salts. Some proton mobility may perhaps be observable in the crystals and melts, e.g. of acid salts. However, a converse role of hydrogen acting as anion in crystalline and molten hydrides seems well established and does present experimentally accessible analogies with the halides of alkali cations. Unfortunately, studies on molten hydrides have as yet seldom dealt in any depth with the questions considered in this book. One not very serious experimental difficulty is the dissociation pressure of hydrogen which must be counteracted for their stabilization. It may however be noted, for example, that the conductivity of molten LiH is reported to be about $30\,\text{ohm}^{-1}\,\text{cm}^{-1}$ measured under an atmosphere of hydrogen at 700°C (Johnson, Wood, and Cavins, 1967).

### 8.3.3. *Other ions of inert gas type*

Ions of the inert gas type include cations of Groups IA and IIA of the Periodic System. In most cases where their behaviour has been studied, anions of Group VII and VI also are found to retain electron shells corresponding with the nearest inert gas atom of the Periodic System. So long as this applies, crystalline and molten assemblies of salts of these ions must show considerable resemblance to those of corresponding inert gases. Forces between these particles are predominantly centrosymmetric, and can be expressed by a combination of Van der Waals and primary electrostatic interactions.

Table 8.1 (Schinke and Sauerwald, 1956) records some of the principal volume and entropy changes on melting inert gas halides.

**Table 8.1. *Thermodynamic melting parameters of inert gas type halides***

| Salt | $\Delta V_f/V_s$ (%) | $S_f$ (e.u.) | Salt | $\Delta V_f/V_s$ (%) | $S_f$ (e.u.) |
|------|------|------|------|------|------|
| LiF | 29·4 | 5·78 | CsCl | 10·5 | 3·5 |
| LiCl | 26·2 | 5·6 | — | — | — |
| LiBr | 24·3 | 4·9 | AgCl | 8·9 | 4·34 |
| | | | AgBr | 8·2 | 3·28 |
| NaF | 27·4 | 5·5 | $CaCl_2$ | 0·9 | 5·8 |
| NaCl | 25·0 | 6·7 | $SrCl_2$ | 4·2 | 3·6 |
| NaBr | 22·4 | 6·0 | $BaCl_2$ | 3·5 | 4·4 |
| NaI | 18·6 | 5·6 | $AlCl_3$ | 83·0 | 36·5 |
| KF | 17·2 | 5·8 | | | |
| KCl | 17·3 | 6·2 | | | |
| KBr | 16·6 | 4·9 | | | |
| KI | 15·9 | 4·3 | | | |
| RbCl | 14·3 | 4·4 | | | |
| RbBr | 13·5 | 3·9 | | | |

It has been suggested (Forland, 1964) that in a total entropy of fusion of 6 e.u. about 2·2 e.u. arises from random distribution of defects, and about 3·8 e.u. from vibrational entropy changes and rotation of paired vacancies. These values should be regarded as illustrative estimates only.

With reference to other peculiarities in Table 8.1 an abnormally small $\Delta V_f/V_s$ for $CaCl_2$ is attributed to the low coordination number in the crystals; not much change in coordination numbers appears to be involved to accompany the positional disordering on melting this salt. The very high $\Delta V_f/V_s$ and $S_f$ for $AlCl_3$ has been attributed to a complete change of structure on melting, from an ionic crystal to a molecular melt containing covalent molecules of $Al_2Cl_6$ (see Biltz and Voigt, 1923). The density of $AlCl_3$ has been measured with some precision in relation to its formation of mixed melts with KCl in which the anion $(AlCl_4)^-$ is formed. Molar volumes (Morrey and Carter, 1968; Boston, 1966) in $g\,cm^{-3}$ indicate (see Table 8.2) a fairly rapid change of volume with temperature and suggest ligand changes as the volume expands. Conversely, it may be expected that when $AlCl_3$ is melted under sufficiently high pressures an ionic melt would be formed, though this has not yet been verified.

**Table 8.2 Relationship between molar volume and temperature for $AlCl_3$**

| $T(°C)$ | $\rho_L$ | |
|---------|----------|---|
| 188·5   | 1·293    | |
| 201·7   | 1·265    | $T_f = 192°C$ |
| 211·2   | 1·238    | |
| 220·8   | 1·218    | |

Whereas the electrical conductivity of molten $Al_2Cl_6$ is zero in conformity with other molecular liquids, addition of even small proportions of KCl steeply raises the conductivity presumably owing to complex ion formation $2KCl + Al_2Cl_6 \rightleftharpoons 2K^+[AlCl_4]^-$ (Boston, Grantham, and Yosim, 1970).

The volume change on melting $Al_2O_3$ is large $(\Delta V_f/V_s = 20\%)$ but no measurements of $S_f$ appear to be available as a clue to the melt structure (Tyrolerova and Lu, 1969).

In the vapour phase aluminium bromide is almost wholly in the form of $Al_2Br_6$ (Johnson, Silva and Cubicciotti, 1968) at moderate densities.

Similar changes of bonding may also be looked for in other parts of the Periodic System when ionic crystals are melted.

## 8.4 Effects of positional disordering on electrostatic compensation

In considering the preceding sections, it is helpful to keep in mind an important effect in the melting of ionic crystals. As a consequence of positional randomization (which always accompanies melting) the geometrical compensation of electrostatic forces acting on any particular ion as a result of its neighbours is reduced on passing from crystal to melt. For example, a cation in an alkali halide with perfect rock salt structure has six neighbouring anions situated at equal distances at the corners of an octahedron. Electrostatic polarizing forces between this cation and any one anion (which would tend to bring them close together), are reduced in the perfect lattice by compensation from the opposing forces from the other anions, having regard to the directions in which they act. As one consequence of this reduced action, *ion contact radii* in crystals are larger than in an isolated ion pair. For ions of simple inert gas structure, this can be verified from comparisons between contact distances $(r_+)$ and $(r_-)$ in the crystals as deduced from X-ray diffraction, and the ion pair distance in ion pairs in a vapour stream of the salt, as established by electron diffraction (see Table 8.3; see Ubbelohde, 1959, 1960). For LiBr and LiI, the distances between ions in the gas were taken from Klemperer, Norris, and Buckler, 1960).

**Table 8.3.** *Shrinkage of ion contact distance on passing from crystal to gaseous ion pairs* $(\mathring{A})$

|    | Cl   | Br   | I    |
|----|------|------|------|
| Li | —    | 0·38 | 0·38 |
| Na | 0·30 | 0·34 | 0·33 |
| K  | 0·35 | 0·35 | 0·30 |
| Rb | 0·38 | 0·37 | 0·40 |
| Cs | 0·50 | 0·57 | 0·54 |

As one extreme for the dilute vapour of a salt, in the gaseous ion pair the electrostatic interactions are not compensated in any way by the actions of similar ions, which in the crystals are situated symmetrically around any ion so that their actions tend to neutralize one another. In the melts a situation arises intermediate between the dilute gas and the perfect crystals. There will normally still be partial compensation, but the absence of long-range order and consequent loss of local symmetry in the melt must imply that at least some $[r_+][r_-]$ distances are smaller than in the crystal.

Although statistical consequences have not yet been analysed in any precise detail, because of their close approach such regions in an ionic melt have structural analogies with 'ion pairs' whose presence has been postulated in other systems (see below). (Energetically, however, an 'ion pair' cannot be specifically identified in a melt unless its formation involves a distinct minimum of potential energy well below the minimum characteristic for the crystal lattice.)

## 8.5 Melting parameters of ionic halides with low polarizabilities

Because the fluorine anion has the lowest polarizability of the halogen series, ionic parameters of molten fluorides have a special interest in relation to melting of ionic crystals. Unfortunately, sufficient purity of these salts is not easy to ensure because of hydrolysis during their formation. Melting points reported for fluorides specially purified by exhaustive treatment with gaseous hydrogen fluoride (Kohma, Whiteway, and Masson, 1968) are listed in Table 8.4.

In situations where a small impurity of dissolved oxide or hydroxide in the fluoride melt is important these parameters may help to test the purity critically.

**Table 8.4. Melting points of fluorides with inert gas type cations (°C)**

| Li | 845 | Mg | 1256–61 |
|----|-----|----|---------|
| Na | 993 | Ca | 1423 |
| K | 852 | Sr | 1473 |
| | | Ba | 1355 |
| | | Cd | 1078 |

## 8.5.1. Halides of high valence cations (lanthanides)

Entropies of fusion (and, where relevant, entropies of precursor transitions in the solid state) are of considerable interest for melting models of halides of the trivalent lanthanide cations (Dworkin and Bredig, 1971), whose polarizability is low, and in which the formation of covalent ligands seems unlikely from valence chemistry considerations. As might be expected, the values can be grouped according to the crystal type (see Table 8.5).

**Table 8.5. Melting parameters of halides of high valence cations. (Melting points in K. Entropies in $cal\ mole^{-1}\ deg^{-1}$)**

| Chlorides | Crystal type A entropies around 7.5 e.u. | | Crystal type B intermediate values | | | Crystal type F entropies around 12 e.u. | | |
|-----------|-----|-----|-----|-----|-----|-----|-----|-----|
| Cation | $S_f$ | $T_f$ | Cation | $S_f$ | $T_f$ | Cation | $S_f$ | $T_f$ |
| Y | 7·6 | 994 | La | 11·5 | 1131 | Sc | 13·0 | 1240 |
| Dy | 6·6 | 924 | Ce | 11·7 | 1090 | | | |
| Ho | 7·3 | 993 | Pr | 11·4 | 1059 | | | |
| Er | 7·4 | 1049 | Nd | 11·6 | 1032 | | | |
| | | | Gd | 11·1 | 875 | | | |
| | | | Tb | $\begin{cases} 5·4 \\ 4·3 \end{cases}$ | $\begin{cases} 855 \\ 783 \\ trans \end{cases}$ | | | |

| Bromides | | | Crystal type all Y except NdBr$_3$ (Pu) | | | Crystal type Y | | |
|----------|-----|-----|-----|-----|-----|-----|-----|-----|
| (F)Gd | 8·6 | 1058 | Pr | 11·7 | 966 | La | 12·3 | 1061 |
| | | | (Pu)Nd | 11·4 | 955 | Ce | 12·4 | 1005 |
| | | | FHo | 10·0 | 1192 | | | |

| Iodides | | | Type Pu | | | Type Pu | | |
|---------|-----|-----|-----|-----|-----|-----|-----|-----|
| | | | Nd with trans | $\begin{cases} 9·4 \\ 3·9 \end{cases}$ | $\begin{cases} 1060 \\ 847 \end{cases}$ | La | 12·7 | 1051 |
| | | | | | | Ce | 12·0 | 1033 |
| | | | | | | Pr | 12·6 | 1011 |
| | | | Type F | | | | | |
| | | | Gd with trans | $\begin{cases} 10·7 \\ 0·14 \end{cases}$ | $\begin{cases} 1203 \\ 1013 \end{cases}$ | | | |
| | | | Tb with trans | $\begin{cases} 11·2 \\ 0·30 \end{cases}$ | $\begin{cases} 1228 \\ 1080 \end{cases}$ | | | |

Clearly, the rather close comparisons of the data with those of the crystals
of

$$FeCl_3 \text{ (type F)} \qquad Y(OH)_3 \text{ (type Y)}$$

$$AlCl_3 \text{ (type A) or} \quad PuBr_3 \text{ (type Pu)}$$

indicate strong model correlations of melting mechanisms. Unfortunately, the
data were obtained by drop calorimetry and thus yield no information about
precursor defect formation in the different crystal types below $T_f$ (see p. 182).
It would also be very useful to know the molar volume changes on melting of
these lanthamide halides, in establishing models for these melts.

### 8.5.2. Ionic analogue for the iodide anion

In constructing model systems for ionic melts, alkali halides constitute a
group which is worked rather hard because of the limited number of salts. It is
useful to note that alkali fluoroborates (see Table 8.6) may provide some
extension of the group (Cantor, McDermott, and Gilpatrick, 1970). The
anion $BF_4^-$ contains 42 electrons which is thus intermediate between $Br^-$ (36)
and $I^-$ (54). The molecular polarizability

$$\alpha = \frac{3}{4\pi} V_m \left[ \frac{\mu^2 - 1}{\mu^2 + 2} \right]$$

suggests that this property decreases with increasing size of the alkali cation;
if there were no covalent ligand interaction it should be independent of the
cation size. So far as is known this complex anion shows negligible dissociation in ionic melts, but this may need further testing.

**Table 8.6 Melting and transition
points in °C of fluoroborates**

| Cation | $T_{tr}$ | $T_f$ |
|--------|----------|-------|
| Li | — | 304 |
| Na | 243 | 408 |
| K | 283 | 570 |
| Rb | 245 | 582 |
| Cs | 169·5 | 555 |

## 8.6 Structural studies on ionic melts

The primary objective of this book is to discuss thermodynamic properties
of the molten state, in terms of its structure. Efforts to achieve this aim for the
class of liquids broadly described as ionic melts reveal how little is still known
about this important sector of physical chemistry. Even in the restricted field
of physical inorganic chemistry, it is often found that the behaviour of any

compound in its molten state shows up intrinsic peculiarities in its chemical bonding that remained obscured in the crystalline state because of the extensive compensation of electrostatic forces in a crystal lattice. To make real progress with the physical chemistry of inorganic melts, diverse kinds of experimental information may be helpful. X-ray diffraction which is so powerful for crystals may miss some of the most significant clues about the melts. In the absence of any single complete approach to research on ionic melts, the following sections aim to bring together results of various methods of study on various compounds. At present these methods are not all of comparable quality; some have been included mainly in the hope that this may stimulate the appearance of improved versions.

### 8.6.1. *Diffraction studies on ionic melts—polarizability effects*

Conclusions about disordered ion packing and local shrinkages of contact radii in ionic melts have been derived from X-ray and neutron diffraction studies on the melting of ionic crystals with inert gas shells, reported by various authors. Data (see Furukawa, 1961, 1962, who gives other details) obtained in the form of radial distribution curves for melts of Group IA halides indicate that on melting; (i) the first coordination number $n$ of ions of opposite charge decreases from 6 to between 4 and 5; (ii) the separation between metal and halogen atoms decreases on passing from crystal to melt, for reasons already discussed. Consideration of various other physical changes on melting suggests that this shrinkage can be primarily attributed to the mutual polarizability between anion and cation, as well as the lessened compensation of electrostatic forces in the melt.

Systematic studies of parameters such as the freezing point $T_f$, the entropy of fusion $S_f$, and the fractional volume change on fusion $\Delta V_f / V_s$ at one time seemed particularly hopeful as leading to quite general models of melting for these salts, whose crystal structure is comparatively simple. However, for reasons indicated, differences of polarizability of different ions somewhat distort any regularity which might be anticipated. This provides one reason why correlations between $T_f$ and the crystal radius ratio $r_+/r_-$ (Pauling, 1960) or ion screening constants (Weyl, 1955) cannot be completely relevant, particularly in describing the melts.

Values of the two principal thermodynamic melting parameters of these salts are given in Table 8.1 (see Ubbelohde, 1959). A steady trend can be seen towards smaller volume increases on melting and smaller entropies of fusion as the atomic number and thus the number of electrons and the mutual ionic polarization increases. It seems likely that positional disordering permits association of ions by polarization forces in the melt, to an increasing extent as the atomic number increases.

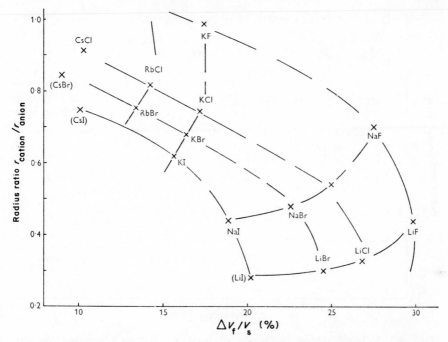

FIG. 8.1. Percentage volume change of inert gas halides on melting as function of radius ratio. Reproduced with permission, from Furukawa, *Faraday Disc.* **32,** 53 (1961)

A semiempirical measure of the influence of ion polarizability $\alpha_{\pm}$ on the volume change on fusion is provided by the correlations found in a plot of $\Delta V_f / V_s \times \alpha_{\pm}^{\frac{1}{3}}$ against the radius ratio $r_+/r_-$ (Furukawa, 1961). In this plot, the polarizability $\alpha_{\pm}$ is taken as $\frac{1}{2}(\alpha_+ + \alpha_-)$ (Figs. 8.1 and 8.2). For alkali fluorides, including LiF, NaF and KF, X-ray studies of the melts are stated to give radial distribution functions with maxima at nearest neighbour separations practically identical with those in the crystals (Zarzycki, 1957). The fluoride would be expected to have the lowest polarizability of any halogen anions and

**Table 8.7. Nearest neighbour and second nearest neighbour coordinaton numbers**

| Salt | Solid | melt | Layer |
|------|-------|------|-------|
| LiF  | 6     | 3·7  | Li/F  |
|      | 12    | 8    | Li/Li |
| NaF  | 6     | 4·1  | Na/F  |
|      | 12    | 9    | Na/Na |
| KF   | 6     | 4·9  | K/F   |
|      | 12    | 9    | K/K   |

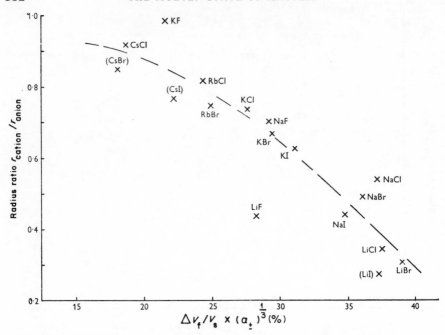

FIG. 8.2. Volume change on fusion of inert gas halides, corrected for ion polarizability, as function of radius ratio. Reproduced with permission from Furukawa, *Faraday Disc.* **32,** 53 (1961)

thus show the least shrinkage attributable to reduced compensation of forces on melting; but if this conclusion is confirmed, the coordination number of the cations and the fluorine anions must decrease sharply on melting (Table 8.7) to account for the observed increases in volume $\Delta V_f / V_s$ (Table 8.1). With other halides, resolution of radial distribution functions given by the ionic melts is less precise (Zarzycki, 1958) and no separation of effects of shrinkage and reduced coordination number can be attempted.

## 8.7 Precursor effects associated with specific ionic crystal types

In many halides with ions of inert gas type, the assumption that a quasi-crystalline melt is formed, with a coordination number of nearest neighbour ion not very different from that in the crystals, appears to give a fairly satisfactory model for the structure of the melt. A significant dfference of melting behaviour is however observed between CsCl and CsBr and CsI. X-ray measurements show a transition in CsCl at 469°C (Menary, Ubbelohde and Woodward, 1951). A volume change is observed in this transition which involves a change from coordination number 8 to coordination number 6 in the crystals at the transition point and $\Delta V_f / V_s$ is only 10% at $T_f$. However even up to the melting point, $T_f$, CsBr and CsI show no transitions in the solid (at any rate up to within 2–3°C of $T_f$). For CsBr $\Delta V_f/V_s = 26 \cdot 8\%$ and for CsI $\Delta V_f/V_s = 28 \cdot 5\%$. This striking difference in behaviour at $T_f$ between otherwise similar salts has led to the plausible suggestion that on melting the transition to coordination number 6 first occurs for these two salts. This would give all three melts similar quasicrystalline structures (Johnson, Agron and Bredig 1955), despite the difference of coordination number in the more tightly packed crystals of the bromide and iodide.

Values of entropies of fusion of halides with polyvalent cations lend considerable general support to quasicrystalline models for the melts closely related to the structure of the crystals. As with the caesium halides, for certain divalent halides when a solid–solid transformation is found below $T_f$, the entropy change at the actual melting point is reduced accordingly. Apparently the melting mechanisms normally include structural changes, which occur well below $T_f$ only in certain salts. A similar comment may be made about the data in Table 8.8.

**Table 8.8** *Specific heats and entropies of transition and fusion of polyvalent halides* (**Dworkin and Bredig, 1963**) *(cal mole$^{-1}$ deg$^{-1}$)*

| Salt | $C_{ps}$ | $C_{PL}$ | $T_f\,(K)$ | $S_f$ (e.u.) | $T_c\,(K)$ | $S_c$ (e.u.) |
|------|------|------|------|------|------|------|
| CaCl$_2$ | 23·6 | 23·6 | 1045 | 6·49 | — | — |
| CaBr$_2$ | 23·0 | 27·0 | 1015 | 6·85 | — | — |
| CaI$_2$[a] | 23·2 | 24·7 | 1052 | 9·5 | — | — |
| SrCl$_2$ | 28·5 | 27·2 | 1146 | 3·39 | 1003[b] | 1·65 |
| SrBr$_2$ | 27·5 | 27·8 | 930 | 2·70 | 918 | 3·16 |
| SrI$_2$ | 22·9 | 26·3 | 811 | 5·80 | — | — |
| BaCl$_2$ | 25·5 | 26·3 | 1223 | 3·17 | 1193 | 3·44 |
| BaBr$_2$ | 21·7 | 25·6 | 1130 | 6·75 | — | — |
| BaI$_2$ | 21·8 | 27·0 | 984 | 6·44 | — | — |

[a] Layer structure.
[b] Continuous transition.

Effects of premelting on diverse experimental methods of evaluation of data must also be allowed for. The fluorite type of structure (which is also the high temperature arrangement in solid SrCl$_2$, SrBr$_2$ and BaCl$_2$) has lattice vacancies equal in number and size to the sites occupied by the cations. This seems likely to permit considerable disordering of cation positions even below $T_f$, and may explain the anomalously high heat capacity of CaF$_2$ in a pre-melting range of temperatures (Chapter 12). Rare earth halides show similar behaviour (see Table 8.9) in the one instance (NdI$_3$) for which solid–solid transformation has been observed (Dworkin and Bredig, 1963).

Salts with comparatively large polyvalent anions have been less investigated than, for example, the fluorides of polyvalent cations. One interesting

**Table 8.9. Specific heats and entropies of fusion of rare earth halides (cal mole$^{-1}$ deg$^{-1}$)$^a$**

| Salt | $C_{p_s}$ | $C_{p_L}$ | $T_f (K)$ | $S_f$ | $T_c (K)$ | $S_c$ |
|---|---|---|---|---|---|---|
| LaCl$_3$ | 34·7 | 37·7 | 1131 | 11·5 | — | — |
| PrCl$_3$ | 32·3 | 32·0 | 1059 | 11·4 | — | — |
| NdCl$_3$ | 35·4 | 35·0 | 1032 | 11·6 | — | — |
| CeI$_3$ | 34·5 | 36·5 | 1033 | 12·0 | — | — |
| PrI$_3$ | 31·3 | 34·2 | 1011 | 12·6 | — | — |
| NdI$_3$ | 30·4 | 36·3 | 1060 | 9·2 | 847 | 4·0 |

$^a$ Where more than one solid form occurs $C_{p_s}$ is given for the form stable at the higher temperatures.

example is K$_2$S for which $S_f = 3·16$ e.u. This comparatively low value for the entropy of fusion resembles that of its antitype salts, CaF$_2$ or SrCl$_2$, but for the sulphide positional randomization of the cations in the persistent lattice of S$^{2-}$ is diffuse instead of occurring at a peak transition, or all at $T_f$ as for MgF$_2$. It takes the form of gradual diffusion of the cations over both octahedrally and tetrahedrally coordinated positions as the temperature is raised from about 550°C to $T_f$ at 948°C. The integrated entropy increase for this diffuse randomisation is approximately 4 e.u.

The peak value of $C_p$ for K$_2$S is 45 cal deg$^{-1}$ mole$^{-1}$ at 1050 K, resembling similar effects for CaF$_2$ at 1423 K and for SrCl$_2$ at 993 K.

Other crystals which may show related changes include Na$_2$O, UO$_2$, ThO$_2$, Li$_2$SO$_4$ and Na$_2$S (Dworkin and Bredig, 1968).

## 8.8  Melting of salts with inert gas type ions—effects of moderate pressures

A less direct source of information about ion packing in the melts can be obtained from the slope of the melting curve at increased pressures. In principle, this can be used in connection with the Clausius–Clapeyron equation to calculate either $\Delta V_f$ or $S_f$ from

$$\Delta V_f/S_f = (dT/dP)_{P=0}$$

Directly obtained parameters tend to yield slopes consistently higher than those calculated from pressure against melting point curves (Clark, 1959; Akella, Vaidya, and Kennedy 1969). Possibly, directly determined values of $\Delta V_f$ (which are experimentally difficult to measure) tend to be too high (see Table 8.10). Effects of pressure on premelting defects may also introduce differences between data obtained by direct calorimetry, and those from thermodynamic phase diagrams; they seem too large to be dismissed.

Normal electrostriction already brings ions into close contact and even penetration of repulsion envelopes, so that enhanced pressures may reveal new types of interaction. For the halides of Na, K, and Rb the most striking effect is the evident tendency of the fluorides to form melts containing the

**Table 8.10. Thermodynamic parameters of melting**

| Salt | $S_f$ (e.u.) | Ref. | $\Delta V_f$ ($cm^3\ mole^{-1}$) | Ref. | $\dfrac{\Delta V_f}{S_f}$ | $\dfrac{dT}{dP_{obs}}$ (deg $bar^{-1}$) |
|------|------|------|------|------|------|------|
| LiCl | 5·6 | a | 5·88 | a | 0·025 | 0·0242 |
| NaF  | 6·2 | b | 5·15 | b | 0·016 | 0·0161 |
| NaCl | 6·7 | a | 7·55 | a | 0·027 | 0·0238 |
| NaBr | 6·0 | a | 8·02 | a | 0·032 | 0·0287 |
| NaI  | 5·6 | a | 8·58 | a | 0·037 | 0·0327 |
| KCl  | 6·2 | a | 7·23 | a | 0·028 | 0·0265 |

References
a Schinke and Sauerwald (1956), see Table 8.1.
b Landon and Ubbelohde (1956).

dimerized ion $MF_2^-$ to oppose the cation $M^+$ (see Pistorius, 1965a,b; 1966). The nitrites and nitrates of monovalent cations confirm the tendency to interlocking association (see section 8.15) (Rapoport, 1966a,b; Rapoport and Pistorius 1966), with a maximum on the curve for $KNO_2$ (Rapoport 1966a). Other polyatomic ions show distinctive differences. $Li(BH_4)$ resembles simple halides (Pistorius, 1974). Poorly conducting halide melts of $HgCl_2$ and $HgI_2$ increase in electrical conductivity on increasing the pressure (Darnell and McCollum, 1971). A similar effect is found for bismuth halides (Darnell, McCollum, and Yosin 1969).

Ultrasonic measurements of compressibilities of ionic melts yield data much as may be expected (Blanc, Diélou, and Petit, 1964).

As stated, melting curves for many inert gas halides obey Simon equations of the form

$$(P-P_0) = A[(T/T_0)^c - 1]$$

$P_0$ differs from zero only for the polymorphic forms of KCl, RbCl and CsCl stable at higher pressures. $T_0$ is the melting temperature at the triple point (see Chapter 2).

Determination of the 'constants' in the Simon melting equation may not always give very accurately defined parameters, either because of experimental uncertainties, or because this equation is not really applicable to ionic melts. By way of example, for AgCN (Pistorius, 1961), in the proposed melting equation

$$(P/P_0) = (T/619)^c - 1$$

but

$$P_0 = 300 \pm 100 \text{ kbar}$$

and

$$C = 1·5 \pm 0·4$$

Table 8.11. *Melting parameters in the Simon equation*

| Salt | $T_t$ (K) | $P_0$ (bar) | A (bar) | c |
|---|---|---|---|---|
| LiCl | 878 | 0 | 14500 | 2·5 |
| NaF | 1265 | 0 | 14300 | 5·5 |
| NaCl | 1073·5 | 0 | 16700 | 2·7 |
| NaBr | 1014 | 0 | 12200 | 2·9 |
| NaI | 928 | 0 | 10100 | 2·8 |
| KCl (I) | 1043 | 0 | 6900 | 5·7 |
| KCl (II) | 1315 | 18950 | 12100 | 4 |
| RbCl (I) | 990·5 | 0 | 6600 | 6 |
| RbCl (II) | 1127 | 7800 | 7200 | 4 |

The lack of precision in these values may also possibly be due to an undetected phase transformation in this salt at high pressures (see Table 8.11).

At higher pressures the relatively greater compressibility of the melt, compared with the crystals, may lead to actual maxima in the pressure/melting curve in some ionic crystals (pp. 17, and see $KNO_2$ above). The melting of NaCN and NaCNS have been observed to 40 kbar (Pistorius and Boyens, 1968). Interesting minor contrasts are observed for these two anions.

### 8.9 Cryoscopy with molten salts

A large number of cryoscopic measurements have been made with molten salts, but unfortunately not always with the considerable refinements necessary to yield accurate derived thermodynamic data. To achieve one physicochemical objective, which is to determine whether salts of polyvalent ions are practically completely dissociated, the technique 'of the common ion' is fairly simple to apply. For example, in principle theoretical freezing point depressions differ quite largely according to whether (i) a sulphate such as $Li_2SO_4$ is completely dissociated when added to another melt to give $2Li^+$ and $(SO_4^{2-})$, or (ii) whether it forms ions of intermediate composition such as $(LiSO_4)^-$ or $(Li_3SO_4)^+$ to a predominant extent. Additions of lithium sulphate to a ·melt which contains an ion in common with (i) introduces a different number of 'foreign particles' into the melt than for case (ii). On this basis, many investigations have been made of the extent of dissociation of salts in dilute solution in other melts (e.g. Kordes, Bergmann, and Vogel, 1951). (These will not be detailed in this book.) In melts of NaCl alkali sulphates are completely dissociated according to this criterion (Riccardi and Benaglia, 1961).

When the thermodynamic objective of cryoscopic measurements with molten salts is to evaluate heats and entropies of fusion, one difficulty is that with many ionic melts freezing point depressions appear to depart from the ideal Raoult law (Chapter 12) for even quite small mole fraction of second component. One difficulty seldom allowed for is that premelting defects may lead to changes of $H_f$ near $T_f$ that are appreciable (see Oldham and Ubbelohde, 1940 refs Chapter 12, for a molecular parallel). Non-ideal behaviour may also be masked by formation of solid solutions, which may appear to reduce freezing point depressions if undetected. When a salt such as NaF is used, in which the small sizes of the ions and tight crystal packing tends to reduce solid solution to a minimum, on adding a range of Group IIA fluorides it is significant that freezing point depressions are found to be not independent of the cation radius even at mole fractions as low as 0·02. Such ionic melts clearly cannot be regarded as giving rise to ideal mixtures, until very low mole fractions indeed are attained (Cantor, 1961). With these mixtures the excess partial molal free energy of NaF is always negative (Fig. 8.3).

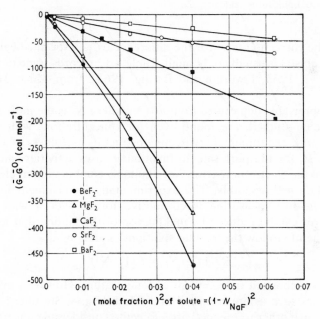

FIG. 8.3. Excess partial molal free energy of sodium fluoride melt with alkaline earth fluorides in dilute solution. Reprinted with permission from Cantor, *J. Phys. Chem.* **65**, 2208 (1961). Copyright by the American Chemical Society

The correct experimental procedure to derive heats of fusion from cryoscopy is to use the complete activity equation for the solvent

$$\ln a = \frac{H_f}{R}\left(\frac{1}{T_f} - \frac{1}{{}^0T_f}\right) + \frac{\Delta C_p}{R}\left(\ln \frac{{}^0T_f}{T_f} - \frac{\Delta T_f}{T_f}\right)$$

**Table 8.12.** *Specific heats* $(cal\ mole^{-1}\ deg^{-1})$
*crystals and melts*

| Salt[a] | $C_{PS}$ | $C_{PL}$ |
|---|---|---|
| LiF | 10·41 | 15·50 |
| NaF | 10·40 | 16·40 |
| NaCl | 10·98 | 16·00 |
| KF | 11·88 | 16·00 |
| KCl | 9·89 | 16·00 |
| AgCl | 14·88 | 16·00 |
| AgBr | 7·93 | 14·90 |
| LiNO$_3$ | 14·98 | 26·60 |
| NaNO$_3$ (I) | 35·70 | 37·00 |
| KNO$_3$ (I) | 28·80 | 29·50 |

[a] (I) is the polymorphic form stable near $T_f$.

in which the correction term for the difference $\Delta C_p$ in specific heats between liquid and crystal is usually only small. Table 8.12 records some of the results (see Blander, 1963; Dworkin and Bredig, 1963), from which $\Delta C_p$ can be estimated.

Since appreciable departures from $a \equiv N_1$ where $N_1$ is the mole fraction of the salt being studied, are found even for concentrations with $N_1 \geqslant 0·05$, difficulties with experimental cryoscopy obstruct its usefulness as a means for evaluating $H_f$ for the pure salt. In favourable cases activities can be determined from the electromotive force of a suitable combination of electrodes or concentration cell assembly. The experimentally inconvenient feature that this may involve measuring electromotive forces over a wide range of temperatures can be reduced by making use of the finding that ionic mixed melts frequently follow the 'regular solutions' equation

$$RT \ln(a/N_1) = B(1 - N_1)^2$$

where $B$ is a constant independent of temperature. However, even so corrected values such as are recorded in Table 8.13 (see Sternberg and Latta, 1960, who give other references) are subject to considerable uncertainty, and it is not clear how far this thermodynamic legerdemain really adds to the knowledge about molten salts.

**Table 8.13.** *Heats of fusion computed from cryoscopic observations* $(cal\ mole^{-1})$

| Salt | $H_f$ |
|---|---|
| KCl | 6239 |
| AgCl | 2868–3089 |
| AgBr | 2194 |
| PbCl$_2$ | 5815 |

Activity coefficients calculated from depressions of freezing point of melts of $NH_4NO_3$ by various added nitrates of univalent and divalent cations show a surprising range (Matsura and Nishikawa, 1960), and again point to the need for careful analysis. Cryoscopic evaluations of $S_f$ uncorrected for activities have been published for a considerable number of simple salts (Darmois and Petit, 1958; Petit and Bourlange, 1957; Petit and Delbove, 1962). The originals should be consulted for details.

Attempts to correlate the entropy of fusion of an ionic crystal with the number of ions required to define its molecular structure in crystal and in melt meet with only partial success (Bourlange, 1959). These are in part hampered by difficulties already indicated above in evaluating entropies of fusion from determinations of freezing point depressions. Major differences in this parameter for example, between sodium molybdate ($S_f = 1\cdot5$) and potassium molybdate ($S_f = 3\cdot2$) appear to be significant, but the smaller difference between this salt and caesium molybdate ($S_f = 3\cdot6$) is based on cryoscopy and may not be real.

A further problem which adds to the uncertainties of cryoscopic evaluations of $S_f$ for ionic crystals stems from the fact that on approaching $T_f$ the specific heat of ionic crystals frequently shows distinct trends increasing above the Dulong and Petit value (e.g. Blanc, 1960). Interpretation of such premelting increases of $C_p$ is discussed in Chapter 12. When they indicate precursor increases in crystal defect concentrations, preceding the jump on passing from crystal to melt (as appears to be the case in silver halides, for example), such premelting phenomena might be regarded as part of the total increase of entropy of fusion. Since such effects vary from crystal to crystal, they add a further source of irregularity which helps to upset similitude rules for the melting of crystals, even those containing ions of the inert gas type.

Molten sodium and molten potassium formates under an inert atmosphere are stated to give good linear cryoscopic plots with a number of halides and nitrates (Leonesi, Piantoni, Borchiesi, and Franzosini, 1968). These experiments indicate for sodium formate $T_f$ 530·7 K, $S_f = 7\cdot6$ e.u.; potassium formate $T_f$ 441·9 K, $S_f = 6\cdot4$ e.u.

## 8.10 Similitude rules for melting of ionic crystals

Despite the similitude of structures of ionic crystals, attempts to establish similitude rules for the transition to their melts have in fact met with only moderate success (e.g. Abraham, Dupuy, Gurion, and Brenet, 1962; and see p. 173). The extent to which a rule of corresponding states can be developed may be indicated by theoretical calculations for melts of inert gas ions (Reiss, Mayer and Katz, 1961). As for molecular crystals of the inert gases, in order to establish a law of corresponding states for molten salts with inert gas ions, it is necessary to represent the potential energy between a pair of ions in terms of a suitably selected single distance encountered in the melt. For this reason, complicating factors already referred to, such as the polarizability of ion atmospheres, perturb accurate application of the quantitative statistical theories involved. A number of corresponding state relationships have nevertheless been worked out in terms of the contact distance (i.e. of closest approach) $r_\pm$ between positive and negative ions as distance parameter. Precise calculations, for example from electron diffraction measurements on molecular beams of salt vapours, show $r_\pm$ to be variable in different states of the same substance, owing to variations in electrostatic compensation (see above). Neglecting such effects, interionic contact distances $r_\pm = r_+ + r_-$ in the crystals may perhaps be used as 'constants' in a tentative derivation of two corresponding state relationships each of which involves $r_\pm$ as recorded in Table 8.14. For univalent salts these relationships are (with $T_f$ in K, $r_\pm$ in a.u. and $\sigma_f$ in dyne cm$^{-1}$)

$$T_f r_\pm = 3 \times 10^{-5} \tag{8.1}$$

$$\sigma_f r_\pm^3 = 3 \cdot 3 \times 10^{-21} \tag{8.2}$$

where $\sigma_f$ is the surface tension $\sigma_{LG}$ of the melt at $T_f$. The fourth and sixth columns of Table 8.14 indicate the degree of similitude achieved by the degree of constancy of the numbers recorded.

For divalent salts parallel results are obtained after division by $Z^2$ where $Z$ is the charge on each ion.

**Table 8.14.** *Corresponding state rules for melting of univalent ionic crystals (for units see text)*

| Salt | $T_f$ | $r_\pm$ | $r_\pm T_f$ | $\sigma_f$ | $\sigma_f r_\pm^3$ |
|------|-------|---------|-------------|------------|--------------------|
| LiF  | 1121  | 2·01    | 1·25        | 251        | 204                |
| Cl   | 887   | 2·57    | 2·27        | 138        | 234                |
| NaF  | 1265  | 2·31    | 2·92        | 202        | 249                |
| Cl   | 1074  | 2·81    | 3·02        | 116        | 257                |
| Br   | 1023  | 2·98    | 3·04        | 99         | 262                |
| I    | 933   | 3·23    | 3·01        | 88         | 297                |
| KF   | 1129  | 2·67    | 3·02        | 142        | 270                |
| Cl   | 1045  | 3·14    | 3·28        | 99         | 306                |
| Br   | 1013  | 3·29    | 3·34        | 89         | 312                |
| I    | 958   | 3·53    | 3·39        | 79         | 348                |
| RbF  | 1048  | 2·82    | 2·96        | 131        | 294                |
| Cl   | 988   | 3·29    | 3·26        | 99         | 352                |
| Br   | 953   | 3·43    | 3·27        | 91         | 367                |
| I    | 913   | 3·66    | 3·34        | 83         | 407                |
| CsF  | 955   | 3·01    | 2·88        | 107        | 292                |
| Cl   | 918   | 3·47    | 3·18        | 90         | 376                |
| Br   | 909   | 3·62    | 3·29        | 85         | 403                |
| I    | 894   | 3·83    | 3·42        | 75         | 421                |

*Table 8.15. Similitude tests for melts of divalent ions*

|      | $T_f$ | $r_\pm$ | $r_\pm T_f / Z^2$ |
|------|-------|---------|-------------------|
| MgO  | 3073  | 2·10    | 1·61              |
| CaO  | 2873  | 2·40    | 1·73              |
| SrO  | 2733  | 2·54    | 1·74              |
| BaO  | 2198  | 2·75    | 1·51              |

The trends which can be seen in Tables 8.14 and 8.15 make it clear that the relationships (8.1) and (8.2) oversimplify the properties of the melts. Nevertheless they point to evident correlations between the interionic potentials and melting parameters. Significantly, these correlations become weaker when the radius ratio $r_+/r_-$ reaches low values. Lithium salts are obviously out of line.

Various empirical attempts to correlate the melting parameters of inert gas halides with other characteristics of the ions have also been examined (Schinke and Sauerwald, 1956). For example, Fig. 8.1 illustrates the correlation between radius ratio, and Fig. 8.2 between the optical polarizability of the ions, with the percentage volume change on melting (Furukawa 1961). Results of this kind support only a broad similitude of melting mechanism for all these halides, without yielding stringent numerical relationships (see also Gopal, 1953, 1955; also Chapter 3).

More refined parameters to be used in formulating similitude relationships have been proposed, without however achieving major advances. It has been pointed out that the anion–cation repulsion contact distances $(r_+) + (r_-)$ show a parallelism with values derived from compressibilities of the molten salts (Luke and Davis, 1967) although close parallelism is not apparent, possibly because a better comparison would be made at truly corresponding temperatures.

A corresponding state analysis based on reduced variables for the interaction potentials between the ions shows that two limiting laws are to be

*Table 8.16. Anion–cation contact distances in condensed phases*

| Halide | From crystal data (Pauling) | From compressibilities of melts at 800°C |
|--------|-----------------------------|------------------------------------------|
| LiCl   | 2·41                        | 2·26                                     |
| Na     | 2·76                        | 2·47                                     |
| K      | 3·14                        | 2·70                                     |
| Cs     | 3·50                        | 2·93                                     |
| LiBr   | 2·55                        | 2·43                                     |
| Na     | 2·90                        | 2·63                                     |
| K      | 3·28                        | 2·85                                     |
| Cs     | 3·64                        | 2·98                                     |
| NaI    | 3·11                        | 2·82                                     |
| K      | 3·49                        | 3·01                                     |

expected. When $(r_+)/(r_-) \sim 1$ the corresponding reduced length to be used is $(r_+) + (r_-)$ but when $(r_-) \gg (r_+)$ (e.g. for lithium salts), the proper length to be used is $(r_-)$ (Woodcock, 1976).

## 8.11 Simple ions forming strongly polarizable systems

A number of simple ions are known whose tendency to polarize is considerably greater than for those with inert gas electron shells. Halides of silver have been much studied. For many other inorganic halides current information is scanty. One criterion used to verify that ionic melts are formed at all from crystals containing such atoms is the electrical conductance. Table 8.17 gives comparisons for halides. The figures quoted should be regarded as indicating orders of magnitude and their accuracy is variable.

**Table 8.17.** *Specific electrical conductance (ohm cm$^{-1}$) of molten halides near $T_t$ as a guide to ionic character (temperatures in °C in parentheses)*

| | | | | |
|---|---|---|---|---|
| HCl $2 \times 10^{-7}$ | | | | |
| NaCl (850) 3·5 | BeCl$_2$ (451) 0·003 | BCl$_3$ v. low | CCl$_4$ v. low | PCl$_3$ v. low |
| KCl (768) 2·2 | MgCl$_2$ (718) high | AlCl$_3$ (200) $5 \cdot 6 \times 10^{-7}$ | SiCl$_4$ v. low | AsCl$_3$ (−16) $10^{-6}$ |
| CuCl (240) 0·2 | CaCl$_2$ (850) 2·1 | ScCl$_3$ (959) 0·5 | TiCl$_4$ v. low | SbCl$_3$ (73) $8 \cdot 5 \times 10^{-7}$ |
| RbCl high | ZnCl$_2$ (370) 0·01 | GaCl$_2$, GaCl$_3$ high | GeCl$_4$ v. low | BiCl$_3$ (266) 0·4 |
| AgCl (456) 3·8 | SrCl$_2$ (900) 2 | — | SnCl$_4$ v. low | — |
| CsCl high | CdCl$_2$ (570) 0·15 | InCl$_3$ high | | |
| | BaCl$_2$ (1000) 2 | LaCl$_3$ (950) 1·3 | PbCl$_2$ (508) 1·5 | |
| | HgCl$_2$ (276) $2 \times 10^{-4}$ | TlCl (431) 1·1 | PbCl$_4$ $8 \times 10^{-7}$ | |

CrCl$_2$, PdCl$_2$, FeCl$_2$ and MnCl$_2$ give melts that are good conductors

This well-known classification of halides into ionic and molecular compounds based on the specific conductances of their melts, involves anomalies in the case of certain crystals with transitional properties (see Biltz, 1924; Biltz and Klemm, 1926).

As one example, the melting point and molar volume of aluminium chloride are abnormally high in comparison with the bromide and iodide. As already discussed (p. 175), an unusually large change in volume of aluminium chloride on melting (see Table 8.1) suggests changes of bonding on passing from crystal to melt. It is further noteworthy that although the conductivity of crystalline aluminium chloride rises steeply on approaching $T_f$ it drops to practically zero on melting (Biltz, 1922; Biltz and Voigt, 1923). The crystal structure of aluminium chloride can be interpreted on an ionic basis (Ketelaar, MacGillavry, and Renes, 1947) but X-ray studies on the melt, as well as other evidence, indicate that in the melt this compound consists predominantly of molecules of $Al_2Cl_6$ (Harris, Wood, and Ritter, 1951; Boston, 1966) (see Table 8.17).

For ions with inert gas shells, to a rough approximation the melts can be regarded as completely dissociated: but for ions originating from other parts of the Periodic System, marked polarization increases ion pair formation and may even involve predominantly covalent bonding in the halides. For wholly covalent or wholly ionic crystals, mechanisms of melting can be described in terms of fairly satisfactory models as discussed above. However an interesting if complex sequence of changes arises when there is a change of valency bias, which may even reach a critical transition owing to the change in nearest neighbour environment on melting. For the silver salts the markedly smaller values of $S_f$ and $H_f$ compared with alkali halides (see Table 8.1) can in part be attributed to stronger polarizations of anions by cations, leading to shrinkage of the contact distance between the ions with an even more marked contraction on melting. For these salts the premelting part of $S_f$ and $H_f$ is likewise much greater than for alkali halides (Chapter 12).

A general pointer to covalent bond formation between at least a fraction of the atoms in melts of uni–univalent halides may perhaps be taken from data calculated by means of the Batschinski 'free volume' equation for the viscosity of the melt

$$\eta = C/(V_L - \Omega)$$

where $V_L$ is the molar volume at any temperature and $C$ is a constant. At $T_f$, for inert gas halides $\Omega > V_L$ (about $1 \cdot 05\ V_L$) whereas for IB halides $V_L > \Omega$ suggesting considerable association (Murgulescu, 1969c).

## 8.12 Formation of ionic melts by autocomplexing—simple structures

This book is not directly concerned with many questions of condensed state valence chemistry. It must be stressed, however, that whether any halide is predominantly ionic or covalent will depend on the nearest neighbour environment in critical cases. Changes of ionic character on melting can involve particularly sensitive instances of the role of the nearest neighbour environment in deciding (say) whether a melt is predominantly ionic. A number of apparent changes of character on melting have been reported in solids whose crystal chemistry suggests they have an intermediate structure, so that neighbours prove to be of controlling importance. As one prominent example, various lines of evidence (e.g. Janz and McIntyre, 1962) indicate that molecules of IIB halides are predominantly covalent. Isolated molecules of mercury halides (Janz and McIntyre, 1962) such as $HgCl_2$, $HgBr_2$ and $HgI_2$ are linear. In the molten mercuric halides, Raman spectroscopy (Janz and James, 1963) confirms that predominantly the linear species X—Hg—Y etc. are present, where X and Y are halogen atoms. From the breadth of the Raman band (e.g. $\sim 100 \, cm^{-1}$ for the stretching mode with the peak at $\sim 313 \, cm^{-1}$) for the melt, pronounced molecular interactions are however inferred. In part, these may be attributed to the heteropolar character of the X—Hg bond, but in addition cooperative molecular rotational interlocking, such as may be expected in molecular clusters (Chapter 13) seems likely in the melt. Melting points of the crystals have low values, characteristic of crystal lattices of covalent molecules, to which positional disordering can be imparted for only small increases of energy. Boiling points of these melts are likewise low. Ratios of activation energies for viscous flow $H_\eta^{\neq}$ to heats of vaporization $H_{vap}$ are characteristic of covalent melts (see Table 8.18).

**Table 8.18.** *Thermodynamic properties of mercuric halides*

| Molecule | $T_f (°C)$ | $T_b (°C)$ | $H_{vap}/H_\eta^{\neq}$ |
|---|---|---|---|
| $HgCl_2$ | 277 | 304 | 3·76 |
| $HgBr_2$ | 238 | 319 | 3·60 |
| $HgI_2$ | 259 | 354 | 2·84 |

It is noteworthy that a marked increase of electrical conductance is observed on melting, with significant premelting effects (Chapter 12) for about 20°C below $T_f$, as is illustrated in Fig. 8.4. Interesting colour changes also are observed on heating the melts. $HgCl_2$ remains colourless, $HgBr_2$ melts to a pale-yellow liquid which, if heated in a sealed cell to 385°C, becomes orange. $HgI_2$ changes from an orange coloured solid to a clear red liquid at $T_f$ and becomes nearly black as the normal boiling point is approached.

Although these melts are predominantly covalent, it seems likely that they contain a small proportion of conducting species, probably formed by autocomplexing reactions:

$$2HgX_2 \rightleftharpoons HgX^+ + HgX_3^- + \Delta H_c \dots$$

Estimates of the fractional ionization of these melts at $T_f$ and at $T_b$ are given in Table 8.19.

The heat $\Delta H_x$ of the autocomplexing reaction is added to the normal activation energy $E_\lambda$ for ion migration in a liquid to yield the temperature coefficient of conductance as recorded in Table 8.20.

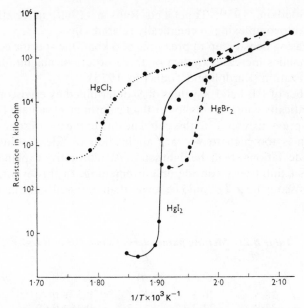

FIG. 8.4. Change of electrical resistance on melting of
mercuric halides. Reprinted by permission of the pub-
lisher, The Electrochemical Society, Inc., from Janz and
McIntyre, *J. Electrochem. Soc.* **109**, 842 (1962)

**Table 8.19. Fractional ionization in molten
mercuric halides**

| Salt | $HgCl_2$ | $HgBr_2$ | $HgI_2$ |
|---|---|---|---|
| at $T_f$ | $3.5 \times 10^{-5}$ | $2.3 \times 10^{-4}$ | $1.8 \times 10^{-2}$ |
| at $T_b$ | $4.0 \times 10^{-5}$ | $3.3 \times 10^{-4}$ | $2.0 \times 10^{-2}$ |

**Table 8.20. Activation energies for ionic conductance in
mercuric halides (kcal mole$^{-1}$)**

| Salt | $HgCl_2$ | $HgBr_2$ | $HgI_2$ |
|---|---|---|---|
| E solid | 13·0 | 22·5 | 25·7 |
| E liq. | 6·15 | 6·2 | −3·00 |

An Arrhenius type of temperature-dependence has been assumed of the
form

$$\lambda = \lambda_0 \exp\left(-E_\lambda / RT\right)$$

The negative value for the iodide is unusual. Calorimetric determinations of
the thermodynamic parameters for some of Group IIB halides are significant

(Janz and Goodkin, 1959; Topel and Ranson, 1960) in relation to the tendency to autocomplexing in chemically related salts.

With increasing density, up to pressures of 6 kbar, the specific conductance of mercury halides increases by two or three orders of magnitude owing to increased ionization (Bardoli and Todheide, 1975).

For a number of IIB halides, values of $S_f$ determined by cryoscopy are seen to be systematically considerably lower than calorimetric values (Table 8.21). As discussed previously (p. 186) this could be due to the use of solutions whose concentration is too high to warrant application of 'ideal' solubility laws, provided the deviations are in the right sense. An alternative explanation is that these crystals exhibit very considerable premelting. In this event, $H_f$ determined by cryoscopy near $T_f$ would be lower than 'integral' calorimetric values (Chapter 12).

**Table 8.21. Melting parameters of some IIB halides[a]**

| Salt | $C_{p_s}$ | $C_{p_L}$ | $T_f$ | $S_f$ | $S_f$ cryoscopy |
|---|---|---|---|---|---|
| CdCl$_2$ | 28·5 | 26·3 | 842·1 | 8·57±0·20 | 5·3 |
| CdBr$_2$ | 22·8 | 24·3 | 841·2 | 9·48±0·09 | 5·0 |
| CdI$_2$ | 21·5 | 24·4 | 661·2 | 7·49±0·20 | 3·7 |
| HgCl$_2$ | 19·2 | 27·2 | 552·7 | 8·40±0·09 | 4·2 |
| HgBr$_2$ | — | — | 511·2 | 8·4 | — |
| BiBr$_3$ | 26·0 | 37·7 | 492·2 | 10·55±0·40 | 4·0 |

$T_f$ in K, $C_p$ and $S_f$ in cal mole$^{-1}$ deg$^{-1}$.

[a] Values were determined by drop calorimetry; uncertainties in $\Delta C_p$ exceed experimental error, except for the last two salts quoted.

For solutions of metallic bismuth in bismuth trichloride at $T_f$, observed freezing point depressions and other evidence point to another kind of reaction between the solute, and the molten salt which acts as solvent (Mayer, Yosim, and Topol, 1960)

$$4Bi + 2BiCl_3 \rightleftharpoons 3Bi_2Cl_2$$

This again brings out the fact that, for molten salts, cryoscopic 'heats of fusion' though of great interest as guides to structure must not be taken as giving direct melting parameters without cautious consideration.

### 8.12.1. Complexing in mixed melt environments

The present book is concerned primarily with mechanisms of melting, and does not extend to the properties of mixed melts except where these throw light on the melting process. For example, various IIB halides form complexes when dissolved in fused nitrate melts (Duke and Garfinkel, 1961) as does silver chloride (Osteryoung, Kaplan, and Hill, 1961). Such behaviour makes autocomplexing probable on melting the pure salts themselves. Owing to the very different environment in the pure melt, the behaviour of these halides in such mixtures might however be quite different. A better clue to the kind of ion formed by autocomplexing may sometimes be derived from the behaviour in strong aqueous solutions, though it must be remembered that the large excess of polar $H_2O$ molecules always still present profoundly modifies the attractive forces involved. For example, precipitated AgI redissolves in strong aqueous solutions of iodides to form the complex ions such as $AgI_3{}^{2-}$ or $AgI_4{}^{3-}$. However in a melt of mixed nitrates of $KNO_3$ and $NaNO_3$ (Golub and Grigorenko, 1961) potentiometric measurements indicate that AgI redissolves to form large clusters as micellar complex ions $[Ag_nI_{n+1}]^-$ instead. These pointers are valuable clues to the behaviour of melts of the pure salts.

### 8.12.2. Autocomplexing leading to ion formation in other melts

Preceding sections of this chapter have dealt with ions of simple structure It may be added that somewhat analogous autocomplexing may explain the electrical conductances of melts of divalent metal carboxylates, laurates, decanoates, stearates, and oleates of Mg, Pb, Zn and Cu (Fogg and Pink, 1959). These conductances are smaller than for most ionic melts by a factor of $10^6$, but are much larger than for most non-ionic liquids. Whilst these molten compounds can hardly be regarded as 'ionic melts' in view of the relatively minute concentrations of ionic species present, the formations of ions in such small concentration may perhaps depend on prior aggregation of the molecules into micelles, e.g.

$$MA_2 \rightleftharpoons MA^+ + A^-$$

$$MA^+ + nMA_2 \rightleftharpoons (M_{n+1}A_{2n+1})^+$$

Temperature coefficients of conductance follow simple Arrhenius expressions with activation energies of the order $10$ kcal g ion$^{-1}$. Dielectric losses are frequency-dependent. The presence of the single ions $A^-$, $M^+$ or $MA^+$ has not been definitely established and another possible alternative is that conductance involves a bond-breaking/bond-making between neighbouring micelles, permitting a hopping process similar to the migration of protons in ice of Grotthus conduction.

Ion cluster formation in molten CuCl is referred to below.

### 8.13 Extended interactions in ionic melts

Models of ionic melts consisting of assemblies of single, well-defined positive and negative ions, as discussed above, are comparatively simple in their valence aspects. An increasing number of crystal compounds is becoming known however in which the melting mechanisms lead to more complex melts with extended interactions near $T_f$; these dissociate more or less rapidly on raising the temperature of the melt above $T_f$, tending towards an unassociated melt at sufficiently high temperatures. As extreme examples, *network melts* are formed at $T_f$ by compounds such as $BeF_2$; their behaviour resembles that of other network liquids discussed in Chapter 10. As links in the network dissociate either thermally, or as a result of the addition of ionic impurities, properties of the melts approximate towards those of simpler ionic melts. Model analogies between molten $BeF_2$ and, say, molten silicates are thus often made (e.g. Baes, 1970).

For compounds whose melts probably contain less extended but still noteworthy network formations, the physical evidence is as yet rather more fragmentary.

### 8.13.1. *Melts of zinc compounds*

Optical evidence suggests marked persistence of tetrahedral or possibly octahedral arrangements of anions in melts of $ZnCl_2$ and $Zn(CN)_2$ (Angell and Wong, 1970). The electrical conductance of molten $ZnCl_2$ when plotted as a function of temperature, appears curved above $T_f = 317°C$, and the nominal activation energy decreases from $27$ kcal mole$^{-1}$ at 320°C to 13 kcal mole$^{-1}$ at 600°C, at which temperature the frequency-dependence of electrical conductivity is no longer marked. The specific conductivity (in reciprocal ohms) is $0·0013$ at 317·8°C and $0·1995$ at 578·2°C. All these facts indicate a fairly rapid break-up of ligand structure with rise of temperature; normal ionic melt models are appropriate but only well above $T_f$ (Bloom and Weeks, 1969). The viscosity of molten $ZnCl_2$ is also much higher than that of simple ionic melts (Bockris, Richards, and Naris. 1965; Grantham and Yosim 1968). Addition of $PbCl_2$ breaks up the network structure (Umetsu, Ishi, Sawada, and Ejima, 1973).

Network or polymer chain association probably accounts for the structure of melts of $SnBr_2$ (Brockner, 1976) and $PbCl_2$ as shown from comparisons of Raman spectra of crystal, melt and vapour.

Another example of autocomplexing is presented by $ZnCl_2$. In the crystal, the zinc ions are packed in holes in a layer-like structure, formed by networks of chloride ions. On melting, clusters or fragments of the original network (see Chapter 10) appear to persist in a prefreezing region of the temperatures. In addition autocomplexing leads to the progressive formation of the ions $Zn^{2+}$ and $[ZnCl_6]^{4-}$ (MacKenzie and Murphy, 1960; Janz, 1962).

Addition of NaCl increases the dissociation of molten $CdCl_2$, most probably by complexing (Koster and Ketelaar, 1969).

The existence of curved plots of electrical conductance as a function of temperature is observed in a number of chemically related halide melts, including $BiX_3$, $ZnI_2$, $InI_3$, $HgX_2$, $SnCl_2$, CuCl (Grantham and Yosim, 1968; Grantham, 1965, 1966, 1968; Boston, Yosim, and Grantham, 1969). It is noteworthy that much of the data can be plausibly recorded on a single 'reduced plot' although no detailed statistical theory of thermal dissociation of networks appears to have been put forward (see Chapter 10).

Increase of pressure appears to inhibit association above about 3 kbar for the halides $BiCl_3$, $BiBr_3$ and $BiI_3$ (Darnell, McCallum, and Yosin, 1969), presumably because the covalent ligand form of these halides occupies a larger volume than the ionic form which is in competition with it. Further evidence for the marked network association of these halides is provided by the data for their electrical conductances at $T_f$. At 1 kbar, the jump in electrical conductivity from $\sigma_S$ to $\sigma_L$ of the bismuth halides is quite small (see Table 8.22) but the conductance itself is only small. $PVT$ data for molten $BiCl_3$ have been measured up to 1200°C and 3·5 kbar (Treiber and Todheide, 1973). The $PVT$ behaviour resembles that of molecular liquids but at high densities the molten chloride appears to be mainly ionic.

**Table 8.22. Conductivity jump on melting bismuth halides**

| | $\sigma_L/\sigma_S$ |
|---|---|
| Bismuth chloride | 1·5 |
| bromide | 1·33 |
| iodide | 3·0 |

Network association in the melts is also shown by their strong tendency to supercool without crystallization, below $T_f$ (see Chapter 16; see also Darnell and McCollum, 1968).

### 8.13.2. Grotthus conduction in network ionic melts

Representative analysis of the data has been proposed in some detail for molten thallium chloride, based on the assumption that electrical conduction of the crystals immediately below $T_f$, owing to premelting (Chapter 12), is wholly due to movement of isolated ions, whereas in the melt above $T_f$ there is superposed a 'Grotthus conduction' through cooperative movements of ions in the network (Spedding, 1973). Since both conductance and mass diffusion data are known for this molten salt, it has been possible to show that immediately above $T_f$ the Grotthus 'network' mechanism predominates. As the melt expands with rising temperature, its covalent (molecular) character becomes more prominent right up to the critical temperature for the liquid. Figure 8.5 illustrates the contribution of the different mechanisms to the electrical conductivity up to $T_{crit}$. In other respects, the polymorphism of

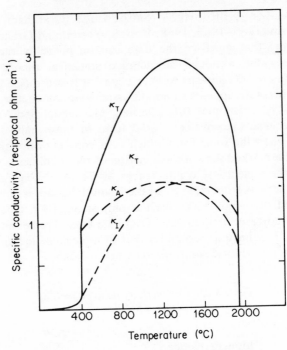

FIG. 8.5. Conductivity of TlCl melt, $\kappa_T$, shown as the sum of the conductivity of individual ions, $\kappa_1$, and the conductivity arising from the Grotthus mechanism of ionic agglomerates, $\kappa_A$. Reproduced with permission from Spedding, *Electrochim. Acta*, **18,** 112 (1973)

crystalline thallium halides (Pistorius and Clark, 1968) resembles that of alkali halides. Presumably the greater freedom of association in the melts is critical. .

## 8.14 Melting of crystals containing polyatomic ions

Crystal chemistry deals with a considerable variety of salts which appear to contain polyatomic ions, when the structures are examined by X-ray diffraction or other methods that do not depend on the chemical stability of such ions. In familiar instances, such as $NH_4^+$ or $NO_3^-$ or $SO_4^{2-}$ chemical bonding between the atoms in the ions is in fact so strong and the distances between the atoms are so short that these charged molecules persist with only minute dissociation on melting, on dissolution in solvents, or on volatilization. This makes it convenient to discuss their behaviour as entities even in disordered states of matter and even at low concentrations. X-ray and neutron diffraction studies (Zarzycki, 1961; Furukawa, 1961) confirm the persistence of ions such as $CO_3^{2-}$ and $SO_4^{2-}$ without serious deformation in the melts, thus supporting other indirect evidence. On the other hand, kinetic data suggest that ions such as $CO_3^{2-}$ both associate and dissociate to a significant extent in carbonate melts (Janz and Lorenz, 1961). The extent to which radial distribution curves do, or even can, give more detailed information about packing in such melts is open to question. Other polyatomic ions such as $(AlCl_6)^{3-}$ and even $(AlF_6)^{3-}$ are crystallographically significant, but dissociate quite readily into simpler species when the crystal symmetry is upset and electrostatic compensation is perturbed, as discussed earlier. No hard and fast distinction can be drawn with regard to the ease of dissociation of polyatomic ions. For practical reasons, a broad differentiation may be made between those ions which do not dissociate at the melting point, to an extent sufficient to modify the entropy of fusion appreciably by introducing new species into the melt, and polyatomic ions whose existence is favoured by crystal symmetry but which dissociate more or less completely on melting. In the discussion that follows 'stable' polyatomic ions will first be dealt with. It may be noted that the stability of a polyatomic anion such as $MO_n$ or $MX_n$, where M is the central atom, and X is a halogen, may be different in its own pure melt, and when the ion is in dilute solution in a foreign salt as solvent. Systematic studies using cryoscopic techniques frequently encounter the problem as to what extent such dissociation occurs in dilute solution in another molten salt, and how far different molten salts exert different 'ionizing tendencies'. For example, molten NaF appears to exert only moderate 'ionizing' tendencies (Seyyedi and Petit, 1959). In this melt the dissociations

$$(B_4O_7)^{3-} \rightleftharpoons 3BO_2^- + BO^+$$

$$P_2O_7^{4-} \rightleftharpoons (PO_4)^{3-} + (PO_3)^-$$

appear to be practically complete at large dilutions, but the dissociation

$$CrO_4^{2-} \rightleftharpoons O^{2-} + CrO_3$$

appears to be only incipient.

### 8.14.1. 'Stable' polyatomic ions

As already indicated in relation to orientational randomization (Chapter 6), polyatomic ions all have repulsion envelopes that are not spherical. Estimates of repulsion envelopes of polyatomic ions are unavoidably subject to a degree of arbitrariness. As one example, assuming $NO_3^-$ is completely randomized with respect to its triad axis in melts, this leads to a disc-like repulsion envelope (Janz and James, 1962). On this basis, comparisons can be made of the fraction of volume of a melt not actually filled by repulsion envelopes of

**Table 8.23. Percentage volumes not occupied by repulsion envelopes in melts at (1.10) $T_f$ and crystals at 298 K**

|  | Chloride | | Nitrate | |
|---|---|---|---|---|
|  | Crystal | Melt | Crystal | Melt |
| Li | 23·9 | 46·7 | 45·8 | 60·0 |
| Na | 36·5 | 56·0 | 55·9 | 61·8 |
| K | 44·0 | 58·8 | 55·6 | 62·6 |
| Rb | 46·7 | 58·3 | 51·0 | 60·5 |
| Cs | 35·6 | 56·5 | 48·5 | 61·9 |

the ions. Data in Table 8.23 give estimates of unoccupied volumes at (1.10) $T_f$ (this temperature is suggested as a kind of substitute for theoretically better based corresponding temperatures which are not however known from experiments). For the crystals, the data refer to 298 K. When the re-entrant parts of the repulsion envelope are not too marked, and intermeshing of the neighbouring ions is not too pronounced, crystal lattice packing may involve repulsion potential barriers for which $V_{rot} \leqslant kT_f$, permitting one or more transitions involving orientational randomization in the crystal at transition temperatures $T_{c_1}, T_{c_2}, T_{c_3} \ldots$ etc. below $T_f$ (some examples for ionic crystals were recorded in Chapter 4).

However, if the barriers opposing randomization of orientation are too high, the onset of orientational randomization may have to wait for the marked lowering of all barriers that accompanies positional randomization on melting. Exactly as for molecular crystals, in such cases

$$S_f = \Delta S_{pos} + \Delta S_{or}$$

Experimental data for polyatomic ions are still somewhat scanty. The entropy of fusion of molten nitrates is well below the 25 e.u. required for 'free' rotation of the nitrate ion (Janz and James, 1962). For NaNO₃ $S_f = 6·03$ at $T_f = 306·2°C$ and $S_c = 1·26$ for a lambda transformation whose peak lies at $T_c = 276°C$ (Mustajoki, 1957). The possibility that complete randomization of orientation is not attained even until well *above* $T_f$ must clearly be considered, as in the case of melts of molecular crystals. However, there appears to be no very direct evidence; indirect evidence from 'cluster formation' is discussed below.

For nitrates with divalent cations, such as Ca, Sr, Ba, direct determination of entropies of fusion is obstructed by thermal decomposition which occurs as $T_f$ is approached. Indirect values calculated for these three cations from heats of solution show an interesting difference recorded in Table 8.24 above.

All these salts have isomorphous fluorite structures, and show no evidence of solid-state transformations below $T_f$. It seems plausible to suppose that in the melts of $Sr(NO_3)_2$ and $Ba(NO_3)_2$ orientational randomization occurs freely, but that in melts of $Ca(NO_3)_2$, with the higher field strength at the

**Table 8.24.** *Melting parameters of nitrates* ($cm^3$ $mole^{-1}$)

| Cation | $T_f$ $(K)$ | $S_f$ $(e.u.)$ | $V_s$ | $V_L$ | $\Delta V_f / V_s$ $(\%)$ | Ref. |
|--------|-------------|----------------|-------|-------|---------------------------|------|
| Li | 524–8 | 11·7 | 31·89 | 38·73 | 21·4 | 1 |
| Na | 580–91 | 6·1 | 40·23 | 44·55 | 10·7 | 1 |
| K | 606–7 | 4·6 | 52·25 | 53·98 | 3·32 | 1 |
| Rb | 583 | 2·2 | 59·47 | 59·33 | − 0·23 | 1, 4 |
| Cs | 687 | 4·7–5 | 61·73 | 69·21 | 12·1 | 1, 4 |
| Ca | (834) | 6·8 | — | — | (10) | 2, 3 |
| Sr | (918) | 11·6 | — | — | (9) | 2, 3 |
| Ba | (865) | 11·5 | — | — | (7) | 2, 3 |

References
1. Schinke and Sauerwald (1960).
2. Kleppa and Hersh (1961).
3. Kleppa (1962).
4. Janz, Solomons, and Gardner (1958).
The volume shrinkage on melting $RbNO_3$ has been redetermined by Cleaver and Williams (1968), who find $V_L - V_S = -0·4$ ml $mole^{-1}$, i.e., rather greater than the value due to Schinke and Sauerwald (1960).

surface of $Ca^{2+}$, 'rotation' of the $(NO_3)^-$ is considerably more restricted even well above $T_f$. This factor may also help to explain reports of glass formation in melts containing $Ca(NO_3)_2$ (Chapter 16). Determination of the molar volume of molten nitrates confirm that 'room to rotate freely' is inadequate in some cases.

### 8.14.2. *Melting of sulphates*

Melting parameters of alkali sulphates (Petit and Bourlange, 1957) have been obtained by cryoscopy. Measurements have suggested that lithium sulphate dissociates without giving rise to either simple ion, e.g. according to the equation

$$2Li_2SO_4 \rightleftharpoons Li_3SO_4^+ + LiSO_4^-$$

(Kordes, Ziegler, and Proeger, 1954). However, this inference appears to have been based on a value of $S_f$ that may be too low. Although the correct value of $S_f$ is still uncertain, freezing point depressions of the salts LiCl, LiF, $BaSO_4$, $PbSO_4$, RbCl, KCl and KBr all agree with the view that the first four salts introduce one dissimilar ion and the last three two dissimilar ions, thus pointing to high concentrations of both $Li^+$ and $SO_4^{2-}$ in molten $Li_2SO_4$ (Riccardi, 1961).

### 8.14.3. *Melting of carbonates*

Direct determination of entropies of fusion and specific heats of alkali carbonates, by drop calorimetry, again have thrown considerable doubt on

**Table 8.25. Comparisons of $S_f$ by cryoscopy and calorimetry for carbonates (e.u.)**

| Salts | $T_f$ (°C) | $S_f$ (e.u.) cryoscopy | $S_f$ (e.u.) calorimetry |
|---|---|---|---|
| $Li_2CO_3$ | 726 | $10\cdot7\pm0\cdot1$ | $10\cdot2$–$11\cdot4$ |
| $Na_2CO_3$ | 858 | $5\cdot9\pm0\cdot2$ | $10\cdot3$ |
| $K_2CO_3$ | 898 | $5\cdot6\pm0\cdot1$ | $8\cdot4$–$8\cdot7$ |

the precision of values of $S_f$ determined from cryoscopy (Janz, Neuenschwander, and Kelly, 1963) (see Table 8.25).

The origin of this discrepancy is not clear. Possibly ion association takes place even in dilute solution in the melts. Some guide to the origin of the trend of (calorimetric) entropies of fusion in alkali carbonates is obtainable (see Table 8.26) from the difference in heat capacities of solid and liquid expressed in the form

$$C_{PS} = a + 2bT$$

$$C_{PL} - C_{PS} = \Delta C + b'T$$

**Table 8.26. Change in heat capacity on melting alkali carbonates**

| Salt | Solid a | b | Change on melting $\Delta C$ | b' |
|---|---|---|---|---|
| $Li_2CO_3$ | $3\cdot82$ | $0\cdot021$ | $27\cdot03$ | $0\cdot029$ |
| $Na_2CO_3$ | $8\cdot14$ | $0\cdot018$ | $25\cdot85$ | $0\cdot025$ |
| $K_2CO_3$ | $19\cdot19$ | $0\cdot013$ | $17\cdot76$ | $0\cdot015$ |

The large increase in heat capacity on melting all these crystals points to greater increases of disorder on melting than could be accounted for merely by positional randomization with respect to the ideal crystal lattice. Vibrational contributions, and other factors such as the gradual dissociation of

**Table 8.27. Calorimetric entropies of fusion for salts with polyatomic anions (e.u.)**

| Anion | Li | Na | K | Ref. |
|---|---|---|---|---|
| $CO_3$ | $10\cdot7$ | $5\cdot9$ | $5\cdot6$ | 1 |
| $SiO_3$ | — | $9\cdot2$ | — | 2 |
| $TiO_3$ | $14\cdot5$ | $12\cdot9$ | — | 2 |
| $NO_3$ | $11\cdot7$ | $6\cdot0$–$6\cdot2$ | $4\cdot6$–$3\cdot8$ | 2, 3 |
| $ClO_3$ | — | $10\cdot2$ | | 2, 3 |

References
1. Janz, Neuenschwander, and Kelly (1963).
2. Kelley (1960).
3. Sokolov and Shmidt (1955, 1956).

association complexes (see the section following) as the temperature of the melt is raised above $T_f$ may help to explain the large heat capacity of the melts, compared with the crystals. A survey including several polyatomic anions suggests that the smallest cation $Li^+$ may tend to favour the largest entropies of fusion (see Table 8.27).

### 8.15 Formation of association complexes on melting

As stated above (p. 176), positional randomization leads to some reduction of the compensation of primary electrostatic forces acting between nearest neighbours in an ionic crystal, owing to the loss of symmetry on melting. For ionic species the shrinkage in repulsion contact distance is very considerably larger than any corresponding effect for molecular crystals. This is one of the factors that upset regularity of melting behaviour; various attempts have been made to establish these factors on the basis of crystal contact radii (Fujiwara, 1952; Kutzelnigg, 1958). Loss of compensation also affects dimensionless ratios such as $H_f/H_{vap}$ which might otherwise show closer similitude.

One striking outcome of the reduced compensation of electrostatic forces on melting is that types of nearest neighbour interactions whose extension in space cannot be arranged on a three-dimensional crystal lattice, and which must therefore be squeezed out when a melt freezes, may become quite prominent in the melt. In particular, *anticrystalline* association complexes may be formed which permit closer approach between anion and cation than in the crystal lattice.

*Table 8.28. Melting points in relation to symmetry of polyatomic anions (°C)*

| Cation anion | Li | Na | K | Rb | Cs | Ag |
|---|---|---|---|---|---|---|
| Cl | 614 | 800 | 770 | 715 | 646 | 455 |
| I | 446 | 651 | 723 | 642 | 621 | 552d |
| $ClO_4$ | 236 | 482 | 610 | — | — | 486d |
| $ClO_3$ | 124 | 250 | 386 | — | — | 230 |
| $NO_3$ | 251 | 318 | 333 | 310 | 414 | 212 |
| $NO_2$ | <100 | 271 | 387 | — | — | 140d |
| $SO_4$ | 860 | 885 | 1076 | 1060 | 1010 | 652 |
| $SO_3$ | (455) | — | — | — | — | — |

A finding which appears to be explained by assuming the formation of association complexes is that freezing points tend to be lower for the less symmetrical polyatomic anions. This hypothesis may be illustrated (see Table 8.28) for example for salts of Group I cations (see Davis, Rogers, and Ubbelohde, 1953).

8.15.1. *Association by closer packing into anticrystalline clusters*

It is not always appreciated how much the requirements of crystal symmetry impose restrictions on the local economy of packing of non-spherical polyatomic ions. In the melt when these requirements are waived the lowering of potential energy achieved by bringing a cation into closest packing with one or more non-spherical anions, for example in a tetrahedral arrangement of oxygen atoms around each cation in molten sodium nitrate

$$
\left[ O-N \begin{matrix} O & O \\ & \\ Na & N-O \\ & \\ O & O \end{matrix} \right]^{-}
$$

may well be considerable in relation to the symmetrical arrangement of ions in crystalline sodium nitrate. Near the melting point each cation is surrounded by 6 anions at equivalent distances in this crystal (see Wells, 1962). It is important to note the geometrical fact that although this and various other association complexes can be devised, whose packing energy may be quite considerably lower than that of the crystal, such more tightly packed complexes or 'clusters' *cannot* be extended indefinitely, so as to occupy space completely, by mere successive additions of more anions and cations. Only a crystal lattice arrangement has packing parameters which remain constant as it extends. Association complexes of the type suggested belong to the class of *anticrystalline* close-packed structures already referred to on p. 151. This feature of melting opens new possibilities for the structure of ionic melts. No complete exploration of such cases has yet been made either theoretically or experimentally. In the light of present knowledge, important aspects of melting mechanisms which lead to the formation of association complexes include the following:

(i) The entropy of fusion for such salts can be regarded as the result of imposing positional disordering and orientational disordering on the crystal, followed by rearrangement of the ions into association complexes. So far as these can be discussed independently, at least to a first approximation, one may write

$$ S_f = \Delta S_{pos} + \Delta S_{or} + \Delta S_{assocn} $$

In this summation $\Delta S_{assocn}$ includes any changes of vibrational entropy on forming local close packed complexes, together with their entropies of mixing. Heat changes relative to the crystal

$$ H_f = \Delta H_{pos} + \Delta H_{or} + \Delta H_{assocn} $$

are positive for the first two terms in $H_f$ since they involve doing work against the interionic forces. Heat changes in the third term $\Delta H_{assocn}$ may be negative since local close-packed clusters of lower potential energy than the crystal are produced. The melting temperature

$$T_f = \frac{H_f}{S_f} = \frac{\Delta H_{pos} + \Delta H_{or} + \Delta H_{assocn}}{\Delta S_{pos} + \Delta S_{or} + \Delta S_{assocn}}$$

tends to be lowered by this third mechanism of melting, which permits small or negative addition to the numerator with appreciable positive addition to the denominator.

This depression of the freezing point by this kind of blurred selfcomplexing, made possible when the constraints imposed by the crystal lattice are removed, can have other consequences for the thermodynamic and transport parameters of these melts.

(ii) The change in volume on melting associated with $\Delta S_{pos}$ and $\Delta S_{or}$ is usually fairly large. In close-packed crystals, the necessary randomization for disorientation would involve too large an increase in repulsion energy unless expansion accompanies disordering. With regard to $\Delta S_{assocn}$, when the energy of association is wholly due to electrostatic effects, the formation of local close-packed association complexes must on the other hand lead to a local shrinkage in volume which would facilitate disorientation.

Increases of pressure should increase the tendency to form complexes with a shrinkage in volume. This may probably explain the effect of pressures of inert gas up to 400 atm (Ar or He) in reducing the electrical conductivity of melts of $AgNO_3$, though more information seems desirable (Copeland and Radak, 1967).

On the basis of considerations such as those above, salts with inert gas type cations which give melts containing association complexes in appreciable proportions should have smaller volume changes on melting than for similar melts which are formed merely by positional and orientational randomization of the crystal structure (see Table 8.29).

**Table 8.29** (see Ubbelohde, 1962). **Volume changes on melting salts with polyatomic anions**

| Salt | $\Delta V_f / V_s$ (%) |
|---|---|
| NaSCN | 7 |
| KSCN | 10 |
| $KHSO_4$ | 4 |
| $KClO_3$ | 10 |
| $Na_3AlF_6$ | 25 |

Despite their great interest, volume changes on melting expressed as the dimensionless fraction $\Delta V_f / V_s$ have only been determined for a few ionic crystals up to the present. The main experimental obstacle appears to be in evaluating the molar volume of the crystal at the melting point by direct measurement. Frequently $(V_s)_f$ needs to be calculated by extrapolating thermal expansion measurements on the crystals up to $T_f$. If the solid near $T_f$ contains an appreciable concentration of defects, i.e. if premelting phenomena are marked (see Chapter 12), extrapolation of the crystal volume to $T_f$ yields values of $(V_s)_f$ that are below that truly occupied by the crystal at the melting point. $(V_L)_f$ can on the other hand usually be calculated quite reliably be extrapolating pyknometric or dilatometric measurements made on the melts down to $T_f$. As the outcome of these different trends, values of $\Delta V_f = (V_L)_f - (V_s)_f$ calculated by such extrapolations tend to be too large. Tables 8.24 and 8.29 record some of these. By comparison with data recorded in Table 8.1 it will be seen that, even with the reservations made, $\Delta V_f / V_s$ tends to be smaller for salts with polyatomic ions than for similar crystals containing only ions with inert gas structures. Much additional information about this important melting parameter is clearly needed to test the generality of this conclusion. If the melts contain complexes somewhat more tightly packed than in the crystals, the expansion in volume on melting $\Delta V_f$ corresponds at least in part with unoccupied space $V_u$ arising from the mismatching of clusters. Estimates (Janz and James, 1962) at temperatures considerably above $T_f$ yield 'unoccupied' volumes $V_u$ considerably higher than $\Delta V_f$. This definitely points to a conglomerate structure for nitrate melts which is of considerable interest. Methods of calculation probably need to be improved considerably however, before definitive structural implications can be quantified.

(iii) If a significant proportion of the ions in salts with abnormally low melting points are in fact associated into complexes, transport mechanisms should be considerably modified compared with those for quasicrystalline assemblies of ions. If small, a cluster might be expected to move only as a whole in ion migration. Free rotation of the nitrate anion for example cannot be assumed (James, 1966). Its rotation in the melt under applied shearing forces might provide relaxation effects additional to the ordinary relative movement of single ions in viscous flow. Up to the present, such transport properties of ionic melts have only been surveyed on a somewhat empirical basis, since detailed theories of micromolecular movements involved in viscous flow are not available from which to interpret experimental results.

Table 8.30 summarizes some of the comparisons made along these general lines (Frame, Rhodes, and Ubbelohde, 1959). Transport properties $P_r$ are assumed to follow Arrhenius equations of the type

$$P_r = P_{r_0} \exp(-E/RT)$$

It will be seen from Table 8.30 that the ratio $E_\eta/E_\sigma$ is uniformly lower for the nitrates than for halides, although for these $E_\sigma$ is comparable or higher. This again points to additional shear relaxation processes in nitrate melts, not available for the halides.

**Table 8.30.** *Activation parameters for transport in ionic melts*[a]

| Anion | Cation $Li^+$ | $Na^+$ | $K^+$ | $Ca^{2+}$ | $Ag^+$ | $Cu^+$ | $Cd^{2+}$ | $Pb^{2+}$ |
|---|---|---|---|---|---|---|---|---|
| $Cl^-$ | 1·15 | 1·54 | 2·30 | 4·1 | 0·99 | 0·85 | 2·30 | (3·9) |
|  | 7·6 | 6·1 | 3·4 | 2·3 | 2·9 | 5·0 | 2·0 | 1·8 |
| $Br^-$ | 1·75 | 1·84 | 2·55 | — | (0·88) | — | — | 4·35 |
|  | 3·4 | 5·8 | 3·1 | — | 6·4 | — | — | 2·1 |
| $I^-$ | — | 1·25 | 2·75 | — | 0·8 | — | — |  |
|  |  | 5·9 | 3·4 | — | 7·3 | — | — |  |
| $NO_2^-$ | — | 2·98 | 2·97 | — |  |  |  |  |
|  | — | 1·34 | 1·42 | — |  |  |  |  |
| $NO_3^-$ | 4·49 | 2·76 | 3·49 |  | 4·57 |  |  | 3·26 |
|  | 0·94 | 1·4 | 1·05 |  | 0·70 |  |  | 1·11 |

[a] The upper figures give $E_\sigma$ in kcal g ion$^{-1}$, the lower the ratio $E_\eta/E_\sigma$, where $\eta$ refers to viscosity and $\sigma$ to electrical conductivity.

Detailed comparisons have been made of ion transport parameters using electrical conductivity $\sigma$ and viscosity $\eta$. Three lines of evidence which point to cluster formation in the molten nitrates are:

(i) the temperature-dependence of viscosity, electrical conductance and density of $AgNO_3$ and $TlNO_3$ are very similar;

(ii) the nominal activation energy for electrical conductance $E_\lambda$ is practically independent of the cation for molten univalent nitrates, whereas a much more marked trend is apparent for molten halides (see Table 8.31);

**Table 8.31.** *E$\lambda$ for molten nitrates and halides (kcal g ion$^{-1}$)*

| Nitrates | | | Halides | | |
|---|---|---|---|---|---|
| $Ag^+$ | 2·9 | | $Li^+$ | 2·02 | |
| $Tl^+$ | 3·35 | | $Cs^+$ | 5·11 | |
| $Li^+$ | 3·59 | | | | |
| $Cs^+$ | 3·69 | | | | |

(iii) the ratio of activation energies for viscous flow and for ion conductance is practically unity for some of these nitrates (see Table 8.32).

**Table 8.32.** *$E_\eta/E_\lambda$ for molten nitrates*

| | |
|---|---|
| $Ag^+$ | 1·06 |
| $Tl^+$ | 1·09 |
| $Li^+$ | 1·42 |
| $Cs^+$ | 1·20 |

As with the other evidence cited, these facts support the general picture of ion association into clusters in these melts, of such a kind that the properties are controlled by anion clusters (Janz, Tomkins, Siegenthaler, and Halasbrahmanyam, 1971; Timidei and Janz, 1968). Whilst these transport properties provide useful diagnostic evidence, this is still only broad and generalized; more precise characterization of the microstructure of melts is to be desired. An extended survey for ionic melts is available (Janz and Reeves, 1965). Viscous flow in molten halides probably involves coupled ion motion (Vasu, 1969). Another basis for using transport properties as a means for detecting appreciable concentrations of association complexes in melts is to look for significant departures from the Nernst–Einstein relationship between diffusion and ionic mobility. If identical hopping probabilities are involved, the coefficient of mass diffusion $D_+$ is directly proportional to the ionic mobility $\sigma_+$ and similarly for the anions. This test to determine whether the structure of an ionic melt is simple i.e. yielding the Nernst–Einstein equation, has not yet been applied at all extensively. Though they provide useful pointers to anomalous melt structures, these and other phenomenological studies of the melting of crystals containing polyatomic anions cannot penetrate very deeply into structural details. Unfortunately, at present the resolving power of X-ray scattering methods gives information about local close-packing which is quite inadequate to test such phenomenological pointers to the formation of association complexes on melting, in any precise way.

Although the examples discussed above have deliberately been referred to melts of polyatomic ions in which association involves simple force fields, this does not exhaust the possibilities. For example, association complexes involving hydrogen bonding are highly likely in melts of $KHSO_4$ (Rogers and Ubbelohde, 1950).

### 8.16 Optical studies of complexing: local ionic environments

Optical measurements of ultraviolet absorption spectra (Rhodes and Ubbelohde, 1959; Cleaver, Rhodes, and Ubbelohde, 1963), yield some additional information about the *local environment* of anions round a cation and *vice versa*.

The fact that the data refer to individual local ionic environment makes them a useful complement to data referring to the average properties of ionic melts. Various electronic transitions leading to optical absorption can be defined for an isolated ion pair in the gas phase. Their modification in the symmetrical environment in a crystal and in the (often unknown) environment in the melt can then be followed systematically. Probably the simplest transition is the electron transfer or electron return process

$$A^+ + B^- \xrightarrow{h\nu} AB$$

For ions with inert gas shells, in the crystal this process is modified to one in which the electron on any anion is excited to a level approximately determined by the nearest shell of cations, of radius $r_0$. Figure 8.6 illustrates the

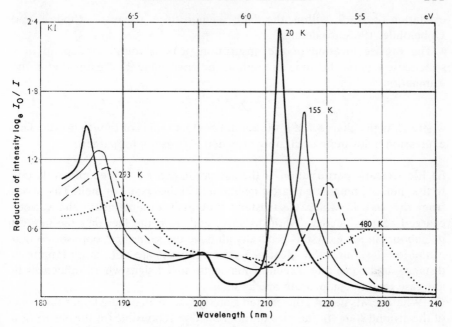

FIG. 8.6. Effect of temperature on absorption peak in KI

effect of temperature on the absorption peak in the case of potassium iodide. As the temperature of the crystal lattice rises, this peak broadens and shifts to longer wavelengths; this can be approximately related to the thermal expansion of the nearest shell of cations, according to the equations

$$E_{max} = E_0 + h^2/4m_0r_0^2$$

$$\frac{dE_{max}}{dT} = -\left(\frac{h^2}{8m_0r_0^3}\right)\frac{dr_0}{dT}$$

where $r_0$ is the (nominal) radius of the nearest shell of cations, $m_0$ is the mass of the electron, and $h$ is Planck's constant.

On melting, if the shell expands still further on average, a finite shift *to the red* $\Delta E_f$ would be expected. This is in fact found for lithium iodide, the one example so far studied by this means.

For polyatomic ions, in addition to electron transfer spectra (which may lie too far to the ultraviolet for convenient observation), other optical transitions may be present at different parts of the absorption spectrum, and may likewise be used to study the changes of *local* ionic environment on melting. For the nitrate ion, one of the absorption peaks in the region 30 000 to 34 000 cm$^{-1}$ in dilute solution and in the melt might be due to an electronic excitation $n_0 \to \pi^*$ within the $NO_3^-$ anion (see Cleaver, Rhodes, and Ubbelohde, 1963, for details).

The precise location of the absorption peak is found to depend in a systematic way on the ionic strength of the environment, as expressed in the summation

$$\sum Z_j / r_j$$

where $Z_j$ is the charge and $r_j$ the crystal contact radius of the $j$th cation. This expression holds quite accurately provided $Z_j/r_j$ is not too large.

(i) Electrostatic perturbation of the excitation process is predominantly due to the shell of immediate nearest neighbours in the crystal or melt. This arises from the way in which electrostatic potentials originating in the various charged centres decrease with distance, and from the fact that the nearest neighbours in ionic crystals generally all have the same sign, opposite to that of the ion they surround. Next nearest neighbours are both more remote in distance, and constitute a shell of ions with mixed signs whose influences in any case tend to compensate one another.

(ii) Whatever its nature, the excited state will have an orbital larger than that of the ground state. In the case of the transition reponsible for the absorption peak arising from charge transfer, in inert gas halides the radius of the excited orbital approximately coincides with the nearest shell of cations surrounding the anion. In other transitions, enlargement of the orbital accompanying electronic excitation seems likely from general considerations of Bohr's correspondence principle.

Insofar as these considerations are applicable, thermal expansion of the shell of nearest neighbours will normally shift the peak energy of absorption to lower energies and lower frequencies. A plausible semiquantitative view is thus that the perturbation of both ground and excited states is lessened by thermal expansion. When the nearest neighbour shell has a charge of opposite sign to the anion, it favours the excitation process and tends to enlarge the excited orbital. Its thermal expansion normally reduces the perturbation and leads to a shift to the red. Detailed theories of this effect may be quite complex, but simple empirical tests of the outcome are obtained by correlating the thermal expansion (solid or melt) with the shift of peak $dE_a/dT$ as the temperature is changed. Not too near a solid–solid or solid–liquid transition, accurate correlation is found, of the form

$$\frac{dE_{max}}{dT} = a\alpha$$

where $a$ is a constant characteristic of the nature and state of the nitrate, and $\alpha$ is the thermal expansion for that state. The important finding is that by contrast, at the transition solid–melt, the nitrates of Group I all show shifts of absorption peak towards *shorter wavelengths*, indicating that the anion and cation shrink locally closer together than in the crystals. This differs for example from LiF (see above). Table 8.33 illustrates some of the findings.

**Table 8.33.** *Correlation between observed optical shifts and thermal changes on melting Group I nitrates*

| Cation | $(dE_2/dT)$ $(cm^{-1}\ deg^{-1})$ melt | $(\Delta E)$ $(cm^{-1})$ on melting |
|--------|------|------|
| Li | — | 120 |
| Na | −1·8 | 200 |
| K | −1·3 | 140 |
| Rb | −1·2 | 280 |
| Cs | −1·5 | 160 |
| $NH_4$ | −2·0 | 330 |
| Ag | −2·0 | 260 |
| Tl | −0·5 | 330 |

Close comparison of the jump $\Delta E$ on melting with the macroscopic volume change, suggests a 'fluctuating cluster or conglomerate' texture for the melt in which more tightly bonded regions or clusters mismatch at their boundaries, where density is lower than the mean (see Chapter 13). At present such models for the texture of nitrate melts are tentative. original papers should be consulted for further details.

### 8.16.1. *Infrared measurements on phase transitions in ionic crystals*

Studies of changes of infrared absorption spectra above and below $T_f$ for the nitrates Li, Na, K, Rb, Cs (Greenberg and Hallgren, 1960) show that melting has only a small effect, generally decreasing the various frequencies which are attributable to various modes of vibration of the $NO_3$ groups for melts of Na, K, Rb and Cs. For Cs the decrease is particularly marked for the $\nu_3$ vibration and may perhaps point to a primary change in coordination number on melting, as already discussed in Section 8.7. For Li, one of the two modes increases slighly in frequency on melting. In general, these changes indicate that interionic forces have only a moderate effect on vibrations of $NO_3$ in the ground state.

For the nitrates, other rather more prominent changes accompany changes of state; relative intensity transmissions are particularly sensitive to melting. This can be attributed to effects of scattering by crystal defects of various kinds in the solids, but other effects may also operate.

Raman and infrared evidence on molten nitrates often shows evidence of 'local ordering' near $T_f$; this fades away at higher temperatures.

Thus for nitrates of Li, Na, K, and AgNO$_3$, low-frequency shift depolarized Raman bands are observed up to about 120°C above $T_f$, corresponding with a microenvironment of the ions which resembles the ionic arrangement in the high-temperature orthorhombic crystal forms. Infrared observations support this conclusion (Clarke, 1969; Williams, Li, and Devlin, 1968), as do Raman spectra (Clarke, 1969). A lattice-like microenvironment is likewise observed for the ions in molten RbNO$_3$ (Devlin and James, 1970) with an increase of frequency in the Raman spectrum, corresponding with the shrinkage in volume on melting this salt (Schinke and Sauerwald, 1960).

Use of Raman laser spectra helps to sharpen the general optical evidence for extensive intermolecular associations in molten nitrates for Ag(NO$_3$) and Tl(NO$_3$). Just below the melting point of AgNO$_3$ ($T_f = 210$°C) at 195°C a band is found at 150 cm$^{-1}$ closely similar to that in the melt, pointing to precursor rearrangements in the crystals (Balasubramanyam and Janz, 1972). Other optical data lead to similar conclusions (James and Leong, 1969, 1970). In molten AgNO$_3$ it has been suggested that association complexes may involve covalent bonding (Vallier, 1968).

KNO$_2$ shows a double infrared peak at room temperature, which fuses into a single peak below $T_f$. This is probably associated with a thermal transformation in the solid at 40°C (Cleaver, Rhodes, and Ubbelohde, 1963; Hazlewood, Rhodes, and Ubbelohde, 1963).

Infrared studies on NaOH show a shift of the OH absorption band to long wavelengths at still higher temperatures (Greenberg and Hallgren, 1961). Comparisons of other physical properties of NaOH with NaCl (see Table 8.34), together with the spectroscopic evidence, point to the presence of clusters or association complexes around $T_f$ in molten NaOH. These gradually dissociate as the temperature of this melt is increased above $T_f$.

**Table 8.34. Comparative properties of NaOH and NaCl in relation to melting**

|  | NaOH | NaCl |
|---|---|---|
| $T_f$ (°C) | 332 | 804·3 |
| $T_b$ (°C) | 1390 | 1413 |
| Specific conductivity (ohm$^{-1}$ cm$^{-1}$) | 2·30 (350°C) | 3660 (827°C) |
| Viscosity (centipoise) | 4·0 (350°C) | 1·30 (841°C) |
| $S_f$ (e.u.) | $\left\{\begin{array}{l}2\cdot69(\text{tr})\\(2\cdot56(\text{f})\end{array}\right.$ | 6·23 |

Optical absorption spectra hardly show any change on melting of Cs$_3$[NiCl$_5$] ($T_f$ 547°C), whereas marked changes are observed on melting Cs[NiCl$_3$], which appears to dissociate thermally ($T_f$ 758°C) (Boston, Brynestad, and Smith, 1967).

Phonon studies on the melting transition and on ionic melts show features broadly in agreement with other known properties of ionic melts. No very discriminating results have yet been reported (Torell, 1975; Hall and Yeager, 1973). Though optical phonon spectra near $T_f$ indicate short range order over small distances, the nature of these clusters is not clearly illuminated using either longitudinal or transverse spectra (Devlin, James and French, 1970).

## 8.17 Atom transfer defects on melting ionic crystals

Formation of simple positional defects has been previously referred to as a mechanism of crystal disorder and of melting (Chapter 5). In crystals containing polyatomic species capable for forming ions, significant positional defects may also include 'internal fragmentation' of these ions in the crystal, with migration of the fragments to defect sites. This kind of thermal disordering may become sufficiently extensive in the crystals to lead to marked premelting (Chapter 12). It is likely to undergo a sharp increase at $T_f$ owing to the decreased compensation of electrostatic forces, and to the expansion in volume which favours dissociation of complex ions.

Examples of various types of behaviour are beginning to be recognized and identified.

### 8.17.1. *Anion dissociation*

In cryolite $Na_3[AlF_6]$ the fragmentation

$$[AlF_6]^{3-} \rightleftharpoons [AlF_4]^- + 2F^-$$

probably accounts for some premelting effects in the solid (Landon and Ubbelohde, 1957), as well as affecting the properties of the melt.

Methods of evaluating the fractional dissociation of such complex ions have frequently been based on cryoscopy. The marked experimental convenience of cryoscopic methods is however counterbalanced by the fact that thermodynamic activities of ionic species depart appreciably from unity at exceptionally low concentrations (see p. 187). With the aid of special hypotheses concerning ion activities, the fractional dissociation $\alpha$ of $[AlF_6]^{3-}$ at $T_f$ for cryolite (1282 K) has been estimated to be $\alpha = 0 \cdot 204$, using freezing point depression methods (Rolin and Bernard, 1962). This gives a value $H_f$ (cryos.) of $26 \cdot 6$ kcal mole$^{-1}$ which differs somewhat from the calorimetric value ($27 \cdot 7$ kcal mole$^{-1}$) in view of the role of the heat of dissociation.

An analogous dissociation of the pyrophosphate anion

$$(P_2O_7)^{4-} \rightleftharpoons (PO_3)^- + (PO_4)^{3-}$$

appears to be practically complete in melts of the sodium salts when in an eutectic mixture (freezing point 725°C).

NaCl 77·8 mole%, $Na_4P_4O_7$ 22·2 mole% (Riccardi, 1962): the high melting point of the pure pyrophosphate (about 990°C) has obstructed study of incipient defect formation in the crystals in a premelting region. Raman spectra also suggest some dissociation of the $HSO_4^-$ in melts of $KHSO_4$ (Walrafen, Irish, and Young, 1962); other evidence indicates ion association as well, e.g. by hydrogen bonding (Rogers and Ubbelohde, 1950).

### 8.17.2. Cation dissociation

In crystals such as $NH_4Cl$ the dissociation (Ubbelohde and Gallagher, 1955) $NH_4Cl \rightleftharpoons NH_3 + HCl$, although not unlikely, has not been studied near $T_f$, mainly because at ordinary pressures this crystal sublimes before melting. At higher pressures considerable premelting followed by dissociation on melting may be anticipated in the ammonium salts from this kind of ion fragmentation.

### 8.17.3. Valence switch defect formation

In crystalline $FeCl_3$, appreciable vapour pressures of chlorine are set up even below the melting point, owing to the valance switch defect formation $FeCl_3 \rightleftharpoons FeCl_2 + \frac{1}{2}Cl_2$. Similar valence switch defects in varying degrees may be expected in salts of other transitional metals, depending on the energy differences between the different valency levels, and on the work of creating defects in the crystals resulting from such dissociation.

For $FeCl_3$ the heat of vaporization is estimated to be $32·9 \pm 0·2$ kcal mole$^{-1}$ of $Fe_2Cl_6$ and the heat of fusion $9 \pm 0·4$ kcal mole$^{-1}$ $FeCl_3$ (Cook, 1962). Premelting is observed at least over the temperature range $297·5°C–T_f$ (ca. 310°C), but owing to the formation of eutectic melts of $FeCl_3/FeCl_2$ (eutectic 297·5°C), attribution of any of the premelting heat uptake to defect formation in the crystals (Chapter 12) is uncertain. For the melt

$$C_p = 30 \text{ cal deg}^{-1} \text{ per } \frac{1}{2}[Fe_2Cl_6]$$

Unfamiliar forms of defect may sometimes be found. When new chemical linkages are actually formed in complexes in any condensed state of the substance (i.e. with quantum mechanical changes in the bonds), the microstructures produced do not necessarily involve a local density higher than that in the ideal crystalline state, particularly if the coordination number in the complex is lower than in the ideal crystal. Because of the valence effects involved, formation of new linkages is much more likely for ions of B metals or transition metals of the Periodic System than for ions with inert gas electron shells (see Murgulescu, 1969a,b).

## 8.18 Other methods of investigating dissociation on melting

Raman spectroscopy can in some cases permit useful discrimination between those ionic crystals which should be regarded merely as crystal compounds, that dissociate completely on melting, and other crystals which give more or less stable complex ions even in the melt. For example, the phase diagram solid–melt for the system $KPO_2/KF$ exhibits maxima at the compositions $KF.KPO_3$ and $2KF.KPO_3$. Raman spectroscopy shows that both melts contain the complex ion $[PO_3F]^{2-}$ which is formed by the reaction

$$(PO_3)^{n-} + nF^- \rightleftharpoons n(PO_3F)^{2-}$$

However, apparently additional fluorine ions do not attach themselves with sufficiently strong binding energy to the phosphate radical to give anions stable in the melt with any higher fluorine content than one atom (Buhler and Bues, 1961). According to this optical criterion, the first maximum on the phase diagram $KPO_3|KF$ corresponds to an ion structure which remains stable on melting, whereas the second apparently involves a crystal compound that dissociates into the ions $F^-$ and $(PO_3F)^{2-}$ on melting.

When the ions carry nuclear spins, nuclear magnetic resonance techniques can give additional information about changes of structure on melting, in favourable cases. For example, melts of the complex thallium salts $Tl_2Cl_4$ and $Tl_2Br_4$ appear to dissociate into the univalent and trivalent ions (Rowland and Bromberg, 1958). On the other hand nuclear magnetic resonance suggests that thallous acetate hardly dissociates on melting. The problem warrants further investigation.

## 8.19 Changes from 'ionic' to 'metallic' structures on melting

With crystals containing atoms of transitional or B metals, joined to electronegative atoms such as O, S, N, P etc., the bonding between such atoms is not always easy to characterize.

Often, but not always, the solids are at least semiconductors of electricity, suggesting covalent or quasimetallic bonding rather than mere ionic juxtaposition in the crystals. In principle, the same bond character need not predominate in the melt as in the crystals; for instance, unlike the crystals, the melts could contain appreciable concentrations of ions (see p. 265 for the analogous situation for other changes on melting of semiconductor solids).

## 8.20 Sublattice melting of ionic lattices

As for all other crystal types, melting of ionic lattices always involves *positional disordering*. However, when certain properties of anion and cation are sufficiently dissimilar, for example properties such as the ion contact radii, the ion charges, and the ion polarizabilities, the sublattice carrying one sign sometimes melts positionally at a much lower temperature than the sublattice with opposite sign. When this occurs, high ionic conductivity is found above the first melting point, with a transport number of practically unity for the ions whose sublattice has undergone fusion. As might be expected, a substantial entropy increase arising from positional disordering of ions of this sublattice is found at the first melting point, with the consequence that when the second melting point is reached (i.e. above which the ionic assembly is wholly molten) the entropy of fusion is then considerably smaller than for related crystals for which both ionic sublattices melt at the same temperature.

Interest in this kind of 'melting by stages' has been stimulated by various technological developments of solid electrolytes.

Examples identified up to the present belong to various groups as follows.

### 8.20.1. *Melting of the cation sublattices*

Univalent ionic crystals may carry the same charge on both ions, but with very different polarizabilities and repulsion radii. Longest known are certain salts of silver which have been frequently studied as examples of solid–solid phase transitions (p. 88). In these cases the positional disordering of the cations above $T_c$ in many ways resembles the melting of the entire crystal lattice except that the framework of the anions remains fixed. In fact, this can permit ionic mobility even higher than when the anion melts likewise, since erratic positions then assumed by the anions may actually decrease the mean free path of the cations. Furthermore, the temperature coefficient of mobility is smaller than in random solutions since there is less interference from thermal scattering. Many recent studies have dealt with the use of these crystals above $T_c$ as 'solid electrolytes'. To meet this demand, properties investigated usually include various mobility parameters, as well as the general positional disordering of the cations, but unfortunately often without determining other properties of thermodynamic significance such as specific heat and entropy changes, or thermal expansion and volume changes. Above a critical pressure the activation energy for cation migration increases (Bradley, Munro, and Ali, 1969). Electrical conductivities are illustrated in Fig. 8.7 (Rickert, 1973). (Useful review articles include Heyne, 1970, Uskke and Bukun, 1972, and Hyde, 1976).

The properties of the double salts $MAg_4I_5$, where M is one of the cations $K^+$, $NH_4^+$, $Rb^+$ or $Cs^+$, show transition from a highly conducting (0.20 ohm $cm^{-1}$ at 20°C) to a low-conductivity phase between $-136°C$ and $-155°C$, according to the cation (see Fig. 8.7) (Owens and Argue, 1967). The lattice structure of $RbAg_4I_5$ has been studied (Geller, 1967).

Salts with organic ions possibly also include examples of this kind of melting behaviour. Up to now, however, only melting to a mesophase, which

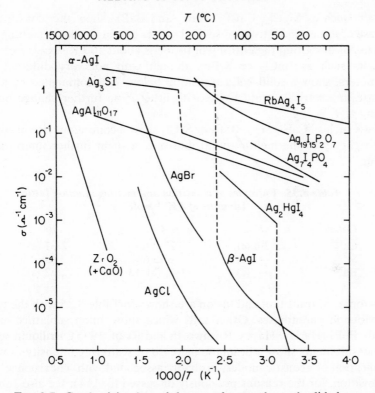

FIG. 8.7. Conductivity data of the most frequently used solid electrolytes. Reproduced by permission of North-Holland Publishing Company

though highly viscous is nevertheless fluid, has been described for molten organic salts. This is in most cases mechanically distinguishable from the cases discussed above, in which one sublattice retains long-range positional order right up to the second melting point, as in the silver and copper salts cited on p. 89. Ionic conductance in the mesophase shows a remarkably low temperature coefficient (Duruz and Ubbelohde, 1976).

8.20.2. *Melting of anion sublattices—ionic crystals carrying doubly charged cations and singly charged anions.*

For many salts of Group II cations, it seems likely that the transport of electricity occurs predominantly by the anion. Recent studies (Derrington, Lindner, and O'Keefe, 1975) indicate that a balance between different factors determines which of three kinds of behaviour is followed as the temperature of the crystals is progressively raised till they are wholly molten.

(a) Salts such as $MgCl_2$, $CaCl_2$, $CaBr_2$ and $BaBr_2$ show the 'conventional' increase of ionic conductivity of several orders of magnitude on melting at $T_f$. For salts of this group the entropy of fusion is about $30\ J\ K^{-1}$.

(b) Salts such as $BaCl_2$ or $SrBr_2$, though with not very different ionic parameters, show a solid–solid transition below $T_f$ accompanied by a large increase in electrical conductivity, with little or no further change on final melting at $T_f$.

(c) Salts such as $CaF_2$, $SrF_2$, $BaF_2$ or $SrCl_2$ show a continuous rise in conductivity right up to the melting point, but only a slight further jump on final melting.

*Table 8.35. Entropies of transition and melting of anion lattice (in units of $J\ K^{-1}\ mole^{-1}$)*

| Cation | $F^-$ | $Cl^-$ | $Br^-$ |
|---|---|---|---|
| $Ca^{2+}$ | 17·6 (c) | 27·2 (a) | 28·7 (a) |
| $Sr^{2+}$ | — (c) | 14·2 (c) | 13·3 (b), 11·3 |
| $Ba^{2+}$ | — (c) | 14·4$_{tr}$ (b), 13·3 | 28·2 (a) |

Entropies of transition and fusion are shown in Table 8.35 with the type of behaviour in parentheses. Other salts which show anion sublattice melting include $PbF_2$ (Harley, Hayes, Rushworth and Ryan, 1975). Brillouin scattering measurements show a dramatic increase of scattering intensity as well as modification of acoustic modes, probably associated with 'coexistence' types of transition, for the reasons previously discussed (p. 114); see also Johnson, investigate hysteresis of properties in cyclic temperature changes for this kind of transition, for the reasons previously discussed (p.    ); see also Johnson, Wiedersich, and Lindberg, 1970.

Theoretical considerations about sublattice melting of ionic lattices have been discussed in terms of a 'mean field' approximation (Welch and Dienes, 1975); at present these discussions are only illustrative.

## 8.21 Molten organic salts—problems of chemical stability of organic ionic melts

As stated above, extended physicochemical studies on the melting of organic salts have until recently been extremely limited. Added to the lack of interest in this sector from physical organic chemists, an outstanding difficulty has been the chemical instability of most of these salts as soon as they are melted, or even when they are dissolved in solvents at temperatures above about 120–140°C. In principle, however, organic salts present some highly intriguing possibilities, giving rise to liquids with unusual force fields between the charged species, and to ions whose shape and size can be modified in a readily controllable way by skilful organic synthesis. Two main subdivisions can be identified with molten organic salts; those whose *cations* are charged organic molecules, such as substituted ammonium salts $[NR_1R_2R_3R_4]^+X^-$ in which the anions $X^-$ are familiar inorganic species, such as $Cl^-$, $I^-$, $NO_3^-$, etc., and those whose *anions* are organic but with simple inorganic cations, particularly the monovalent ions of the alkali metals. A third subdivision of organic salts in which both anion and cation are organic does not appear to have been investigated at all extensively.

Thermal properties of salts for which both anion and cation are large and quasi spherical would be of particular interest if stable melts can be achieved. Information for inorganic or organic salts is as yet, however, too scanty to warrant discussion (see for example, Lind, Abdel Rahim, and Rudich, 1966; Rudich and Lind, 1969, 1970).

*A priori*, one obvious source of the widely experienced instability of organic salts, as soon as they are melted, stems from the decreased compensation of electrostatic forces on passing from ordered crystal to disordered melt. Reference has already been made to the fact that local anion–cation interactions will normally be considerably stronger in certain regions of ionic melts than in the crystalline state of the same substance. As a consequence, if the organic ion is capable of ligand rearrangements, involving its decomposition, quite often these chemical changes will await dissolution or melting since they are catalysed by the strong local field exerted by the nearest neighbour of opposite charge, particularly when this field is no longer compensated through lattice symmetry. For example, an alkali cation $M^+$ may achieve rearrangement of two carboxyl anions.

$$2M^+ + 2[COOR]^- \rightarrow 2M^+ + CO_3^{2-} + R_2CO$$

to yield a ketone, or a simple anion may favour splitting off of unsaturated molecules from substituted ammonium cations, such as

$$[R_1R_2R_3NC_2H_5]^+ \rightarrow C_2H_4 + [R_1R_2R_3NH]^+$$

Such ligand rearrangements are in a sense unavoidable chemical consequences of positional disordering of the lattice on melting. If the main instability of organic salts were in this sense *intrinsic*, there would be little hope of investigating these salts by means of any precise measurements on the molten states, except perhaps by finding salts whose melting points are freakishly so low that intrinsic breakdown reactions are kinetically only slow.

However, quite recently it has been established that at least some salts with organic anions are intrinsically quite stable in the molten state, and usually decompose in well-known ways on melting, because trivial but commonly present *extrinsic* impurities begin to exert their full influence on the kinetics of decomposition, as soon as the system becomes fluid. This discovery is important since if it can be extended to a variety of organic salts (as seems likely), a large and novel sector of ionic melts is made available for physicochemical studies, both with regard to thermodynamic parameters, which are the main concern of the present book, and with regard to many other important fluid properties, where the unusual structure of molten organic salts seems likely to confer unusual behaviour.

It is not necessary to elaborate here chemical aspects of extrinsic impurities whose presence would obstruct accurate measurements on molten organic salts. Some of the major features of this discovery have been investigated with particular reference to molten carboxylates (Hazlewood, Rhodes, and Ubbelohde, 1966; Duruz, Michels, and Ubbelohde, 1971). With some hindsight it is perhaps not surprising to find that organic ions are often attacked fairly readily by molecular oxygen, by the hydroxyl anion $OH^-$, or by other ions with strong electrostatic fields such as transitional metal cations. Fortunately, all these extrinsic sources of instability can be eliminated or suppressed (e.g. by complexing) fairly readily. The relevant organic ionic melts can then be studied over quite extended ranges of temperatures, up to regions where intrinsic ligand rearrangement can no longer be avoided.

The group of salts most extensively subjected to precise physical measurements up to the present includes alkali carboxylates $M^+[COOR]^-$ in which the organic radical R has included alkane radicals $C_nH_{2n+1}$ with $n$ up to 7. Both normal and iso radicals have been studied. They show similarities but also marked differences in the molten salts.

The following sections will serve as introductions to this new sector of ionic melts research. It is hoped that this will, before long, include other portions of the great catalogue of organic compounds, particularly all those salts whose melting point is sufficiently low to avoid intrinsic cracking of organic bonds. At present, this would appear to require melting points somewhat below 500°C.

## 8.22 Thermodynamic parameters in the melting of alkali carboxylates

Alkali carboxylates $M^+[COOC_nH_{2n+1}]^-$ can readily be prepared in characteristic homologous series, although some preparative difficulties are found with the lithium and caesium salts. When $n$ remains fairly small (say, $n \leq 7$) the ratio of ionic charge to the total number of atoms in the crystals is not too different from that of typical inorganic salts whose melting has been widely studied. For example, in alkali nitrates this ratio is $2:5$, in alkali butyrates it is $2:14$, and in alkali caproates it is $2:23$. Again, the repulsion envelope calculated for the butyrate anion is not very different from that of the nitrate anion (Fig. 8.8). One might thus look for considerable similarity of melting behaviour between, say, nitrates and butyrates, with gradual trends as the strong electrostatic forces emanating from the cation and the charged

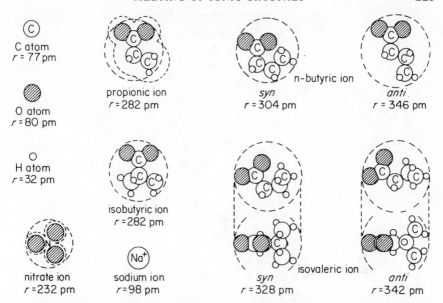

FIG. 8.8. Models and scaled repulsion envelopes of atoms and ions referred to in text. Covalent radii for C and H; weighted mean between covalent and ionic radius for O, Hazelwood, Rhodes, and Ubbelohde (1966); nitrate in model from James and Janz (1962). Short broken circle segments represent Van der Waals repulsion envelopes

carboxylate end of the anion are progressively diluted, on moving up the homologous series. In fact, these expectations have proved almost wholly amiss. Organic ionic melts in this series of compounds show properties often widely different from inorganic ionic melts, for reasons that can be understood in terms of their crystal structure. Although this crystal structure has not been established in complete detail for most of the salts under discussion, the general arrangement adopted by the ions in all cases seems quite clear. It helps to maximize the neutralization of opposing electrostatic charges by forming electrostatic *layers* separated in the crystals by the alkane tails of the carboxyl anions. This general arrangement may be illustrated for potassium caproate in Fig. 8.9A. A diagrammatic version useful for discussing the effect of increasing thermal energy on the crystals and their melts is illustrated in Fig. 8.9B. Melting of these distinctive crystal structures shows various features which can be followed, plausibly, from the preceding discussions of melting of molecular crystals and of ionic crystals, since to some extent both kinds of behaviour control the processes of fusion. Significant experimental features are as follows:

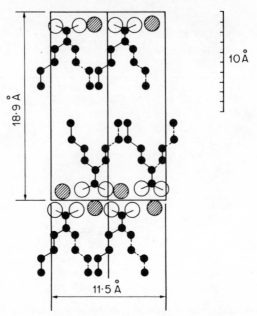

FIG. 8.9A. Alternative configurations of n-alkane chains between 'electrostatic sandwiches' for solid potassium caproate (ions and atoms not to scale). Reproduced by permission of North-Holland Publishing Company

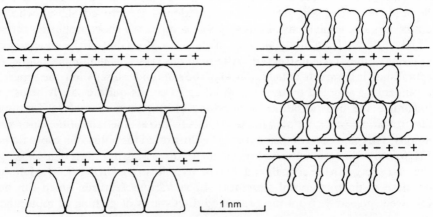

FIG. 8.9B. Sketch illustrating possible arrangements of repulsion envelopes to alkane chains between 'electrostatic sandwiches' (left) sodium n-butyrate; (right) sodium isovalerate

(i) The melting points are not particularly sensitive to changes from one monovalent alkali cation to another. In this respect, the crystals resemble for example the nitrates, as illustrated in Table 8.24.

(ii) As the alkali tail of the carboxyl anion is increased up a homologous series, at first the melting points increase; but for quite short chain lengths the melting process separates into two stages, with an intermediate 'mesophase' separating the lower from the higher melting points (Michels and Ubbelohde, 1972) (see Fig. 8.10) for Na, K, Rb and Cs.

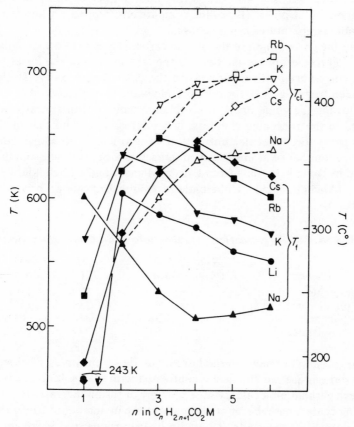

FIG. 8.10. Melting $T_f$ and clearing $T_{clear}$ points for alkali salts of n-alkanoic acids

(iii) On viewing the molten salts through crossed polarioid screens, it is readily verified that the mesophases are optically anisotropic (Duruz and Ubbelohde, 1972). They can indeed be termed ionic 'liquid crystals', though this terminology is not very helpful since these organic ionic mesophases differ in various important ways from much longer familiar molecular 'liquid crystals' (Chapter 14). Here, it is sufficient to note that these organic ionic mesophases, though highly viscous (about 3 poise), can be shown to satisfy even sophisticated mechanical definitions of the liquid state (Ubbelohde, Michels, and Duruz, 1972; Michels and Ubbelohde, 1974) with some exceptions.

(iv) Above the higher melting points, these ionic melts become optically isotropic; their viscosity also drops to values similar to those of inorganic melts, about 0·03 poise. (Relevant comparisons depend on how far really comparable temperatures can be chosen.)

(v) When the chain length of the alkane carboxyl anions becomes sufficiently long to lead to mesophase melting, neither the upper nor the lower melting points change very much on increasing the chain length progressively. This can be seen from Fig. 8.10 for the normal alkane carboxylates.

(vi) Following the guide lines for correlating melting with crystal structures, discussed in the preceding chapters, it is interesting to find that at the first melting point the molecular entropy of fusion is low, notwithstanding the quite large number of atoms involved in each ion. Some results obtained are recorded in Table 8.36 (data from Duruz, Michels, and Ubbelohde, 1971 and Fjortoft, Michels, and Ubbelohde (unpublished data private communication)).

**Table 8.36. Melting parameters of sodium salts of n-alkane carboxylic acids**

| | Solid → Meso | | Meso → Isotropic | |
|---|---|---|---|---|
| Alkane | $T_c$ (°C) | $\Delta S$ (e.u.) | $T_c$ (°C) | $\Delta S$ (e.u.) |
| $n = 1$ (acetate) | none | | 329 | 11·0 |
| $n = 2$ (propionate) | none | | 288 | 3·7 |
| $n = 3$ (butyrate) | 252 | 2·8 | 324 | 0·35 |
| $n = 4$ (isovalerate) | 188 | 2·17 | 280 | 0·32 |

(vii) The absence of any marked trend in these entropies of first fusion beyond sodium acetate should be contrasted with the melting behaviour of the parent paraffin molecules, whose entropy of fusion increases steeply with increasing chain length, because of the rapid rise in number of configurational isomers formed in the melts of paraffin hydrocarbons (p. 154) of increasing chain length.

These basic findings about the entropy of fusion suggest that the increase of disorder on first melting is only small, even compared with melting inert gas crystals. To elucidate this further, data may be examined concerning the volume changes and concerning the molecular packing as revealed by X-ray diffraction. Volume data are illustrated for sodium n-butyrate in Fig. 8.11.

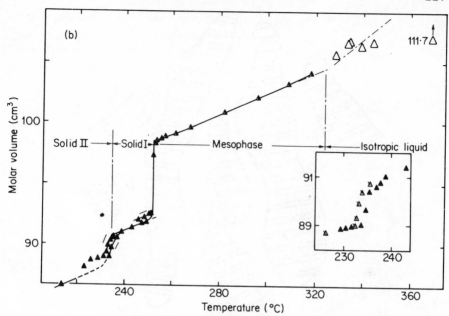

FIG. 8.11. Molar volume of sodium n-butyrate

Comparative data for homologues show fractional increases in volume on melting that are intermediate between the inert gases and sodium nitrate. X-ray diffraction reveals the interesting finding that 'long spacings' (about 14·08 Å for sodium butyrate) in the crystals persist into the mesophase at the first melting point $T_{clear}$ at which the melt becomes optically anisotropic.

A structural model to explain these melting parameters can be illustrated with reference to Fig. 8.9 and 8.12. In the crystals, as well as in the mesophase melts, the strong electrostatic forces anchor the individual molecules to rafts or layers of ions. This anchorage presses the alkane tails of the anions roughly parallel, in the crystals as well as the mesophase. There is thus no facility for any large increases of configurational entropy on first melting, since the strong electrostriction suppresses configurational options that do not pack well together. This is in striking contrast with the corresponding paraffins.

The texture of these mesophase melts can be illustrated as in Fig. 8.9, which represents the ionic layers sandwiching the hydrocarbon tails. These layers are compressed quite strongly together, but as the temperature rises the increasing vibrational and torsional energy leads to *lateral* expansion of the organic filling of each sandwich. Figure 8.13 illustrates how, for sodium n-butyrate, the long spacing actually contracts at the melting point, and continues to contract thereafter. Apparently, the usual thermal expansion in

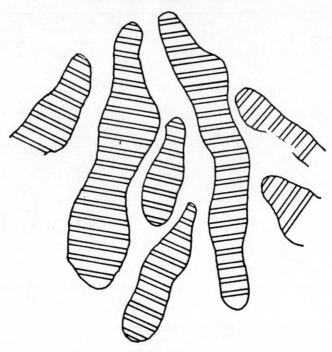

FIG. 8.12. 'Striped domains: possible texture of stacked 'electrostatic sandwiches' in the ionic mesophases

volume of the melt (Fig. 8.12) must be attributed mainly or solely to lateral expansion of the electrostatic sandwiches. If it is assumed that no other structures play an appreciable part in the mesophase, the lateral area $A$ of the ionic layers can be calculated from the equation

$$V/l = A$$

Taking $V$ as the molar volume in $cm^3$ and $l$ the long spacing in $A$ units, the lateral packing area per carboxyl group/cation is

$$V/Nl = A/N$$

On first melting, it seems likely that positional and possibly orientational disorder occur for the carboxyl ends of the anions. Further details of this novel class of ionic melts have been reported elsewhere (Ubbelohde, 1976).

Melting in stages of alkali salts of n-alkane carboxylic acids with longer alkane chains and thus a greater dilution of electrostriction of melts have been studied for $C_n H_{2n+1} CO_2^-$ with $n = 12$ (laurate), $n = 14$ (myristate), $n = 16$

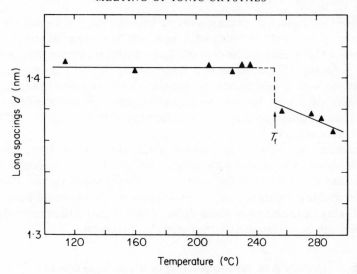

FIG. 8.13. Plot of long spacings against temperature for sodium
n-butyrate

(palmitate), and $n = 18$ (stearate), with substantially similar conclusions
(Skoulios and Luzzatti, 1961; Gallot and Skoulios, 1966a,b, 1967).

### 8.22.1. *Other organic anions*

In addition to the carboxylates discussed in the previous sections, a series of
alkali phenyl sulphonates derived from phenyl sulphonic acid (see Table 8.37)
has been reported (Gal, Halmos, and Eadey, 1966), but with little critical
detail concerning either structure or thermodynamic parameters, and no
guidance about their stability or suitability for precise physical measurements.

**Table 8.37.** *$T_t$ values for a series of alkali phenyl sulphonates*

| Cation | $T_t$ (°C) |
|--------|-----------|
| Li⁺ | 520 (decomposes) |
| Na⁺ | 415 |
| K⁺ | 375 |
| Rb⁺ | 305 |
| Cs⁺ | 232 |
| NH₄⁺ | 275 |

### 8.23 Melting of salts with organic cations

Up to the present comparatively little detail has been published concerning
salts with organic cations. The principal aim appears to have been to extend

the descriptive variety of organic salts of known molecular structure. Quite often merely the melting point of a new salt has been recorded, so that information about intrinsic or extrinsic factors concerning the stability of the melt is generally absent, or at best fragmentary. Melting points and other observations with a microscope do not necessarily represent the conditions for stability of a bulk sample. Some known data are surveyed in usefully compact form (Reinsborough, 1968; Gordon, 1969; Angell, 1971) (see Tables 8.38 and 8.39).

Substituted ammonium ions provide a rich diversity of salts, which are known both with simple and with complex anions. There is reason to believe that the intrinsic stability of the melts is not very good (Gordon, 1969), but since the melting points are often remarkably low a moderate liquid range above $T_f$ may be found, over which physical and in particular thermodynamic properties can be determined with reasonable precision.

**Table 8.38  Melting points of salts with organic anions**
**(Reinsborough, 1968)**

|  | $T_f$ | $H_f$ (kcal mole$^{-1}$) |
|---|---|---|
| Ethylammonium nitrate | 8 | |
| Tri-isopentylammonium iodide | 100 | |
| Tetra-n-butylammonium iodide | 144 | |
| perchlorate | 208 | |
| Tetra-n-pentylammonium thiocyanate | 50·5 | |
| Tetraisopentylammonium iodide | 132 | |
| perchlorate | 118 | |
| Tetra-n-hexylammonium iodide | 104 | |
| Pyridinium chloride | 146 | 2·3 |
| N-Methylpyridinium chloride | 138 | |
| bromide | 150 | |
| N-Ethylpyridinium bromide | 114 | |

**Table 8.39.  Transport properties of molten salts (organic cations)**

| Organic cations | $T$ (°C) | $K$ (ohm$^{-1}$ cm$^{-1} \times 10^2$) |
|---|---|---|
| Methylpyridinium bromide | 160 | 6·3 |
| Methylanilinium chloride | 150 | 1·7 |
| Tetra-isopentylammonium iodide | 150 | 0·17 |
| AgNO$_3$ | 210 | 70 |

Tables of melting points of substituted ammonium salts, where known, are given by Gordon (1969). In many cases these lie below 200°C; little attempt has been made to characterize the thermodynamic properties in most cases (see Table 8.40).

**Table 8.40. Comparative properties of molten chlorides of organic cations (Kisza and Hawranek, 1968)**

| Cation | $T_f$ (°C) | Molar conductance ($ohm^{-1}\, cm^2\, mole^{-1}$) at $1\cdot05\, T_f$ | Activation energy for conductance ($kcal\, mole^{-1}$) | Surface tension $\sigma_{LG}$ ($dynne\, cm^{-1}$) |
|---|---|---|---|---|
| $CH_3NH_3^+$ | 225 | — | — | 59·9 |
| $C_2H_5NH_3^+$ | 109 | — | 4·28 | 46·35 |
| Isopropyl $NH_3^+$ | 153 | 3·47 | 6·52 | 36·08 |
| n-Amyl $NH_3^+$ | 216 | 8·6 | 3·18 | 24·87 |
| Dimethyl $NH_2^+$ | 167 | 31·13 | 2·74 | 49·46 |
| Diethyl $NH_2^+$ | 224 | 12·47 | 3·60 | 37·18 |

The low surface tension of these ionic melts is particularly noteworthy. Data on the transport parameters appear to need further elucidation (Gatner and Kisza, 1969).

Fusion parameters have been determined for a number of quaternary n-alkyl ammonium compounds (see Table 8.41), whose melts can be studied for short periods above $T_f$ without serious decomposition (Coker, Ambrose, and Janz, 1970). In the data listed below, it is noteworthy that the volume changes on melting $\Delta V_f/V_s$ are smaller than for alkali halides, probably because the structures of the melts are closer to those of the crystals for the quaternary ammonium salts. Attention may also be drawn to the very large entropy changes found for the salts with the larger organic cations. There is evidence that this arises from disordering of the kink-block type of the flexible chains in the cation (see p. 158) at transition temperatures which may or may not coincide with the positional disordering which occurs on melting.

The structural basis of the thermodynamic changes occurring has been followed in considerable detail for the salt $[(n\text{-}myl)_4N]^+CNS^-$, which undergoes melting after a transition in the sequence (Coker, Wunderlich, and Janz, 1969)

$$\text{Crystal I} \xrightarrow[\text{about } 38°C]{T_c} \text{Crystal II} \xrightarrow[49\cdot5°C]{T_f} \text{Melt}$$

Ultraviolet absorption shifts indicate a sharp transition in the crystals at $T_c = 38°C$, with an entropy change of about 17 e.u. The specific electrical conductance ($ohm^{-1}\, cm^{-1}$) changes from $0\cdot3 \times 1 \times 10^{-3}$ at this transition. The total volume change from 25°C, traversing the transition and on into the melt, is about $4\cdot8\%$. Most of this occurs at $T_c$, after which there is a decrease in volume followed by a slight jump ($0\cdot4\%$) at $T_f$.

At $T_f$, the specific electrical conductance changes from about 5 to $122 \times 10^{-3}$, with a steep premelting rise (see p. 327). As with other quaternary ammonium salts the entropy of fusion $S_f = 14$ e.u. is quite large.

**Table 8.41. Melting parameters of quaternary n-alkyl ammonium salts[a]**

| Cation n-alkyl group | | $T_{transition}$ | $S_{transition}$ | $T_f$ | $S_f$ | $\Delta V_f/V_s$ |
|---|---|---|---|---|---|---|
| Chlorides | 4 | — | — | 314 | 15·6 | — |
| | 5 | 281 | 2·3 | 295 | 1·1 | — |
| | 7 | 215 | — | 265 | — | — |
| Bromides | 4 | 370 | 9·2 | 395 | 9·3 | 8·5 |
| | 5 | — | — | 376 | 26·4 | 9·5 |
| | 6 | 315 | 9·1 | 377 | 10·1 | 4·9 |
| | 7 | 343 | 5·1 | 369 | 23·3 | — |
| Iodides | 4 | 392 | 17·1 | 419 | 5·5 | 10·2 |
| | 5 | 345 | 10·2 | 412 | 22·9 | 10·9 |
| | 7 | 356 | 6·4 | 396 | 22·5 | 7·3 |
| Nitrates | 4 | — | — | 392 | 9·0 | — |
| | 5 | 366 | 8·2 | 387 | 17·7 | 3·0 |
| | 6 | 323 | 16·5 | 345 | 12·3 | — |
| Perchlorates | 6 | 335 | 16·5 | 383 | 11·4 | 11·9 |
| | 7 | 354 | 3·1 | 399 | 19·0 | — |

[a] Temperatures in K; entropies in e.u., cal mole$^{-1}$ deg$^{-1}$.

Like other substituted ammonium cations, (quaternary n-amyl N)$^+$ is considered to undergo a kink-block randomization of the conformations of the n-amyl groups—between 1000 and 10 000 conformations may be presumed above $T_c$. This randomization appears to require considerable time, so that $\Delta H_c$ and $\Delta H_f$ may be increased if the cooling is sufficiently short (by up to 8%). The ability of neighbour amyl groups to kink and unkink may even be frozen-in if cooling does not allow sufficient time, leading to a kind of hysteresis.

With n-hexylammonium perchlorate three transitions are found in the crystals, before melting. Conformational disorder of the cation and orientational disorder of the $(ClO_4)^-$ may be involved (Andrews and Gordon, 1973).

## REFERENCES

M. Abraham, J. Dupuy, J. Guion, and J. Brenet (1962) *C.R. Acad. Sci. Paris* **254**, 4290.

J. Akella, S. N. Vaidya, and G. C. Kennedy (1969) *Phys. Rev.* **185**, 1135.

J. T. S. Andrews and J. E. Gordon (1973) *J. Chem. Soc. Faraday Trans. I* **69**, 546.

C. A. Angell and J. Wong (1970) *J. Chem. Phys.* **53**, 2053.

C. A. Angell (1971) *Fused Salts in Ann. Rev. Phys. Chem.* **22**, 429.

C. F. Baes (1970) *J. Solid State Chem.* **1**, 159.

K. Balasubramanyam and G. J. Janz (1972) *J. Chem. Phys.* **57**, 4085.

B. Bardoli and K. Todheide (1975) *Ber. Bunsenges. Phys. Chem.* **79**, 490.
A. J. Barker (1972) *J. Phys. Chem.* **5**, 2276.
L. W. Barr and D. K. Dawson (1971) *Proc. Brit. Ceram. Soc.* **19**, 152.
W. Biltz (1922) *Z. Anorg. Allg. Chem.* **121**, 257.
W. Biltz (1924) *Z. Anorg. Allg. Chem.* **133**, 312.
W. Biltz and A. Voigt (1923) *Z. Anorg. Allg. Chem.* **26**, 39.
W. Biltz and W. Klemm (1926) *Z. Anorg. Allg. Chem.* **152**, 267.
M. Blanc (1960) *Ann. Phys. (Paris) Series 3* **5**, 615.
M. Blanc, L. Diélou and G. Petit (1964) *C.R. Acad. Sci., Paris* **258**, 491.
H. A. Bloom and I. A. Weeks (1969) *J. Chem. Soc. London* 2028.
J. O'M. Bockris and T. Emi (1970) *J. Phys. Chem.* **74**, 159.
J. O'M. Bockris, S. R. Richards, and L. Naris (1965) *J. Phys. Chem.* **69**, 1627.
C. R. Boston, L. F. Grantham, and S. J. Yosim (1970) *J. Electrochem. Soc.* **117**, 28.
C. R. Boston (1966) *J. Chem. Eng. Data* **11**, 262.
C. R. Boston, S. J. Yosim, and L. F. Grantham (1969) *J. Chem. Phys.* **51**, 1669.
C. R. Boston, J. Brynestad, and G. P. Smith (1967) *J. Chem. Phys.* **47**, 3193.
C. Bourlange (1959) *Ann. Phys. Paris Ser. 13* **4**, 1017.
R. S. Bradley, D. C. Munro, and S. I. Ali (1969) *High Temp. High Pressures* **1**, 103.
W. Brockner (1976) private communication.
K. Buhler and W. Bues (1961) *Z. Anorg. Allg. Chem.* **308**, 62.
S. Cantor (1961) *J. Phys. Chem.* **65**, 2208.
S. Cantor, D. P. McDermott, and L. O. Gilpatrick (1970) *J. Chem. Phys.* **52**, 4600.
S. P. Clark, Jr. (1959) *J. Chem. Phys.* **31**, 1526.
J. H. R. Clarke (1969) *Chem. Phys. Lett.* **4**, 39.
B. Cleaver, E. Rhodes, and A. R. Ubbelohde (1963) *Proc. Roy. Soc., Ser. A* **276**, 437, 453.
B. Cleaver and J. F. Williams (1968) *J. Phys. Chem. Solids* **29**, 877.
B. Cleaver, E. Rhodes, and A. R. Ubbelohde (1961) *Discuss. Faraday Soc.* **32**, 22.
T. G. Coker, J. Ambrose, and G. J. Janz (1970) *J. Am. Chem. Soc.* **92**, 5293.
T. G. Coker, B. Wunderlich, and G. J. Janz (1969) *Trans. Faraday Soc.* **65**, 3361.
C. M. Cook, Jr. (1962) *J. Phys. Chem.* **66**, 219.
J. L. Copeland and S. Radak (1967) *J. Phys. Chem.* **71**, 4360.
G. Darmois and G. Petit (1958) *Bull. Soc. Chim. Fr.* 511.
A. J. Darnell and W. A. McCollum (1968) *J. Phys. Chem.* **72**, 1327.
A. J. Darnell and W. A. McCollum (1971) *J. Chem. Phys.* **55**, 116.
A. J. Darnell, W. A. McCollum and S. J. Josin (1969) *J. Phys. Chem.* **73**, 116.
W. J. Davis, S. E. Rogers, and A. R. Ubbelohde (1953) *Proc. Roy. Soc., Ser. A* **220**, 14.
C. E. Derrington, A. Lindner, and M. O'Keefe (1975) *J. Solid State Chem.* **15**, 171.
J. P. Devlin and D. W. James (1970) *Chem. Phys. Lett.* **7**, 237.
J. P. Devlin, D. W. James, and R. French (1970) *J. Chem. Phys.* **53**, 4394.
Y. Doucet (1959) *J. Chim. Phys.* **56**, 578.
F. R. Duke and H. M. Garfinkel (1961) *J. Phys. Chem.* **65**, 1627.
J. J. Duruz and A. R. Ubbelohde (1972) *Proc. Roy. Soc. Ser. A* **330**, 1.
J. J. Duruz and A. R. Ubbelohde (1976) *Proc. Roy. Soc., Ser. A* **347**, 301.
J. J. Duruz, H. Michels, and A. R. Ubbelohde (1971) *Proc. Roy. Soc., Ser. A* **322**, 281.
A. S. Dworkin and M. A. Bredig (1963) *J. Phys. Chem.* **67**, 697.
A. S. Dworkin and M. A. Bredig (1968) *J. Phys. Chem.* **72**, 1277.
A. S. Dworkin and M. A. Bredig (1971) *High Temp. Sci.* **3**, 81.
P. G. T. Fogg and R. C. Pink (1959) *J. Chem. Soc., London* 1735.

T. Forland (1964) in B. R. Sundheim, *Fused Salts*, McGraw-Hill, New York.

J. P. Frame, E. Rhodes, and A. R. Ubbelohde (1959) *Trans. Faraday Soc.* **55**, 2039.

S. Fujiwara (1952) *Bull. Chem. Soc. Japan* **25**, 347.

K. Furukawa (1961) *Discuss Faraday Soc.* **32**, 53.

K. Furukawa (1962) *Rep. Progr. Phys.* **25**, 95.

S. Gal, T. Meissel, Z. Halmos, and L. Erdey (1966) *Mikrochim. Acta* 903.

B. Gallot and A. Skoulios (1967) *Kolloid-Z. Z. Polym.* **213**, 144.

B. Gallot and A. Skoulios (1966b) *Kolloid-Z. Z. Polym.* **210**, 143.

B. Gallot and A. Skoulios (1966a) *Kolloid-Z. Z. Polym.* **209**, 164.

K. Gatner and A. Kisza (1969) *Z. Phys. Chem. (Leipzig)* **241**, 1.

S. Geller (1967) *Science* **157**, 310.

A. M. Golub and F. F. Grigorenko (1961) *Russ. J. Inorg. Chem.* **6**, 1189.

R. Gopal (1953) *J. Indian Chem. Soc.* **30**, 10.

R. Gopal (1955) *Z. Anorg. Allg. Chem.* **279**, 229.

J. E. Gordon (1969) *Applications of Fused Salts in Organic Chemistry*, ed. D. B. Denny, Marcel Dekker, New York.

L. F. Grantham and S. J. Yosim (1968) *J. Chem. Phys.* **72**, 762.

L. F. Grantham (1965) *J. Phys. Chem.* **43**, 1415.

L. F. Grantham (1966) *J. Phys. Chem.* **44**, 1509.

L. F. Grantham (1968) *J. Phys. Chem.* **49**, 3835.

J. Greenberg and L. J. Hallgren (1960) *J. Chem. Phys.* **33**, 900.

J. Greenberg and L. J. Hallgren (1961) *J. Chem. Phys.* **35**, 180.

C. Hall and E. Yeager (1973) Vol. 2, p. 399 *Techniques of Electrochemistry* ed. Yeager and Salkind, Wiley.

R. T. Harley, W. Hayes, A. J. Rushworth, and J. F. Ryan (1975) *J. Phys. C* **8**, L 530.

R. L. Harris, R. E. Wood, and H. L. Ritter (1951) *J. Am. Chem. Soc.* **73**, 3150.

J. Hazlewood, E. Rhodes, and A. R. Ubbelohde (1963) *Trans. Faraday Soc.* **59**, 491.

J. Hazlewood, E. Rhodes, and A. R. Ubbelohde (1966) *Trans. Faraday Soc.* **62**, 3101.

L. Heyne (1970) *Electrochim. Acta* **15**, 1251.

D. W. James (1966) *Aust. J. Chem.* **19**, 993.

D. W. James and W. H. Leong (1969) *J. Chem. Phys.* **51**, 640.

D. W. James and W. H. Leong (1970) *Trans. Faraday Soc.* **66**, 1948.

D. W. James and G. J. Janz (1962) *Electrochem. Acta* **7**, 427.

G. J. Janz (1962) *J. Chem. Educ.* **39**, 59.

G. J. Janz and J. Goodkin (1959) *J. Phys. Chem.* **63**, 1975.

G. J. Janz and D. W. James (1962) *Electrochim. Acta* **7**, 427.

G. J. Janz and D. W. James (1963) *J. Chem. Phys.* **38**, 902.

G. J. Janz and M. R. Lorenz (1961) *J. Electrochem. Soc.* **108**, 1052.

G. J. Janz and J. D. E. McIntyre (1962) *J. Electrochem. Soc.* **109**, 842.

G. J. Janz and R. D. Reeves (1965) in *Advances in Electrochemistry and Electrochemical Engineering*, ed. Delahay and Tobias, Wiley–Interscience.

G. J. Janz, C. Solomons and H. J. Gardner (1958) *Chem. Rev.* **58**, 461.

G. J. Janz, E. Neuenshchwander, and F. J. Kelly (1963) *Trans. Faraday Soc.* **59**, 841.

G. J. Janz, R. P. T. Tomkins, H. Siegenthaler, K. Balasubrahmanyan, and S. W. Lurie (1971) *J. Phys. Chem.* **75**, 4025.

J. W. Johnson, P. A. Agron, and M. A. Bredig (1955) *J. Am. Chem. Soc.* **77**, 2734.

J. W. Johnson, W. J. Silva, and D. Cubiciotti (1968) *J. Phys. Chem.* **72**, 1664, 1669.

W. V. Johnston, H. Wiedersich, and G. W. Lindberg (1970) *Chemistry of Extended Defects in Non-Metallic Solids*, ed. Le Roy Eyring and M. O'Keefe, North-Holland, Amsterdam.

C. E. Johnson, S. E. Wood, and E. J. Cairns (1967) *J. Chem. Phys.* **46**, 4168.
K. K. Kelley (1960) *Bur. Mines Bull.* 584.
J. A. A. Ketelaar, C. H. Macgillvary, and P. A. Renes (1947) *Rec. Trav. Chim. Pays-Bas* **66**, 501.
J. A. A. Ketelaar (1975) *Z. Phys. Chem.*, *N.F.* **98**, 453.
J. A. A. Ketelaar (1976) *Silicates Industriels* **3**, 105.
A. Kisza and J. Hawranek (1968) *Z. Phys. Chem.* (*Leipzig*) **237**, 210.
W. Klemperer, W. G. Norris, and A. Buckler (1960) *J. Chem. Phys.* **33**, 1534.
O. J. Kleppa (1961) *J. Phys. Chem. Solids* **23**, 819.
O. J. Kleppa and L. S. Hersh (1962) *Discuss. Faraday Soc.* **32**, 99.
H. Kohma, S. G. Whiteway, and C. R. Masson (1968) *Can. J. Chem.* **46**, 2968.
E. Kordes, G. Ziegler, and H. J. Proeger (1954) *Z. Elektrochem.* **58**, 168.
E. Kordes, W. Bergmann, and W. Vogel (1951) *Z. Elektrochem.* **55**, 600.
J. Koster and J. A. A. Ketalaar (1969) *J. Chim. Phys. Physicochim. Biol* (1969) Num. Spéc 171–6. Eng.
A. Kutzelnigg (1958) *Monatsh. Chem.* **89**, 459.
G. J. Landon and A. R. Ubbelohde (1956) *Trans. Faraday Soc.* **52**, 647.
G. J. Landon and A. R. Ubbelohde (1957) *Proc. Roy. Soc.*, *Ser. A* **240**, 160.
D. Leonesi, G. Piantoni, G. Berchiesi, and P. Franzosini (1968) *Ric. Sci.* **38**, 702.
H. A. Levy and M. D. Danford (1964) in M. Blander, *Molten Salt Chemistry*, 1964, Wiley, New York.
J. E. Lind, H. A. Abdel Rehim, and S. W. Rudich (1966) *J. Phys. Chem.* **70**, 3610.
K. D. Luke and H. T. Davis (1967) *Ind. Eng. Chem. Fundamentals* **6**, 194.
J. D. MacKenzie and W. K. Murphy (1960) *J. Chem. Phys.* **33**, 366.
S. Matsura and M. Nishikawa (1960) *Sci. Pap. Coll. Gen. Educ. Univ. Tokyo* **10**, 29.
S. W. Mayer, S. J. Yosim, and L. E. Topol (1960) *J. Phys. Chem.* **54**, 238.
J. W. Menary, A. R. Ubbelohde, and I. Woodward (1951) *Proc. Roy. Soc.*, *Ser. A* **208**, 158.
H. J. Michels and A. R. Ubbelohde (1972) *J. Chem. Soc.*, *London* 1879.
H. J. Michels and A. R. Ubbelohde (1974) *Proc. Roy. Soc.*, *Ser. A* **338**, 447.
J. R. Morrey and D. G. Carter (1968) *J. Chem. Eng. Data* **13**, 94.
I. Murgulescu (1969a) *Rev. Roum. Chim.* **14**, 965.
I. Murgulescu (1969b) *J. Chim. Phys. Physicochim. Biol.* **66**, 43.
I. Murgulescu (1969c) *J. Chim. Phys. Physicochim. Biol. Num. Spéc.* 43.
A. Mustajoki (1957) *Ann. Acad. Sci. Fenn. Series* AVI, Paper 7.
R. A. Osteryoung, C. Kaplan, and D. L. Hill (1961) *J. Phys. Chem.* **65**, 1951.
B. B. Owens and G. R. Argue (1967) *Science* **157**, 308.
G. Petit and C. Bourlange (1957) *C.R. Acad. Sci.*, *Paris* **245**, 1788.
G. Petit and F. Delbove (1962) *C.R. Acad. Sci.*, *Paris* **254**, 1388.
C. W. F. T. Pistorius (1961) *J. Inorg. Nucl. Chem.* **19**, 367.
C. W. F. T. Pistorius (1965a) *J. Phys. Chem. Solids* **26**, 1543.
C. W. F. T. Pistorius (1965b) *J. Chem. Phys.* **43**, 155.
C. W. F. T. Pistorius (1966) *J. Chem. Phys.* **45**, 3513.
C. W. F. T. Pistorius (1974) *Z. Phys. Chem.*, *N.F.* **88**, 253.
C. W. F. T. Pistorius and J. C. A. Boyer (1968) *J. Chem. Phys.* **48**, 1018.
C. W. F. T. Pistorius and J. B. Clark (1968) *Phys. Rev.* **173**, 692.
L. L. Quill (1950) *Chemistry and Metallurgy of Miscellaneous Materials* (*Thermodynamics*), McGraw-Hill, p. 76.
E. Rapoport (1966a) *J. Chem. Phys.* **45**, 2721.
E. Rapoport (1966b) *J. Phys. Chem. Solids* **27**, 1349.

E. Rapoport and C. W. F. T. Pistorius (1966) *J. Chem. Phys.* **44,** 1514.
V. C. Reinsborough (1968) *Rev. Pure and Appl. Chem.* **18,** 281.
H. Reiss, S. W. Mayer, and J. L. Katz (1961) *J. Chem. Phys.* **35,** 820.
R. Riccardi (1961) *Gazz. Chim. Ital.* **91,** 1479.
R. Riccardi (1962) *Gazz. Chim. Ital.* **92,** 34.
R. Riccardi and C. Benaglia (1961) *Gazz. Chim. Ital.* **91,** 315.
H. Rickert (1973) in *Fast Ion Transport in Solids,* ed. W. van Gool, North-Holland, Amsterdam.
E. Rhodes and A. R. Ubbelohde (1959) *Proc. Roy. Soc., Ser. A* **251,** 156.
S. E. Rogers and A. R. Ubbelohde (1950) *Trans. Faraday Soc.* **46,** 1051.
M. Rolin and M. Bernard (1962) *Bull. Soc. Chim. Fr.* **423,** 42.
T. J. Rowland and J. P. Bromberg (1958) *J. Chem. Phys.* **29,** 626.
S. W. Rudich and J. E. Lind (1969) *J. Phys. Chem.* **73,** 2099.
S. W. Rudich and J. E. Lind (1970) *J. Chem. Phys.* **50,** 3035.
H. Schinke and F. Sauerwald (1956) *Z. Anorg. Allg. Chem.* **287,** 313.
H. Schinke and F. Sauerwald (1960) *Z. Anorg. Allg. Chem.* **304,** 26.
A. Seyyedi and M. G. Petit (1959) *J. Phys. Radium* **20,** 832.
A. F. Skoulios and V. Luzzatti (1961) *Acta Cryst.* **14,** 278.
V. A. Sokolov and N. E. Shmidt (1955) *Izv. Sekt. Fiz-Khim. Anal. Inst. Obschei Neorg. Khim. Akad. Nauk SSSR* **26,** 123.
V. A. Sokolov and N. E. Shmidt (1956) *Izv. Sekt. Fiz.-Khim. Anal. Inst. Obschei Neorg. Khim. Akad. Nauk SSSR* **27,** 17.
P. L. Spedding (1973) *Electrochim Acta* **18,** 111.
S. Sternberg and L. Marta (1960) *Rev. Chim. Acad. Rep. Populare Roum.* **5,** 281.
A. Timidei and G. J. Janz (1968) *Trans. Faraday Soc.* **64,** 202.
L. E. Topel and L. D. Ranson (1960) *J. Chem. Phys.* **64,** 1339.
L. M. Torell (1975) *J. Acoust. Soc. Am.* **57,** 876.
G. Treiber and K. Todheide (1973) *Ber. Bunsenges. Phys. Chem.* **77,** 1079.
P. Tyrolerova and W. K. Lu (1969) *J. Am. Ceram. Soc.* **52,** 77.
A. R. Ubbelohde (1959) in *The Structure of Electrolytic Solutions,* Wiley, New York, p. 401.
A. R. Ubbelohde (1960) *Proc. Chem. Soc.* 332.
A. R. Ubbelohde (1962) *Rev. Chim. Acad. R.P. Roum.* **7,** 625.
A. R. Ubbelohde (1965) *Melting and Crystal Structure,* Oxford University Press.
A. R. Ubbelohde (1976) *Rev. int. Hautes Temp. et Refract.* **13,** 5.
A. R. Ubbelohde and K. J. Gallagher (1955) *Acta Cryst.* **8,** 71.
A. R. Ubbelohde, H. J. Michels, and J. J. Duruz (1972) *J. Phys. E* **5,** 283.
Y. Umetsu, Y. Ishii, T. Sawada, and T. Ejima (1973) *J. Japanese Inst. Met.* **37,** 997.
E. A. Uskke and N. G. Bukun (1972) *Elektrokhimiya* **8,** 163.
J. Vallier (1968) *J. Chim. Phys. Phys. Chim. Biol.* **65,** 1762.
G. H. Vasu (1969) *Rev. Roum. Chim.* **14,** 167.
G. E. Walrafen, D. E. Irish, and T. F. Young (1962) *J. Chem. Phys.* **37,** 662.
D. O. Welch and G. J. Dienes (1975) *J. Electron Mater. (U.S.A.)* **4,** 973.
A. F. Wells (1962) *Structural Inorganic Chemistry,* 3rd edn., Oxford University Press.
W. A. Weyl (1955) *J. Phys. Chem.* **59,** 147.
K. Williams, P. Li, and J. P. Devlin (1968) *J. Chem. Phys.* **48,** 3891.
L. V. Woodcock (1976) *Proc. Roy. Soc., Ser. A* **348,** 187.
G. Zarzycki (1957) *J. Phys. Radium* **18,** 65A.
G. Zarzycki (1958) *J. Phys. Radium* **19,** 4, 13A.
G. Zarzycki (1961) *Discuss. Faraday Soc.* **32,** 38.

# 9. MELTING OF METALS

## 9.1 Metallic crystals as models of melting

By metals is commonly meant condensed states of matter that conduct electricity readily, by electron rather than ion transport, and whose electrical conductivity decreases with rising temperature. Some authors regard mechanical properties such as plasticity and work hardening as even more significant for the metallic state than the collective electronic properties. However, in this book only rather indirect aspects about such mechanical aspects of metals will be discussed.

Crystals of metals have been favourite targets for contrivers of model mechanisms of melting for many years. This arises from the fact that their crystal structure often contains atoms with radially symmetrical force fields and that close-packed assemblies are formed by the majority of metals. An apparent simplicity results, which is however probably deceptive; refined methods of measurement reveal a complex diversity of behaviour. In addition to the other structural changes occurring, melting may affect specific metallic (i.e. electronic) characteristics which depend upon the collective band of conduction electrons, with respect particularly to the entropy, and enthalpy. Information from changes of melting of near-metallic crystals (i.e. semimetals and semiconductors) brings out yet other ways in which complexities in melting mechanisms may arise.

In principle, a rough separation may be attempted between different kinds of contributions to the entropy of melting, as has already been illustrated for other types of crystals. For metallic crystals

$$S_f = \Delta S_{vib} + \Delta S_{pos} + \Delta S_{electronic}$$

Complexities arise especially in the ways that liquid metals show variations in the second and third terms of this approximate summation.

This book emphasizes features and problems of melting in relation to crystal structure, and one major objective is to interpret the thermodynamic parameters of crystals and melts in terms of known crystal structures. In the upshot, much research on the electrical properties of conducting crystals and their melts has yielded two lines of information about melting that are distinctive. One of these stems from the actual *changes* in electrical properties at the melting point, on passing from crystal to melt. As is detailed below, in many cases these changes are relatively small; quite close similitude might be inferred between the structures of metallic crystals and their melts, though the possibility of extended precursor effects in the crystals and in their melts affects this conclusion, which still seems undecided. A second line of information can be important whenever the liquid state considered *per se* shows 'anomalous' electrical properties. Detailed theories concerning 'anomalous' electrical properties of melts are often only qualitative, but what can be regarded as normal behaviour seems well established so that informative 'anomalies' can be picked out with assurance. Review articles on liquid electrical conductors are further referred to in what follows (see also Guntherodt, 1970; Aleksseev, Andreev, and Prokhorenko, 1972; Regel, Glazov, and Alvazov, 1974).

Many metals assume close-packed structures in the crystals. Hexagonal close packed (A3) and face centred cubic (A1) have 12 nearest neighbours, and body centred cubic (A2) has 8 nearest neighbours. These high coordination numbers can be explained in resonance theories of valency (Pauling, 1960) on the basis that the atoms in the crystals form a multiplicity of equivalent resonance bonds with several nearest neighbours. In some metallic elements more than one type of nearest neighbour bond can be identified (see Table 9.1). This generally leads to 'anomalous' behaviour on melting; a change to more uniform nearest neighbour linkages sometimes results in the melt, e.g. for Sb (Krebs, 1969). On the other hand, the greater packing freedom which results from the expansion in volume on melting may favour formation of 'hybrid' regions in the melt; notionally these can be referred to more than one crystal structure (see Chapter 4).

## 9.2 Vibrational effects in the melting of metals—solid–solid transformations

It might be thought that the close-packed crystal structure of many metals would be associated with a particularly simple vibrational spectrum. Probably for this reason, vibrational theories of melting such as the Lindemann theory (Chapter 3) at one time tended to concentrate on metallic crystals in the search for illustrative examples. This is unfortunate; there can be no doubt however that interatomic vibrations play a large part in determining various physical properties of metals. The importance of vibrations for the thermodynamic parameters as well as the complexity of their role is borne out by the fact that at least 26 metals transform at ordinary pressures from a low temperature close-packed structure to a high-temperature b.c.c. phase. These transformations occur often quite close to the melting point and thus may lead, *inter alia*, to a hybrid reference state for positional melting as well as hybrid domain topology of the melt (see below). They appear to have a vibrational origin. In general terms, competition between the rival crystal structures can be attributed to the somewhat lower Debye temperature $\theta_D$ in the b.c.c. polymorph (Grimvall and Sjodin, 1975) in a way discussed on p. 73. Anharmonicity of the vibrations does not appear to be a prime factor causing the change.

The entropy of transformation $\Delta S_{trans}$ in general depends on the changes of nearest neighbour environment. A suggested regularity (Cho, 1974) is

$$\text{h.c.p.} \to \text{f.c.c.} \qquad \Delta S_{trans} = 0\cdot17 \text{ e.u.}$$

$$\text{f.c.c.} \to \text{b.c.c.} \qquad \Delta S_{trans} = 0\cdot51 \text{ e.u.}$$

$$\text{h.c.p.} \to \text{b.c.c.} \qquad \Delta S_{trans} = 0\cdot68 \text{ e.u.}$$

but closer examination (Gschneider, 1975) indicates that other factors such as the number of valence electrons also contribute to both crystal transformation and fusion. This can be seen from actual experimental values.

**Table 9.1.** *Thermodynamic parameters of metallic crystals and their melts in relation to periodic system*[a]

| | $CN$ in crystal | $S_f$ (e.u.) | $\dfrac{\Delta V_f}{V_s}$ (%) | $C_p$ at $I_f^{-1}$ cal atom deg$^{-1}$ | | $V_s$ (cm$^3$) | $10^3 \times \alpha$ (deg$^{-1}$) at $T_f$ | |
|---|---|---|---|---|---|---|---|---|
| | | | | S | L | | S | L |
| Li | 8 | 1·53 | 1·65 | 15·3 | 15·4 | 13·68 | 0·18 | — |
| Na | 8 | 1·70 | 2·5 | 19·0 | 20·2 | 24·74 | 0·22 | 0·275 |
| K | 8 | 1·70 | 2·55 | 20·8 | 22·8 | 47·15 | 0·25 | 0·29 |
| Rb | 8 | 1·68 | 2·5 | 23·0 | 25·2 | — | 0·27 | 0·34 |
| Cs | 8 | 1·65 | 2·6 | 24·4 | 26·4 | — | 0·29 | 0·365 |
| Be | — | — | — | — | — | — | — | — |
| Mg | 12 | 2·25 | 3·1 | — | — | 15·29 | 0·078 | 0·122 |
| Al | 12 | 2·70 | 6·0 | 8·4 | 8·7 | 11·38 | 0·099 | 0·122 |
| *Ga[c] | 1+6 | 4·42 | −3·2 | 21·1 | 22·5 | — | 0·054 | 0·126 |
| In | 4+8 | 1·82 | — | — | — | — | — | — |
| Si[b] | 4 | 6·47 | −9·6 | — | — | 11·1 | 0·23 | 0·145 |
| Ge[b] | 4 | 5·93 | — | — | — | — | — | — |
| Ti | 12 | 1·79 | — | 12·4 | 11·7 | — | 0·126 | 0·150 |
| Sn[b] | — | 3·32 | — | — | — | — | — | — |
| Pb | 12 | 1·98 | 3·5 | 11·6 | 12·5 | 19·58 | 0·12 | 0·13 |
| *Bi[c] | 3+3 | 4·78 | −3·35 | 12·6 | 14·0 | — | 0·040 | 0·12 |
| Sb | 3+3 | 5·25 | −0·95 | 7·1 | 7·9 | 18·80 | 0·033 | 0·10 |
| Zn | 6+6 | 2·48 | 4·2 | 12·6 | 14·0 | 9·95 | 0·113 | 0·154 |
| Cd | 6+6 | 2·57 | 4·7 | 11·6 | 12·0 | 14·05 | 0·126 | 0·165 |
| Hg | 6+6 | 2·37 | 3·7 | 29·0 | 28·3 | — | 0·171 | 0·182 |
| Fe | 12 | 2·01 | — | 5·8 | 5·2 | — | 0·057 | — |
| Co[b] | 12 | 2·12 | — | — | — | — | — | — |
| Ni | 12 | 2·45 | — | 4·6 | — | — | — | — |
| Cu | 12 | 2·29 | 4·1 | 5·5 | 5·7 | 7·95 | 0·057 | 0·097 |
| Ag | 12 | 2·22 | 3·8 | 6·05 | 6·7 | 11·6 | 0·042 | 0·069 |
| Au | 12 | 2·29 | 5·1 | 5·5 | 5·2 | 11·5 | 0·093 | 0·150 |

[a] Data from Kubaschewski, 1949; McGonigal, Kirshenbaum, and Grosse, 1962; Sing and Kumer, 1968. Minor discrepancies are found between different sets of data.
[b] See Lumsden 1952.
[c] Asterisks indicate contraction on melting.

Table 9.1 groups metals according to the Periodic System, i.e. with regard to the (probable) number of valence electrons.

Data for solid–solid transformations and for fusion of other rare earth metals (Jayaraman, 1965b) are illustrated in Tables 9.2 and 9.3. Other correlations between the valence state of the metal atoms and their thermodynamic properties are also discussed by Jayaraman (1965b).

**Table 9.2. Entropy of transformation of metallic crystals**

| Type h.c.p. $(A3) \rightarrow b.c.c.$ $(A2)$ | $\Delta S_{trans}$ $(e.u.)$ |
|---|---|
| Li | 0·134 |
| Sc | 0·60 |
| Ti | 0·88 |
| Y | 0·68 |
| Zr | 0·83 |
| Dy | 0·60 |
| Tl | 0·18 |

| f.c.c. $(A1) \rightarrow h.c.c.$ $(A2)$ | |
|---|---|
| Ca | 0·31 |
| Mn | 0·32 |
| Fe | 0·12 |

### 9.2.1. Melting parameters of rarer metals

**Table 9.3. Crystal transformations $(f.c.c. \rightarrow b.c.c.)$ and melting of rare earth metals ($^{\circ}C$)**

| Atom | $T_c$ | $\Delta S_{trans}$ $(e.u.)$ | $T_f$ | $S_f(e.u.)$ |
|---|---|---|---|---|
| La | 864 | 0·67 | 920 | 1·34 |
| Ce | 725 | 0·71 | 797 | 1·16 |
| Pr | 792 | 0·71 | 935 | 1·37 |
| Nd | 862 | 0·63 | 1024 | 1·32 |
| Sm | 917 | 0·63 | 1072 | 1·53 |
| Eu | — | — | 826 | 2·23 |
| Gd | 1264 | 0·61 | 1312 | 1·53 |
| Tb | 1317 | 0·8 | 1356 | 1·56 |
| Dy | 1392 | — | 1407 | 2·04 |
| Ho | 1442 | — | 1461 | 2·25 |
| Er | — | — | 1497 | 2·67 |
| Tm | — | — | 1545 | 2·31 |
| Yb | 798 | 0·4 | 824 | 1·64 |
| Lu | — | — | 1652 | 1·71 |

(For the metals Sm, Pr, Nd, the solid–solid transformation itself may be of hybrid type (see Chapter 4).

## 9.2.2. *Role of crystal similitude*

At this point, it is convenient to cite some entropies of fusion in relation to the crystal type, to illustrate the role of crystal similitude without implying that $\Delta S_{vib}$ contributes appreciably, or that $\Delta S_{pos}$ and $\Delta S_{electronic}$ can be disregarded in the melting of metals (Gschneider, 1975). Table 9.4 groups the metals according to the crystal types in which they crystallize; it may be contrasted with Table 9.1 in some cases, pointing to the need for further research, e.g. on crystal structures near $T_f$, and on melting of hybrid crystals (see below).

**Table 9.4. Melting parameters in relation to crystal coordination numbers**

| h.c.p. (A3) metals | $S_f$ (e.u.) | $\dfrac{\Delta V_f}{V_s}$ | f.c.c. (A1) metals | $S_f$ (e.u.) | $\dfrac{\Delta V_f}{V_s}$ |
|---|---|---|---|---|---|
| Mg | 2·32 | 3·6 | Al | 2·76 | 6·5 |
| Zn | 2·53 | 4·3 | Co | 2·19 | 3·5 |
| Cd | 2·49 | 3·8 | Ni | 2·42 | 5·4 |
| Ho | (2·31) | (7·4) | Cu | 2·30 | 4·2 |
| Er | 2·65 | (9·0) | Pd | 2·25 | 5·9 |
| Tm | 2·22 | (6·9) | Ag | 2·19 | 3·8 |
| *b.c.c. (A2) metals* | | | Pt | 2·30 | 6·6 |
| | | | Au | 2·24 | 5·2 |
| Li | 1·58 | 2·2 | Pb | 1·91 | 3·6 |
| Na | 1·67 | 2·5 | | | |
| K | 1·66 | 2·5 | Mn | 2·31 | 1·7 |
| Rb | 1·68 | 2·4 | Yb | 1·51 | — |
| Cs | 1·66 | 2·6 | Nb | 2·34 | — |
| Tl | 1·72 | 2·2 | Mo | 2·69 | — |
| Ca | 1·84 | — | Sm | 1·53 | 3·6 |
| Sr | 2·10 | — | Eu | 2·02 | 4·8 |
| Ba | 1·85 | — | Gd | 1·52 | 2·0 |
| Sc | 1·86 | — | Tb | 1·59 | 3·1 |
| Cr | 1·62 | — | Dy | 1·57 | (4·5) |
| Fe | 1·82 | 3·5 | Yb | 1·67 | 5·1 |
| | | | W | 2·31 | — |
| | | | U | 1·45 | 2·2 |

b.c.c. metals with exceptionally low or negative volume changes on melting include

| | $S_f$ (e.u.) | $\dfrac{\Delta V_f}{V_s}(\% \ e.u.)$ |
|---|---|---|
| La | 1·24 | 0·6 |
| Ce | 1·22 | −1·1 |
| Pr | 1·36 | 0·02 |
| Pu | 0·74 | −2·4 |
| Nd | 1·32 | 0·9 |

### 9.2.3. *Correlation of vibrational properties with $T_f$*

Some metals also give other smooth correlations of vibrational properties with $T_f$ (see p. 80). For example, it has long been known that a close connection can be found between vibrations and the viscosity of the melts in the form of a correlation between $T_f$ and the viscosity $\eta_f$ at the melting point of a number of molten metals (Andrade, 1934) in the equation

$$\eta_f = 5 \cdot 7 \times 10^{-4} (MT_f)^{\frac{1}{2}} (1/V)^{\frac{2}{3}}$$

where $M$ is the atomic weight and $V$ the atomic volume. For metals with cubic structure in the solid state, calculated values of $T_f$ agree within 10% of those observed. Coefficients of self-diffusion $D_f$ can be evaluated from $\eta_f$ and the Stokes–Einstein relation

$$D_f = RT/6\pi r \eta_f$$

but it is notable that $r$ must be taken as the ionic and not the atomic radius (Mackenzie and Hillig, 1958). The extent to which such correlation of transfer parameters can be uniquely interpreted in terms of vibrational properties and by a simple structural model for molten metals which consist of positive ions interpenetrated by a plasma of electrons is, however, not clear. No alternative models for melts of metals appear to have received detailed analysis in relation to viscous relaxation of shearing stresses. However, significant findings are that grain boundary diffusion is already high in the solid state. Also, on melting the diffusion coefficient may increase by several orders of magnitude (Lange, 1966).

The role of cooperative defects (p. 127) in the melting of metals calls for further study with particular reference to their effects on transport parameters such as diffusion or viscosity. Data in Tables 9.2, 9.3, and 9.4 suggest that comparisons between melts originating from different crystal types could be revealing with respect to these transport properties. As an empirical finding probably related to this problem, it is intriguing to note that at their melting points the ratio of velocities $U$ of longitudinal waves $U_S/U_L$ is approximately $1 \cdot 26$ for all the metals Zn, Cd, Sn, Pb, Bi, Hg, Na, K, Rb, Cs and Ga, whereas for waves of dilation $U_S/U_L \sim 1 \cdot 0$; values for individual metals fluctuate considerably about this mean. Improved experimental methods of measuring mechanical properties may perhaps resolve difficulties thrown up by sonic measurements, e.g. of compressibilities of molten IIA metals (McAlister, Crozier, and Cochran, 1974). At present the accuracy of sonic measurements hardly warrants more detailed discussion (Schaafs, 1956). Measurements of the propagation of longitudinal sound have been made in molten Sn and Pb using a frequency of $5 \text{ Mc s}^{-1}$. Down to $T_f$ results can be fitted by a straight-line equation with no marked anomalies (Gordon, 1959). Calculations of velocities of sound in liquid metals have been compared with experimental data to test the Percus–Yevick theory of liquids (Siegel, 1975). In providing a norm by means of which to single out those molten metals which are not simple this approach can probably prove useful.

Proposed uses of diverse parameters based on acoustic measurements, to characterize 'restructuring' in metals on melting, and as their temperature rises above $T_f$, include a Gruneisen parameter $\Gamma$ (Kor and Pandey, 1973) and an 'ultrasonic' parameter $\gamma_G$ (Knopoff and Shapiro, 1970). $\Gamma = (\alpha/c_p)(\beta V)$ where $\beta$ is the adiabatic compressibility and $\gamma_G = (\alpha/c_p)(v_H)^2$ where $v_H$ is the velocity of ultrasound. With more data such parameters may prove diagnostically useful in differentiating between different textures of molten metals.

### 9.3 Thermodynamic parameters of melting of metals

Tables 9.3 and 9.4 record a number of melting parameters. These tables, as well as Table 9.1, serves to illustrate some general features in the melting of metals. Further peculiarities of behaviour may be found, particularly when accurate data are examined with regard to premelting and prefreezing effects.

On comparing with values for inert gas atoms (Table 6.1), it is notable that metals in general have considerably lower entropies of fusion. The volume increase on melting is also much smaller for most metals than for inert gas crystals, or even for many ionic crystals with inert gas ions (Table 8.1). These findings suggest that the melting mechanisms of metals may involve different entropy factors, not prominent in the other types of crystals considered. Plausibly, these may be related to the intermetallic bonding forces which can be much more versatile (through resonance effects) than in other crystal types. Consequences for the domain topology of melts are discussed in what follows. Furthermore, properties of electron bands in metallic crystals may undergo significant changes on melting; unfortunately, electron band theory by its nature tends to give an unduly smooth representation for disordered condensed states of matter.

### 9.3.1. *Specific heats of metals near $T_f$*

The specific heats $C_p$ of solid and molten metals, and the entropies of fusion, clearly call for a more searching interpretation than can be given by elementary vibrational theories, even making allowances for anharmonicity. It does not seem likely that evaluation of $C_v$ from $C_p$ will greatly help with the problem to be considered (see various papers by Grimvall and co-authors 1975–1977).

On approaching $T_f$ atomic heats can rise far above Dulong and Petit values; as discussed below, this rise is particularly notable for the alkali metals whose structure is often considered to be specially simple but which may be contrasted with Cu, Ag, Au in their atomic heat. Again, at the melting points, the jump in entropy generally is notably lower than for even inert gas crystals, and the specific heats of the melts remain practically as high as for the solid state. Corresponding statements can be made about thermal expansions, to which the vibrational energies must make a major contribution.

### 9.3.2. Entropies of fusion

As stated, a peculiarity to be noted from Tables 9.1, 9.3, 9.4 is that entropies of fusion are low compared with values calculated for inert gas crystals on the basis of simple theories of positional disordering (Chapter 5). This may perhaps be due to precursor effects on either side of $T_f$, as discussed elsewhere (p. 334). However, because of the high coordination number and resonance bonding to nearest neighbours in metals, types of crystal defect may need to be reckoned with that are not found with other kinds of crystals forces. It has been estimated (Borelius, 1970) that in the metals Cu, Ag, Au and Pb, defect sites require only about 5 kcal (g atom)$^{-1}$ for their creation, compared with vacancy creation energies of about 24 kcal mole$^{-1}$. In addition to simple positional defects as described, for example, for ionic crystals or inert gas crystals, it seems possible that the overall entropy of fusion of metals might include to an exceptional degree cooperative positional disorder as part of the mechanism of melting.

Although $\Delta V_f / V_s$ is not large for the majority of metals, the number of 'abnormal' crystals for which $\Delta V_f / V_s$ is negative is very small; it is interesting that many of them have semi-metallic or near-metallic properties. For these, the ratio of electrical resistivity $\rho_L / \rho_S$ is less than unity (see Table 9.5), unlike most metals. The range of known instances has been somewhat extended by including intermetallic compounds, such as $Mg_2Pb$ for which both these parameters change in the same anomalous direction (Knappwost, 1959) on melting. There is considerable evidence that melts of this crystal compound contain undissociated clusters of $Mg_2Pb$ as well as positionally randomized atoms of Pb and Mg (see Chapter 13).

Dislocations in metal crystals have been extensively studied, but mainly in connection with theories of strength and work hardening. Only rather elementary forms of dislocations have been considered up to the present. A theory of dislocation melting has indeed been examined for certain simple metals but only on a tentative basis (see Chapter 11). One significant feature of this model of dislocation melting is that when the number of atoms on dislocation lines becomes appreciable in relation to the total number of lattice sites, the vibrational entropy may be appreciably different for melt and for crystal. A tendency to undergo dislocation melting in certain crystals could also help to contribute to the phenomenon of superplasticity observed in some alloys (Sherby, 1969). The difference $S_{vib}$(liquid)$-S_{vib}$(solid) may then make a substantial contribution to $S_f$, thereby increasing contributions from positional disorder. Any factor which raises $S_f$ without demanding corresponding increases in $H_f$ tends to favour the existence of the liquid over a wider range of temperatures, relative to the crystals. Although no quantitative theory of melting of metal crystals has been put forward, various lines of evidence indicate that the liquid range is in fact exceptionally long, e.g. for the alkali metals. Some comparative figures are given in Table 2.6. Other aspects of the melting mechanisms are discussed in the sections following.

It might be sought to attribute these varied thermodynamic peculiarities of metallic crystals to (only partially known) contributions from the electronic band energy to the entropy. These cannot take any very simple form however. If one writes tentatively

$$S_f = \Delta S_{positional} + \Delta S_{vib} + \Delta S_{electronic}$$

the total value of $S_f$ is already so low that no major contributions to $S_f$, from $\Delta S_{vib}$, e.g. from dislocation melting, can be looked for. No major contributions to $\Delta S_{electronic}$ appear to have been yet identified from electronic or magnetic free energy of metals (Allgaier, 1969; Grimvall, 1975b) except perhaps in the solid–solid transitions in iron (Grimvall, 1976). Positional defects call for closer examination, in the light of all the foregoing.

## 9.4 Positional defects in melting of metals

Various authors have attempted to attribute peculiarities in the melting of metals to the formation of positional vacancies. Evidence is often rather indirect (see discussion of premelting, Chapter 12). Even though some impressive correlations between various melting parameters can be interpreted in terms of vacancy formation (see Gorecki, 1974), direct evidence is hard to establish incontrovertibly for even simple lattice holes. The proposed formula for their concentration $n/N$ in the crystals at $T_f$

$$(n/N)_{T_f} = 0.0037$$

must still be regarded as semiempirical.

Systematic plots of specific heats and of thermal expansions of some 20 crystals, mostly metals, have been discussed by Borelius (1963, 1970), who has stressed that the process of melting should always be considered to include quite extended premelting and postmelting. Figure 9.1 illustrates the way in which excess specific heat may build up in metallic sodium, as one example (Borelius, 1963). With respect to structure and energy, the problem is to find sufficiently precise descriptions of departures from regularity in the crystals, as well as in melts of metals, that could account for such extended anomalies. Known types of vacancy would not seem to fit theoretical models

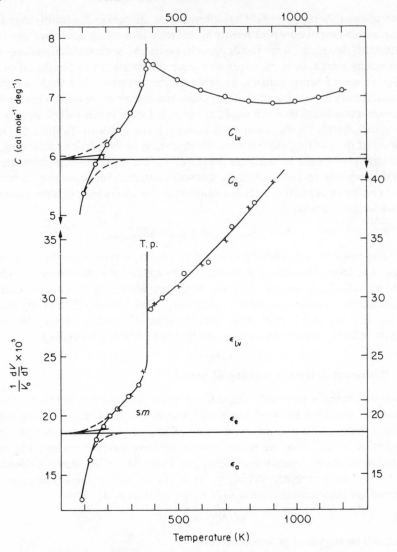

FIG. 9.1. Specific heat and expansion coefficient of sodium. Reproduced
with permission from Borelius, *Solid State Physics* **15,** 1 (1963)

for vacancy melting of crystals such as are discussed in Chapter 11. Furthermore, for melts of metals the precursor problems are far from symmetrical (as
discussed in Chapters 12, 13).

One way of resolving the apparent dilemma is to look for less usual kinds of
defects capable of giving extensive premelting effects in the crystals, and
perhaps also of prefreezing effects in the melts. Defect structures that could

help to spread the total increase of disorder over the wide range of temperatures observed are discussed in Chapters 12 and 13 in relation to premelting and prefreezing generally. Clearly, if the total excess entropy $S_{liquid} \rightarrow S_{solid}$ includes precursor effects in significant proportions, it becomes less baffling to account for the thermodynamic peculiarities that are found in the melting of metals. For these, in contrast to other types of crystals, additional possibilities could arise from the nature of the interatomic (resonance) bonding in crystalline metals, which varies according to their location in the Periodic Table. In Group IA and Group IIA metals all the bonds are equivalent. Just as for the inert gases, unitary positional defects can be created by vacancy formation and/or insertion of atoms. However, in addition, small domains of b.c.c. and f.c.c. packing can probably coexist near a solid–solid transformation point. Resonance bonding between atoms on opposite sides of any interface separating two domains will not be very different from the resonance bonding between atoms within any one domain, since these ligand energies are not very sensitive to bond angles when the coordination number is high.

In terms of parameters discussed previously, a reasonable hypothesis is thus that the *domain interface energies* $\sigma_{12}$ in the hybrid solids (Chapter 4) may be exceptionally small for transitions in these metals. By analogy, the solid–melt interface energy $\sigma_{SL}$ (Chapter 12) may also be small compared with other types of crystals. When this is so, statistical thermodynamic theory of the kind originally outlined by Frenkel (p. 314) indicates that premelting should be extensive. The peculiar thermodynamic properties of crystals of the alkali metals already referred to (p. 243) could thus be attributed to their having an unusually extensive system of cooperative defects in the crystals. This may at the same time help to explain their plasticity, which is a well known peculiarity of metals. It is probably also significant that near $T_f$ vibration amplitudes as observed at crystal–gas interfaces by LEED are much greater than inside the bulk of the crystals (Goodman and Somorjai, 1970). A similar situation is likely to apply at domain interfaces, with consequent decrease of free energy of the interface and reduction of $\sigma_{12}$ or $\sigma_{SL}$ (cf. also a model of domain texture for molten metals by McLachlan 1969). One main problem in adopting any such interpretation of the observation that $C_p \gg 3R$ in the crystals near $T_f$ is that a rather different cause must then be sought for the parallel experimental finding $(C_p)_L \gg 3R$ near $T_f$. The actual entropy jump $S_f = S_L - S_S$ is unusually small, and perhaps this could be due to the fact that in melts of alkali metals, positional correlation of the atoms is retained in 'clusters' which are only gradually dispersed as the temperature rises above $T_f$. Not all these clusters need be quasicrystalline since rearrangements to *anticrystalline* packing need not involve large enthalpy changes. Inspection of Fig. 9.1 and of similar plots for other alkali metals confirm the fact that the temperature range for prefreezing (if that is the correct explanation) is much more extended than for premelting.

Further development of structural explanations for the unusual thermodynamic parameters of some metal crystals, recorded in Tables 9.1 and 9.3, could be greatly helped by improvement of methods for detecting and evaluating cooperative positional defects and domain topology especially in the liquid state. Unfortunately, conditions generally do NOT favour their detection from diffraction spectra of various kinds. Other methods of detecting and studying 'cluster' anomalies are being applied with special reference to molten metals. Thus from heat capacity (Fitterer, 1967) and resistivity data (Davies and Llewellyn Leach, 1970), it seems likely that molten 'white tin' contains clusters of grey tin up to about 200°C above $T_f$.

The complexities encountered in attempting to refine structural descriptions of molten metals may be illustrated by reference to some current investigations. For simple metals, such as sodium, these confirm that structural/entropy contributions are only a subsidiary factor in determining electronic properties (Ziman, 1970; Stroud and Ashcroft, 1972; Jones, 1973). Changes in thermo-electric power (Marwaha and Cusack, 1965) and in magnetic susceptibility (Duprée and Ford, 1973) observed on melting of varius metals are only small and not very helpful in elucidating relationships between melting and crystal structure. Unlike the alkali metals, metals from other parts of the Periodic System mostly have a more complex ligand structure, often with one distance of closest approach but with a second set of neighbours at a not much greater distance; quasicovalent bonds are formed by this second set. At the melting point, these covalent bonds may suffer breakage, with collapse to a closer packing and formation of more uniform resonance bonds with neighbours to a higher coordination number. In consequence, the volume change on melting $\Delta V_f / V_s$ is small, or even negative, as for the crystals of Ga, Bi and Sb. Anomalous entropies of fusion for these crystals and for Si and Ge can be attributed to various peculiarities of the liquid structures without as yet definitive inferences (Grimvall, 1977).

It may also be noted that a number of crystalline elements are *semiconductors*, such as Ge, Si or Te. Their crystals are made up of covalent bonds; on melting these bonds undergo progressive breakage, with a collapse in volume (10% for Si and 5% for Ge) and increase in metallic conduction. On melting semiconductors, structural changes can thus have major relevance for the changes in electronic properties which may occur on passing from crystal to melt. Phenomenologically it is meaningful to differentiate crystalline semiconductors into examples such as Ge, Si, which change on melting to characteristic metallic behaviour (in properties such as the temperature coefficient of electrical conductivity, the Hall effect, and thermoelectric power), and other examples, such as certain tellurides including elemental Te and $Bi_2Te_3$, $Sb_2Te_3$, SnTe, GeTe, PbTe, whose melts retain the semiconductor properties of the crystals (Busch and Guntherodt, 1966; Krebs, 1969). Some properties which are particularly sensitive to detailed bond structure, such as the thermal conductivity, show that the jump at $T_f$ may occur with extensive premelting and postmelting in cases such as $Sb_2S_3$. Persistence of a proportion of covalent structures in regions in the melt is indicated (Fedorov and Machuev, 1969, 1970). Tentative statistical calculations of the contribution to the configurational entropy have been put forward, assuming a kind of random breaking of the covalent network on melting. The original should be consulted for details (Chakraverty, 1968). Semiconductor melts are similar to other network melts in which bond breaking continues to make (moderate) changes in the physical properties for a long range of temperatures above $T_f$ (Chapter 10). Thus, for Te, neutron diffraction shows that molecular bond structures found in the crystals persist in the melts both above and below $T_f$ (Tourand, 1975). Other features such as chain vibrational modes persist in Se but not in Te, indicating that the polyatomic networks may be flexible in the melt (Axmann, Gissler, Killmar, and Springer, 1970).

## 9.5 Melting of metals at high pressures and temperatures

More information about each of the thermodynamic parameters of metals at enhanced pressures could be of value in deciding between alternative explanations of the peculiarities discussed above. Unfortunately, vital information is seldom available as yet. Most work so far has been limited to phase diagrams. When the melting point curves of metals are explored at high pressures, since $\Delta V_f$ is normally positive, this increases $T_f$ in most cases, in accordance with the Clausius–Clapeyron equation. Unexpected polymorphic transitions become apparent, however. Clearly, in some metals equalization of the free energies $G_S$ and $G_L$ of solid and liquid at $T_f$ may involve quite a complex balance of contributory factors, instead of a single positional disordering. As one example, for caesium two transitions are found at higher pressures (Newton, Jayaraman, and Kennedy, 1962) (Fig. 9.2).

Possibly one crystal transition $I \rightarrow II$ involves a rearrangement from h.c.p. to cubic close packed: a second transition $II \rightarrow III$ has been tentatively attributed to an electron shell collapse at higher pressures ($6s \rightarrow 5d$), but the evidence is by no means conclusive (Kennedy, Jayaraman, and Newton, 1962). Less pronounced anomalies appear in the plots of $T_f$ vs. pressure for the other alkali metals, (Luedemann and Kennedy, 1968).

A maximum has likewise been reported in the melting curve of barium, which may involve similar effects (Jayaraman, Klement, and Kennedy, 1963). It is noteworthy that, in cerium, a polymorphic transition probably involves an electronic promotion ($4f \rightarrow 5d$) for the smaller specific volumes occupied at higher pressures. A critical limit for this transformation has been reported (Poniakovskii, 1958). These effects are significant in underlining the importance of changes of electronic enthalpy and entropy, at any rate in certain types of metallic crystals.

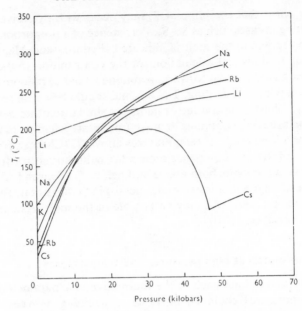

FIG. 9.2. Fusion curves of alkali metals. Reproduced from Newton, Jayaraman, and Kennedy, *J. Geophys. Res.* **67,** 2559 (1962), copyrighted by American Geophysical Union

Unlike most other metals, cerium contracts on melting at ordinary pressures, but as the pressure rises $T_f$ passes through a minimum at 675°C. There are two transitions in solid Ce, followed by melting of a b.c.c. crystal with

$$S_f = 1 \cdot 16 \text{ e.u.}; \; \Delta V = -0 \cdot 228 \text{ cm}^3 \text{ mole}^{-1}; \; dT/dP = -4 \cdot 7 \text{ deg kbar}^{-1}$$

(Jayaraman, 1965a).

For germanium, and for other semimetals whose lattice collapses on melting, the plot of temperature *vs.* pressure slopes downwards (Vaidya, Akella, and Kennedy, 1969). Perhaps not surprisingly, the nominal Debye temperature $\theta_D$ calculated for such metals from a Lindemann formula for the melting point differs very greatly from the Debye temperature calculated from the specific heat. Melting mechanisms are simply not amenable to vibrational interpretation for these crystals.

### 9.6  General similitude rules in the melting of metals

Up to the present, the number of metals for which critical temperatures have been evaluated is too small to make tests of conventional 'corresponding state' rules really useful. Various other correlations of properties have however been proposed. Similitude rules based on similarities of crystal structure were briefly referred to above. The exceptionally close similarity of thermodynamic properties such as $V$ and $C_p$ between so many metal crystals and their melts, as recorded in Tables 9.1 and 9.4 above, at first suggests that these melts might perhaps constitute good general examples of quasicrystalline liquids. Thus, at ordinary pressures, a variety of properties of metallic melts can in fact be approximately equated with properties of the crystals extrapolated to the increased volume of the melt. For molten sodium, the measured isothermal compressibility at $T_f$ of $19{\cdot}1 \times 10^{12}$ c.g.s. (Pochapsky, 1951) is practically the same as the value $18{\cdot}9 \times 10^{12}$ c.g.s. extrapolated for the solid to negative pressures sufficient to expand the crystal to $U_L$. Again, the increase of electrical resistivity of sodium from $6{\cdot}8$ to $9{\cdot}67$ ohm cm has been attributed mainly to the increase in volume on melting ($\Delta V_f / V_s = 2{\cdot}6\%$) (Addison, Creffield and Pelham, 1969).

At their melting points, many molten metals show a degree of similitude of surface energy $(\sigma_s)_f$ in following a linear correlation between $\log(H_{vap}/V_L)$ and $\log(\sigma_f)^{\frac{2}{3}}$. As with other similitude relationships whose basis is semiempirical, this correlation rule can serve to identify marked departures from normality. A plot (Fig. 9.3) suggests that metals such as Zn, Cd, Hg and Mg

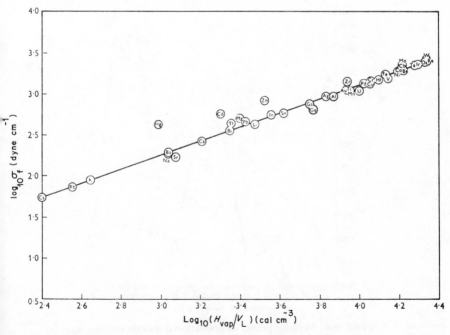

FIG. 9.3. Correlation between surface energy of molten metals at $T_f$, and $H_{vap}/V_L$.
Reproduced by permission of the American Nuclear Society

show significant departures from such a correlation (Strauss, 1960), pointing to possible anomalies of structure of these melts.

Empirical correlations between $T_f$ and the interatomic bonding in metal crystals have been reported from time to time. Since $T_f = H_f/S_f$, correlations with strength of bonding can be expected only for crystals with similar mechanisms of melting, for which $S_f$ is therefore the same. One interesting test is to plot (Fig. 9.4) the periodic variation of $T_f$ for the transitional metals. Such plots have, as is well known, been used to discuss bonding relationships for a great diversity of elements, ever since the Periodic System of the elements was first devised. Periodicity plots can highlight the influence of other factors, such as the communal electronic energy of metals, particularly when the crystal structures being compared are closely similar. This makes it interesting to find a periodic relationship between $T_f$ and the main quantum number of atoms of the transitional metals, with lanthanum as an exception (Szabo and Lakatos, 1952). Using a comparable criterion, $T_f$ for the rare element technetium appears to show an anomaly resembling manganese in this respect (Anderson, Buckley, Hellawell, and Hume-Rothery, 1960). For metallic melts of the rare earths periodic regularities are found (Jayaraman, 1965b) in relation to the quantum number of the atoms.

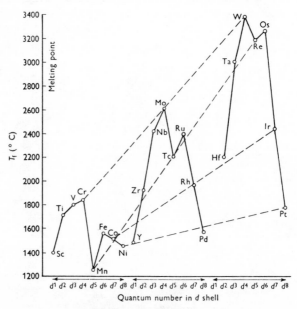

FIG. 9.4. Melting points of transitional metals in relation to atom quantum numbers. Reproduced by permission of Springer-Verlag

## 9.7 Structural information about molten metals

Many X-ray diffraction studies of liquids have in fact been made on liquid metals, partly because of the technical facility of such measurements, and partly because the spatial distribution of atoms in molten metals are sometimes thought to be particularly simple. A favoured model assumes that the melt consists of a random close-packed lattice with about 10% vacancies with respect to the crystal lattice. On this basis the average number of nearest neighbours in the melt is about $10 \cdot 85$ at $T_f$ (Furukawa, 1959, 1960). Transport properties, such as the coefficients of self-diffusion in the melts, and their viscosity, can be reduced to a single model with fair consistency. This approach to positional melting remains highly important in indicating broad structural features common to the melting mechanisms of most metals. However, as already indicated (p. 120) X-ray diffraction studies of melts definitely tend to obscure or smudge important but more refined aspects of their structure, particularly with respect to *anticrystalline* packing. A common assumption that liquid metals adopt quasigaseous distributions seems unwarranted. The possibility of substantial fluctuations of structure and substantial concentrations of *anticrystalline* clusters near $T_f$ would tend to be missed in the usual X-ray procedures. A 'conglomerate' structure of liquid metals can be inferred when prefreezing effects are steeply prominent near $T_f$ (Chapter 13). However when an anomaly merges smoothly into average behaviour, special techniques may be needed even to establish it. As is well known, radial distribution functions as ordinarily evaluated from X-ray diffraction data, though very informative in the first approximation, are comparatively insensitive to local variations of spacing of nearest neighbours in the melts and may accordingly give an unduly simplified model for the liquid state, with little information about domain topology. Some recent studies of molten metals with RDF give little evidence for any structural anomaly. These include:

transitional metals Pd, Pt, Zr, Ti, V, Cr, Mn, Fe, Cd, Ni, Cu (Waseda and Tamaki, 1975; Waseda and Obtani, 1975)
Hg (Ruppersberg and Reiter, 1972)
Alkaline earth metals Mg, Ca, Sr, Ba (Waseda, Yokoyama, and Suzuki, 1974; Waseda and Yokoyama, 1975) (but see also Cusack, 1967; Alekseev, Andreev, and Prokhorenko, 1972).

More sensitive means of calculation have however been made which refer to analyses of X-ray, electron or neutron scattering from melts of monatomic metals (Richter and Breitling, 1961, 1962, 1966, 1967, 1968a,b, 1970). In such melts, regions or domains of 'spherical close packing' are found in which the nearest neighbour distance is close to that in the crystals as represented in conventional radial distribution curves. In addition, it is claimed that microregions of the order $10^{-6}$–$10^{-7}$ cm across are present, e.g. with the layer

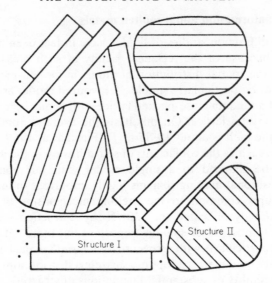

FIG. 9.5. Domain model structure for melts of gold, silver, lead, thallium and the alkali metals. Reproduced by permission of Dr Riederer-Verlag, Stuttgart

lattice (l.l.) packing (see also Richter and Steeb, 1958; Brozel, Handtman, and Richter, 1962a). Figure 9.5 illustrates a domain model structure proposed for Au, Ag, Pb, Tl and the alkali metals (Richter and Breitling, 1970).

Using sufficiently sensitive methods of determining radial distribution curves, it is found that the layer lattice, or other non-uniform regions in different molten metals, show differing degrees of prominence. In Sn and Ga they are evident from 'shoulders' on the radial scattering curves (Brozel, Hardtmann, and Richter, 1962b). In Ag and Au they become relatively more prominent at higher temperatures. The view of the authors is that the heat of disaggregation of spherical close-packing is well below that for the layer lattices, which are assumed to involve covalent bonding. Consequently, such layer regions become relatively more important at higher temperatures, though with marked distortion owing to the smallness of the regions involved. Mercury at room temperature corresponds with tin at 750°C in this respect.

Dual 'structures' in the melts have also been inferred from diffraction data for molten Bi (Lamparter and Steeb, 1976). The original paper should be consulted for details.

As a further argument that liquid metals particularly near $T_f$ may be far from quasigaseous in the spatial distribution of their atoms, it may be added that the unit cell of some metals such as manganese involve quite complex structures (Wells, 1962). Such assemblies differ only to a comparatively minor extent from much smaller unit cells of simpler structure. Presumably they are differentiated from these, and are a preferred structure in the crystals, simply because lower electronic energy favours the more complex arrangements. However, small regions of such privileged packing may be missed in the averaging process of evaluating RDFs for the molten state. In the same way, clustering in molten metals may be due to electronic factors promoting certain types of micellar rearrangement. This could result in an electronic contribution to $S_f$, whose magnitude is at present difficult to estimate.

Studies of NMR confirm that with respect to this property there is no marked change in local structure on melting many metals. Exceptions are Bi and Ga, whose lattice structure is, on other grounds, known to collapse on fusion (Knight, Berger, and Heine, 1959) (see Table 9.1).

## 9.8 Melting of metals with non-equivalent bonds

Most metals crystallize in close-packed structures, h.c.p. or f.c.c. with coordination number 12, or b.c.c. with coordination number 8. Melting of such crystals involves mainly positional disorder, and any changes of electronic properties can, at least in part, be correlated with this increasing disorder. As stated, a number of metallic crystals are known, however, for which X-ray diffraction and other methods of study show that not all the bonds are equivalent. Attempts have been made to make a distinction of classification between 'true metals' such as Li, Na, K, Cu, Ag, Au, Mg, Ca, Al, Co, Ni, Ti, and 'semimetals' in which not all the bonds between the atoms in the crystals are equivalent, including Si, Ge, Sn, Sb, Bi, Te, Se, Ga. Like many such semiempirical classifications, this one provides useful groupings, but leaves a middle group of 'meta metals' such as Zn, Cd, Hg, In, Pb whose allegiance is difficult to define (Schneider and Heymer, 1958). One advantage of this particular classification is, however, that it provides a base for discussion and for interpreting premelting and prefreezing anomalies (Chapters 12, 13) that tend to be prominent particularly for the semimetals (see Alekseev, Andreev, and Prokhorenko, 1972). Unlike those of true metals, it is claimed that melts of Ga, Bi and Ge show marked variations of X-ray diffraction intensities with temperature. Side maxima appear just above $T_f$ and their intensity rapidly decreases above $T_f$ (see Hendus, 1947; Glocker, 1949). In a similar way, chain structures in the melts of Te at $T_f$ (142°C) change to 'metallic' arrangements of the atoms within 10°C or so above $T_f$ (Mokrovskii and Regel, 1955).

As another distinctive exception to normal behaviour, electrical resistance and viscosity of melts of tellurium show rapid decreases as the temperature rises above $T_f$ (see below).

## 9.9 Changes of electronic properties of metals on melting

As indicated in section 9.3 metal crystals appear to show exceptionally low entropy and volume increases on fusion, compared with crystals of the inert gases. This striking finding suggests that some additional mechanism of melting is effective for metals, not available to the inert gases, favouring the liquid state for smaller increases in volume and enthalpy. Various structural possibilities have been discussed in preceding sections. One model feature to be included for consideration is the plasma of 'metallic' electrons surrounding the ions in the metal. Energy and entropy contributions from this plasma might change sufficiently on melting to make a substantial contribution to the overall change from solid to liquid, in crystals of metals. By degrees, a diversity of information is being obtained about these changes. In the outcome, electronic contributions do not on the whole appear to account for the particularly low entropy and volume changes on melting metals. Studies of these properties around $T_f$ in some cases yield useful information about structural and vibrational changes in these substances. Changes in various electronic properties often involve complex physical theory for their interpenetration. (When such theory is only remotely concerned with thermodynamic aspects of melting, more detailed discussion should be sought, for example in various review articles; see Cusack, 1967; Alekseev, Andreev, and Prokhorenko, 1972; Busch and Guntherodt, 1974.)

### 9.9.1. *Changes of electrical resistivity*

On melting, the loss of long-range correlation of positions and the general increase in positional disorder may be expected generally to increase scattering of the conduction electrons. Amplitudes of vibration of each atom about a mean position also increase, leading to a jump in vibrational scattering at $T_f$. Finally, the Fermi level of the electrons may change as a consequence of the volume change and (for non-cubic crystals) the increase of average symmetry on melting, altering both the electronic energy and entropy.

A very wide range of electrical properties of conductors can in principle be measured. It must be stressed that many of these properties prove to be not very sensitive to structural changes on melting. Unfortunately, for some of the most readily measured changes, such as those of electrical resistivity, complex mathematical interpretations may restrict the information at present obtainable from them concerning mechanisms of melting (see Cusack, 1963). Only the most prominent or decisive results in this direction are referred to in what follows. In certain cases, the behaviour of the crystals is anomalous, but this becomes smoothed out on melting (e.g. Cd, Ge). In other cases, the molten state is anomalous for reasons not fully understood (e.g. Pb, Bi, Tl) (Busch and Guntherodt, 1974).

Attempts have been made to separate the different contributions to the entropy of fusion (see above):

$$S_f = \Delta S_{pos} + \Delta S_{vib} + \Delta S_{electr}$$

using measurements of changes of resistivity and thermoelectric power on melting (Cusack and Enderby, 1960). A theory due to Borelius (1958) suggests $\Delta S_{vib} = 0$. On the other hand calculations due to Mott (1934) (see also Knappwost, 1954; Knappwost and Thieme, 1956) correspond with the assumption $\Delta S_{pos} = 0$ and $\Delta S_{electr} = 0$ and lead to the result $\rho_L/\rho_S = (\theta_S/\theta_L)^2$ where $\theta_L$, $\theta_S$ are the respective characteristic temperatures of the melt and solid, treated as Einstein vibrators. However, present evidence suggests that no single entropy contribution is sufficient to explain the findings for all metals.

A dual contribution to the change on melting of electrical resistivity and of other properties such as thermoelectric power appears to be operative at any rate for metals of Group 1A and 1B (Ziman, 1961). Furthermore, observed effects depend on the position of a metal in the Periodic System. Transport properties of molten transitional metals may be quite different (Guntherodt and Kunzi, 1973). Essentially, for simple metals, a 'thermal' term arising from scattering of the electrons by phonons in the crystal or melt is augmented by a 'structural' term which is nearly independent of temperature, and which is large, particularly in metals with strongly distorted Fermi surfaces. This duality makes a direct correlation between melting mechanisms and the change of electrical parameters in melting much less clear cut than if either contribution were insignificant (see Table 9.5).

For the transitional metals and rare earths the change on melting is clearly quite small. For uranium, the only actinide so far studied, the change is practically zero on melting (see Schneider and Heymer, 1958; Cusack and Enderby, 1960; Cusack, 1963).

On the basis of Mott's theory the drop in $\theta$ would have to range from 20% to 40% on melting if the whole of the change in electrical conductance recorded in Table 9.5 were due to vibrational effects (MacDonald, 1959). Measurements at constant pressure of the change of resistance of rubidium at $T_f$ indicate that perhaps up to 25% of the increase in resistance on melting is due to increased scattering of the electron waves by positional defects in the molten metal. For transitional metals, the change in electrical resistivity and of magnetic susceptibility is small because most of the resistivity in the crystals results from holes in the $d$ band, which do not depend on the structure factor (Mott, 1972; Ten Bosch and Bennemann, 1975). Only part of this small change at $T_f$ is to be attributed to the change in vibrational content on melting (Cusack and Enderby, 1960). Insofar as molten metals can be adequately represented as quasicrystalline, theories referred to above can be regarded as

*Table 9.5. Changes of metallic resistivity on melting*

| Metal | $(\rho_L/\rho_S)_t$ | Metal | $(\rho_L/\rho_S)_t$ |
|---|---|---|---|
| Li | 1·64 | Be | — |
| Na | 1·45 | Mg | 1·78 |
| K | 1·56 | Ca | — |
| Rb | 1·60 | Sr | — |
| Cs | 1·66 | Ba | 1·62 |
| Cu | 2·04 | Zn | 2·24 |
| Ag | 2·09 | Cd | 1·97 |
| Au | 2·08 | Hg | 3·74 |
| Al[a] | 2·20 | Ge | 0·053 |
| Ga | 0·45c $\rangle$ axes<br>1·46a<br>3·12b | Sn | 2·10 |
|  | | Sb | 0·61 |
| In | 2·18 | Bi | $\|0·35\}$<br>$\perp 0·47\}$ |
| Ti | 2·06 | Tl | 2·06 |
| Si | 0·034 | Mn | 0·061 |
| Se | ~1·000 | Fe | 1·01 |
| Te | $\|0·091\}$<br>$\perp 0·048\}$ | Ni | 1·33 |
|  | | Co | 1·09 |
| Mo | 1·23 | Pt | 1·40 |
| La[b] | 1·08 | W | 1·08 |
| Ce[c] | <1·05 | | |
| Pr[c] | <1·05 | | |
| Nd[c] | <1·05 | | |

[a] Marty (1958) gives $\rho_L/\rho_S = 2·33$.
[b] Kreig, Genter, and Grosse, 1969.
[c] Gaibullaev, Regel, and Khusarov, 1969.

giving at least a first approximation. Relatively simple structures are observed, e.g. for melts of transition metals (Meyer, Stott, and Young, 1976) and alkali metals such as Li (Guntherodt, Kunzl, Muller, Evans, 1972). However, the possibility of marked cluster structure (see above and Chapter 13) in certain metallic melts introduces a new feature and could demand a rather different theoretical approach. Phenomenological analysis of the change in resistivity on fusion has been proposed in terms of a break-up of the crystalline metal into microcrystalline regions. In molten alkali metals estimates of the size of these scattering regions range from about $10^3$ atoms (Oriani, 1951) to about $10^2$ atoms (Gerstenkorn and Sauter, 1951). The suggested antithesis between 'microcrystalline' and other molten metals to account for scattering of electrons in liquid metals appears to be at present unresolved (Cusack, 1963).

The metals Zn, Cd and Sb show a remarkably long range of temperatures where resistivities in the liquid state remain practically constant (Fig. 9.6(a), (b), (c)), almost certainly because of compensation by more than one effect. This contrasts with the behaviour of metals whose melts have simpler

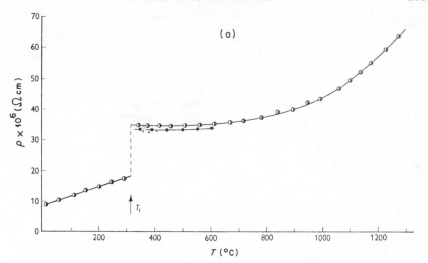

FIG. 9.6(a). Temperature dependence of specific resistance $\rho$ of Zn (solid and liquid). Reproduced by permission of the American Institute of Physics

structure, such as alkali metals, whose resistivities rise uniformly for both solid and liquid as the temperature rises (Regel, 1958). This anomalous behaviour of IIB metals can probably be attributed to the consequences of cluster formation near $T_f$ in the melts, further discussed in Chapter 13. Figure 9.6(a), (b), (c) includes values obtained by different authors (see Regel, 1958).

Liquid tin has become a favourite object of study, possibly because of the ease of obtaining a highly purified sample. Some authors claim there is an

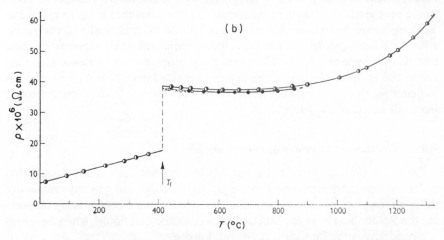

FIG. 9.6.(b). Temperature dependence of specific resistance $\rho$ of Cd (solid and liquid). Reproduced by permission of the American Institute of Physics

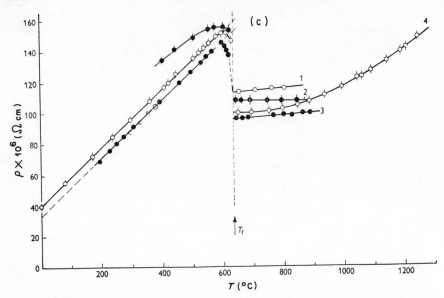

FIG. 9.6(c). Temperature dependence of specific resistance $\rho$ of Sb (solid and liquid). Reproduced by permission of the American Institute of Physics

abrupt change of slope in its resistivity plot around 320°C (Davies and Llewellyn Leach, 1970). Other data indicate continuous changes, with a two-component model for the melt (Pokorny and Astrom, 1975).

Closer characterization of the Fermi surfaces of crystalline and molten metals might help to put the interpretation of temperature changes of electronic properties on firmer foundations. On the basis that a liquid metal has 'no long-range order', the Fermi surface should be spherical and the relaxation time isotropic. Liquid mercury, for example, might be expected to show a Hall coefficient of $-77{\cdot}4 \times 10^{5}$ e.m.u. as an atom with two valence electrons giving rise to 'free' electrons. The experimental value ($-74{\cdot}6 \times 10^{-5}$ e.m.u.) is sufficiently close to support this view, but awaits further confirmation (Kendall and Cusack, 1959).

### 9.9.2. Thermoelectric power of molten metals

Owing to changes in electron band structure that accompany crystal disordering, the thermoelectric powers $\mu_S$, $\mu_L$ of solid and molten metals are different. In principle, these can either be investigated directly, or by measuring the Peltier heat $\pi$ at the junction between solid and liquid, when a current is passed across the interface. As is well known

$$\pi/T_f = \mu_S - \mu_L$$

Estimates of the Peltier coefficient made by passing current up and down the temperature gradient in the metal, and noting the positions of the solid and liquid interface in the two cases have led to the following values (Bardeen and Chandrasekhar, 1958):

| Metal | $\pi$(mV) |
|-------|-----------|
| Bi | −14 |
| Cd | +3 |
| Zn | +3 |

(current flowing from solid to liquid, negative

Changes in absolute thermoelectric power, in $mV/deg^{-1}$ have been evaluated for various Group I metals, as given in Table 9.6 (Cusack and Enderby, 1960).

**Table 9.6.** *Changes in absolute thermoelectric power on melting Group I metals*

| Li | 6·0 | Cu | 9·5 |
|----|-----|----|-----|
| Na | 0·0 | Ag | 0·0 |
| K | 2·3 | | |
| Rb | 3·8 | Au | 0·0 |
| Cs | 5·8 | | |

Discussions of these interesting differences of behaviour on passing from crystal to melt are still far from definitive. An experimental method to maximize the differences of thermoelectric power between molten and crystalline metals is to investigate thermocouples with limbs in each of these states (Ubbelohde, 1966). These reveal interesting precursor effects in some metals (see below).

### 9.9.3. *Hall effect of liquid metals*

Studies show (Busch and Tièche, 1963; Busch and Guntherodt, 1974) that many liquid metals follow a free electron model with respect to the Hall effect. This can be tested using the formal relation

$$(R_H)_{calc} = \left(\frac{1}{Ne}\right) M/\rho n_v$$

where $\rho$ is the density, $n_v$ the number of free electrons per atom, and $M$ the atomic weight. For the metals listed, measured values of the Hall effect $R_H$ at $T_f$ correspond with values of $n_v$ within ±10% of the number of estimated valency electrons. However, as is seen from Table 9.7, molten Pb, Tl and Te present exceptions. This would appear to indicate structural peculiarities in the liquid state of these metals.

For Cd, the Hall coefficient changes from positive to negative at $T_f$ (Busch and Guntherodt, 1974). The Hall coefficients of molten Pb, Bi and Tl are notably smaller than for the free electron model.

Examination of the detailed results for crystal and melt suggest considerable premelting and prefreezing in Zn, and some prefreezing in Te and premelting in Sb (see Chapters 13 and 12).

It is also noteworthy that the quantity

$$(R_H)_S/(R_H)_L - 1$$

is negative for Zn, positive for Sb, Bi and Te. (For Pb, present information is incomplete near $T_f$.) Marked departures from unity in Table 9.7 may point, for example to domain or cluster formation in the melt, modifying the overall electronic change on melting. More information and better precision seem to be required before a complete analysis of these results can be usefully further discussed.

Table 9.7. *Ratio of experimental to nominal Hall effects at $T_t$ in metallic melts*

| Metal | $\eta_v$ nominal | Ratio |
|---|---|---|
| Na, K, Rb, Cs | 1 | 0·98–1·00 |
| Cu, Ag, Au | 1 | 1·00–1·02 |
| Tl | 1 | 0·70–1·00 |
| Zn | 2 | 0·70 |
| Cd | 2 | 0·98 |
| Hg | 2 | 0·96 |
| Al | 3 | 1·00 |
| Ga[a] | 3 | 0·99 |
| In | 3 | 1 |
| Ge | 4 | 1·06 |
| Sn | 4 | 1·06 |
| Pb | 4 | 0·38–0·88 |
| Sb | 5 | 1·14 |
| Bi | 5 | 0·60–0·95 |
| Te | [a] See 3·3 | |

[a] See Cusack, Kendall, and Marwaha, (1962).

### 9.9.4. Changes in magnetic susceptibility on melting

It appears to be significant for the nature of the phenomena that no examples have been reported of ferromagnetic behaviour persisting up to the melting point. However, paramagnetic susceptibilities could probably give useful information about melting mechanisms of metals, especially when these are influenced by longer-range rather than by nearest neighbour interactions. Values of the magnetic susceptibility often show only small changes at $T_f$ (Nakagawa, 1956, 1957; see Busch and Guntherodt, 1974).

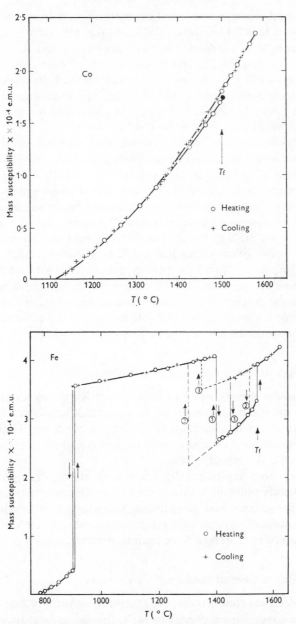

FIG. 9.7. Change of magnetic susceptibility of iron on melting. 1, Liquid supercooled, $\delta$-phase not super-cooled. 2, Both liquid and $\delta$-phase supercooled. 3, Liquid supercooled to Y-region. Reproduced by permission of the Physical Society of Japan

More marked changes are found on melting pure iron, and also certain iron-rich alloys. In solid Fe, the crystal forms that are stable in various ranges of temperatures have magnetic susceptibilities as indicated in Fig. 9.7. The f.c.c. phase $\gamma$-iron (CN 12) has $\chi$ much smaller than either the $\alpha$-$\beta$ or $\gamma$-phase (both b.c.c. CN 8). On melting the atoms behave as though they rearranged their positions to a perturbed crystalline version of the $\gamma$-phase (see Cusack, 1963). Similar basic changes of coordination number on melting have been previously referred to for CsCl (p. 182).

Structural influences on melting which may likewise involve collective electronic factors are suggested by the finding (Ferrier, 1961) that the heat of fusion of $\delta$-iron is 3·3 kcal g atom$^{-1}$ whereas for $\gamma$-iron it is 4·4 kcal g atom$^{-1}$. Changes of magnetic susceptibility for iron-nickel alloys whose melting does not involve any change of primary structure can be accounted for wholly in terms of the increase in volume on fusion (Nakagawa, 1957). Changes of the Knight shift $\kappa$ on melting metals (McGarvey and Gutowsky, 1953; Cusak, 1963) have been given somewhat conflicting interpretation (see Oriani, 1959). The suggestion is that changes in the characteristics of the electronic wave functions arising from loss of long-range order and acquisition of a different nearest neighbour distribution are as important as the macroscopic volume change on melting, but the effects cannot yet be used to give discriminating information about mechanisms of melting.

### 9.9.5. Changes in X-ray band edges on melting

Band edges for the absorption bands of soft X-rays by atoms show some dependence on the environment of nearest neighbours, which can modify to a perceptible extent excitation levels for the electrons. Thus it has been suggested that the band edge in molten gallium is slightly displaced relative to that for the solid (Cauchois, 1949).

On passing from crystalline to amorphous states, changes in X-ray $L$ absorption bands point to a broadening of levels accompanying positional disordering for gallium and germanium (Lucasson, 1960). However, quantitative evaluation of possible changes in communal electron levels on melting of metals has not yet emerged from such techniques.

### 9.9.6. Changes in thermal conductivity on melting

Like the electrical conductivity, the thermal conductivity $\lambda$ of many metals decreases on passing from crystal to melt. Insofar as heat transport can be attributed to the metallic electrons this would be expected. A fraction of the energy transported must however be due to lattice vibrations. Discrimination between changes of electron scattering and phonon scattering on melting metals has not yet been attempted. Table 9.8 (see Cusak, 1963) illustrates the

**Table 9.8. Decrease of thermal conductivity on melting (ratios)**

| Metal | $\lambda_S/\lambda_L$ | Metal | $\lambda_S/\lambda_L$ |
|---|---|---|---|
| Na | 1·3 | Te | 2·04 |
| Zn | 1·5 | Sn | 1·82 |
| Cd | 2·4 | Pb | 1·83 |
| Al | 1·47–3·7 | Sb | 1·1[a] |
| Ga | 1·06[a] | Bi | 0·5 |

[a] For Ga and Sb it has been claimed that $\lambda$ increases on melting, roughly proportional to the changes expected from the Wiedemann–Franz Law (Dutshak and Panasyuk 1967). This would imply heat conduction predominantly by the electrons.

order of changes observed. Scatter in some of the published data indicates caution in using numerical values without careful scrutiny of their reliability.

Partly for experimental reasons, the Lorenz number is more conveniently determined with precision than the thermal conductivity $\lambda$. Future developments of this procedure will probably reduce or banish some of the anomalies claimed for liquid metals (Busch and Guntherodt, 1974).

## 9.10 Melting of semiconductors

Attention has already been drawn to the observation that for semimetals, like Bi or Sb, there is an increase of electrical conductivity on melting (Table 9.5). This can plausibly be attributed to a partial collapse in the melt of covalent 'non-metallic' structures that are present in the crystals. Thus bismuth has a layer structure, with each metal atom equidistant from three nearest neighbours at 3·17 Å (Wells, 1962). In metallic antimony, the general arrangement is isostructural with bismuth. Each atom has three nearest neighbours at 2·87 Å together with three next nearest neighbours relatively more distant than in bismuth at 3·37 Å. On melting, more symmetrical packing is taken up, though there may be a gradual change of clustering in local regions in the melt near $T_f$ (Chapter 13) as the temperature of the melt is raised. In the resonance theory of metal binding (Pauling, 1960) a high coordination number favours delocalization of the valence electrons and enhances metallic character. No quantitative theory appears to have been put forward in these terms but the theory provides a qualitative guide. Thus semiconducting crystals of elements such as germanium likewise show an increase of electrical conductance on melting. The more metallic behaviour of liquid germanium may be associated with the observation that in it the predominant bonding has coordination number 8, as in many metals, whereas in the crystal a tetrahedral 'network' structure is maintained with coordination 4 (Hendus, 1947; Keyes, 1951; see also Fritzche, 1952; Tanuma, 1954, for changes in other physical properties on melting).

Selenium and especially tellurium present even greater possibilities of structural diversification on melting, since in the crystals the atoms are arranged in spiral chains. These chains might be expected to dissociate in the melt to give greater delocalization of valence electrons and presumably a more metallic liquid. For Te, $T_f = 445°C$, $S_f = 5·80$ e.u. The solid has semiconducting properties with somewhat unusual temperature coefficients. At $T_f$, the resistivity drops sharply and continues to drop as the temperature rises still further (see Busch and Tièche, 1962, 1963) In many respects, both Te and $Bi_2Te_3$ behave as molten metals (Shadrichev and Smirnov, 1970).

Molten tellurium shows electronic properties that can be attributed to a mixture of spiral chains of atoms, of various lengths, together with isolated tellurium ions and free electrons. The conductivity $1/\rho$ is given by a composite expression (Busch and Tièche, 1962, 1963). Just above the melting point the behaviour is wholly metallic.

The Hall coefficient drops quite sharply on melting and continues to decrease as the temperature rises, but passes through an inversion point around 575°C. Supercooled liquids show no break of electrical properties at the freezing point. Thermoelectric power measured against tungsten shows a steep discontinuous drop on passing from melt to crystal at $T_f$. These changes of properties on melting are all consistent with the model for molten tellurium which contains spiral chains of atoms together with some single ions and electrons. As the temperature of the melt rises above $T_f$ the proportion of chains dissociated into atoms increases.

### 9.10.1. *Melting of semiconducting crystal compounds*

Melting of crystal compounds is discussed in this book only insofar as this helps to throw light on mechanisms of melting generally. Theoretical considerations about effects of melting on the properties of semiconductors indicate a distinction between two main classes of behaviour (Gubanov, 1957). For the collapsing network type of crystal structure (Class I) the coordination number increases on melting. As explained above, accompanying the increase of density, the short range environment of the atoms undergoes fundamental changes, and the electrical conductivity becomes metallic in character. However, despite the great interest of this type of melting, theories are not very successful up to the present. The radical change of bond character on melting and the electronic properties of the liquid can be referred to a (hypothetical) metallic crystal of the same atomic composition. This point of view is particularly suggestive, in relation to the possibility that even in the crystalline state the metallic packing may perhaps be attainable as a polymorph at sufficiently high temperatures and pressures. Because of the radical change in structure, the normal relation for change of resistivity of metals on melting (section 9.9.1) does not apply.

Other suggestions include the speculation (Kontorova, 1959) that the change from semiconductor to metallic properties of crystals such as Ge and

Si involves a 'rotational' melting of $sp^3$ hybridization, which holds the atoms in an open network structure in space. On melting, the crystal collapses to increase the coordination number from 4 to a value more characteristic of close-packed spheres; the $sp^3$ orbitals are considered to 'rotate' between the various nearst neighbour atoms. Quantitative formulation of this concept of defect bonding does not yet appear to have been provided. It has obvious analogies with the Pauling resonance model for metals.

### 9.10.2. *Environmental collapse on melting*

This noteworthy type of collapse-melting is frequently observed for network crystals with a diamond or sphalerite lattice structure, and includes Ge, Si, GaSb, InSb, and others for which the coordination number is 4 in the crystal and 8 (as a limit) in the melt. Marked premelting has been observed in the thermoelectric power of compounds such as InSb, or GaSb, but no critical assessment has been made which would prove that this is due to lattice defects (see Chapter 12).

### 9.10.3. *Changes of electrical conductance on melting Class II semiconductors*

In a second type of semiconducting solid (Class II) the coordination number does not change to any large extent on melting. The short-range order changes only slightly, and the density decreases by a few per cent, as in other instances of positional melting in crystals. Semiconductor properties are retained. Examples include $Tl_2S$, $Cu_2S$, $Ag_2S$, $Bi_2O_3$, $V_2O_5$, $Sb_2S_3$, FeS, CuS, as well as Cu, $GaSe_2$ and $CuAsSe_2$ (Mal'sagov, 1970). It is interesting to note that the (low) conductivity of $CuAsSe_2$ actually falls on melting, owing to a decrease in the role of doping impurities once the lattice is molten. $CuInSe_2$ is considered to melt to an ionic semiconductor in which the bonds are dissociated.

Effects of changes in defect concentration on semiconductor behaviour in various substances are well studied, so long as these remain solid. In the most elementary formulation, when a voltage gradient is applied, electric current in the crystals is carried by conduction electrons or by 'positive holes'. Either type of carrier requires activation, associated with an activation energy $\varepsilon$, to a partly filled energy band in the crystal. Because of this requirement, conduction increases with rise in temperature. $\varepsilon$ gives a measure of the height of the conduction band above ground level.

Changes in the conductivity of a semiconductor on melting the crystal could throw light on the change in structure on passing from solid to liquid, in various ways. In the simplest formulation, the conductivity $1/\rho$ is determined by the number of carriers per $cm^3$ $n_+$ or $n_-$ and the mobility $\mu$ which is a measure of the scattering power of the crystal. $(1/\rho) = ne\mu$ where $e$ is the electronic charge.) The number of carriers is related to the number of defect sites in the crystal, either intrinsic (thermal defects in equilibrium) or owing to impurities. On melting, when positional disordering intervenes the concentration of defects undergoes a radical change. Other factors change likewise, so that no final attributions of the overall effect can yet be made. By way of example, for wüstite (FeO prepared by melting the oxide in equilibrium with metallic iron under purified $N_2$) an increase in $1/\rho$ by a factor of about 3 occurs on melting. Plots of $\log \rho$ $vs.$ $1/T$ do not give a straight line between the melting point (137°C) and about 1460°C, above which $\varepsilon$ is approximately 3660 cal mole$^{-1}$. The solid is known to be a p-type semiconductor with the lattice deficient in iron. It contains cation vacancies which are compensated for charge deficiency by the insertion of two $Fe^{3+}$ ions per vacancy. On melting, the ratio Fe/O increases. If the melt can be regarded as quasi-crystalline, the higher conductivity could be attributed to a change in the number of carriers owing to decrease in $\varepsilon$ or alternatively to a change in their mobility $\mu$. Relative attribution of the observed increase in $1/\rho$ on melting between these two factors has not yet been made.

### 9.10.4. Effects of pressure

The changes of electronic properties brought about by pressure changes on melting semiconductors could, in principle, be due to changes in the contribution from defects. Semiconductor activation energies $\varepsilon$ are likely to depend on pressure, though little systematic information is available. A further complication is that since positional defects mostly involve an increase in volume for their formation, their concentration in crystal and in melt will be pressure-dependent. These considerations could help to account for changes in melting behaviour of a semiconductor such as bismuth telluride (Ball, 1962) under pressure. Lattice-collapse processes are also likely to depend on pressure. This could provide an alternative basis of interpretation of the remarkable results observed for $Bi_2Te_3$. Again, around 300 000 atm a sharp drop is found in the resistivity of indium antimonide, InSb. This effect is probably due to the transformation to a second 'metallic' structure of this semiconducting crystal, since the change occurs with a shrinkage in volume of not less than 20% and can be followed to a triple point (solid solution I, solid II melt, at about 335°C) (Jayaraman, Newton, and Kennedy, 1961). The decrease in resistivity accompanying collapse has analogies with the change on passing from grey to white tin.

### 9.11 Melting of intermetallic compounds $AB_n$

In relation to the various mechanisms of melting for crystals, illustrative information is obtainable from related properties of molten intermetallic compounds $AB_n$. In principle, these melts could assume a range of structures. One extreme would be a fully randomized distribution of A and B atoms on interchangeable sites in a quasicrystalline model for the melt. This resembles mixed melts of the rare gases or of isotopic molecules. At the other extreme intermetallic compounds could give melts with a high degree of local order, similar to ionic melts with polyatomic ions. If specific covalent bonds are formed between the A and B atoms, it may even be possible to identify polyatomic crystal fragments in the melts. When alloys are used, concentration fluctuations or clusters are more readily revealed than in one component systems. For example, clusters of diameter about 10 Å are observable by small angle X-ray scattering on Al–In alloys (Hoehler and Steeb, 1975).

One example of marked persistence of local ordering (incomplete randomization) is found in alloys between strongly electropositive metals and mercury, such as KHg and NaHg (Schumann, 1962). X-ray radial distribution functions of these melts indicate marked persistence of at least local order. Positional defects are presumably formed, as in ionic melts, but these are not readily evident from X-ray data. For molten Na/Hg, 8 Hg atoms surround each Na as nearest neighbours (3·27 Å) with 4 Na atoms as next nearest neighbours (3·45 Å). The volume increase on melting is about 6%.

For KHg each potassium atom has approximately the same number of nearest neighbours as in the solid. Other properties of these amalgams also indicate strong persistence of nearest neighbour ordering on melting. These include high heats of mixing of the liquids, of the order 4 kcal g atom$^{-1}$ at the maximum. Entropies of mixing are negative, which likewise points to charge transfer or ionic binding between the atoms (Kubaschewski and Catterall, 1956). Marked shrinkage occurs on forming the systems K–Hg (24·5%) and Na–Hg (8·8%) in the liquid state; these volume changes hardly differ from those for the same processes between the crystalline solids.

Other electrical and thermodynamic properties likewise support the assumption of some persistence of compounds of K with Hg, Tl or Pb in the melts, and even in the vapour (Itemi and Shimoji, 1973). Interesting questions about what determines the principal contributions to the entropy of fusion of such crystals cannot as yet be completely answered.

Many other pairs of metals can be found which form crystal compounds and for which various thermodynamic and physical properties of the melts indicate marked persistence of local ordering. Generally, this behaviour is found when the electronegativity of the two components differs markedly. Such crystal compounds are usually formed with marked evolution of heat, and show a contraction in volume even on mixing the melts. They may be contrasted with conventional solutions which in the limit may be regarded as random-mixed on a statistical basis (Bauer and Sauerwald, 1961). Prominent examples of such intermetallic compound formation are conveniently displayed in terms of heats of mixing, though in some cases other properties may show even more marked evidence. Typical examples of prominent atom pairs showing this behaviour are illustrated by a schematic table which also records some percentage contractions in volume.

For discussing melting, the importance of these results is that they illustrate the persistence in the molten state of local order, in very varying degree.

Physical evidence for numerous other systems likewise points to persistence of non-random structures in melts of intermetallic compounds, such as K–Tl (Lantratov and Tsarenko, 1959).

## REFERENCES

C. C. Addison, G. K. Creffield, and R. J. Pulham (1969) *J. Chem. Soc., London* 1482.
V. A. Aleksseev, A. A. Andreev, and V. Ya. Prokhorenko (1972) *Sov. Phys. Usp.* **15**, 139.
R. S. Allgaier (1969) *Phys. Rev.* **185**, 227.
E. Anderson, R. A. Buckley, A. Hellawell, and W. Hume-Rothery (1960) *Nature, Lond.* **188**, 48.
E. N. da C. Andrade (1934) *Phil. Mag.* **17**, 497.
A. Axmann, W. Gissler, A. Killmar, and T. Springer (1970) *Faraday Discuss. Neutron Spectrosc.* Meeting on the Vitrous State, Bristol, England.
D. L. Ball (1962) *Inorg. Chem.* **1**, 805.
J. M. Bardeen and B. S. Chandrasekhar (1958) *J. Appl. Phys.* **29**, 1372.
G. Bauer and F. Sauerwald (1961) *Wiss. Z. Univ. Halle Math. Nat.* **X/5**, 1029.
G. Borelius (1958) *Solid State Phys.* **6**, 65.
G. Borelius (1963) *Solid State Phys.* **15**, 1.
G. Borelius (1970) *Physica Scripta* **1**, 141.
R. Brozel, D. Handtmann, and H. Richter (1962a) *Naturwiss.* **49**, 129.
R. Brozel, D. Handtmann, and H. Richter (1962b) *Z. Phys.* **168**, 322.
G. Busch and H. J. Guntherodt (1966) *Phys. Kond. Materie* **5**, 31.
G. Busch and H. J. Guntherodt (1974) in *Solid State Physics Advances*, Vol. 29, ed. Ehrenreich, Seitz and Turnbull, Academic Press, p. 235.
G. Busch and Y. Tièche (1962) *Report of the International Conference on Physics of Semiconductors*, Physical Society, London, p. 237.
G. Busch and Y. Tièche (1963) *Phys. kond. Materie* Germany **1**, 78.
J. A. Cahill and A. D. Kirshenbaum (1962) *J. Phys. Chem.* **66**, 1080.
Y. Cauchois (1949) *J. Chim. Phys. Phy. chim. Biol.* **46**, 307.
B. K. Chakraverty (1968) *J. Phys. Chem. Solids* **30**, 454.
S. A. Cho (1974) *J. Solid State Chem* **11**, 234.
N. E. Cusack (1963) *Rep. Progr. Phys.* **26**, 361.
N. E. Cusack (1967) *Contemporary Physics* **8**, 583.
N. E. Cusack and J. E. Enderby (1960) *Proc. Phys. Soc., Lond.* **75**, 395, 400.
N. E. Cusack, P. W. Kendall,and A. S. Marwaha (1962) *Phil. Mag.* **7**, 1745.
H. A. Davies and J. S. Llewellyn Leach (1970) *Phys. Chem. Liquids* **2**, 1.
R. Duprée and C. J. Ford (1973) *Phys. Rev.* **8B**, 1780.
Ya. I. Dutschak and P. V. Panasyuk (1967) *Sov. Phys. Solid State* **8**, 2244.
V. I. Fedorov and V. I. Machuev (1969) *Sov. Phys. Semicond.* **3**, 228.

V. I. Fedorov and V. I. Machuev (1970) *Sov. Phys. Solid State* **11**, 2179.
A. Ferrier (1961) *C. R. Acad. Sci., Paris* **254**, 104.
G. R. Fitterer (1967) *Trans. Quart. Am. Soc. Metals* **60**, 15.
H. Fritzche (1952) *Science* **115**, 571.
K. Furukawa (1959) *Nature, Lond.* **184**, 1209.
K. Furukawa (1960) *Sci. Rep. Res. Inst. Tohuku Univ.* **12A**, 368.
F. A. Gaibullaev, A. R. Regel, and Kh. Khusanov (1969) *Sov. Phys. Solid State* **11**, 1138.
H. Gerstenkorn and F. Sauter (1951) *Naturwiss.* **38**, 158.
R. Glocker (1949) *Ergebn. Exakt. Naturw.* **22**, 186.
R. M. Goodman and G. A. Somorjai (1970) *J. Chem. Phys.* **52**, 6331, 6325.
T. Gorecki (1974) *Z. Metall.* **65**, 427.
R. B. Gordon (1959) *Acta Metall.* **7**, 1.
G. Grimvall (1975) *Physica Scripta* **11**, 381.
G. Grimvall (1975) *Physica Scripta* **12**, 173.
G. Grimvall (1976) *Physica Scripta* **13**, 59.
G. Grimvall (1977) *Inst. Phys. Conf. Series No. 30.*
G. Grimvall and I. Ebbsjo (1975) *Physica Scripta* **12**, 168.
G. Grimvall and S. Sjödin (1975) **10**, 340.
K. A. Gschneider (1975) *J. Less-Common Metals* **43**, 179.
A. I. Gubanov (1957) *Sov. Phys.-Tech. Phys.* **2**, 2335.
H. J. Guntherodt (1970) Schweizer Archiv. **36**, 179.
H. J. Guntherodt and H. U. Kunzi (1973) *Phys. Kondens. Mat.* **16**, 117.
H. J. Guntherodt, H. U. Kunzi, R. Muller, and R. Evans (1972) *Phys. Lett. (Netherlands)* **54A**, 155.
H. Hendus (1947) *Z. Naturforsch.* **2A**, 505.
J. Hoehler and S. Steeb (1975) *Z. Naturforsch.* **30A**, 775.
T. Itami and M. Shimoji (1973) *Phil. Mag.* **28**, 85.
A. Jayaraman (1965a) *Phys. Rev.* **137A**, 179.
A. Jayaraman (1965b) *Phys. Rev.* **139A**, 690.
A. Jayaraman, W. Klement, Jr., and G. C. Kennedy (1963) *Phys. Rev. Lett.* **10**, 387.
A. Jayaraman, R. C. Newton, and G. C. Kennedy (1961) *Nature, Lond.* **191**, 1288.
H. D. Jones (1973) *Phys. Rev.* **8A**, 3215.
P. W. Kendall and N. E. Cusack (1960) *Phil. Mag. Ser. 8* **5**, 100.
G. C. Kennedy, A. Jayaraman, and R. C. Newton (1962) *Phys. Rev.* **126**, 1363.
R. W. Keyes (1951) *Phys. Rev. Lett.* **84**, 367.
A. Knappwost (1954) *Monatsh. Chem.* **85**, 3.
A. Knappwost (1959) *Z. phys. Chem. (Frankfurt am Main)* **21**, 358.
A. Knappwost and F. Thieme (1956) *Z. Elektrochem.* **60**, 1175.
W. D. Knight, A. G. Berger, and V. Heine, (1959) *Ann. Phys. (New York)* **8**, 173.
L. Knopoff and J. N. Shapiro (1970) *Phys. Rev.* **1B**, 38, 93.
T. A. Kontrova (1959) *Sov. Phys. Solid State* **1**, 1610.
S. S. Kor and S. K. Pandey (1973) *Acustica* **29**, 63.
H. Krebs (1969) *J. Non-Cryst. Solids* **6**, 455.
G. Krieg, R. B. Genter, and A. V. Grosse (1969) *Inorg. Nucl. Chem. Lett.* **5**, 819.
O. Kubaschewski (1949) *Trans. Faraday Soc.* **45**, 931.
O. Kubaschewski and J. A. Catterall (1956) *Thermodynamic Data of Alloys*, Pergamon Press, London.
P. Lamparter and S. Steeb (1976) *Z. Naturforsch.* **31A**, 99.
W. Lange (1966) *Z. Metallk.* **57**, 653.
M. F. Lantratov and E. V. Tsarenko (1959) *Zh. Fiz. Khim.* **33**, 1792.

A. S. Lashko (1959) *Zh. Fiz. Khim.* **33**, 1730.
I. Lauerman and G. Metzger (1961) *Z. Phys. Chem.* (*Leipzig*) **216**, 37.
A. Lucasson (1960) M.Sc. Thesis, Fac. Sci. Univ. Paris, Masson & Cie, Paris, 545.
H. D. Luedermann and G. C. Kennedy (1968) *J. Geophys. Res.* **73**, 2795.
J. Lumsden (1952) *Institute of Metals Monograph II.*
S. P. McAlister, E. D. Crozier, and J. F. Cochran (1974) *Can. J. Phys.* **52**, 1847.
D. K. C. MacDonald (1959) *Phil. Mag.* **4**, 1283.
B. R. McGarvey and H. S. Gutowsky (1953) *J. Chem. Phys.* **21**, 2114.
P. J. McGonigal, A. D. Kirshenbaum, and A. V. Grosse (1962) *J. Phys. Chem.* **66**, 737.
J. D. MacKenzie and W. B. Hillig (1958) *J. Chem. Phys.* **28**, 1259.
D. McLachlen (1969) *Proc. Nat. Acad. Sci.* **62**, 337.
Mal'sagov (1970) *Fiz. Tekh. Polnpov.* **4**, 1417.
W. Marty (1958) *Brown Boveri Rev.* **45**, 549.
A. S. Marwaha and N. E. Cusack (1965) *Phys. Lett.* **22**, 556.
A. Meyer, M. J. Stott, and W. H. Young (1976) *Phil. Mag.* **33**, 381.
N. P. Mokrovskii and A. R. Regel (1955) *Zh. Tekh. Fiz.* **25**, 2093.
N. F. Mott (1934) *Proc. Roy. Soc., Ser. A* **146A**, 465.
N. F. Mott (1972) *Phil. Mag.* **26**, 1249.
Y. Nakagawa (1956) *J. Phys. Soc. Japan* **11**, 855.
Y. Nakagawa (1957) *J. Phys. Soc. Japan* **12**, 700.
R. C. Newton, A. Jayaraman, and G. C. Kennedy (1962) *J. Geophys. Res.* **67**, 2559.
R. A. Oriani (1951) *J. Chem. Phys.* **19**, 93.
R. A. Oriani (1959) *J. Chem. Phys.* **31**, 557.
L. Pauling (1960) *Nature of the Chemical Bond*, 3rd edn., Cornell University Press, Ithaca.
T. E. Pochapsky (1951) *Phys. Rev.* **84**, 553.
M. Pokorny and H. U. Astrom (1975) *J. Phys. F* **5**, 1327.
E. G. Poniakovskii (1958) *Sov. Phys. Dokl.* (*U.S.A.*) **3**, 498.
A. R. Regel (1958) *Sov. Tech. Phys.* **3**, 489 (Engl).
A. R. Regel, V. M. Glazov, and A. A. Aivazov (1974) *Sov. Phys. Semicond.* **8**, 335.
H. Richter and S. Steeb (1958) *Naturwiss* **45**, 512.
H. Richter and G. Breitling (1961) *Z. Naturforsch.* **162**, 187.
H. Richter and G. Breitling (1962) *Z. Phys.* **168**, 69.
H. Richter and G. Breitling (1966) *Fortschritte Phys.* **14**, 71.
H. Richter and G. Breitling (1967) *Z. Naturforsch.* **22A**, 655, 658.
H. Richter and G. Breitling (1968) *Z. Naturforsch* **23A**, 2063.
H. Richter and G. Breitling (1968) *Z. Naturforsch* **20**, 1062.
H. Richter and G. Breitling (1970) *Z. Metallk.* **61**, 628.
H. Richter and G. Breitling (1966) *Z. Naturforsch* **21A**, 1710.
H. Ruppersberg and H. Reiter, (1972) *Acta Cryst.* **28A**, 233.
W. Schaafs (1956) *Acustica* **6**, 387.
A. Schneider and G. Heymer (1958) *Symposium on Physical Chemistry of Metallic Solutions and Intermetallic Compounds 1958*, N.P.L. Teddington, Middlesex, Paper 4a.
H. Schumann (1962) *Z. Anorg. Allg. Chem.* **317**, 204.
E. V. Shadrichev and I. A. Smirnov (1970) *Sov. Phys. Solid State* **11**, 1557.
O. D. Sherby (1969) *Sci. J.* **5**, 75.
E. Siegel (1975) *Phys Chem. Liquids* (*London*) **4**, 233.
M. Sing and R. Kumar (1968) *Proc. Symp. Non-Ferrous Metals Technol.* **3**, 261.
S. W. Strauss (1960) *Nucl. Sci. Eng.* **8**, 362.

D. Stroud and N. W. Ashcroft (1972) *Phys. Rev.* **5B**, 371.
Z. G. Szabo and B. Lakatos (1952) *Naturwiss.* **39**, 486.
S. Tanuma (1954) *Sci. Rep. Tohuku Univ. Res. Inst.* **6**, 159.
A. Ten Bosch and K. H. Bennemann (1975) *J. Phys. F* **5**, 1333.
G. Tourand (1975) *Phys. Lett. (Netherlands)* **54A**, 209.
H. U. Tschimer (1969) *Z. Metallk.* **60**, 46.
A. R. Ubbelohde (1966) *Proc. Roy. Soc., Ser. A* **293**, 291.
S. N. Vaidya, J. Akella, and G. C. Kennedy (1969) *J. Phys. Chem. Solids*, **30**, 1411.
Y. Waseda and M. Obtani (1975) *Z. Phys.* **21B**, 229.
Y. Waseda and S. Tamaki (1975) *Phil. Mag.* **32**, 273.
Y. Waseda and K. Yokoyama (1975) *Z. Naturforsch* **30A**, 801.
Y. Waseda, K. Yokoyama, and K. Suzuki (1974) *Phil. Mag.* **30**, 1198.
A. F. Wells (1962) *Structural Inorganic Chemistry*, Clarendon Press, Oxford.
J. M. Ziman (1961) *Phil. Mag.* **6**, 1013.
J. M. Ziman (1970) *Proc. Roy. Soc. Ser. A* **318**, 401.

# 10. NETWORK MELTING

## 10.1 Independent defects leading to network melting

Preceding chapters have discussed evidence about the melting of each of the principal types of crystals, which shows that

(i) The transition from crystal to melt is always discontinuous, but generally with precursor phenomena over a range of temperatures in both crystal and melt.

(ii) The transition may or may not involve a difference in specific volume but it always involves an increase of entropy on passing from crystal to melt.

(iii) This increase of entropy can be attributed to one or more ways of introducing disorder in the ideal crystal. Melting always involves the introduction of positional disorder in the ideal crystal lattice. Other kinds of disorder may also occur at the melting point. Some classes of crystals show examples in which orientational disorder need not await actual fusion, but takes place at solid–solid transitions with transition points $T_{c_1}$, $T_{c_2}$ ... below the eventual melting point $T_f$.

Establishing the above conclusions has given a powerful impulse to the development of statistical theories of melting.

A key feature in many statistical theories is the way in which the energy increase $\varepsilon$ required to introduce a specified defect into the crystal lattice changes as the fraction $n/N$ of defect sites is increased. For many types of crystals (with coordination number 6 or over) $\varepsilon$ decreases fairly steeply as $n/N$ increases so as to produce defect sites that are nearest neighbours. This leads to a kind of autocatalytic rise in $n/N$ with rising temperature, which is finally superseded by a jump to the molten structure as soon as this becomes thermodynamically more favourable. Statistical theories with this general characteristic are well developed and are discussed in the following chapter. In principle, it is however perfectly possible to discuss a notionally simpler alternative model for melting behaviour in a crystal in which $\varepsilon$ remains independent of $n/N$. The nearest practical realizations of such a model appear to be shown by a number of covalent bonded crystals, with $CN \leqslant 4$. In these the defect energy may refer to the breaking of primary interatomic bonds, which could be largely independent of bonding at neighbouring molecules. Of course the basic assumption of $\varepsilon$ independent of $n/N$ may not be even nearly correct for the majority of examples. Some discussion of the behaviour of network models is nevertheless of relevance. It helps to bring out the fact that one source of difficulty in matching experimental data to statistical theories of melting is that the process is not symmetrical about $T_f$. Whereas it is usually adequate to discuss the progressive introduction of defects into a unique crystal lattice (but see p. 183), in the melt there is greater freedom for alternative arrangements that can be regarded as highly defective versions of rival crystal lattices. In addition, molecular packing into *anticrystalline* clusters may become prominent in certain cases. In fact, the model for melting may be drawn up as in Fig. 10.1.

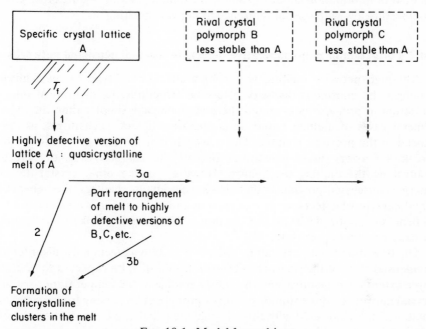

FIG. 10.1. Model for melting

The occurrence of processes (2) and (3a, b) if prominent, may be detectable in prefreezing 'anomalies' discussed in Chapter 13 as well as in rate processes controlling spontaneous nucleation and crystal growth discussed in Chapter 15.

In relation to these introductory considerations it is useful to have in mind some typical network crystals. One-dimensional networks involve covalent chains as in 'infusible white precipitate' with groups linked to two neighbours, represented by

$$-Hg-NH_2^+-Hg-NH_2^+-$$

separated by bromine anions (Weiss, Nagorsen, and Weiss, 1953). The chains may be flexible as in the case of S, Se or Te (Chapter 7). Two-dimensional networks include layer lattices such as graphite, molybdenum disulphide, and boron nitride (see Wells, 1962). Melting mechanisms for many of the substances of this important crystal type are not at all well investigated, partly on account of technical difficulties in handling melts, e.g. carbon at high temperature. Three-dimensional networks are much better known. They include the technologically important silicate and oxide glass-forming substances, and their analogues, also discussed in Chapter 16. They also include hydrogen-bonded networks. For example there is an enormous amount of information about a prominent example—water—some of which is discussed below with special reference to network melting.

## 10.2 Formation of non-cooperative defects in three-dimensional networks

Although network melting is a comparatively rare type of mechanism amongst the enormous diversity of crystal structures, its theoretical interpretation in principle is as stated thermodynamically simpler than the more general cases of melting based on cooperative defect mechanisms, as discussed in the previous chapter. The reason is that the energy $\varepsilon$ required to break a network bond constitutes one of the characteristics of defects controlling the melting mechanism. However, unlike other crystal defect energies, discussed previously and also in subsequent chapters, the network-breaking type of defect energy $\varepsilon$ at present under discussion is assumed not to depend steeply (if at all) on the fraction $n/N$ of defect (broken bond) sites already present in the crystal.

The transition from a crystalline (well-ordered) to a positionally disordered condensed state of the network substance would then be expected to be to a large extent *uncooperative*, and thus fairly gradual as the temperature rises. In crystal networks a discontinuous jump of properties on passing from crystal to melt, which is associated with the self-enhancing fall of $\varepsilon$ in many other types of disorder as $n/N$ increases, would not be expected to be nearly as prominent. Figure 10.2 illustrates the theoretical trend of excess heat uptake owing to increase of defect concentration with temperature in a crystal for the ideal case where $\varepsilon$ is wholly independent of $n/N$ in a so-called Schottky transition (Ubbelohde, 1952). This plot would only apply to actual network melting in the limiting case where $\varepsilon$ remains wholly independent of $n/N$. In practice, cooperative action between neighbours in the solid always modifies to some extent and sharpens the rise towards a peak shown in Fig. 10.2. Truly ideal network melting with $\varepsilon$ independent of $n/N$ should lead to a continuous change from 'solid' to 'liquid' (see also p. 28). In practice, on the 'liquid' side of $T_f$ rearrangements make further varied contributions to the enthalpy and entropy of fusion, usually enhancing the discontinuous nature of the transition. Melting may be particularly difficult to delimit sharply on a structural basis. Crystals of diverse ligand types with network characteristics might be

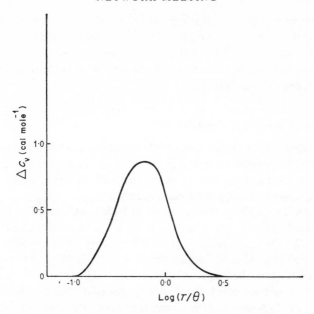

FIG. 10.2. Excess specific heat resulting from formation of non-cooperative defects in thermal equilibrium
in crystals, as a function of the characteristic temperature $\theta = \varepsilon/k$

expected to show exceptionally large ranges of prominent premelting (see
Chapter 12) and prefreezing (see Chapter 13).

Despite the simplicity of the basic assumption about independent defect
formation, development of more elaborate statistical thermodynamics for
network melting is hampered in various ways. No really satisfactory probability theory appears to be available from which to calculate the sizes of
separated domains i.e. of regions wholly cut off by bond rupture from all
attachment to neighbouring domains, as a function of the coordination
number in the ideal crystal, and the average fraction of defect sites $n/N$.
Thermodynamic properties for network melting which might be correlated
with such basic probability calculations are often hardly investigated. Transport properties such as molecular diffusion and viscosity are however much
more generally known, though it is not evident whether liquid flow in network
melts partakes principally of a sequence of bond-breaking/bond-making
events, or whether it involves mainly separated domains moving as whole
entities. Accordingly, existing information concerning transport properties in
network melts gives somewhat uncertain information about structure and
thermodynamic parameters.

For specific molecular species it must be considered whether, with poly-valent atoms, the postulated crystal defects involving bond rupture are likely to occur only once at each network site, or whether multiple ruptures can occur. This is often not yet known. A brief review of existing information is given for purposes of illustration.

### 10.2.1. *Network melting in Elements*

It seems likely that most crystals of the elements in which the linkages are covalent will exhibit network melting, but data on the high melting solids are scanty and not always very precise. For boron $T_f \sim 2350$ K and $S_f = 5 \cdot 1$ e.u. (Stout, 1973, observed using drop calorimetry).

For graphite, $T_f \sim 4000$ K. The entropy of fusion, estimated by a somewhat indirect method (Bundy, 1963) is $S_f \sim 6 \cdot 2$ e.u. A more direct evaluation is $T_f \sim 4300$ K, $S_f \ 7 \cdot 08$ e.u. (Haaland, 1976). Probably liquid carbon is far from monatomic. For this solid the liquid range is very short $(T_b/T_f \sim 1 \cdot 0)$ and it sublimes readily. For this melt $\Delta V_f/V_s \sim 45\%$ (Haaland, 1976) and there appears to be a maximum on the P–T diagram (Korsunkaya, Kamentskaya, and Aptekar 1972) pointing to an unusual network structure of the melt.

For germanium with its diamantane structure the collapse in volume result-ing from bond rearrangements on melting (Chapter 9) appears to be pro-gressive above $T_f$ (Krebs, Lazarev and Winkler, 1967).

For (red) phosphorus melting occurs with molecular rearrangement to give the (colourless) melt (Krebs and Fischer, 1970). Molecular bond rearrange-ments also seem likely for the melting of S, Se and Te (Haisty and Krebs, 1969a,b) as well as for some of their compounds. When the liquid range for a compound is short, as for $MoS_2$ (Cannon, 1959), this makes melting mechanisms particularly difficult to study.

### 10.3 Binary oxides and halides

Information about network melting is most extensive for these classes of crystals, possibly because fusion seldom involves major bond rearrangements.

### 10.3.1. *Three-dimensional oxide networks*

Some of the properties of the oxides $B_2O_3$, $SiO_2$ and $GeO_2$ illustrate the general range of melting phenomena to be expected. With $B_2O_3$, the low melting point makes measurements most readily accessible. Table 10.1 records some correlations of the properties related to melting (Mackenzie, 1956, 1958, 1959) of this group of oxides.

For BeO, $T_f$ lies between 2428°C and 2431°C but the transformation in the crystals shows large hysteresis (2095°C heating, 2058°C cooling; Latta, Duderstadt, and Fryxell, 1970). This broad hysteresis indicates a pronounced

**Table 10.1.** *Properties of crystals and melts with network characteristics*

| Bond disruption energy M–O | $B_2O_3$ | $SiO_2$ | $GeO_2$ |
|---|---|---|---|
| kcal $(g \ atom)^{-1}$ | 110 | 106 | 104 |
| $T_f$ (°C) | 450 | 1720 | 1115 |
| $1/\rho$ $(ohm^{-1} \ cm^{-1})$ | $10^{-6}$ | $10^{-5}$ | $10^{-5}$ |
| $\eta$ (poise at $T_f$) | $10^5$ | $10^7$ | $10^7$ |
| $E_\eta$ $(kcal \ mole^{-1})$ | 40 | 180 | 180 |
| $\eta(T_f + 200°C)$ | $10^3$ | $10^5$ | $10^5$ |
| $E_\eta(T_f + 200°C)$ | 30 | 150 | 160 |

tendency for the crystals to break up into domains with cooperative interface defects below $T_f$; enthalpy data do not appear to be available to verify this.

A characteristic feature which has made basic studies of network melting particularly difficult is that it can be so readily catalysed. Bond-breaking is frequently highly sensitive to the presence of certain types of impurity which promote break-up of the network. This can be readily understood, since primary covalency bonds such as Si—O—Si require much less energy for disruption if they can be hydrated to —Si—OH OH—Si, or broken in other ways by insertion of cations. Again, in transport processes the migration of impurity defect sites may be quantitatively more important, in a number of experimental examples, than the migration of true broken bond (free valency) defect sites which are analogous with free radicals. Thus in a bond-breaking/bond making sequence in the case of $B_2O_3$, it seems likely that the break-up of the network is strongly promoted by dissolved $H_2O$. These melts have a very low electrical conductivity and thus are not ionic. At higher temperatures, defect formation may perhaps involve the introduction of doubly linked oxygens, as in $B_2O_3$. The structural changes may be represented by Fig. 10.3.

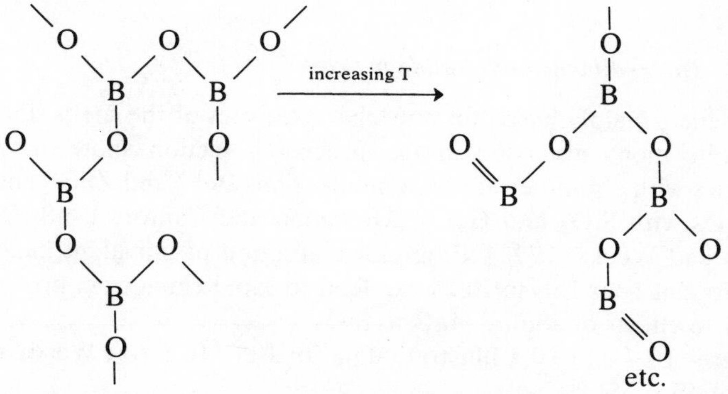

FIG. 10.3. Structural changes in the melting of $B_2O_3$

On cooling molten $GeO_2$, the viscosity–temperature relationship shows a characteristic steep increase on approaching $T_f$, over the value extrapolated from higher temperatures using a linear Arrhenius relationship $\log \eta = A_\eta + E_\eta/RT$. This anomalous behaviour has been attributed to a viscosity mechanism, whose activation energy involves the breaking of a Ge—O—Ge bond in the melt ($\varepsilon$ ranges from 160 to 80 kcal mole$^{-1}$). Molten $GeO_2$ is assumed to have a network structure with broken links (Mackenzie, 1958). It is remarkable, however, that the electrical conductivity $\sigma$ of the melt, which is quite low, yields a linear relationship $\log \sigma = A_\sigma + E_\sigma/RT$ with activation energy $E_\sigma$ of only 46·6 kcal mole$^{-1}$. This discrepancy between activators for the two processes has been plausibly explained on the basis that electrical conductance is due to calcium ions present as impurities in the melts, which migrate by a process different from that which controls the viscosity. The problem is to choose the most appropriate structural model to describe different properties of the melt. On the assumption of bond disruption, even at 300°C above $T_f$, the activation energy would be still 80 kcal mole$^{-1}$ for viscous flow. An alternative explanation of the steep curvature of the log $\eta$ plot could involve cluster formation (see Chapter 13), but has hardly been considered.

Physically it seems difficult to explain why the energy required to break covalent bonds in these molten oxides shows such a wide spectrum. The viscosity of pure molten silica near $T_f$ (e.g. $2·9 \times 10^6$ P at 1720°C) and the very large activation energy for viscous flow (approximately 175 kcal mole$^{-1}$ in the range 1200–1400°C, decreasing to about 150 kcal mole$^{-1}$ in the range 1800–2000°C) have been interpreted on the basis that in these network melts the unit flow in the transition state is a single $SiO_2$ molecule, formed by activation of the network bonds to the breaking point. Some support for this view is claimed from the finding that the activation energy drops to about 45 kcal mole$^{-1}$ if the bonds are permanently disrupted by dissolving metal oxides such as calcium oxide in the silica (Bockris and Lowe, 1954).

### 10.3.2. Three-dimensional halide networks

Evidence, mainly based on flow characteristics of the melts, indicates that the oxide melts discussed in the preceding section show marked model analogies with a number of halide melts. Thus $BeF_2$ and $ZnCl_2$ show marked analogies with $SiO_2$ and $GeO_2$ (Moynihan and Cantor, 1968; Baes, 1970; Bloom and Weeks, 1971). Progressive addition of univalent halides such as NaCl to $ZnCl_2$ or LiF to $BeF_2$ can lead to rapid changes in flow parameters similar to effects of adding MgO to $SiO_2$.

Figures 10.4 and 10.5 illustrate data for $BeF_2$ (Cantor, Ward, and Moynihan, 1969).

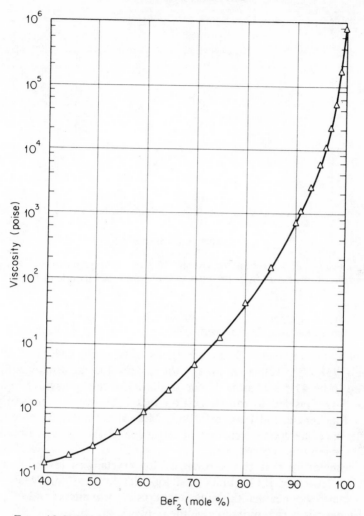

FIG. 10.4. Viscosities in the BeF$_2$—LiF system at 600°C.
Reproduced by permission of the American Institute of Physics

For BeF$_2$ the partial molal entropy in two component melts with LiF has
been evaluated (Holm and Kleppa, 1969). To complete the information about
network melting, what is also required is studies of enthalpy, specific volume,
and their derivatives for the one-component networks free from these catalytic additives.

Molten Zn(CN)$_2$ shows analogies with ZnCl$_2$ which may make it useful for
model purposes since its melting may be more accessible to experimentation
(Angell and Wong, 1970).

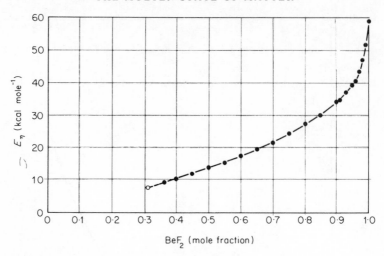

FIG. 10.5. Energy of activation for viscous flow *vs.* composition in the $BeF_2$—LiF system. Reproduced by permission of the American Institute of Physics

### 10.3.3. *Two-dimensional oxide networks*

The melt of tetragonal $P_2O_5$ has a high viscosity, of the order of $5 \cdot 15 \times 10^6$ poise at $T_f$ (580°C). Activation energy for viscous flow, as usually defined, is estimated to be $41 \cdot 5$ kcal mole$^{-1}$. The similarity of this behaviour to molten $B_2O_3$ or $SiO_2$ has led to the suggestion that $P_2O_5$ yields a network melt (Cormia, Mackenzie, and Turnbull, 1963). More structural information seems desirable since the crystal structure of tetragonal $P_2O_5$ suggests two-dimensional linking in this network melt.

It is noteworthy that this substance also crystallizes in an alternative 'molecular' lattice of $P_4O_{10}$ units held together by Van der Waals forces. There is some evidence that this molecular crystal form melts at 423°C; but its melt polymerizes very rapidly to give the network structure.

It seems characteristic of many more or less rigid planar networks that their liquid range appears to be extremely short, and may even be non-existent at ordinary pressures. Clearly, the crystal chemistry of such species would repay further study, including investigations at enhanced pressures.

### 10.4. Water as a network melt

From the preceding sections, it should be evident that in some respects ice may be treated as a characteristic network crystal, and water a network melt. This approach is however not very profitable, for the following two main reasons.

(a) In liquid water, domains of alternative packings are likely to be present with an unusually large variety of structures. With reference to Fig. 10.1 processes of type (2) and (3) are prominent and crystal polymorphs of many distinguishable structures A, B, C, etc. probably make a marked contribution to the melt near $T_f$. Effects of enhanced pressure in modifying this contribution also seem particularly marked. Accordingly, it is not very helpful to discuss the melting of ice in terms of a single bond-breaking defect operating on a three-dimensional network.

(b) Hydrogen bonds between neighbouring molecules can undergo angular distortion involving much smaller energy requirements than for angular distortion of primary covalent bonds. This again makes the assumption of a single defect energy $\varepsilon$ for the melting of water, not a good starting point in discussing structural consequences. Aspects of the unusually complex structure of water around $T_f$ are discussed further in relation to prefreezing phenomena (Chapter 13) and nucleation (Chapter 15).

Recent reviews of the structure of water which have a bearing on the melting of ice in relation to its crystal structure include Ives and Lemon, 1967; Safford and Leung, 1971.

# REFERENCES

C. A. Angell and J. Wong (1970) *J. Chem. Phys.* **53,** 2053.
C. F. Baes (1970) *J. Solid State Chem.* **1,** 159.
H. A. Bloom and I. A. Weeks ( 1971) *Trans. Faraday Soc.* **67,** 1420.
J. O'M. Bockris and D. C. Lowe (1954) *Proc. Roy. Soc. Ser. A* **226,** 423.
F. P. Bundy (1963) *J. Chem. Phys.* **38,** 618.
P. Cannon (1959) *Nature, Lond.* **183,** 1612.
S. Cantor, W. T. Ward, and C. T. Moynihan (1969) *J. Chem. Phys.* **50,** 2874.
R. L. Cormia, J. D. Mackenzie, and D. Turnbull (1963) *J. Appl. Phys.* **34,** 2245.
D. M. Haaland (1976) *Carbon* **14,** 357.
R. W. Haisty and H. Krebs (1969a) *J. Non-Cryst. Solids* **1,** 399.
R. W. Haisty and H. Krebs (1969b) *J. Non-Cryst. Solids* **1,** 427.
J. L. Holm and O. J. Kleppa (1969) *Inorg. Chem.* **8,** 207.
D. J. G. Ives and T. H. Lemon (1967) *Roy. Inst. Chem. Rev.* **1,** 62.
I. A. Korsunkaya, D. S. Kamenetskaya, and I. L. Aptekar (1972) *Fiz. Metal Metalloved.* **34,** 942.
H. Krebs, V. B. Lazarev, and L. Winkler (1967) *Z. Anorg. Allg. Chem.* **352,** 277.
H. Krebs and P. Fischer (1970) *Trans. Faraday Soc.* **1A** (Bristol Symposium).
R. E. Latta, E. C. Duderstadt, and R. E. Fryxell (1970) *J. Nucl. Mat.* **35,** 350.
J. D. Mackenzie (1956) *Trans. Faraday Soc.* **52,** 1564.
J. D. Mackenzie (1958) *J. Chem. Phys.* **29,** 605.
J. D. Mackenzie (1959) *Res. Rep. G.E.C. Schenectady N.Y. 59-RL-2231M.*
C. T. Moynihan and S. Cantor (1968) *J. Chem. Phys.* **48,** 115.
G. J. Safford and Pak. S. Leung (1971) in *Techniques of Electrochemistry Vol. II,* John Wiley and Sons, 1970-71.

N. D. Stout (1973) *High Temp. Sci.* **5,** 241.
A. R. Ubbelohde (1952) *Introduction to Modern Thermodynamic Principles,* 2nd edn., Oxford University Press.
A. Weiss, G. Nagorsen, and A. Weiss (1953) *Z. Anorg. Allg. Chem.* **274,** 151.
A. F. Wells (1962) *Structural Inorganic Chemistry,* Oxford University Press.

# 11. STATISTICAL THEORIES OF MELTING AND CRYSTAL STRUCTURE

## 11.1 Melting of crystals with simple units of structure

In considering reasons why melting has given rise to such a diversity of models, it should be remembered how often it is that the liquid state remains comparatively ill-characterized in modern theory. This is in striking contrast with the gaseous state, and even the crystalline (solid) state of matter. The present book emphasizes the structural changes which accompany melting, starting with the crystals as reference basis. Such an approach helps to make it clear that the increase of entropy on passing from crystalline to the molten states of matter is likely to involve a variety of mechanisms. A classification of theories about melting in terms of several different kinds of structural changes involved is almost essential, even if merely used as an introduction to more general treatments of the molten state.

As discussed in the preceding chapters, major differences might be expected in the melting of crystals of different ligand types, because their melts have such different properties. However, by far the greatest number of crystals in Nature belong to the molecular class. What follows refers to this ligand type unless otherwise stated. Even for crystals of the simplest possible structure, in which the units are atoms exerting radial forces on their neighbours, a surprising variety of physical theories has been put forward in the attempt to characterize the basic difference between solid and its melt. Some of the leading features of the statistical thermodynamics involved have been usefully summarized (Lennard-Jones, 1940). One of these which has given rise to some confusion is the 'communal entropy' attributed to liquids.

### 11.1.1. *Communal melting entropy*

As in all thermodynamic theories of melting starting with a suitable model, the requirement is to evaluate the available energy. Energy $A$ mole$^{-1}$ of either state is given by $A = U - TS = -kT \log f$ where $f$ is the partition function of the system, defined by

$$f = \frac{1}{N!} [(2m\pi kT)^{\frac{3}{2}}/h^3]^N \int \exp[-W(q_1, q_2, \ldots)/kT] \, dq_1 \, dq_2 \ldots$$

$$(11.1)$$

where $W(q_1, q_2, \ldots)$ is the potential energy of the atoms of a system in a configuration defined by the coordinates $q_1, q_2, \ldots$ etc., and the factor $N!$

arises from the fact that the atoms are indistinguishable *inter se*. (Molecules whose internal structure can be disregarded for the present purpose, and which are quasispherical may be treated as atoms.)

One method of evaluating $f$ is to suppose each atom is confined to a cell by forces arising from its neighbours. The field acting on it is calculated on the assumption that these neighbours are in their equilibrium positions. The intergral in $f$ is then given approximately by an expression of the form $(v_T^*)^N$ where $v_T^*$ is termed the available or 'free' volume defined as $v_T^* = \int e^{-\phi/kT} dv$. This is an integral over an atomic cell, $\phi$ being the potential energy of the field within it measured relative to the value at the centre of the cell, and gives a weighted average of the volume. The parameter $\phi$ takes a high positive value when the atoms enter each other's repulsive force fields, so that the integrand rapidly approaches zero. The parameter $v_T^*$ is a finite small fraction of the actual volume occupied by each atom.

If $\Phi$ is the energy of the system when the atoms are in their equilibrium positions, for the solid, for $N$ atoms

$$A = \Phi_0 - NkT[\log v_T^* + \log(2\pi mkT)^{\frac{3}{2}}/h^3] \tag{11.2}$$

and the entropy

$$S = -\left(\frac{\partial A}{\partial T}\right)_v$$

is given by

$$S = Nk\left[\log v_T^* + \log(2\pi mkT)^{\frac{3}{2}}/h^3 + \frac{T}{v_T^*}\left(\frac{\partial v_T^*}{\partial T}\right)_v + \frac{3}{2}\right] \tag{11.3}$$

Thus the internal energy is given by

$$U = A + TS = \Phi_0 + \frac{NkT^2}{v_T^*}\left(\frac{\partial v_T^*}{\partial T_v}\right) + \frac{3}{2}NkT \tag{11.4}$$

By comparison, for the so-called 'Einstein' solid consisting of $N$ vibrators of characteristic frequency $\nu_E$, the available energy $A$ is given by

$$A = -NkT \log(kT/h\nu_E)^3 + \Phi_0 \tag{11.5}$$

and the entropy per mole

$$S = Nk\left[\log(kT/h\nu_E)^3 + 3 - \frac{3T}{\nu_E}\left(\frac{\partial \dot\nu_E}{\partial T}\right)_v\right] \tag{11.6}$$

The main difference between (11.3) and (11.6) is the extra contribution $(\frac{3}{2})Nk$ allowed for the potential energy of vibration; this is introduced automatically in the Einstein model of the simple crystal. Comparisons of these expressions show that $v_T^*$ may be defined as

$$v_T^* = (kT/2\pi m\nu_E^2)^{\frac{3}{2}} \tag{11.7}$$

For a dilute gas of $N$ atoms, $W$ in (11.1) is zero for most of the configuration space and the available energy is simply

$$A = -kT \log[(2\pi mkT)^{\frac{3}{2}}/h^3] \times V^N/N!$$

where $V$ is the total volume. Writing $V = Nv^*$ and applying Stirling's theorem this reduces to

$$A = -NkT[\log v^* + \log(2\pi mkT)^{\frac{3}{2}}/h^3 - NkT] \tag{11.8}$$

and the entropy

$$S = Nk[\log v^* + \log(2\pi mkT)^{\frac{3}{2}}/h^3 + \tfrac{5}{2}] \tag{11.9}$$

When this is compared with $S$ in (11.3) for the solid, apart from the last 'expansion term', which in any case is small, the main difference is an increase in entropy of $Nk = R$ owing to the sharing of the whole volume by the atoms in the gas.

It has been suggested (Hirschfelder, Stevenson, and Eyring, 1937) that this increase of entropy, which is a statistical consequence of considering the whole of the volume as 'communal' or available to all the atoms, should be regarded as occurring at $T_f$, i.e. on melting the crystal. On this basis, the 'communal' entropy of melting would be $Nk = R$ cal mole$^{-1}$ deg$^{-1}$ or about 1·98 e.u.

In principle, the assumption that the whole 'communal' increase occurs at $T_f$ despite the small difference in density between solid and melt is open to criticism. A more gradual take up of communal entropy in the melt above $T_f$ can be argued on statistical grounds (Brusset, Kaiser and Perrin, 1968) for molecules such as $N_2$, $C_2H_6$, $C_3H_8$, $CH_2Cl_2$ and $CHCl_3$, and is also apparent in probability calculations on the solid–liquid transition in hard-sphere fluids (Hoover and Ree, 1968).

Furthermore a communal entropy of melting of $R$ e.u. does not match with experience. Only very few of the crystals reviewed throughout this book (e.g. Tables 9.1 and 9.3) have entropies of fusion as low as 2 e.u., and any relation to crystal structure is erratic. Even the inert gas crystals have values of $S_f$ that are higher, and other sources of entropy increase on fusion must therefore be sought.

The cell theory predicts the existence of solid and liquid phases, qualitatively, when allowance is made for the correlation between pairs of molecules (Barker, 1956). However quantitative developments (in their present form) do not match experiment closely, probably because this model fails to yield the solid state parameters adequately.

## 11.2 Disorder theories of fusion. Positional melting

As has been stated, the great diversity of values found for the entropy of fusion of different crystals arises from the fact that several kinds of structural changes may be contributing. Numerous X-ray studies have made it evident that whatever else may also happen to a crystal on melting, the regular repetition of lattice positions over long sequences in three dimensions, which is characteristic of any crystal structure, disappears on changing to a melt.

As indicated in the survey of experimental information in preceding chapters, positional disordering thus appears to be a *universal mechanism of melting*. However it is a sole mechanism only when all other sources of entropy change on passing from crystal to melt are absent. This universality has favoured the development of statistical theories of positional melting with direct reference to the inert gases as model crystals. Despite the fact that the models of positional disordering used are necessarily somewhat artificial, they can serve to illustrate some important features of this universal mechanism of melting in a very clear way.

## 11.3  Melting of inert gas crystals to quasicrystalline melts. The Lennard-Jones–Devonshire model

In this model, only one kind of positional disorder is considered, for simplicity. Perfectly ordered crystalline argon is an f.c.c. lattice in which each atom has twelve nearest neighbours. An important type of positional disorder involves the insertion of atoms interstitially, at sites on an interpenetrating lattice, the two lattices being related as are the $Na^+$ and $Cl^-$ lattices in crystalline rock salt.

In the Lennard-Jones–Devonshire model (1939) (see de Boer, 1952) atoms in the normal lattice sites are termed A and those in interstitial sites B. The total number of A and B sites is equal to the number of atoms $N$. Six adjacent B sites surround any A site, and *vice versa*. (More generally, according to the type of crystal lattice, one may assume each A atom to be surrounded by $Z$ adjacent B sites and *vice versa*.)

To place a single atom on a B site when all the rest are on A sites requires a considerable increase of energy, owing to the repulsive field of the neighbouring atoms. This energy increase per displaced atom would diminish if several atoms changed from A to B sites simultaneously, and would become zero in a state of complete disorder in which all sites have equal probability of being occupied. Comment has previously been made on how any *cooperative* character usually decreases the energy of defect formation in solids; in any model theory of melting the way in which a *cooperative* factor is introduced proves to be of particular significance. In the particular model described here, the partition function for an assembly of $N$ atoms in a state of perfect order is $f^N$ where $f$ is the contribution of each atom arising from vibration in its cell and is a function of the volume $V$ and temperature $T$:

$$f = \left(\frac{2\pi mkT}{h^2}\right)^{\frac{3}{2}} v_T^* \exp[-\Phi_0/NkT]$$

$v_T^*$ the 'free volume', and $\Phi_0$ the energy of the system with all atoms in their positions of equilibrium, are as stated previously. In disordered states of the crystals, keeping the position of each cell fixed, an additional factor must be introduced into the partition function to allow for positional disorder of the atoms.

Writing $Q = N_A/N$, $(1-Q) = N_B/N$ for the fraction of atoms in each kind of site, then provided their distribution is random over the two kinds of site, i.e. provided that domain formation or segregation of defects can be disregarded, the number of occupied B sites around any occupied A site is $Z(1-Q)$. Similarly the number of occupied A sites round any occupied B site is $ZQ$. The energy of interaction of each pair of atoms in adjacent A and B sites is taken as $W$, and will be a function of the distance between these sites and therefore of the volume of the solid as a whole. This confers an average energy to an atom on an A site of $W_A = ZW(1-Q)$ and on a B site $W_B = ZWQ$, so that the work of transfer A → B is

$$\Delta W = W_B - W_A = ZW(2Q - 1)$$

Clearly $\Delta W = ZW$ in a state of complete order ($Q = 1$) and $\Delta W = 0$ in a state of complete disorder ($Q = \tfrac{1}{2}$).

The total energy of interaction owing to disorder is

$$N_A W_A = N_B W_B = ZNWQ(1 - Q)$$

The number of ways of distributing the atoms at random is

$$Y(Q) = \frac{N!}{(N-N_A)!N_A!} \times \frac{N!}{(N-N_B)!N_B!} = \left\{ \frac{N!}{(NQ)![N(1-Q)!]} \right\}^2$$

It is usual to make the simplifying assumption (which is by no means evident) that the vibrations remain unchanged as a result of the disorder. On this basis it is sufficient to multiply the complete partition function $f^N$ (see Kirkwood, 1938) by the disorder factor $Y(Q) \exp[-ZNWQ(1-Q)/kT]$. At constant $V$, $T$, and $W$ this factor has a maximum for a definite

$$\frac{ZW(2Q-1)}{2kT} = \log Q - \log(1-Q)$$

or $(2Q-1) = \tanh ZW(2Q-1)/4kT$. This equation is always satisfied by $Q = 1/2$ and when $ZW/4kT \geqslant 1$ there will be another root $Q_{max} > 1/2$. When this root exists it corresponds to the maximum value of the partition function. One may write for the overall free energy minimum $A = A' + A''$ where $A'/NkT = \log f$ as for the fully ordered solid, and

$$A''/NkT = \frac{ZWQ_{max}(1-Q_{max})}{kT} - 2\{(1-Q_{max})\log(1-Q_{max})\}$$

Other principal thermodynamic functions can likewise be separated into an 'ordered' part and a contribution resulting from positional disorder, with the expressions

$$U = U' + U''$$

$$S = S' + S''$$

$$P = P' + P''$$

where

$$U' = NkT^2 \left[ \frac{\partial}{\partial T} \log f \right]_v$$

$$U'' = ZNWQ_{max}(1 - Q_{max})$$

$$S' = NkT \frac{\partial}{\partial T} \log f + N \log f$$

$$S'' = -2(1 - Q) \log(1 - Q) - 2Q \log Q$$

$$P' = -\left( \frac{\partial A'}{\partial V} \right)_T = NkT \left[ \frac{\partial}{\partial V} \log f \right]_T$$

$$P'' = -\left( \frac{\partial A''}{\partial V} \right)_T = -NZ \left( \frac{dW}{dV} \right) Q(1 - Q) - \left( \frac{\partial A}{\partial Q} \right)_{V,T} \left( \frac{\partial Q}{\partial V} \right)_T$$

$W$ for this model is assumed to be a function of $V$ alone*. The second term in $P''$ vanishes in view of the minimum of property $A''$ with respect to $Q$ for given volume and temperature.

Denoting by $n''$ the number of pairs of atoms in adjacent A and B sites, since $n'' = ZNQ(1 - Q)$, then $P'' = -n'' \, dW/dV$. The effect of this new term $P''$ on the equation of state profoundly modifies the isotherm, as can be seen from Fig. 11.1. Corresponding plots of the available energy are given in Fig. 11.2. With an assembly homogeneous throughout its volume the curve ABC would be followed. Following a well-known argument in statistical thermodynamics, the total free energy can be reduced by following the straight line AC which refers to a two-phase system. In this mixture of two phases the minimum A in Fig. 11.2 corresponds to a state of comparative order, the point C to nearly complete disorder.

In relation to matters discussed in Chapters 12 and 13, it is interesting to note that even this simple model requires some premelting in the solid and some prefreezing in the melt as these two phases approach $T_f$. In the present theory, structural characterization (particularly of the melt) is, however, only rudimentary. The theory assumes the melt is purely *quasicrystalline* and makes no allowance for possible *anticrystalline* clusters in this phase which

---

* For a modified evaluation of $W$ in this equation see Lennard-Jones (1940).

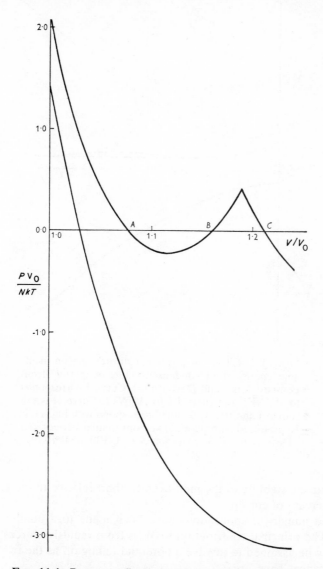

FIG. 11.1. Pressure effects resulting from disordering in
condensed states of matter (from Lennard-Jones and
Devonshire). The lower curve gives $P'$, the pressure for a
state of order. The upper curve gives the sum of $P'$ and $P''$.
Reproduced with permission from Lennard-Jones and
Devonshire, *Proc. Roy. Soc. Ser. A.* **169,** 317 (1939)

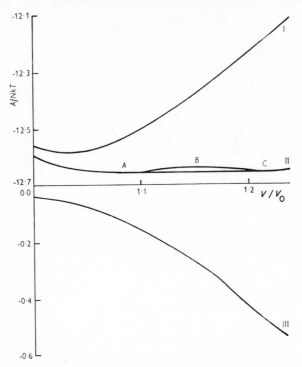

FIG. 11.2. Effects of positional disordering on available energy $A$ of condensed states of matter (from Lennard-Jones and Devonshire). Curve I corresponds to $A'/NkT$ and curve III to $A''/NkT$. Curve II is the sum of I and III. A, B and C correspond with Fig. 11.1. Reproduced with permission from Lennard-Jones and Devonshire, *Proc. Roy. Soc. Ser. A.* **170**, 464 (1939)

may favour the stability of the melt even further, relative to the crystal, by a further increase of entropy.

Detailed numerical applications have been made to a number of simple crystals. The principal contribution to $W$ is from repulsive forces, which for argon may be assumed to involve a potential falling off as the twelfth-power of the distance. Thus

$$W = W_0\left(\frac{V_0}{V}\right)^4$$

and the complete equation of state is

$$\frac{PV}{NkT} = V\frac{\partial(\log f)}{\partial V} + 4ZW_0\left(\frac{V_0}{V}\right)^4\frac{Q(1-Q)}{kT}$$

with $Z = 6$. $W_0$ can be evaluated by matching $T_f$ at any given pressure. This leads to a value $S_f = 1 \cdot 70R$ compared with the experimental $1 \cdot 66R$. The coefficient of thermal expansion $\alpha$ of the argon melt is calculated to be $0 \cdot 0040 \text{ deg}^{-1}$ compared with an observed value of $0 \cdot 0045 \text{ deg}^{-1}$. Approximately, the model with the empirical rule for $\phi_0$ discussed in the following paragraph indicates $\alpha T_f = 1/3$ in rough agreement with experiment (see also p. 67).

Detailed examination of the results suggests the semiempirical correlation $kT_f = 0 \cdot 7\phi_0$ where $\phi_0$ is an energy parameter which can be directly related to the interatomic forces. This rule can be extended with fair agreement to other crystals containing atoms, or even molecules with quasispherical repulsion envelopes, such as Ne, $N_2$, CO, $CH_4$, and $H_2$. The rule can also be useful as a diagnostic norm from which to measure deviations observed in cases such as crystalline oxygen. $T_f$ for this substance lies well below the calculated value, indicating that other entropy effects additional to positional disorder contribute to its melting. This particular model mechanism also gives values of $V_L$ and $V_S$ in fair agreement with experiment, for the atoms and molecules stated.

Physical models are acceptable only so far as they are serviceable. Despite the crudity of the model, several important features characteristic of melting appear quite clearly in this theory.

(i) *Lattice forces and melting*. Melting parameters are directly related to the intermolecular forces, which determine the work of creating positional disorder in the crystal, and to the expansion in volume $\Delta V_f = V_L - V_s$ on passing from crystal to melt.

An even simpler statistical theory of melting neglects attractive forces altogether, and regards the molecules as impenetrable hard spheres. This is briefly referred to in section 11.4 below. The Lennard-Jones specification of what happens at the melting point raises in another way the question already considered previously (p. 25)—to what extent at very high external pressures a continuous transition might be found between crystal and liquid. If extrapolation is permissible the Lennard-Jones–Devonshire model would indicate that above a temperature $(T_{crit})_{fusion} = 1 \cdot 1\phi_0/k$ the isotherms lose the sigmoid-shape characteristic of a change of phase, reminiscent of the way isotherms of a fluid lose their sigmoid-shape above the liquid–gas critical temperature. Presumably melting on this model would become continuous, since any sharp change $S \rightarrow L$ would be difficult to describe in structural terms above $(T_{crit})_{fusion}$. For argon $(T_{crit})_{fusion} = 132$ K, $(P_{crit})_{fusion} = 2400$ atm, on the above theory; yet experimentally no critical melting is found even up to 6000 atm. Further theoretical studies of effects of pressure on positional melting are desirable to clear up this discrepancy. In any case, the characterization of interatomic repulsions in terms of a single force-field parameter $\phi_0$ may not be valid at these higher pressures.

(ii) *A model for a strictly quasicrystalline fluid.* The features of the model and the conclusions about positional melting to which it leads are all characteristic of a crystalline solid changing into a strictly quasicrystalline assembly in the melt. This conforms with the definition of such melts in which specific 'sites' can be attributed to 'ordered' and to 'disordered' atoms. Accordingly, it would not be easy to extend the Lennard-Jones–Devonshire treatment to melts that are for one reason or another far from quasicrystalline in character. Specific examples have been discussed above for melts of flexible molecules (p. 154). *Anticrystalline* character proves to be of particular importance in prefreezing phenomena of certain melts discussed in Chapter 13.

(iii) *Premelting of crystals.* The model quite naturally introduces an increasing proportion of disordered sites up to $T_f$ in the solid. In argon, this increase has not reached a stage where it becomes markedly autocatalytic, so that premelting is not quantitatively very obvious in the properties usually chosen for experimental measurement. Other properties such as mass diffusion could however be more revealing even when the fraction of defect sites $n/N$ below $T_f$ remains small. By suitable adjustment of the force fields, other hypothetical crystals can be constructed to show marked premelting phenomena below $T_f$. Experimentally observed premelting in crystals generally is discussed in Chapter 12.

(iv) *Structural and other origins of thermal expansion of melts.* Since a quasicrystalline structure is assumed for the melt, part of its thermal expansion and specific heat must be attributed to the creation of crystal defects, and part to the usual expansive consequences of anharmonic lattice vibrations in crystals.

However, for anticrystalline melts, many other factors also contribute to their thermal expansion. Classical thermodynamics shows the overall thermal expansion to be given by $-\alpha/\beta = (\partial S/\partial V)_T$ where $\alpha$ is the coefficient of thermal expansion and $\beta$ is the compressibility. From this it can be seen that in unusual liquids various other sources of entropy increase may also have to be allowed for as the volume of a liquid expands, in addition to the conventional vibrational effects, and the positional effects discussed in the Lennard-Jones theory. This highlights the fact that surprisingly little systematic study has yet been made of *thermal expansion* of melts above $T_f$, to determine whether they are normal or abnormal with respect to the simple model of melting.

(v) *Structure of supercooled melts.* A further consideration is to determine what the model suggests will happen to the structure of a strictly quasicrystalline melt when such an assembly is supercooled down to the temperature of spontaneous crystallization, $T_N$. Empirically (as detailed in Chapter 15) for many liquids of simple structure it seems likely that

$$T_N/T_f \sim 0.8$$

Extrapolation of the thermal properties of supercooled molten argon down to $T_N$ has not yet been discussed on an order–disorder model of the Lennard-Jones–Devonshire type. It seems possible that refinements to theories of spontaneous nucleation, to allow for appreciable changes suggested by the model on cooling the melt below $T_f$ before it crystallizes spontaneously, would normally be swamped by other more powerful influences of temperature on the kinetics of nucleus formation; but these refinements should nevertheless provide a correction term.

The Lennard-Jones–Devonshire model refers to positional melting only. Various attempts to refine this kind of treatment of positional melting have been made, without however attaining a close degree of correspondence with experiment, commensurate with the additional complications introduced in the theoretical development (see Domb, 1951, who gives other references).

As an alternative to calculating the increase in entropy on melting, resulting from positional disorder, from a crystal lattice model, the change of coordination number has been discussed as an additional parameter to characterize entropy and volume changes. In an f.c.c. solid the coordination number is 12 in the fully ordered state, whereas in the melt it could sink to 11 or 10·5 or even less as a result of the formation of holes (Rice, 1938, 1939). However, this approach, though it has some experimental support, does not lend itself to calculations of entropy of disorder at all readily, and awaits further development.

## 11.4 Solid–liquid transitions in hard-sphere assemblies

Hard spheres are molecules for which the repulsive forces rise from zero to infinity when they make contact and which have no attractions. Transitions to a solid from a liquid in which only repulsive forces operate have been discussed by Rushbrooke (1967).

Computerized simulation of the behaviour of assemblies of 'hard' spheres as the volume containing them shrinks progressively leads to the important conclusion that a solid–liquid transition appears even when attractive forces between the molecules are wholly neglected. More realistic interaction potentials have also been assumed by a number of authors when using such models to calculate melting and other thermodynamic parameters of the inert gas crystals. As yet it seems too soon to test many of the outcomes from such computerized methods, using sensitive criteria such as those listed on pp. 295 seq. (see Crawford, 1974a,b; Street, Raveche, and Mountain, 1974). Repulsion forces appear to be much 'softer' between alkali metal atoms than between inert gas atoms to account for the much greater volume change on freezing. $(\Delta V_f / V_s)$ for alkali metals is about 2·5% compared with 15% for inert gases (Hansen and Verlet, 1969; Hansen and Schiff, 1973) but see p. 246 for an alternative explanation.

## 11.5 Cooperative positional disorder leading to melting

In addition to positional disordering which takes the form of individual lattice defects such as vacancies, or the occupation of interstitial lattice sites, positional disordering can also be 'cooperative' over a number of atoms (Oldham and Ubbelohde, 1940). A characteristic feature of cooperative defects is that the energy of their formation per unit of structure in a lattice can often be greatly reduced, relative to isolated defects. However this kind of crystal disorder is subject to the condition that the appropriate sequence of neighbouring sites must be displaced cooperatively from the ideal lattice positions. Imposition of this restrictive condition on siting of the defects lessens the contribution of such defects to the statistical positional entropy of the crystals. Modifications of the vibrational entropy through their presence can however also modify the vibrational entropy so as to favour the melt.

### 11.5.1. *Dislocation melting (elementary model)*

Within recent years, much work has been done on a particular kind of cooperative defect known as a lattice dislocation. This simple kind of communal defect by no means exhausts possibilities of cooperation between neighbouring units of structure in a crystal. Various properties of elementary dislocations are however fairly well known, as a result of their importance for studies of mechanical properties, especially for metallic crystals whose structure is often fairly simple. As has repeatedly been stressed in this book, the relative importance of different mechanisms of melting must be expected to vary with different types of crystal structure. Even the relative importance of different types of positional disorder will vary from one lattice arrangement to another. For metals, numerical estimates suggest cooperative positional disorder may perhaps make a major contribution to melting (e.g. Mott, 1952; and see p. 244). Attempts have been made to provide more detailed calculations of the contribution of dislocations to the melting mechanisms of simple metals, taking into account the increase in concentration of dislocations on fusion (Mizushima, 1960; see also Ookawa, 1960).

A model proposed for cooperative positional melting of a simple cubic lattice may be discussed, by way of illustration, for a fixed lattice constant $a$, and a density of dislocations at any temperature considered as uniform throughout the crystal. The excess free energy resulting from the presence of a density of dislocations $\rho$ is to be calculated. For simplicity, the lattice is divided into a number of cubic domains, all of equal volume. Each domain is given a line element of dislocation, assumed parallel to one of the three (100) crystal axes. Thus the dislocation density determines the size of each domain. This cannot become excessively small because of the specification that it is to contain a dislocation line element. When neighbouring domains have their line elements passing through their common domain boundary, they are regarded as forming one dislocation branch. If the boundary between neighbouring domains is passed only by the line element of either domain, this is assumed to extend to the centre of the other domain, where it meets a node, in order to complete the connection of all the branches into a network. Such extensions are, however, neglected for simplicity in the calculations which follow.

Thus a branch which is always straight may be composed of 1, 2, or more line elements. Some domains have nodes, others not. Any node is associated with a minimum of 3, up to a maximum of 6, branches. In these terms, any configuration of the dislocation network can be completely defined except that the Burgers vectors remain to be determined.

The dislocation line density $\rho$ (total length of dislocation lines per unit volume of the lattice) is thus associated with an edge length of a domain $\rho^{-\frac{1}{2}}$. Each domain contains $\rho^{-\frac{3}{2}}a^{-3}$ atoms, and there are $n = N_0 a^3$ domains per g atom, where $N_0$ is Avogadro's number. The total length of the dislocation lines is $N_0 a^3 \rho$.

11.5.1.1. *Energy arising from dislocations.* Somewhat harsh simplifications are introduced to solve the difficult problem of evaluating the excess energy $\Delta E$ resulting from dislocations in terms of the excess per plane traversed, which in turn is subdivided into a term $\varepsilon_{\text{core}}$ attributable to the core energy, and $\varepsilon_{\text{elas}}$ attributable to the elastic deformation energy of the atoms surrounding the core. The parameter $\varepsilon_{\text{elas}}$ depends logarithmically on the ratio between the mean distance between the lines, i.e. $\rho^{-\frac{1}{2}}$ and the radius $r_0$ of a core, assuming its cross-section to be a circle. An approximate expression is

$$\varepsilon_{\text{elas}} = \tfrac{1}{4}\pi G_0 a^3 \log \rho^{-\frac{1}{2}} / \pi^{\frac{1}{2}} r_0$$

The absolute value of the Burgers vector is taken to be $a$ and the difference between different types of dislocation is neglected. The form of this expression is chosen to make $\varepsilon_{\text{elas}}$ zero, when the area of a domain wall $\rho^{-1}$ becomes equal to the area of cross-section $\pi r_0^2$ of the core. $G_0$ is the rigidity modulus of the lattice. The effect of the value of Poisson's ratio is neglected for simplicity.

Core energies are difficult to evaluate. Rather forced model calculations give

$$\varepsilon_{\text{core}} = \log[(r_0/r')4\pi] G_m a^3$$

or more generally $\varepsilon_{\text{core}} = \alpha G_m a^3$ in which the parameter $\alpha$ is a fraction of unity when the core radius is of the order $a$. $G_m$ may be taken as the geometric mean of two elastic coefficients, and $r'$ is a length (much smaller than $r_0$) which depends on the lattice constant, and slightly on the ratio of the two elastic coefficients. This leads to the final energy contribution:

$$\Delta E = N_0 a^3 \rho \left[ G_m \alpha - \frac{1}{8\pi} G_0 \log(\pi \rho r_0^2) \right]$$

A maximum contribution must be attained when the lattice is filled with core, making

$$\rho_{max}^{-\frac{1}{2}} = \pi^{\frac{1}{2}} \rho_0$$

and

$$\Delta E_{max} = \left( \frac{\alpha}{\pi} \right) \left( \frac{a}{r_0} \right)^2 G_m V_0$$

$$V_0 = N a^3$$

This calculation can be even further simplified by noting that

$$r_0 \sim a$$

11.5.1.2. *Entropy increase resulting from dislocations.* Calculation of the energy increase $\Delta E_{max}$ resulting from the presence of cores introduces difficulties of mechanical theory. Entropy increases must involve both the consequences of random selection of configurations of the dislocation networks, and consequences of modifying the vibrational spectrum of the crystal through the presence of the dislocations, i.e. $S_f = \Delta S_{config} + \Delta S_{vib}$. Approximately, for the very crude model selected, it is found that $\Delta S_{config}/R = a^3 \rho^{\frac{3}{2}} \log C$ where $C$ is the order of 20.

If $\nu$ is the Einstein frequency characterizing the perfect crystal, the vibrational entropy of the perfect crystal at temperatures $T \gg h\nu/k$ is taken to be $S_{vib} = 3R \ln kT/h\nu$.

Dislocations introduce atoms in the cores vibrating with a different frequency $\nu'$ in the direction parallel to the Burgers vector (atoms outside the cores are assumed to be unaffected). This makes

$$\Delta S_{vib} = qa^2 \rho R \log \nu/\nu'$$

where $q$ is the number of atoms in any plane, belonging to the core.

The usual summation of energy and entropy contributions, owing to the presence of these cooperative defects, leads to a free energy $A$:

$$A = V_0 a^2 \rho [\alpha G_m - G_0/8\pi \log \pi \rho r_0^2)]$$
$$- RT[a^3 \rho^{\frac{3}{2}} \log C + qa^2 \rho \ln \nu/\nu']$$

Inspection of this expression yields a critical temperature $T_f$, such that below $T_f$, $A$ is a minimum when $\rho = 0$. Above $T_f$, $A$ is a minimum when $\rho = \rho_{max}$. At $T_f$

$$H_f = \frac{a}{\pi}\left(\frac{a}{r_0}\right)^2 G_m V_0$$

$$\frac{S_f}{R} = \left[\pi^{-\frac{3}{2}}\left(\frac{a}{r_0}\right)^3 \log C + \log(\nu/\nu')\right]$$

$$T_f = H_f/S_f$$

These expressions have clearly been derived by means of simplifications too drastic to be of any precise numerical significance. Even so, insertion of plausible values for $r_0 \sim a$, $a/\pi \sim 0.05$, $\nu/\nu' = \sqrt{2}$ and $C = 20$ leads to the interesting conclusion, with this model, that both vibrational and positional disorder contribute to $S_f$, which is given in e.u. by

$$S_f = 1.08_{config} + 0.69_{vib}$$

This expression for melting by way of cooperative positional defects may be contrasted with the Lennard-Jones–Devonshire model based on isolated defects, which gives a configurational entropy of fusion disregarding any changes of vibrational entropy. It may also be contrasted with the Lindemann model since the only vibrations that control melting in it are those associated with dislocation cores, not with the undisturbed regions of the crystal.

### 11.5.2. Dislocation melting (refined model)

More sophisticated theories of dislocation melting have tended to concentrate on dominant terms in thermodynamic functions for melting. Only f.c.c. or h.c.p. close-packed crystals seem readily amenable to calculations either by analytical (e.g. Eckstein, 1967) or computerized procedures. For crystals of the inert gases, it has been computed that (in e.u.)

$$\Delta S_{config} = 0.50 \pm 0.15$$

$$\Delta S_{vib} = 0.76 \pm 0.17$$

i.e. with a vibrational contribution relatively larger than for the previously discussed theory (Kristensen, Jensen and Cotterill, 1974). An analytical treatment highlights the behaviour of normal glide dislocations (Kuhlman Wilsdorf, 1965). Model computational calculations of the generation of crystal defects by way of dislocations have been discussed, though only in a rather abstract way (Jensen, Kristensen, and Cotterill, 1972; Kristensen, Jensen, and Cotterill, 1974; Cotterill, Kristensen, and Jensen, 1974). The most common and most important dislocations lie about the close-packed plane of h.c.p. or f.c.c. crystals, and the critical temperature $T_{crit}$ at which the calculated free energy of formation of dislocations becomes equal to or smaller than zero is identified as the melting point. The original publications should be consulted for details, in particular concerning experimental criteria to help in identifying for which crystal types which theories of dislocation melting are most appropriate. For example, none of the publications cited deals with contributions to the entropy of fusion from orientational or configurational entropy. This tends to limit such comparisons of theoretical models with experimental melting data to crystals with atomic units of structure, such as metals (Chapter 9).

It is also important to note that if the process crystal → melt occurs at unusually low temperatures as a result of developing cooperative defects, such as dislocations, this must become apparent in certain transport parameters near $T_f$, for example mass diffusion or ion mobility. Activated diffusion along dislocations may be expected to exhibit chain sequence of behaviour (Ubbelohde, 1957) unlike the point-defect diffusion by way of more or less independent hopping from site to site. Comparisons between experimentally determined self-diffusion parameters and mobilities calculated from NMR measurements (Bladon, Lockhart, and Sherwood, 1971) suggest that some diffusion processes may indeed be cooperative in certain crystals. More experimental evidence, particularly with atomic crystals, would be welcome (see also Hooper and Sherwood, 1976).

## 11.6 Positional disorder theories with more than one mechanism of melting

The Lennard-Jones–Devonshire theory of positional disorder as a mechanism of melting is based on a simple central force law (the 6–12 potential) and thus implies a law of corresponding states. Quantities such as the entropy of fusion $(S_f)_{trip}$ at the triple point, and the ratio of the triple-point temperature $T_{trip}$ to the critical temperature $T_{crit}$ should be the same for all solids conforming to this behaviour. For many solids very approximately $S_{f(trip)} = S_f$, and $T_{trip} = T_{(atm)}$. Since the boiling point at 1 atm pressure $T_b$ is approximately proportional to $T_{crit}$ for such substances, melting should be governed by the two similitude rules

$$T_b / T_f = \text{constant}$$

and

$$S_f = \text{constant}$$

insofar as corresponding state theories can be applied. However, empirical data reviewed earlier show that neither of these relationships applies generally, though they often offer fairly good approximations within a narrowly chosen class of crystal structures.

Departures from this simple formulation of positional melting must be expected whenever more than one mechanism of disordering is feasible in the crystals. In particular for polyatomic units of structure, orientational disordering presents such a mechanism, which experience has shown to be important, though erratic in incidence. Orientational disordering can generate thermal transitions in the solid below $T_f$, or may accompany the melting process, in ways that can be recognized empirically, as described in Chapter 6. When the magnitudes of barriers opposing orientational disorder are made to depend upon the degree of positional disordering, extension of the Lennard-Jones–Devonshire model of positional melting can be developed, on somewhat harshly simplified lines, incorporating an elementary kind of orientational order–disorder process (Pople and Karasz, 1961a,b). In each of the two kinds of positions $\alpha$ and $\beta$ with reference to the idealized crystal lattice, one of two orientations may also be attributed to each molecule, as an additional parameter, leading to different minima of potential energy $\alpha_1$, $\alpha_2$ and $\beta_1$, $\beta_2$. These minima are regarded as linked with solid angles within which the molecular axes lie. A jump of molecular axes from one position to the next has to overcome a potential barrier.

Just as the simple positional theory of melting introduces (see above) a repulsive energy $W$ for molecules on neighbouring $\alpha$ and $\beta$ sites, to allow for the higher energy of $\beta$ sites when these are mainly surrounded by $\alpha$ sites, allowance is made for orientational disorder introduced by each pair of occupied neighbouring $\alpha_1\alpha_2$ sites, of $\beta_1\beta_2$ sites, by introducing a second repulsive energy $W'$. $Z'W'$ is the energy required to turn one molecule from $\alpha_1$ to $\alpha_2$ with the rest of the lattice in orientation $\alpha_1$.

As has been stressed in connection with phenomena such as the 'lubrication' of orientational disorder by impurities in solid solution, where one molecule is turned into the unfavourable $\alpha_2$ position local strains are set up. Normally these lower the energy required to pass from an $\alpha$ to a $\beta$ position. Such coupling between the energy required to create two different kinds of defect can be allowed for by the simple (if somewhat artificial) device of assuming that the energy $W$ of neighbouring $\alpha$ and $\beta$ sites is independent of orientation, thus favouring their being neighbours.

With this additional disorder, the new partition function for the solids $Z = f^N \Omega$ where

$$\Omega = \sum \exp(-1/kT)[N_{\alpha\beta}W + N_{\alpha_1\alpha_2}W' + N_{\beta_1\beta_2}W']$$

Following the earlier treatment, $Q$ represents the fraction of molecules on $\alpha$ sites. Similarly, $S$ represents the fraction of molecules with $\alpha$ orientations. Using the Bragg–Williams simplification for the summations that lead to $\Omega$ (see Lennard-Jones and Devonshire, 1939)

$$\Omega(Q,S) = Y(Q,S)\exp - N[ZWQ(1-Q) + Z'WS'(1-S)(1-2Q+2Q^2)]/kT$$

Since there are

> $NQS$ molecules in $\alpha_1$ positions,
> $NQ(1-S)$ molecules in $\alpha_2$ positions,
> $N(1-Q)$ molecules in $\beta_1$ positions,
> $N(1-Q)(1-S)$ molecules in $\beta_2$ positions,

the number of ways $Y(Q, S)$ of arranging the molecules for given $Q, S$ is

$$Y(Q, S) = \left\{ \frac{N!}{[NQ]![N(1-Q)]!} \right\}^2 \times \left\{ \frac{NQ!}{[NQS]![NQ(1-S)]!} \right\}$$

$$\times \left\{ \frac{[N(1-Q)]!}{[N(1-Q)S]![N(1-Q)(1-S)]!} \right\}$$

Using Stirling's theorem to maximize $N^{-1} \log \Omega$, two conditions are

(A)  $\log \dfrac{Q}{(1-Q)} = \left[ \dfrac{ZW}{2kT} - \dfrac{Z'W'S(1-S)}{kT} \right](2Q-1)$

(B)  $\log \dfrac{S}{(1-S)} = \left[ \dfrac{Z'W'}{kT}(1-2Q+2Q^2) \right](2S-1)$

If $W' = 0$, condition (A) reduces to the corresponding result for the Lennard-Jones–Devonshire theory. The ratio

$$y = \frac{Z'W'}{ZW}$$

is a measure of the relative energy barriers for the orientational and positional disordering of a molecule.

As in the parent theory which deals with positional melting only, a solution of these equations can always be found for $Q = S = \frac{1}{2}$ at sufficiently high temperatures. Both kinds of disorder then occur together at $T_f$. For large values of $ZW/kT$ values of $S$ and $Q$ maximizing $N^{-1} \ln \Omega$ may differ from $\frac{1}{2}$. Stepwise determination of values of $Q$ and $S$ that maximize the partition function show that these vary as a function of $ZW/kT$ and the parameter $y$. As $ZW/kT$ increases, at first a range of solids is found for which $S = \frac{1}{2}$, but $Q \neq \frac{1}{2}$. Over this interval, separation of orientational from positional disorder is to be expected. Such a separation would correspond experimentally with thermal randomization of orientations in the crystal with a peak transition temperature $T_c$ below $T_f$. Melting when it finally occurs then refers only to randomization of positions. The overall maximum of the partition function corresponds with $Q \neq \frac{1}{2} \neq S$ for sufficiently large values of $y$. Interestingly enough, if $y$ is increased still further, $S$ leaves the value $\frac{1}{2}$ for $ZW/kT$ smaller than $Q$, which would seem to indicate orientational but not positional ordering persisting above $T_f$ as an equilibrium state. This *may* correspond with liquid crystal behaviour (Chapter 14) though the structure of liquid crystals is usually not quasicrystalline. More generally, it may correspond with the failure to achieve rotation in the melt near $T_f$ (p. 130). However this statistical model is too artificial to make further exploration of this extension of algebraic theory profitable with such simple order–disorder calculations. Nevertheless, it helps to stress the general importance of considering whether molecules are 'free to rotate' in their melts (p. 149). Tests for this should be much more frequently applied than at present, e.g. in discussing transport parameters of melts.

Exactly as for the simple positional melting, excess functions $A''$, $S''$, $P''$ can be calculated to describe the consequences of these two kinds of disorder, when $y \leqslant 0.325$. Unfortunately, numerical evaluations of such a highly simplified model cannot be expected to be very accurate. In addition to other omissions made to achieve simplification, allowance should also be made for the fact that many crystals have more than two positions of minimum orientational energy, which increases $S_f$ accordingly. In a straightforward extension of the Pople–Karasz procedure, allowance has been made for $D$ possible orientations of molecules in a molecular crystal, with $D \geqslant 2$ (Amzel and Becka, 1969). Improved representation of actual behaviour is achieved, with close agreement between observed and calculated entropies of orientational transition, in cases where this is separable from the melting process ($T_c/T_f < 1$). Illustrative examples are given in Table 11.1.

As in the parent Lennard-Jones–Devonshire model, one serious weakness of this theory lies in the unrealistic representation of the molten state. When disorder of orientation is followed separately from positional disorder, as stated it is possible mathematically to arrive at solutions of the equations for which $T_c/T_f > 1$. However, the lack of freedom to rotate, which does indeed appear to prevail in many melts, is unlikely to take such a simple lattice-like

**Table 11.1. Orientational and positional melting**

| Molecule | $T_f$ | $T_c/T_f$ | $D$ | $S$ (e.u.) trans (expt.) |
|---|---|---|---|---|
| $CCl_4$ | 250·3 | 0·90 | 6 | 4·79 |
| $CF_4$ | 89·5 | 0·85 | 6 | 4·63 |
| $CBr_4$ | 53·2 | 0·88 | 6 | 4·98 |
| Adamantane | 54·3 | 0·38 | 6 | 3·87 |
| $WF_6$ | 275·2 | 0·96 | 14 | 7·81 |
| $ReF_6$ | 291·7 | 0·93 | 14 | 7·49 |
| $OSF_6$ | 306·4 | 0·90 | 14 | 7·29 |
| $IrF_6$ | 317·1 | 0·86 | 14 | 7·08 |
| $PtF_6$ | 334·8 | 0·83 | 14 | 6·77 |
| $MoF_6$ | 290·6 | 0·91 | 14 | 7·22 |
| Bicyclo-2,2,2-octane | 447 | 0·37 | 20 | 6·66 |
| Bicyclo-2,2,1-heptane | 360 | 0·36 | 36 | 7·59 |

form. Instead, packing into *anticrystalline* clusters or groups seems more likely in most cases.

Effects of pressure on positional as well as orientational defects are not allowed for in the models discussed so far. As explained, increased pressure could have interesting consequences for the merging of the two transitions, as well as for the possibility of critical melting. Unfortunately, data are not yet available. (A useful critique of the Pople–Karasz equations presents these in reduced form, Smith, 1965, without however lifting the limitations referred to above.)

### 11.7 Conglomerate models of melting

Theories of melting based on a quasicrystalline model for the liquid phase lend themselves to fairly straightforward treatment by conventional statistical mechanical procedures, as in the examples outlined. At the other extreme, freezing of 'hard sphere' liquids could be regarded as developing quasigaseous models of melting. For crystals of molecules with very low defect energies, such as helium or hydrogen, this may be adequate. The general question remains open, whether a more realistic model for any melt than either the quasicrystalline or quasigaseous might be represented by a conglomerate of domains. Not all of these need even have the same packing. Every crystal lattice is geometrically well defined but nothing compels a melt to develop its increased disorder with reference to only a single crystal lattice, especially when there are several alternatives with almost the same free energy to which different domains may be referred.

Particularly near $T_f$, small domains or clusters of molecules may be formed whose local arrangement differs from the most stable crystal polymorph from which the melt is formed and also from a quasigaseous condensate. It may be presumed that domains representing disordered versions of less stable crystal polymorphs must also be present in the melt. In addition, other causes of such clustering may be quite varied. For example, in ionic melts association complexes may be formed. 'Non-crystallizable' atomic packings may occur in molten metals. In melts of rigid organic molecules, interlocked clusters may be favoured by a marked departure from a spherical repulsion envelope. Re-entrant portions on this repulsion envelope appears to facilitate *anticrystalline* close-packings once long range positional order is lost. In melts of long chain compounds, which contain many configurational isomers, a micellar mechanism for introducing disorder may operate. Some of the consequences of clustering for thermodynamic behaviour have already been discussed, in relation to the volume required for free rotation in the melt (p. 149), and in relation to abnormally low melting points in ionic melts with non-spherical ions (p. 205). Other aspects are described in relation to prefreezing phenomena (Chapter 13).

To describe the melt accurately, some kind of summation to include all the principal varieties of domains is needed, but up to the present no fully satisfactory statistical thermodynamics has been proposed for any model of an assembly of clusters, both crystalline and anticrystalline swimming in a quasigaseous environment. Starting with quite different considerations, related concepts have been discussed by Eyring, Ree, and Hira (1958).

### 11.7.1. Swimming domain models for melts

Another perhaps more familiar approach is to treat any kind of clusters in a quasigaseous environment in 'chemical' terms following a mass action formulation, as though they were polymers.

The concentration of such polymers will be controlled by an equilibrium constant $K_n = [A]^n/[A_n]$ so that if a fraction $x$ of the total species is in the polymerized state and fraction $(1-x)$ in the monomer state

$$K_n = (1-x)^n/xV^{n-1}$$

where $V$ is the volume occupied by one nominal mole of fluid. The heat of association which controls the expression $d(\log K_n)/dT = H_n/RT^2$ is given by $H_n = nH_c$. For any type of cluster to be of significant importance in a melt the heat of clustering $H_c$ per unit attached must be of the same order of magnitude as the heat of fusion $H_f$, and the volume occupied cannot on average be very different from that in the crystal. (If $H_c$ were much larger than $H_f$, this kind of packing would predominate even above $T_f$; if it were much smaller it would be trivial.)

Such a 'conglomerate' formulation of association equilibrium is unlikely to be useful if $(1-x)$ falls below, say, 70%. Though very crude it can be used to analyse transport phenomena in melts in the prefreezing region (Chapter 13). As a general model for melts it suffers from the present weakness that no independent general physical methods are yet available for assessing the average size of clusters $n$. Until experimental and mathematical techniques are developed for dealing with cooperative fluctuations whose structure can be specified, advances in conglomerate models of condensed states of matter will remain seriously hampered.

### 11.7.2. Conglomerate melting of hydrogen-bonded crystals

As discussed in Chapter 13 reasons for postulating a conglomerate structure for melts, particularly in the region of $T_f$, have become increasingly prominent for a variety of melts, in view of the *prefreezing* anomalies they exhibit. Although such anomalies are not conditioned by the existence of any singular form of intermolecular attraction, some types of intermolecular forces particularly favour cluster formation. For example, hydrogen bonding combines fairly strong attractive forces between molecules with a diversity of configurations of atoms in which these attractions are more or less equally effective. Clusters involving hydrogen bonding can thus readily satisfy the requirements of being formed through molecular packings not referable to any regular crystal lattice, but having much the same molecular volume and heat of formation as any fully regular crystal stable in the same range of temperature. Water has prompted much the most numerous researches on melting of hydrogen-bonded crystals, including quite detailed calculations of cluster formation. Figure 11.3 illustrates schematically the kind of conglomerate structure proposed for this melt. Statistical computations of the extent of cluster formation follow the procedure proposed for 'significant structures' (Eyring, Ree, and Hirai, 1958) and lead to distributions in the melt (Nemethy and Scheraga, 1962) as summarized in Table 11.2. More than one type of cluster is included in the calculation, to include tetra-, tri- and di-hydrogen-bonded water molecules. The product of cluster size (column 2) and concentration (column 3) expresses the fraction of $H_2O$ molecules bound into some kind of cluster.

These values illustrate semiquantitatively the progressive breakdown in a liquid with a marked tendency to molecular bonding. Some authors would

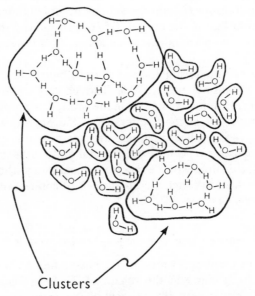

Clusters

FIG. 11.3. Schematic representation of cluster formation in water. Reproduced by permission of the American Institute of Physics

### Table 11.2.  Cluster formation in liquid $H_2O$

| $T$ (°C) | Size of cluster (number of units) | Concentration $\times 10^2$ (moles per mole $H_2O$) | Fraction of hydrogen bonds unbroken |
|---|---|---|---|
| 0 | 90·6 | 0·84 | 0·53 |
| 10 | 71·5 | 1·02 | 0·49 |
| 20 | 57·0 | 1·24 | 0·46 |
| 30 | 46·5 | 1·47 | 0·43 |
| 40 | 38·4 | 1·72 | 0·41 |
| 50 | 32·3 | 1·98 | 0·39 |
| 60 | 27·8 | 2·24 | 0·37 |
| 70 | 24·9 | 2·43 | 0·36 |
| 80 | 22·9 | 2·57 | 0·34 |
| 90 | 21·8 | 2·64 | 0·33 |
| 100 | 21·0 | 2·68 | 0·325 |

include water among the network class (Chapter 10). Concentrations of clusters near $T_f$ involve volume fractions too large to permit use of the simple blocked volume theory of prefreezing anomalies discussed in Chapter 13. Evaluation of various transport properties can be made following a 'lattice-like' model (Frank and Wen, 1957; Frank, 1958) but will not be discussed here.

# REFERENCES

L. M. Amzel and L. N. Becka (1969) *J. Chem. Phys. Solids* **30**, 521.
J. A. Barker (1956) *Proc. Roy. Soc. Ser. A* **240**, 265.
P. Bladon, N. C. Lockhart, and J. N. Sherwood (1971) *Mol. Phys.* **20**, 577.
J. de Boer (1952) *Proc. Roy. Soc. Ser. A* **215**, 41, cf. 21.
H. Brusset, L. Kaiser, and F. Perrin (1968) *J. Chim. Phys. Phys. chim. Biol.* **65**, 260.
R. M. J. Cotterill, W. D. Kristensen, and E. J. Jensen (1974) *Phil. Mag.* **30**, 245.
R. K. Crawford (1974a) *J. Chem. Phys.* **60**, 2169.
R. K. Crawford (1974b) in *Rare Gas Solids*, ed. M. L. Klein and J. A. Venables, Academic Press, London.
C. Domb (1951) *Phil. Mag.* **42**, 1316.
B. Eckstein (1967) *Phys. Status Solidi* **20**, 83.
H. Eyring, T. Ree, and N. Hirai (1958) *Proc. Nat. Acad. Sci., Wash.* **44**, 683.
H. S. Frank (1958) *Proc. Roy. Soc., Ser. A* **247**, 481.
H. S. Frank and W. Y. Wen (1957) *Discuss. Faraday Soc.* **24**, 133.
J. P. Hansen and L. Verlet (1969) *Phys. Rev.* **184**, 151.
J. P. Hansen and D. Schiff (1973) *Mol. Phys.* **25**, 1281.
J. O. Hirschfelder, R. Stevenson, and H. Eyring (1937) *J. Chem. Phys.* **5**, 896.
R. M. Hooper and J. N. Sherwood (1976) *J. Chem. Soc. Faraday Trans.* **72**, 2872.
W. G. Hoover and F. H. Ree (1968) *J. Chem. Phys.* **49**, 3609.
E. J. Jensen, W. D. Kristensen, and R. M. J. Cotterill (1972) *Phil. Mag.* **27**, 623.

J. G. Kirkwood (1938) *J. Chem. Phys.* **6**, 70.
W. D. Kristensen, E. J. Jensen, and R. M. J. Cotterill (1974), *Phil. Mag.* **30**, 229.
D. Kuhlman Wilsdorf (1965) *Phys. Rev.* **140A**, 1599.
J. E. Lennard-Jones (1940) *Proc. Phys. Soc., Lond.* **52**, 729.
J. E. Lennard-Jones and A. F. Devonshire (1939) *Proc. Roy. Soc., Ser. A* **169**, 317; **170**, 464.
S. Mizushima (1960) *J. Phys. Soc. Japan* **15**, 70.
N. F. Mott (1952) *Proc. Roy. Soc., Ser. A* **215**, 1.
G. Nemethy and H. Scheraga (1962) *J. Chem. Phys.* **36**, 3382.
A. Ookawa (1960) *J. Phys. Soc. Japan* **15**, 2191.
J. W. Oldham and A. R. Ubbelohde (1940) *Proc. Roy. Soc., Ser. A* **176**, 50.
J. A. Pople and K. E. Karasz (1961a) *J. Phys. Chem. Solids* **18**, 28.
J. A. Pople and K. E. Karasz (1961b) *J. Phys. Chem. Solids* **20**, 294.
O. K. Rice (1938) *J. Chem. Phys.* **6**, 476.
O. K. Rice (1939) *J. Chem. Phys.* **7**, 883, *cf.* 889.
G. S. Rushbrooke (1967) *Discuss. Faraday Soc.* **43**, 7.
G. W. Smith (1965) *J. Chem. Phys.* **42**, 4229.
W. B. Street, H. J. Raveche, and R. D. Mountain (1974) *J. Chem. Phys.* **61**, 1960 and following paper in journal.
A. R. Ubbelohde (1957) *Discuss. Faraday Soc.* **23**, 128.

# 12. PREMELTING IN CRYSTALS

## 12.1 Precursor effects in the transition from crystal to melt

As already stated, classical thermodynamics assumes complete independence between two phases which are in equilibrium at the melting point. Geometrically, this implies intersection of their $P$, $V$, $T$ surfaces so that there is a jump in the tangent angles on passing from one phase to another (p. 9). On this classical basis, precursor effects on either side of a transition point or melting point can apparently only be due to impurity shifts discussed in the following paragraph.

On the other hand, structural theories of melting inevitably involve some kind of defect formation. The origin of precursor effects must be given a different attribution according to the type of theory developed, though this essential consideration is seldom spelt out clearly. For example, vibrational theories presumably would require increasing anharmonicity on approaching $T_f$. In more sophisticated theories such effects need not be uniform over the entire phonon spectrum. Theories of positional disordering require an increase of appropriate defects; for orientational disorder $S_{or}$ these may possibly be collected into domains, as in hybrid crystals. Theories of dislocation melting would demand measurable changes of mechanical and acoustic properties even below $T_f$. More attention to the origins of these different kinds of defect formation could well be rewarding. Statistical thermodynamics shows that formation of defects must begin to occur in most crystals even at temperatures well below their change to another more stable state. Precursor effects are thus definitely required, though they may not be very prominent in all cases. No real conflict is introduced by this requirement of statistical thermodynamics, since the classical postulate of phase-independence is satisfied so long as there is a discontinuous jump in tangent slopes at the transition point itself. Accordingly, precursor effects may be looked for with the expectation that their identification may throw considerable light on actual mechanisms of fusion, since different mechanisms entrain different precursor effects.

To save confusion, it may be pointed out here that in phase transitions precursor effects are seldom symmetrical about a transition point. Even in solid–solid transitions, defect formation involves different energy requirements in the two lattices, with consequent differences in the defect concentration $vs.$ temperature curves for the two crystalline phases. However, in solid–melt transitions, there is much greater freedom for alternative molecular packings in the melt. This implies that *prefreezing* effects are often likely to be more prominent, and may seldom even resemble premelting in the same transition. For this, and for other reasons connected with the thermodynamics of the supercooled molten state, the present chapter is restricted to consideration of premelting only in crystals.

## 12.2 Theoretical possibilities for incipient disordering of a crystal lattice on approaching $T_f$

### 12.2.1. Two-phase effects (trivial premelting)

Much lack of clarity may be found, particularly in the older literature, about premelting phenomena. Even the nomenclature is not uniform in all publications on the subject. One usage refers to a phenomenon which is rather trivial in its theoretical aspects, but which is of considerable importance in the accurate experimental evaluation of thermodynamic parameters. When a melt contains even a small amount of an impurity *which is not soluble in the crystals*, as the liquid freezes the concentration of this impurity rises progressively, so that the instantaneous equilibrium temperature $T_f$ between crystals and (impure) melt is progressively lowered as more solid is frozen. Mathematical treatments of this kind of spreading of the equilibrium temperature between solid and liquid are discussed below. Its occurrence does not imply any kind of anomaly in the solid or melt near $T_f$. Usage of the term 'premelting' to describe this kind of gradual depression of the freezing point is unfortunately rather widespread in the literature. It should NEVER be employed without a descriptive term to indicate that two phases are present throughout. A term such as 'heterophase premelting' would perhaps be suitable, but has unfortunately been applied by Frenkel to a more basic statistical fluctuation phenomenon discussed below. Possibly 'two-phase premelting' or 'freezing point depression spread' will prove acceptable in evaluations of melting parameters by scientists who are aware of the confusions introduced by using 'premelting' without precise specification of what is implied. The present author would recommend 'two-phase premelting' or 'heterophase premelting' whenever the observed precursor effects arise from the coexistence of two distinct phases.

### 12.2.2. Homophase premelting

The suggestion that *precursor effects in the crystal lattice itself* might be responsible for some premelting phenomena on approaching $T_f$ appears to have first been made in relation to the fusion of crystals of long-chain paraffins and their derivatives (Ubbelohde, 1938). Since the precursor effects take place in the crystal before it melts, the term 'homophase premelting' is reasonably precise in referring to such phenomena.

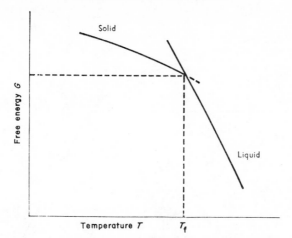

FIG. 12.1. Classical intersection of free energy curves for solid and liquid at $T_f$. Note the sharp angle of intersections and absence of superheating of the solid

In view of certain experimental difficulties which have obstructed some observations of homophase premelting, it is useful to examine the theoretical possibilities before discussing actual instances.

When premelting is due to a change *in the crystals* as they approach $T_f$, this implies that the classical thermodynamic treatment of the phase change

$$\text{solid} \rightarrow \text{liquid}$$

is no longer adequate. As stated above, in this treatment each phase is regarded as having a wholly independent free-energy surface and the melting curve is determined by the intersection of these independent surfaces, as illustrated in Fig. 12.1 (see Ubbelohde, 1950). Premelting in the crystals shows that these two surfaces are not wholly independent, however, since $G_S$ 'knows' that it is about to intersect with $G_L$. To meet the requirement that there is an actual jump at $T_f$, homophase premelting must imply intersection at a somewhat reduced angle, preceded by some enhanced curvature in the solid, as illustrated in Fig. 12.2. it may be noted (see p. 395) that whereas careful experimentation permits observations on the supercooled liquid for a considerable range of temperatures below $T_f$, no clear evidence has been obtained of extensive superheating of the solid above $T_f$. This means that the dotted portion of the curve for the solid in Fig. 12.1 is somewhat hypothetical. Probable reasons for such behaviour are discussed in Chapter 15 in relation to the kinetics of melting and freezing.

An enhanced rate of change of curvature for the solid implies that its heat content $H_S$ changes abnormally fast on approaching $T_f$, as is even more

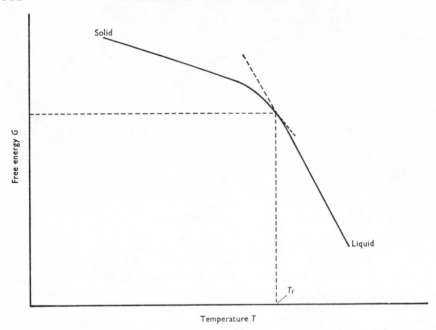

FIG. 12.2. Intersection at an angle of free energy curves for the solid and liquid at $T_f$, modified to allow for premelting in the crystals below $T_f$

clearly apparent from specific heat plots. Unless there is an equally rapid change of the molar heat content of the liquid phase $H_L$, the heat of fusion $H_f = H_L - H_S$ should change steeply with temperature. Marked changes of curvature of properties of the melt near $T_f$, referred to as prefreezing (Chapter 13), are not necessarily found in the same crystal types for which premelting is most prominent. In consequence, rapid changes in $H_f$ may permit a rather sophisticated thermodynamic test for premelting in the crystals. This requires very precise determinations of the freezing point depression near to the ideal limiting freezing point $^0T_f$, on adding impurities not soluble in the crystal. Measurements using this procedure have been described for molecular crystals (Oldham and Ubbelohde, 1940) and for ionic crystals (Rhodes and Ubbelohde, 1959), and see p. 187.

### 12.2.3. *Phonon premelting*

If vibrational breakdown of the crystal lattice were a valid explanation of melting (see Chapter 3) it might be expected that as $T_f$ is approached properties dependent on vibration, such as the lattice specific heat and thermal expansion, would begin to show precursor anomalies. Vibrational anharmonicity which has sometimes been adduced to interpret premelting anomalies in various types of crystal lattice (see below) could be regarded in this sense but the evidence is not very firm. It is not always easy to discriminate between vibrational anharmonicity and other effects such as positional defect formation as causes of premelting, particularly when only one or two lattice characteristics have been determined near $T_f$. Other properties such as comparisons between X-ray estimates and bulk thermal expansion, or molecular diffusion using isotopically tagged molecules, or ionic conductance in ionic crystals, can however be particularly useful in providing discriminatory information. Computer studies on the melting of assemblies of hard spheres indicate that as the vibrational energy increases the longitudinal frequencies increase whilst transverse frequencies at long wavelength ultimately decrease. This effect is most marked for the most densely packed (111) direction and at sufficiently high amplitudes these planes shear into a new configuration and the vibrational energy is redistributed. This matches experimental findings with Ne (Dickey and Paskin, 1969; Leake, Daniels, Skalyo, Frazer, and Shirane, 1969). Vibrational 'melting' on this model involves atoms sliding past one another, i.e. not undergoing head-on obstruction as in the primitive Lindemann model.

In solid–solid transformations precursor phonon anomalies are sometimes readily identified (Chapter 4). Strong vibrational anharmonicity thus almost certainly contributes to the marked precursor phenomena which accompany solid–solid lambda transitions such as orientational randomization in crystals (see Ubbelohde, 1963). When such 'phonon-fading' effects merge with the solid–liquid transition (Chapter 9, p. 238) they may begin to appear in premelting anomalies.

In the crystal compound $In_2Bi$ $(T_f = 89°C)$ a steep increase in phonon attenuation (longitudinal 10 Mc s$^{-1}$) is observed within 1·5°C of $T_f$ (Saunders, Pace, and Alpar, 1967).

## 12.2.4. *Defect premelting*

With many crystals, various types of defect can be envisaged, for which the characteristic temperatures given by $\theta = \varepsilon/k$ where $\varepsilon$ is the work of creating the particular defect, and $k$ is Boltzmann's constant, lie well above $T_f$. At any lower temperature the fraction present in thermal equilibrium will be of the order

$$n/N \sim \exp(-\varepsilon/kT)$$

where $N$ is Avogadro's number. The contribution from such defects to an excess molar specific heat is $C_{exc} = \varepsilon(dn/dT)$ and to an excess thermal expansion $V\alpha_{exc} = v(dn/dT)$ where $v$ is the volume increase on creating a defect and $V$ is the molar volume. Such defects clearly cannot contribute appreciably to changes of enthalpy or volume unless $dn/dT$ is appreciable. Thus they are not easily detected experimentally from such excess functions, below $T_f$.

Other physical properties, such as the electrical conduction of ionic crystals, can often be used to detect much smaller concentrations of defects than any precursor excess of enthalpy or volume. In this respect, all crystals must show homophase premelting, since all are subject at least to positional defects below $T_f$. One question is to decide for which kinds of crystal structures defect formation is sufficiently marked, before the crystal melts, to affect appreciably even insensitive properties such as heat content and specific heat, or specific volume and thermal expansion. Though no general theoretical guide can be given at the present time, it is clearly desirable to classify homophase premelting phenomena according to the type of crystal under examination. So far as the present experimental information permits, this is done in sections which follow. With regard to general theoretical considerations about defect concentrations, it may be noted that the Lennard-Jones–Devonshire model for positional melting, involving interstitial defects (Chapter 11) indicates a mole fraction of about $0.05$ of defect sites at $T_f$ in crystalline argon ($Q = 0.95$). If present this should be fairly easily detectable; the value calculated is however sensitive to the numerical assumptions made in this theory. Whatever the details assumed, in view of the universal contribution of positional defects to melting processes, some premelting should be regarded as a normal feature.

As against this expectation, in calorimetric determinations on crystals of the simple molecules Ar, CO, $N_2O$, HCl, and HBr, any premelting is stated to be less than experimental error (Clusius and Staveley, 1941). More sensitive methods of measurement are apparently required to establish defect formation in such crystals (see below).

### 12.2.5. *Fluctuation premelting*

In general, premelting phenomena must in part be attributed to the exponential increase in concentration of crystal defects as $T_f$ is approached. As the fraction of defects increases, various factors may be particularly helpful in lowering $\varepsilon$ in some crystals. These include the presence of impurities *in solid solution* which may facilitate (lubricate) orientational transformations below $T_f$, and which may also influence other cooperative interactions of defects that lower $\varepsilon$.

A somewhat different approach to pretransition phenomena generally, including premelting, has been discussed by Frenkel (1939). In this fluctuation theory, the basic thermodynamic concept is that fluctuations lead to a statistical distribution of 'micro-regions' of a phase B in a phase A below the transition temperature $T_c(A \to B)$ (e.g. the melting point) and of 'microregions' of A in B above $T_c$. This concept involves the theory of cooperative fluctuations (see Ubbelohde, 1937). It may be noted that particularly extensive experimental support for the general concept of cooperative fluctuations has been derived from studies of the spontaneous transformation of phases supercooled below their transition temperature, in relation to the formation of crystallization nuclei (Chapter 15).

In Frenkel's theory, if $T_c$ is the temperature of transition or of melting, the concentration of microregions at any temperature $T_c + \Delta T$ is calculated in terms of the parameter

$$\beta = -\frac{S_f \Delta T}{kT_c}$$

and in terms of the volume $gv_b$ of the microregions, and their surface tension $\sigma_{AB}$ with respect to the surrounding medium. (For the liquid–solid transition, this parameter becomes $\sigma_{SL}$ and it controls nucleation in a way discussed in Chapter 15.) $v_b$ is the volume per molecule in the solid.

With special reference to premelting, the excess specific heat below $T_f$ attributable to cooperative fluctuations in an assembly of $N$ molecules is given by

$$C_{ex} = NH_f(dJ/dT)$$

where

$$J = \int_{g_0}^{\infty} g \exp[-\beta g - Y_g^{\frac{2}{3}}]\, dg$$

$$Y = (\sigma_{SL}/kT_f) \times (36\pi)^{\frac{1}{3}} v_b^{\frac{2}{3}}$$

As with other theories of microphase fluctuations, the physical difficulty with this theory is to arrive at a satisfactory means of evaluating $\sigma_{SL}$ for different types of crystal structures which give rise to different kinds of melts (see also Lukasik, 1957). Even for macrosurfaces, direct evaluation of $\sigma_{SL}$ involves complex difficulties (Grange, Landers, and Mutaftschiev, 1976).

A related difficulty not properly discussed by Frenkel is that the configuration of atoms in microregions in condensed phases needs to be accurately defined particularly when the host phase is a crystal lattice. Even the incomplete evidence at present available about the structure of melts indicates that attribution of a 'liquid' structure to fluctuation regions in the crystal is much too indefinite to permit theoretical progress. For very small regions of 'liquid' the packing is likely to fluctuate substantially from the average for bulk liquid, even well away from $T_f$. A microregion of 'liquid' surrounded by a crystalline matrix may differ even more from the statistical average for liquid in bulk, so that the term cannot be given any very precise meaning. Again, some premelting effects are much too large to be attributable to a *heterophase* defect such as is postulated by Frenkel (Ivlev and Mal'tseva, 1971).

A more definite as well as more consistent picture would involve the coexistence of microdomains of small but detectable size of each type of molecular assembly in the other, around $T_f$. In the case of solid–solid transformations, this postulate has received experimental support from the discovery of 'hybrid' single crystals (Chapter 4). However, detectable coexistence of domains of two structures in a hybrid crystal lattice only seems likely when these structures are closely similar (see Ubbelohde, 1962) because otherwise the compression and tension energies involved become rather large.

In theories of premelting resulting from cooperative fluctuations these energies have hitherto been neglected. For example, the Frenkel theory of phase fluctuations merely allows for the interfacial energy term $\sigma_{SL}$. As already stated, the solid–melt transition in reality is far from symmetrical with regard to fluctuations near $T_f$.

An alternative statistical model is to accept the structure of a melt as in a sense quasicrystalline, with cooperative defects present in very high concentrations (Chapter 11, p. 294). On this basis the corresponding (defective) crystalline counterpart of the melt at $T_f$ is a related arrangement with much. fewer cooperative defects. This approach to premelting would have the advantage that micro domains of 'liquid' in the crystal lattice, with plausible energies of formation, do not have to be specified. No *heterophase* defects are involved. Such a structure for a melt has not yet been fully tested experimentally, even for substances for which positional disordering to produce a quasicrystalline liquid gives a formally adequate description of the melting process. In more complex melting mechanisms, when the melt cannot be regarded as quasicrystalline, for example because mechanisms other than positional disordering are operative, this approach may altogether fail to give any adequate representation of fluctuations with regard to melting, especially on the prefreezing side of $T_f$ (Chapter 13).

### 12.2.6. *Corresponding state theories for precursor effects*

As already stated in preceding chapters (p. 55) no extended corresponding state theorems can be expected to apply to fusion, particularly when more than one mechanism of melting operates. However, for closely similar crystal lattices greater uniformity may be looked for. Thus it has been found that for close-packed metals such as Ag, Cu, Au, Al, etc. a single reduced plot of thermal expansion $\alpha_T/\alpha(0\cdot5T_f)$ *vs.* $T/T_f$ applies in all cases, with good linearity up to $T/T_f = 0\cdot45$. This behaviour has been attributed to similitude in the role of lattice vibrations for the thermal expansion as well as for other thermal properties of these close-packed metals (Pathak and Vasavada, 1970; see Catz, 1955). One advantage of this reduced linear plot is that it gives a foundation from which to calculate excess functions. Above $T/T_f = 0.45$ the data continue to show similitude though the plot is curved and a logarithmic plot of the excess $\Delta\alpha_T$ vs. $1/T$ points to vacancy defect formation to account for the excess. For all the metals Au, Ag, Al, Cu, Pb, Pt, Mg, Ni the energy of defect creation on this basis is given by $\varepsilon \sim 8.85\, kT_f$ (see however p. 245).

## 12.3 Surface melting of single crystals

The influence of the parameter $\sigma_{SL}$ for premelting, as introduced in fluctuation theory, does not exhaust its significance. When the melting of single crystals is considered, it seems plausible that since $\sigma_{SG}$ is known to be different for faces with different indices, $\sigma_{SL}$ may also exhibit differences near $T_f$. This must mean that different crystal faces have slightly different melting points, when their surfaces are of infinite extent. The number of layers beyond the outermost, for which properties can still be distinguished as different from the bulk interior, probably differs markedly according to the type of force fields involved in the crystal and its melt. The premelting reported in heat capacity data, when gallium is heated very slowly through $T_f$, may perhaps involve surface effects (Jach and Sebba, 1954). A single crystal surrounded by its melt at a controlled temperature $T_f$ may be able to adjust the extent of various faces until true solid–melt thermodynamic equilibrium is reached at each of them. The physical concepts concerning surfaces of crystals are not really well defined, however, and experimental work in this field is difficult; up to the present, advances have been very limited. Furthermore, minute concentrations of heteroatoms can affect surface phenomena in a variety of ways. For single crystals of a number of metals, there is some evidence that not all the faces are wetted by the melts (see refs. in Stranski, 1942). This could point to differing values of $\sigma_{SL}$ for different crystal faces. However, since numerous examples are known of differing absorption of impurities on different crystal faces, surface contamination could perhaps also account for the phenomena reported.

For Pb and Bi, vibration amplitudes perpendicular to and in the surface near $T_f$ are much greater than in the bulk (Goodman and Somorjai, 1970).

## 12.4 Superheating of solids above $T_f$

Melts may be supercooled tens or even hundreds of degrees below $T_f$, without the occurrence of spontaneous crystallization (Chapter 15). This is because nuclei are necessary for growth of the crystals to proceed. On the other hand it has frequently been verified that solids cannot be heated appreciably above $T_f$ without melting. The generally accepted reason is that nucleation of the disordered liquid phase can take place readily at the solid–gas interface, which is in any case in a state of strain with respect to the three-dimensional lattice so that growth of the liquid layer proceeds. Since transfer of heat to a crystal normally occurs through this interface, from the hotter melt outside it, the liquid–solid layer normally cannot be raised appreciably above $T_f$ without accelerated melting.

A different situation might arise when only the internal portions of a crystal are superheated, e.g. by absorption of radiation or by dielectric loss effects, without the introduction of a free interface with the gas. For example, it has been claimed that ice has been overheated internally by both these methods, by at least $0.3°C$ (Käss and Magun, 1961). Trick methods of internal heating of crystals can lead to large superheating in the case of $SiO_2$ ($+300°C$) and $P_2O_5$ ($+50°C$) (Cormica, Mackenzie, and Turnbull 1963).

## 12.5  Trivial two-phase melting—the Raoult test

As already indicated, the depression of the freezing point by an impurity NOT soluble in the crystals makes it difficult to obtain valid measurements of various physical properties, particularly extremely close to the ideal freezing point of the pure one-component material $^0T_f$, and even if the amount of impurity still present is extremely small. It is usual to assume that if the mole fraction of impurity is $n/N$ just before any crystals begin to separate, the measured freezing point $(T_f)_{lim}$ will be given by a Raoult expression

$$^0T_f - (T_f)_{lim} = Kn/N \tag{12.1}$$

where $K$ is the (mole fraction) constant expressing the depression of freezing point.

As freezing proceeds, the concentration of impurity in the residual melt rises. At a temperature $T$ below $(T_f)_{lim}$ it will have risen to a concentration $n/(N-N_x)$ where $N_x$ refers to the moles frozen. Still assuming Raoult's law

$$^0T_f - T = \frac{Kn}{N-N_x} = \frac{N(^0T_f - (T_f)_{lim})}{N-N_x} \tag{12.2}$$

If $\Delta^0P_r$ is the total change of the property (e.g. volume or enthalpy) being measured, when all the melt has frozen, the change measured down to $T$ is

$$\Delta P_r = \Delta^0P_r N_x/N \tag{12.3}$$

Measurements usually lead to the evaluation of an excess property $X$ on heating the solid to $T_f$, such that

$$X = \Delta^0P_r - \Delta P_r = \Delta^0P_r \left(\frac{N-N_x}{N}\right) \tag{12.4}$$

for example, the excess specific heat $C_{\text{premelting}}$

$$C_{\text{premelting}} = H_f \frac{dX}{dT}$$

or the excess thermal expansion $\alpha$

$$V_S \times \alpha_{\text{premelting}} = V_f \frac{dX}{dT}$$

owing to the formation of melt below $(T_f)_{\text{lim}}$.

The steep rise in these excess properties follows a hyperbolic form, as can be seen by combining (12.2) with (12.4) in the form

$$\frac{(^0T_f - T)}{(^0T_f - (T_f)_{\text{lim}})} = \frac{\Delta^0 P_r}{X} \tag{12.5}$$

From (12.5), a plot of the reciprocal of *any* excess property against $T_f$ should give a straight line whose slope is determined by the mole fraction of impurity $n/N$ initially present when the whole sample is molten. Values of excess properties on approaching $T_f$ can usually be determined with fair accuracy by graphical means. When the plots indicated are used, in many cases the mole fraction of impurity $n/N$ calculated to account for the observed premelting is less than experimental estimates of what unavoidably remains present even after the use of refined techniques of purification. A trivial origin of premelting must be assumed in such cases. Plots based on equation (12.5) are currently used to 'correct' calorimetric determinations of enthalpies, when these traverse $T_f$. Although this procedure is harmless enough in evaluating integrated thermodynamic functions, it may also conceal genuine homophase premelting. By way of illustration of what probably is two-phase premelting for a particular sample of Bi, equation (11.2) gives an impurity of 0·023 atom %, whereas analysis shows 0·014 atom % (Carpenter and Hearle, 1932). These quantities are too close together to permit confident assessment of any homophase premelting using this particular method of analysis. Studies of the decrease of electrical resistivity of Bi as it approaches $T_f$ lead to a similar conclusion (Knappwost, 1954).

When it is suspected that homophase premelting is genuine, a useful plot takes into account the general exponential trend of defect concentrations in crystals (see Plester, Rogers, and Ubbelohde, 1956) according to which $n/N = C \exp(-\varepsilon/kT)$ where $C$ is a constant. For $n$ units of structure (atoms or molecules) $\Delta P_r = n\delta p_r$ where $\delta p_r$ is the contribution per unit of structure. On this basis a plot of $\log \Delta P_r$ against $1/T$ should be linear and the slope should indicate the energy required to create defects.

A further trivial origin of distortion of cryoscopic observations can arise from the presence of undetected impurities not soluble in the crystals but soluble in the melts. If there are $n_x$ moles of unsuspected impurity in $N$ moles of the solvent melt, the known addition of $n$ moles of added impurity will produce an overall lowering of freezing point of

$$K\left(\frac{n_x+n}{n_x+n+N}\right) = {}^0T_f - T_f \qquad (12.6)$$

The cryoscopic constant calculated from such experimental data will appear. to have a value $K'$ where

$$\left(\frac{n}{n+N}\right)K' = {}^0T_f - T_f$$

in cases when the correct value ${}^0T_f$ for the ideally pure crystal is used to evaluate $\Delta T_f$ as in (12.6). Approximately, $K'/K = (n_x + n)n$. Experimentally this situation can arise, for example, from an undetected impurity such as $H_2O$ or $CO_2$ (Rhodes and Ubbelohde, 1959) or thermal decomposition products (Ricci, 1944) introduced in the course of preparing a two-component mixture to determine $T_f$. If, however, $n_x$ is already present, originally measured freezing point depressions refer to $(T_f)_{lim}$ and not to the ideal freezing point ${}^0T_f$. With this starting point the presence of undetected impurity has only a marginal effect on evaluations from freezing point depressions.

Diverse NMR techniques have been proposed for evaluating the actual proportion of 'liquid' when the premelting is due to a trivial two-phase situation (Burnett and Muller, 1967). Whilst these could prove helpful, effects of homophase cooperative defects on 'spin echo' in NMR are not known and could confuse the conclusions from such measurements.

## 12.6 Homophase premelting in different types of crystal lattice

Theoretical possibilities for incipient disordering of a crystal lattice on approaching $T_f$, discussed in the sections under 12.2, indicate that published investigations of homophase premelting effects can usefully be classified and examined according to the type of crystal structure involved. When the methods of observation refer solely to thermodynamic measurements, for example of the excess specific heat $\Delta C_{ex}$ or excess volume expansion on approaching $T_f$, these excess functions can only be detected if they are relatively large. Any genuine changes within the crystal lattice are easily obscured by two-phase premelting. This confusing possibility must always be investigated with particular care. When expansions are measured directly on the crystal lattice by X-ray diffraction, or when various non-thermodynamic methods are used for studying precursor effects in a crystal below $T_f$, homophase premelting is much more readily authenticated. Reference to various experimental methods of study is made in connection with the substances for which they have been used as examples in what follows.

12.6.1. *Premelting in molecular crystals*

*Argon*. Premelting of argon has been followed from calculations on accurate equations of state established from melting data at various pressures, from $T_f = 83 \cdot 8$ K to $T_f = 320$ K (Crawford, Lewis, and Daniels, 1976). Vacancy defect concentrations $n/N \leqslant 10^{-3}$ are indicated (see Flynn, 1972), though very near $T_f$ the equations are subject to some uncertainty.

*Krypton*. Comparisons (Losee and Simmons, 1968) between bulk and X-ray lattice densities indicate a mole fraction $n/N \sim 3 \times 10^{-3}$ at $T_f$.

*Molecular hydrogen*. Nuclear spin resonance could provide a very useful non-thermodynamic method for studying premelting in crystals of compounds containing hydrogen, provided that normal behaviour can be estimated with sufficient precision to assess excess functions adequately, as $T_f$ is approached. Measurements of relaxation times for proton magnetic resonance in crystalline $H_2$ and crystalline HD show that appreciable self-diffusion of these molecules takes place near $T_f$ (Bloom, 1957). These measurements can be represented fairly well by a linear relationship between the logarithm of the relaxation time and the reciprocal of the absolute temperature, but the data do not permit critical assessment of the extent to which the increased mobility near $T_f$ can be regarded as anomalously high. For premelting studied in hydrogen at pressures up to 200 atm a new phase transition has been reported close to $T_f$ (Manzheli, Udovidchenko, and Esel'son, 1975). It would be of particular interest to test the Lennard-Jones–Devonshire theory (Chapter 11) which one would expect to be acceptable for these molecular crystals, and to verify its calculated values for incipient defect formation on approaching the melting point.

*Long-chain paraffins and their derivatives*. A variety of physical measurements indicates marked premelting for normal paraffins and their derivatives, though a discontinuous jump always occurs, as it must do under ordinary pressures at $T_f$.

Specific heat measurements (Ubbelohde, 1938) first drew attention to homophase premelting in view of the fact that premelting in carefully purified paraffins was greater than could be attributed to two-phase effects. This was further tested by deliberate addition of appreciable amounts of contaminants. Any likely impurities remain in solid solution where they could facilitate homophase premelting (see p. 133) but would not give rise to two-phase formation of liquid. Closely corresponding studies have been made using dilatometric techniques (van Hook and Silver, 1942). Figure 12.3 illustrates plots of specific heats as a function of reduced temperature $T/T_f$, which also demonstrate the accelerated rise in $C_p$ as the melting point is approached for a number of normal paraffins from $C_8$ to $C_{16}$ (Finke, Gross, Waddington, and Huffman, 1954). Excess entropy and specific heat functions of $C_n$-alkanes with $16 \geqslant n \geqslant 12$ appear to undergo a particularly steep increase above about

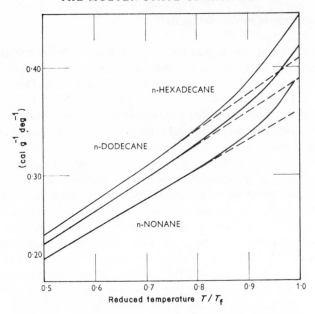

FIG. 12.3. Premelting in paraffins. Plots of $C_p$ for crystals against the 'reduced' temperature $T/T_f$ for the n-alkanes $C_8-C_{16}$. Reprinted with permission from Finke, Gross, Waddington and Hoffman, *J. Am. Chem. Soc.* **76**, 333 (1954). Copyright by the American Chemical Society

$0.9 \ T_f$ (Silverman, 1970). Cryoscopic studies have also been made to examine the change of $H_f$ as the crystals approach $^0T_f$ (Oldham and Ubbelohde, 1940).

Various aspects of structural behaviour and physical properties around $T_f$ have been investigated (Daniel, 1953). The dielectric response of normal alcohols $C_nH_{2n+1}OH$ with $n$ from 6–12 indicates premelting over a range of 10° or more; the extent of heterophase effects has not, however, been tested very carefully (Brot, 1955).

X-ray studies of the change of structure in the premelting region for long chain paraffins indicate that it involves principally increasing randomization of orientation about the (stretched) length of the molecule. A number of the paraffins actually take up hexagonal packing below $T_f$. For chains with carbon atoms $n = 24$–44, the change takes place by way of an abrupt transformation, whereas it is gradual for other chain molecules. Impurities in solid solution may, however, modify details of this behaviour (Müller, 1932). Any very marked changes of crystal packing as observed by X-ray diffraction on approaching $T_f$ appear to run parallel to thermodynamic anomalies in the premelting region. In an example studied in some detail, n-hexatriacontane $C_{36}H_{74}$ (Vand, 1953), the transformation from a near-orthorhombic

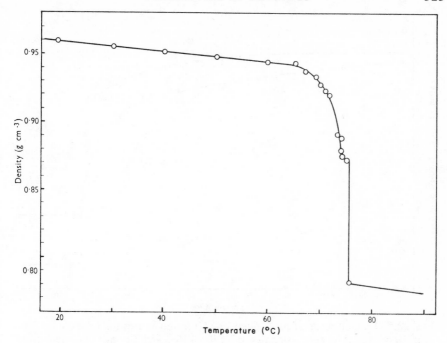

FIG. 12.4. Precursor density change in n-hexatriacontane $C_{36}H_{74}$ in a transformation 20°C below $T_f$. Reproduced by permission of the International Union of Crystallography

structure, stable at ordinary temperatures, to a hexagonal form in which the chains are 'rotating' about their long axes, occurs in the solid. It is significant that in this crystal the transition temperature orthorhombic → hexagonal (at 73·5°C) is only about 2°C below $T_f$ at 75·5°C. The solid → solid transformation is preceded by a precursor increase in thermal expansion extending over more than 10°, as illustrated in Fig. 12.4. In the melt, the distance between stretched chains is nearly the same as in the hexagonal form, but the chains appear to be arranged in a roughly square array, Whether such precursor effects can be clearly detected as preceding a transition temperature $T_c$ in the crystals, if $T_c$ lies below $T_f$, or whether they are merged and appear as smeared precursor effects in a premelting region below $T_f$, but without any clear resolution of a solid–solid transition, depends to some extent on comparatively minor features in the intermolecular force fields (see p. 141). Instances where premelting becomes very marked may sometimes be attributed to unresolved transitions merging in this way into the melting process.

Nuclear magnetic resonance spectra and infrared absorption between 700 and 750 cm$^{-1}$ can both be used to give information about progressive disordering in the premelting range of temperatures, for various even-number

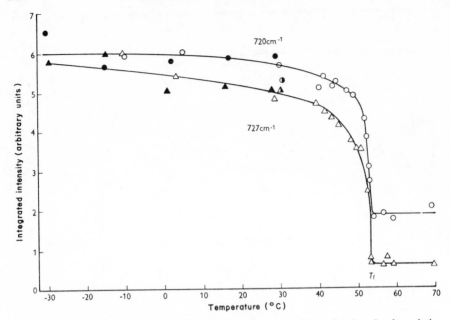

FIG. 12.5. Premelting decrease of intensity of infrared absorption bands of myristic acid. Reproduced by permission of the National Research Council of Canada from the *Canadian Journal of Chemistry* (**41**, 1188–1196 (1963)

fatty acids (Barr, Dunnell, and Grant, 1963). Nuclear magnetic resonance spectra show that the fraction of protons involved in very rapid and extensive motion increases as $T_f$ is approached. Infrared spectra show two peaks of the rocking mode of $CH_2$ progressively less resolved on approaching $T_f$, as is illustrated for myristic acid in Fig. 12.5.

Using graphical methods, each of the two methods of observation can be used to yield estimates of the fraction $(1-x)$ of molecules in the crystals subjected to 'liquid type disordering'. Figure 12.6. shows values based on infrared data; NMR results are similar but not identical. This discrepancy may partly be attributed to the fact that a different measure of disorder is involved in the two methods. Lack of crystalline order becomes apparent rather sooner in the infrared studies than does freezing of the rapid motion of protons in NMR investigations.

Increasing impurity appears to enhance premelting observed by these techniques. A probable explanation is that impurity molecules in solid solution (usually kindred long chain molecules, see Oldham and Ubbelohde, 1940) distort the crystal lattice around them, thereby 'lubricating' and facilitating thermal disorder generally. This lowering of crystal barriers as a result of impurities in solid solution has been repeatedly observed in various

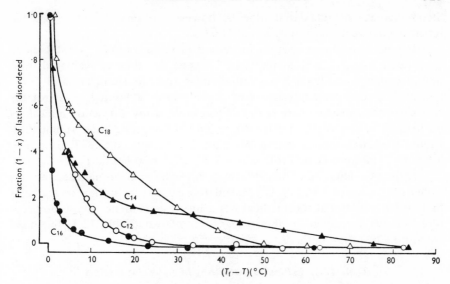

FIG. 12.6. Fraction $(1-x)$ of lattice disordered through premelting, on approaching $T_f$ for n-fatty acids. Reproduced by permission of the National Research Council of Canada from the *Canadian Journal of Chemistry* **41**, 1188–1196 (1963)

phenomena (see p. 100), and tends to be particularly important for molecular crystals.

In the dumb-bell molecule $Cl_3C-CH_3$ (Rubin, Levedahl, and Yost, 1944) a transformation in the crystals at $224\cdot3$ K lies close to $T_f$ at $240\cdot2$ K. Complex premelting is observed between 224 K and $T_f$.

*Quasispherical molecules.* Gradual merging of orientational randomization in the crystal with the positional melting point seems likely to occur in precursor phenomena in a number of cases. For example, in benzene, X-ray diffraction studies indicate randomization about the hexad axis within 9°C of $T_f$ (Cox and Smith, 1954) (see p. 151). In cyclohexane and a number of its derivatives orientational changes are apparent from X-ray studies rather further below its melting point, in the range −40 to −8°C (Hassel and Sommerfeldt, 1938). Only a slight increase of the barriers opposing this randomization, for example, in other derivatives of these molecules, could bring $T_c$ so close to $T_f$ as to change a distinct solid–solid transformation in the crystal into what merely appears as a marked premelting. In view of this consideration, it would be interesting to investigate the effects of increase of pressure on such merging of two kinds of disorder in the crystals, but no studies at enhanced pressures appear to have been reported up to the present. Practically no change is found in the spin-lattice relaxation time as determined by NMR (Andrew and Eades, 1953) at $T_f$ on melting the crystals.

Steep increase of relaxation time is, however, observed in the crystals, especially above a transition point at 186 K.

Attempts to evaluate vacancy formation in various organic crystals have been made using critical comparisons between densities as evaluated from bulk occupancy and from X-ray lattice measurements (Baughan and Turnbull, 1971). Unfortunately the accuracy of these is limited since rotator crystals in their plastic phase seldom give usable X-ray diffraction intensities up to high values of the Debye angle, and furthermore their sublimation vapour pressures are inconveniently high. For cyclooctane at $T_{trip}$, $n/N \sim 3.6 \times 10^{-3}$ and for succinonitrile $n/N \sim 3.3 \times 10^{-4}$ at 4° below $T_{trip}$.

Less direct evaluation of defect formation can be made as indicated above, from a plot of $\log (T^2 \Delta C)$ vs. $1/T$, calculating excess specific heat on the basis that lattice vibrational specific heat has a linear dependence upon temperature. Results are given for their suggestive interest (Table 12.1) but need confirming by other methods.

**Table 12.1. Estimated premelting in molecular crystals**

| Molecule | $T_{trip}$ $(K)$ | $S_f$ (e.u.) | $10^4 \times$ vacancy mole fraction |
|---|---|---|---|
| Benzene | 278·69 | 8·46 | 9 |
| 1, 2-Diethylcyclohexane | 239·8 | 2·02 | 22 |
| Krypton | 115·8 | 3·39 | 30 |
| Perfluoropiperidine | 274·1 | 2·46 | 33 |
| Cyclooctane | 288·0 | 2·0 | 36 |
| cis-1,2-Dimethylcyclopentane | 219·5 | 1·8 | 46 |
| Hexamethylethane | 374·0 | 4·8 | 53 |
| Pentaerythritylfluoride | 367·4 | 3·35 | 60 |
| 2,3,3-Trimethylbutane | 248·6 | 2·17 | 76 |

An ingenious procedure for measuring thermodynamic properties of crystals very close to $T_f$ has been examined with crystals of benzene, ice and cyclohexane (Staveley and Parham, 1952). Determinations are made of the adiabatic compressibility $\beta_S$ measured for various mixtures of crystals and melt at $^0T_f$. For benzene and ice, the results indicate compressibilities up to one-third greater than for the same crystals a few degrees below $^0T_f$. For cyclohexane, the compressibility of the solid at $^0T_f$ appears to be actually higher than that of the melt. These findings are of value for suggesting further lines of investigation, but are subject to reservations in view of various practical difficulties, such as the fact that most melts liberate bubbles of dissolved gas on freezing. However, the elaborate precautions taken to eliminate such disturbances support the view that values of $\beta_S$ some degrees below $^0T_f$ may be considerably less than in the premelting region. Presumably this enhanced compressibility very near $T_f$ is to be attributed to anomalously high defect concentration in the crystals, with consequent anomalous expansion of the molar volume, preceding the jump on passing to the even more highly disordered melt at $T_f$. It should be noted that this anomalous trend also affects the calculation of specific heats at constant volume from specific heats observed at constant pressure, in the premelting region itself.

High-pressure studies using piezothermal analysis promise to yield useful discrimination concerning the diverse origins of precursor effects (Ter Minassian and Pruzan, 1976, 1977) when sufficient data are available for comparative estimates.

*Premelting in crystal–mesophase transitions.* Premelting effects appear to be fairly prominent in a number of crystal–mesophase transitions, probably because the disordering of structures involves defect energies that are not large compared with $kT_f$, thus permitting appreciable mole fractions $n/N$ of defects in the crystal lattice on approaching $T_f$. A variety of physical methods of measurement have been used to study these effects. When rod-like molecules are involved, orientational disordering in the crystals about the rod-axis may be expected. 'Soft' orientational modes of vibration have been found by both infrared and Raman spectroscopy, in $p$-azoxyanisole; their frequency decreases steeply on approaching $T_f$ (Bulkin and Prochaska, 1971). In melting leading to nematic crystals, translational disordering parallel to the long axis may also be prominent (Bulkin, Grunbaum, and Santoro, 1969). Using deuterated $p$-azoxyanisole, the neutron diffraction peaks remain sharp up to $T_f$ but a steep decrease of the integrated Bragg intensity point to a 'softening' of torsional oscillations on approaching $T_f$ (Riste and Pynn, 1973).

## 12.6.2. *Homophase premelting in ionic crystals*

In favourable cases ionic crystals might be expected to show marked premelting, since the energy of creating positional defects can involve characteristic temperatures $\theta = \varepsilon/k$ not too large compared with $T_f$. Reference has been previously made (p. 185) to the fact that thermodynamic calculated parameters will show a discrepancy from values directly observed if the crystals or melts are subject to marked precursor changes of structure.

For NaCl, the pressure coefficient of the freezing point $dT/dP = 0.0212°C\,bar^{-1}$, which should be compared with that calculated by Clausius–Clapeyron from estimated volume and entropy changes on freezing, $dT/dP = V_L - VS/S_S = 0.028°C\,bar^{-1}$ (Akella, Vaidya, and Kennedy, 1969). Either or both of these volume and entropy changes may be modified by precursor effects very close to $T_f$, in some methods of evaluation.

Lithium fluoride (Douglas and Dever, 1954) exhibits a marked precursor rise in $C_p$ for as much as 400°C below $T_f$ (Fig. 12.7). For this salt $S_f$ is 5.77 e.u. Correction of values of $C_p$ to yield $C_v$ could give a useful check on the interpretation that premelting effects occur in the crystals. With KCl, determination of the adiabatic elastic constants coupled with data on the linear

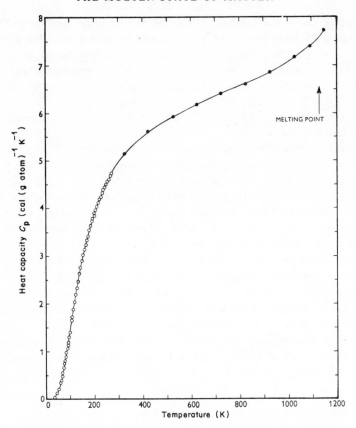

FIG. 12.7. Premelting rise of $C_p$ for lithium fluoride. Reprinted
with permission from Douglas and Dever *J. Am. Chem. Soc.* **76,**
4828 (1954). Copyright by the American Chemical Society

thermal expansion have yielded values of $C_v$ up to $T_f$ (Enck, 1960). Unlike
values of $C_p$ for this salt, values of $C_v$ show no precursor rise larger than
experimental error.

With the simple alkali halides it may be quite difficult to attribute the excess
thermal functions that are observed to any single cause. Comparisons of
X-ray with dilatometer and interferometer data for thermal expansion have
led to the claim that for alkali halides observed effects may be due solely to
vibrational anharmonicity (Lawrence, 1974). For specific heats of KBr, KCl,
and NaCl the major part of any excess functions on approaching $T_f$ may
similarly be due to anharmonicity, though some vacancy creation may also
contribute (Leadbetter and Settatree, 1969). However, it is difficult to dis-
entangle contributions from various causes to any overall excess function. They
may even interact for example in the sense that defect formation lowers the

cohesion and thereby the overall expansion in lattice spacing, quite apart from any local increases in bulk (Laredo, 1969). Furthermore, premelting enhancement at ends of vertical dislocation lines and at grain boundaries is suggested by observations on thermal etching of alkali halides just below $T_f$ (Hämalainen, 1968).

A discrepancy between the thermal expansion of $NaClO_3$ observed by X-ray lattice and by macroscopic methods has been reported (Deshpande and Mudholker, 1960). This discrepancy has been attributed to the role of cooperative defects as $T_f$ is approached, in this instance decreasing the macroscopic expansion more than the lattice expansion. For AgCl (see Nicklow and Young, 1963) X-ray evaluations of thermal expansion show steep increases above the value attributable to lattice vibrations, above about 400 K. Determinations up to 710 K ($T_f = 728$ K) can be interpreted by assuming the formation of Frenkel point defects. A useful confirmatory check

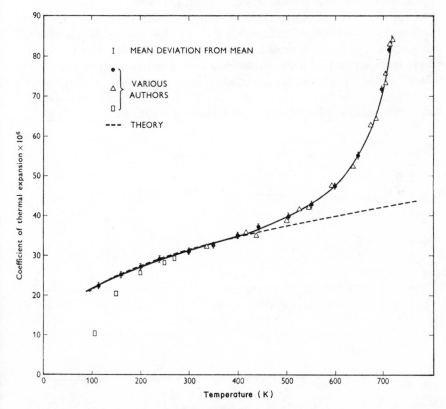

FIG. 12.8. Premelting in the thermal expansion of AgCl. Reproduced by permission of *The Physical Review*

is that the activation energy agrees with corresponding estimates of pre-melting rises observed from two other quite different physical measurements *viz*, in ionic conductivity measurements, and in values of $C_p$. Figure 12.8 illustrates the observed rise above the theoretical estimate for the vibrational contribution to thermal expansion. Similarly for the volume premelting of AgBr the activation energy calculated agrees with values for premelting ionic conductivity (Weiss and Meyer, 1964).

For NaCl, the elastic moduli have been measured right up to $T_f$, using crystal oscillations induced piezoelectrically (Hunter and Siegel, 1942). Marked premelting effects are reported as is illustrated by the value for the adiabatic compressibility $\beta$ (Fig. 12.9). The fact that it apparently drops in this region is attributed by the authors to a correction term

$$\beta_{\text{isothermal}} - \beta_{\text{adiabatic}} = \alpha^2 VT/C_p$$

in which the volume thermal expansion $\alpha$ increases rapidly in the premelting region, whilst the other parameters change only slowly.

As already indicated (Chapter 3) these measurements also verify that solid NaCl remains mechanically fully stable at $T_f$, i.e. that melting does not for this salt involve asymptotic approach to mechanical breakdown. In KCNS, cryo-scopic studies point to a transition (probably orientational) in the crystals within $1 \cdot 5°$ of $^0T_f$. Impurities in solid solution cause this to merge with a premelting region; erratic results are obtained for the cryoscopic constant $K$ unless this is allowed for (Rhodes and Ubbelohde, 1959).

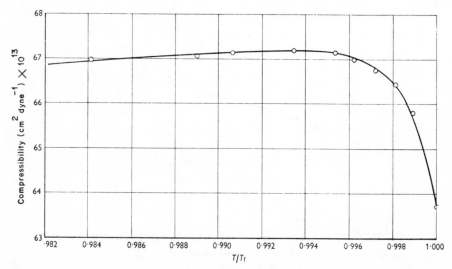

FIG. 12.9. Adiabatic compressibility of sodium chloride on approaching $T_f$. Reproduced by permission of *The Physical Review*

Many ionic crystals show premelting rises of electrical conductance. This is a particularly sensitive property since any reduction of barriers impeding ion migration is readily detected. In NaCl and KCl there is a steep and reproducible rise in electrical conductivity from about 4° below $T_f$, and the mole fraction ($\sim 10^{-4}$) would seem to be about ten-times larger than could be attributed to two-phase impurity premelting (Allnatt and Sime, 1971). A rapid increase of dislocations may occur near $T_f$. Though such a useful property, electrical conductance has not been studied at all systematically near $T_f$; some of the trends may be illustrated by data for caesium salts (Fig. 12.10) (Harpur and Ubbelohde, 1955; Harpur, Moss and Ubbelohde, 1955).

In ionic crystals with large radius ratio, defect formation can become particularly prominent through a considerable positional randomization of the smaller ion, before the lattice framework maintained by the larger ion of opposing sign undergoes melting by positional randomization at $T_f$ see pp. 217 seq.

In $Li_2SO_4$ the jump in electrical conductivity on melting $\sigma_L/\sigma_S \sim 1\cdot 30$ at

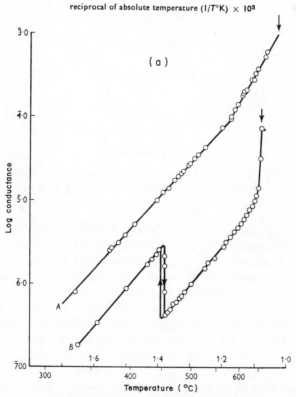

FIG. 12.10(A). Electrical conductance of caesium fluoride (A) and chloride (B) on approaching $T_f$.

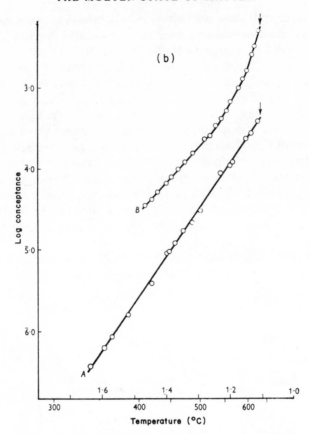

FIG. 12.10(B). Electrical conductance of caesium
bromide (A) and iodide (B) on approaching $T_f$

$T_f = 860°C$, which is even smaller than in $Ag_2SO_4$ for which $\sigma_L/\sigma_S = 7.1$ at
$T_f = 657°C$. Disordering in the cation lattice may become prominent on
approaching $T_f$ (Kvist, 1967). As previously pointed out, in silver iodide
positional randomization of the small cation $Ag^+(r_+ = 1.26$ a.u.) amongst all
possible lattice sites is complete at a lambda transition in the crystals, with
$T_c = 146°C$, whereas the anion lattice only melts at $T_f = 555°C$. For other
silver salts, even though no complete cation randomization occurs below $T_f$,
various properties point to extensive premelting disorder. Values of the
electrical conductivity likewise show steep premelting rises, as illustrated in
Fig. 12.11. For AgCl studies have been published by Ebert and Teltow
(1955). Calculation of defect concentrations from the excess properties may
involve (long-range) interactions between defects when the mole fraction
becomes at all appreciable; estimates of number fractions tend to be too high
if such interactions are neglected (Fouchaux, 1970). An empirical rule

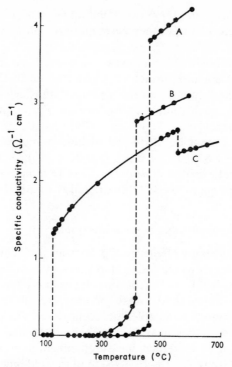

FIG. 12.11. Specific conductance of silver halides on approaching $T_f$. A, silver chloride; B, silver bromide; C, silver iodide (with transition in the solid)

correlating energies of defect formation in ionic crystals is $\varepsilon$ (e.v.) = $2 \cdot 14 \times 10^{-3} T_f$ (Barr and Dawson, 1971).

Premelting in ionic crystals containing polyatomic ions may of course involve rapid growth of more than one type of crystal defect on approaching $T_f$. Thus, extensive premelting is found in $Li_2BeF_4$ (Douglas and Payne, 1969) and the volume increase at $T_f$ is remarkably small ($\Delta V / V_S$ 2–3%), pointing to a complex interplay of defects.

## 12.6.3. *Homophase premelting in network crystals*

As already discussed (p. 282) in some crystal lattices the forces between the units of structure form a network, e.g. of covalent bonds or of hydrogen bonds as in ice. In such crystals defect formation may follow distinctive patterns.

(a) Bond-breaking energies between any pair of units remains more nearly independent of any other bonds, even when the mole fraction $n/N$ of breaks is quite large. In this event, the onset of melting is much more gradual than in cases where the defect energy $\varepsilon$ decreases quite steeply as $n/N$ and the crystal volume increase.

(b) The coordination number in networks is small and the energy required to introduce bond angle deformation is likewise small. Distorted networks may be easily formed. Interpretation of premelting effects may not be simple in such crystals. A much studied example is ice. Proton magnetic resonance (PMR) gives evidence of a 'liquid like' component at $-12°C$ with $n/N \sim 3 \times 10^{-3}$, but this cannot be attributed to hetero-impurities since the mole fraction needed ($\sim 1 \cdot 6 \times 10^{-2}$) is much too great. A quasiliquid behaviour of the surface molecules of the crystals, or of $H_2O$ molecules at intergranular surfaces, has been suggested (Bell, Myatt, and Richards, 1971).

### 12.6.4. *Homophase premelting in metallic crystals*

A considerable diversity of electrical properties can be used in the study of premelting phenomena of metals, in addition to the other types of physical measurement already referred to for other types of crystals. Furthermore, the convenience with which solid metals of exceptionally high purity can be obtained as polycrystalline samples or even in the single crystal form has encouraged careful studies near $T_f$. By themselves, and in the absence of other data on the same materials, measurements of any single property such as heat capacity (Grønvold, 1974, 1975) tend to yield indecisive conclusions.

### 12.6.5. *Macroscopic vs. lattice thermal expansion*

One line of study is to measure macroscopic thermal expansions with sufficiently high accuracy to evaluate any difference from lattice expansions as determined by X-rays. If formation of liquid inclusions by two phase premelting can be eliminated, the difference between macroscopic and lattice thermal expansions can be attributed to the presence of crystal defects in thermal equilibrium. On this basis, for metallic silver of 99·999% purity the mole fraction of defects (about 90% single defects) at $T_f$ has been estimated as $n/N \sim 1 \cdot 7 \times 10^{-4} (\pm 0 \cdot 5)$ with $n/N \sim 9 \cdot 4 \times 10^{-4}$ for aluminium (Simmons and Balluffi, 1960). Similar work on Pb and Al indicates a vacancy concentration $<1 \cdot 5 \times 10^{-4}$ for Pb and possibly about $3 \times 10^{-4}$ for Al at $T_f$ (Feder and Nowick, 1958). A mole fraction concentration of lattice vacancies of about $10^{-3}$ seems likely for many simple crystals at $T_f$ (Barron, 1956).

### 12.6.6. *Specific heats*

The specific heats of many metals show anomalous increases above the Dulong and Petit value ($3R$ per g atom) as $T_f$ is approached (see p. 243).

Varying degrees of critical assessment have been made to evaluate possible two-phase effects, so that the significance of the evidence is at present very patchy. Anomalies for the Group B metals Sn and Cd (Khomyakov, Kholler, and Zhvanko, 1952) can perhaps be attributed to homophase premelting if trivial liquid formation can be excluded. For metallic Mo a vacancy concentration at $T_f = 2890$ K has been calculated to be $2 \cdot 9 \times 10^{-2}$ with $\varepsilon = 42 \cdot 9$ kcal g atom$^{-1}$ (Kirillin, Shandlin, and Chekhowsky, 1968). In general, the possibilities can be checked and defined in a much more concrete way when more than one method of investigation is applied to the same crystals. By way of illustration a favourite target for investigation of precursor effects is the melting of alkali metals. Metallic sodium shows marked premelting as well as prefreezing (Chapter 13) in its specific heat (Ball, 1950). Elastic constants have been measured from 20°C to 95°C (Fritsch, Geipel and Prasetyo, 1973). Data for the specific heat (Martin, 1967) and for thermal expansion (Feder and Charbnau, 1966; Sullivan and Weymouth, 1964) point to a high concentration of defect sites $\sim 3 \times 10^{-3}$ at $T_f$, but the exact value depends on the norms assumed when calculating excess functions as well as on the type of defect assumed. Even after allowing for the effects of impurities, the atomic heat of solid potassium shows an anomalous rise over many degrees below $T_f$. A slight prefreezing rise is also observed for the melt (Carpenter and Steward, 1939). Both anharmonicity and vacancy defects have been proposed to explain the precursor effects in the thermal expansion of potassium (Montfort and Swenson, 1965; Stokes, 1966; Buyers and Cowley, 1969). Premelting increases are found in the electrical resistances of the alkali metals Li, Na, K on approaching $T_f$ (McDonald, 1953). These have been attributed to the formation of vacant lattice sites. Energies of defect formation are estimated to be about 9 kcal g atom$^{-1}$ for each of these metals, in fair agreement with activation energies of self-diffusion in Li and Na of about 10 kcal g atom$^{-1}$. Plausible correlation has been made between excess properties determined from the specific heats (Carpenter, 1953), using methods outlined at the beginning of this chapter. At $T_f$ these quite different methods of investigation indicate a proportion of defect sites of about 1/10 in these crystals (but see p. 245). Solid solution of Na in K and of K in Na occurs to a small extent near $T_f$ for the purified metals (McDonald, Pearson and Toele, 1956) but this does not modify the conclusions about authentic homophase premelting stated above.

## 12.6 X-Ray diffraction

Intensities of X-ray scattering from small particles of finely dispersed sodium show a steep decrease as $T_f$ is approached. This can probably be attributed to homophase premelting; unfortunately the method of investigation used makes an accurate quantitative assessment difficult (Kulp, Shaw, and Speizer, 1955). Individual particles may be regarded as crystals, in which an increase in positional disorder by the production of lattice vacancies and interstitial atoms is added to the root mean square displacement $\bar{u}_x^2$ of the atoms, measured normal to any diffracting plane, arising from ordinary thermal vibrations. However, the root mean square value of the observed abnormally rapid increase in $\overline{u_x^2}$ on approaching $T_f$, as deduced from X-ray diffraction intensities, does not permit discrimination between consequences of a steep increase in amplitude of vibration of all the atoms, owing to the anharmonicity of the potential well in which they vibrate, and alternatively a steep increase in the fraction which undergo a finite jump displacement to defect sites, such as interstitial positions.

## 12.8 Electromagnetic effects

X-ray data do, however, help to confirm that the premelting is due to a change in the actual crystals, even if they do not discriminate between abnormal vibrational amplitudes of all the atoms and specific positional displacements of some of the atoms. Measurements of premelting anomalies in the electrical resistance or in the specific heat are in principle subject to the same difficulty. Only the detailed trend with temperature can serve to discriminate between these alternative modes of atomic displacement from the ideal lattice position as $T_f$ is approached.

Marked decreases in the Hall coefficient of Zn, and increases in its electrical resistivity, are observed in premelting regions of temperature extending many decades below $T_f$ (see Cusack, 1963; Busch and Tièche, 1963) (Fig. 12.12). Antimony also exhibits marked premelting effects. Premelting decreases in electrical resistance of tellurium have been reported below $T_f$, as a precursor to the discontinuous decrease at the melting point (Blum, Mokrovskii, and Regel, 1952). Similar observations have been reported for the crystal compound $Mg_2Pb$ (Knappwost, 1952). Like other premelting effects, each incidence calls for a separate assessment to decide whether two-phase formation of liquid below $T_f$ or authentic homophase defect formation in the crystals is responsible. With the magnesium compound the effect is largest for the purest specimens (Horn, 1951).

In metallic bismuth, the electrical resistivity decrease found on melting varies with the direction of measurement in the crystals, for which $\rho_0/\rho_{90°}$ increases from 1·28 at room temperature to 1·34 at $T_f$ ($\rho_0$ is the resistivity parallel to the trigonal axis). However, careful studies on single crystal rods, 0·25-in. diameter and 5-in. long showed no precursor decrease in a premelting region below $T_f$, for bismuth of 99·99% purity (Hurle and Weintroub, 1960).

Metallic gallium shows various unusual properties, such as the drop in specific resistivity on melting (Chapter 9). Single crystals are obtained fairly

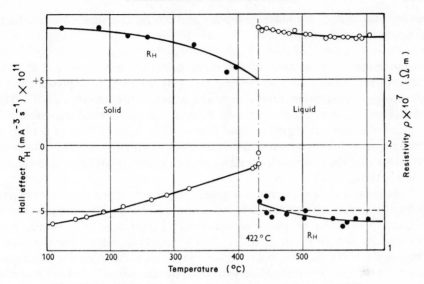

FIG. 12.12. Hall effect $R_H$ and resistivity $\rho$ of zinc around $T_f$

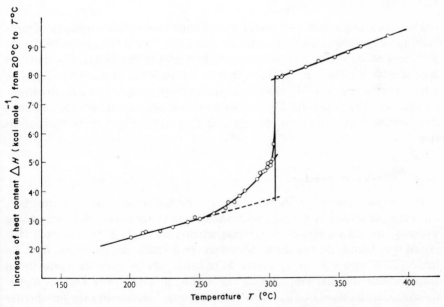

FIG. 12.13. Increase of heat content of NaTl on approaching $T_f$. Reproduced by permission of *Z. anorg. allg. Chemie* **286**, 97 (1956)

easily, since it does not nucleate readily and can thus be supercooled without difficulty.

Electromagnetic properties in the three principal axes $a$, $b$, and $c$ show marked anisotropy. As the melting point at 29·75°C is approached, the anisotropy shows premelting anomalies within a few degrees of $T_f$. This behaviour can depend, in an erratic manner, on mechanical treatment and content of impurities (Epelboin and Erny, 1955). With gallium of 0·1 p.p.m. impurity specific heat data reveal pronounced precursor properties both in the crystals and in the supercooled melt (prefreezing rise in $C_p$) (Bottard, Bros, and Lafitte, 1969).

Intermetallic crystal compounds appear to present opportunities for crystal defect formation with particularly low values of $\varepsilon/k$, and this can give rise to marked premelting in the thermodynamic parameters such as the heat content. By way of illustration Fig. 12.13 records the values for the crystal compound NaTl. Similar observations have been recorded for analogous 'electron transfer' intermetallic crystal compounds such as LiIn (and probably also LiGa). LiCd in addition to showing a premelting rise gives indications of a transformation of higher order immediately below $T_f$ (Schneider and Hilmer, 1956).

## 12.9 Thermal conductance

Thermal conductances of metal compounds show clear instances of premelting in certain cases. For example (Pashaev, 1962) the alloys 49% Cd–51% Sn and 70% Cd–30% Sn show marked rises in thermal conductivity on approaching $T_f$. These increases cannot be attributed to formation of liquid in a trivial fashion below $^0T_f$, since the liquid compounds have markedly lower conductivity than the solids. Possibly a complex interaction between phonons and defects in the crystal on approaching $T_f$ is responsible for the anomalous rise.

## 12.10 Mössbauer effect

The Mössbauer effect (intensity of recoilless resonance absorption of $\gamma$-rays) can, in principle, be used with solids with suitable isotopes undergoing $\gamma$-decay, to study changes in thermal movement as $T_f$ is approached. It is stated that liquids do not show this effect. In metallic tin, the decay of $^{119}$Sn from 120 K upwards to $T_f$ shows Mössbauer effects corresponding to an amplitude of thermal vibration of the atoms, as given by a Debye–Waller formula, with $\theta_D = 142$ K, with considerable anharmonicity at higher temperatures. However, some 5° below $T_f$ a steep decrease is observed towards the zero value for the liquid. This premelting effect in tin is attributed to greatly enhanced diffusion rates of the atoms throughout the crystal lattice neat $T_f$ (Boyle, Bunbury, Edwards, and Hall, 1961), presumably owing to a steep increase in the proportion of crystal defects. A similar premelting increase of diffusion has been reported for indium (Eckert and Drickamer, 1951). Near $T_f$ steady state creep of metallic tin increases very steeply. Below about 215°C a point defect migration process with activation energy $22 \cdot 5$ kcal g atom$^{-1}$ accounting for the observations. Between this temperature and $T_f$ creep mechanisms involving cooperative defects appear to contribute to the overall process (Zama, Lang, and Brotzen, 1960).

From all the examples quoted, it can be seen that homophase premelting is apparent in particular instances of each of the types of crystal examined. Certain metals show effects of particularly valuable theoretical significance. However, to make further progress a more complete understanding of melting mechanisms is obviously necessary.

# REFERENCES

J. Akella, S. N. Vaidya, and G. C. Kennedy (1969) *Phys. Rev.* **185**, 1135.
A. R. Allnatt and S. J. Sime (1970) *Trans. Faraday Soc.* **67**, 674.
E. R. Andrew and R. G. Eades (1953) *Proc. Roy. Soc. Ser. A* **216**, 398.
A. F. Ball (1950) *J. Res. Nat. Bur. Stand.* **45**, 23.
L. W. Barr and D. K. Dawson (1971) *Proc. Ceram. Soc. G.B.* **19**, 152.
M. R. Barr, B. A. Dunnell, and R. F. Grant (1963) *Can. J. Chem.* **41**, 1188.
T. H. K. Barron (1956) *Nature, Lond.* **177**, 23.
R. H. Baughan and D. Turnbull (1971) *J. Phys. Chem. Solids* **32**, 1375.
J. D. Bell, R. W. Myatt, and R. E. Richards (1971) *Nature Lond.* **230**, 91.
M. Bloom (1957) *Physica (Utrecht)* **23**, 767.
A. I. Blum, N. D. Mokrovskii, and A. R. Regel (1952) *Izv. Akad. Nauk SSSR, Ser. Fiz.* **16**, 139.
E. Bottard, J. P. Bros, and M. Lafitte (1969) *J. Chim. Phys. Phys. chim. Biol.* **66**, 166.
A. J. F. Boyle, D. S. P. Bunbury, C. Edwards, and H. E. Hall (1961) *Proc. Phys. Soc. Lond.* **77**, 129.
C. Brot (1955) *C.R. Acad. Sci., Paris* **240**, 1989.
B. J. Bulkin and F. T. Prochaska (1971) *J. Chem. Phys.* **54**, 635.
B. J. Bulkin, D. Grunbaum, and A. Santoro (1969) *J. Chem. Phys.* **51**, 1602.
L. J. Burnett and B. H. Muller (1967) *Bull. Am. Phys. Soc.* **12**, 360.
G. Busch and Y. Tièche (1963) *Phys. Kond. Materie* **1**, 78.
W. J. L. Buyers and R. A. Cowley (1969) *Phys. Rev.* **80**, 755.
L. G. Carpenter (1953) *J. Chem. Phys.* **21**, 2244.
L. G. Carpenter and T. F. Hearle (1932) *Proc. Roy. Soc. Ser. A* **136**, 243.
L. G. Carpenter and C. J. Steward (1939) *Phil. Mag.* **27**, 551.
L. Catz (1955) *Proc. Phys. Soc.* **68**, 957.
E. G. Cox and J. A. S. Smith (1954) *Nature, Lond.* **173**, 75.
K. Clusius and L. A. K. Staveley (1941) *Z. Phys. Chem.* **49B**, 1.
R. L. Cormica, J. D. MacKenzie, and D. Turnbull (1963) *J. App. Phys.* **34**, 2239.
R. K. Crawford, W. F. Lewis, and W. B. Daniels (1976) *J. Phys. Chem. Solid State* **9**, 1381.

N. E. Cusack (1963) *Rep. Progr. Phys.* **26**, 361.
V. Daniel (1953) *Phil. Mag. Suppl.* **2**, 450.
V. T. Deshpande and V. M. Mudholker (1960) *Acta Cryst.* **13**, 483.
J. M. Dickey and A. Paskin (1969) *Phys. Lett.* **30A**, 209.
T. B. Douglas and J. L. Dever (1954) *J. Am. Chem. Soc.* **76**, 4824.
T. B. Douglas and W. H. Payne (1969) *J. Res. Nat. Bur. Stand.* **73A**, 479.
L. Ebert and J. Teltow (1955) *Ann. Phys. Leipzig* **15**, 268.
R. E. Eckert and H. G. Drickamer (1951) *J. Chem. Phys.* **20**, 13.
F. D. Enck (1960) *Phys. Rev.* **119**, 1873.
I. Epelboin and M. Erny (1955) *C. R. Acad. Sci., Paris* **241**, 1118.
R. Feder and H. P. Charbnau (1966) *Phys. Rev.* **149**, 464.
R. Feder and A. S. Nowick (1958) *Phys. Rev.* **109**, 1959.
H. L. Finke, M. E. Gross, G. Waddington, and H. M. Huffman (1954) *J. Am. Chem. Soc.* **76**, 333.
C. P. Flynn (1972) *Point Defects and Diffusion*, Oxford University Press.
R. D. Fouchaux (1970) *J. Phys. Chem. Solids* **31**, 1113.
J. Frenkel (1939) *J. Chem. Phys.* **7**, 538.
G. Fritsch, A. Geipel, and A. Prasetyo (1973) *J. Phys. Chem. Solids* **34**, 1961.
R. M. Goodman and G. A. Somorjai (1970) *J. Chem. Phys.* **52**, 6325.
G. Grange, R. Landers and B. Mutaftschiev (1976) *Surface Sci.* **54**, 445.
F. Grønvold (1974) *Rev. Chim. Mineral.* **11**, 569.
F. Grønvold (1975) *Acta Chem. Scand* **29A**, 945.
M. Hämalainen (1968) *J. Crystal Growth* **2**, 181.
W. W. Harpur and A. R. Ubbelohde (1955) *Proc. Roy. Soc., Ser. A* **232**, 310.
W. W. Harpur, R. L. Moss, and A. R. Ubbelohde (1955) *Proc. Roy. Soc., Ser. A* **232**, 196.
W. V. Hashkovskii and P. G. Strel'kov (1937) *Phys. Z. Sowjet* **12**, 23.
O. Hassel and A. M. Sommerfeldt (1938) *Z. Phys. Chem.* **40B**, 391.
F. H. Horn (1951) *Phys. Rev. Lett.* **84**, 855.
L. Hunter and S. Siegel (1942) *Phys. Rev.* **61**, 84.
D. J. J. Hurle and S. Weintroub (1960) *Proc. Phys. Soc., Lond.* **76**, 163.
V, I. Ivlov and G. K. Mal'tseva (1971) *Sov. Phys.* **12**, 1810.
J. Jach and F. Sebba (1954) *Trans. Faraday Soc.* **50**, 226.
M. Käss and S. Magun (1961) *Z. Kristallogr.* **116**, 354.
K. G. Khomyakov, V. A. Kholler, and S. A. Zhvanko (1952) *Vestn. Mosk. Univ.* **7**, No. 3, *Ser. Fiz. Mat.* **2**, 41.
V. A. Kirillin, A. E. Shandlin, and V. Ya Chekhowsky (1968) *Proc. 4th Symposium Collge park A.S.M.E.*, p. 54.
A. Knappwost (1952) *Z. Elektrochem.* **56**, 594.
A. Knappwost (1954) *Z. Phys. Chem. (Leipzig)* **2**, 320.
B. A. Kulp, C. H. Shaw, and R. Speizer (1955) *Ohio State Univ. Res. Foundation Tech. Rep. No. 3.*
A. Kvist (1967) *Z. Naturforsch* **22**, 208.
E. Laredo (1969) *J. Phys. Chem. Solids* **30**, 1037.
J. L. Lawrence (1974) *Acta Cryst.* **30A**, 294.
A. J. Leadbetter and G. R. Settatree (1969) *J. Phys. Chem. Solid State* **2**, 385.
J. A. Leake, W. B. Daniels, J. Skalyo, B. C. Frazer, and G. Shirane (1969) *Phys. Rev.* **181**, 1251.
D. L. Losee and R. O. Simmons (1968) *Phys. Rev.* **172**, 934, 944.
S. J. Lukasik (1957) *J. Chem. Phys.* **21**, 177.
D. K. C. McDonald (1953) *J. Chem. Phys.* **21**, 177.

D. K. C. McDonald, W. B. Pearson and L. T. Toole (1956) *Can. J. Phys.* **34**, 389.
V. G. Manzheli, B. G. Udovichenko, and V. B. Eselson (1975) *Fiz. Nizke Temp.* (*U.S.S.R.*) **1**, 799.
D. L. Martin (1967) *Phys. Rev.* **154**, 571.
C. E. Montfort and C. A. Swenson (1965) *J. Phys. Chem. Solids* **26**, 291.
A. Müller (1932) *Proc. Roy. Soc. Ser. A* **138**, 514.
R. M. Nicklow and R. A. Young (1963) *Phys. Rev.* **129**, 1936.
J. W. H. Oldham and A. R. Ubbelohde (1940) *Proc. Roy. Soc., Ser. A* **176**, 50.
B. P. Pashaev (1962) *Sov. Phys. Solid State* **3**, 1773.
P. D. Pathak and N. G. Vasavada (1970) *J. Phys. C* **3**, 144.
D. W. Plester, S. E. Rogers, and A. R. Ubbelohde (1956) *Proc. Roy. Soc., Ser. A* **235**, 469.
E. Rhodes and A. R. Ubbelohde (1959) *Trans. Faraday Soc.* **55**, 1705.
J. E. Ricci (1944) *J. Am. Chem. Soc.* **66**, 658.
T. Riste and R. Pynn (1973) *Solid State Commun.* **12**, 409.
T. R. Rubin, B. H. Levedahl, and D. M. Yost (1944) *J. Am. Chem. Soc.* **66**, 279
G. Saunders, N. G. Pace, and T. Alper (1967) *Nature, Lond.* **216**, 1298.
A. Schneider and O. Hilmer (1956) *Z. Anorg. Allg. Chem.* **286**, 97.
S. M. Silverman (1970) *J. Phys. Chem. Solids* **31**, 2371.
R. O. Simmons and B. W. Balluffi (1960) *Phys. Rev.* **119**, 600.
L. A. K. Staveley and D. N. Parham (1952) *Compt. Rendu Reun. Ann. Soc. Chimie Physique, Paris,* **2**, 366.
R. H. Stokes (1966) *J. Phys. Chem. Solids* **27**, 51.
I. N. Stranski (1942) *Naturwiss.* **30**, 425.
G. A. Sullivan and J. W. Weymouth (1964) *Phys. Rev.* **136A**, 1141.
L. Ter Minassian and P. Pruzan (1976) *Rev. Sci. Instrum.* **47**, 66.
L. Ter Minassian and P. Pruzan (1977) *J. Chem. Thermodynamics* **9**, 375.
A. R. Ubbelohde (1937) *Trans. Faraday Soc.* **33**, 1203.
A. R. Ubbelohde (1938) *Trans. Faraday Soc.* **34**, 282.
A. R. Ubbelohde (1950) *Quart. Rev. Chem. Soc.* **4**, 362.
A. R. Ubbelohde (1962) *Proc. Roy. Ned. Akad. Sci. Lett.* **B65**, 5.
A. R. Ubbelohde (1963) *Z. Phys. Chem.* (Frankfurt am Main) **37**, 183.
V. Vand (1953) *Acta Cryst.* **6**, 797.
A. van Hook and L. Silver (1942) *J. Chem. Phys.* **10**, 686.
K. Weiss and H. A. Meyer (1964) *Z. Phys. Chem.* (Frankfurt am Main) **42**, 211.
W. A. Zama, D. D. Lang, and F. R. Brotzen (1960) *J. Mech. Phys. Solids* **8**, 45.

# 13. PREFREEZING PHENOMENA IN LIQUIDS

## 13.1 Asymmetry of precursor phenomena on either side of $T_f$

In phenomena of premelting the crystals begin to show increases of various kinds of disorder to a detectable extent even before the jump in properties at their actual melting point $T_f$. As a notional counterpart it might perhaps be expected that corresponding prefreezing phenomena might become apparent in the melts on approaching $T_f$ downwards. Of course, classical thermodynamics which disregards structural considerations and treats solid and melt as wholly independent gives no support to this expectation; but when the similarities in structure and force fields in these two types of condensed phase are kept in mind, precursor effects might perhaps be looked for. More refined theoretical consideration suggests, however, that correspondence between prefreezing and premelting need not be close. The most prominent prefreezing effects need not be expected only in those melts of crystal types that are particularly prone to disordering in the crystalline solid. Although it is convenient to discuss various prefreezing anomalies together with premelting, the distribution of crystal types showing marked departures from normal average behaviour is different above and below $T_f$.

### 13.1.1. *Non-inversion of prefreezing with premelting*

In elementary theories of cooperative fluctuations such as that of Frenkel (Chapter 12), reasons for this lack of symmetry are not obvious. In principle, calculation of pretransition anomalies on the basis of theories of cooperative fluctuations are equally valid on either side of $T_f$. Indeed, in the prefreezing range of temperatures topological description of a crystal nucleus or other kind of cluster surrounded by melt (i.e. with a structure less specifically organized) is more readily made plausible than the inverse description of a microregion of 'liquid' surrounded by crystal lattice (see Chapter 12). In both cases, the controlling parameters introduced in the elementary fluctuation theory are the solid–liquid interfacial energy $\sigma_{SL}$ and the minimum size of the distinctive microregion or cluster $g_0$ that is physically meaningful. As an example, for benzene and $o$-cresol, theoretical calculations have been explored with various values of $\sigma_{SL}$. Using as a plausible estimate $\sigma_{SL} \sim 0 \cdot 1 \sigma_{LG}$, and taking the interfacial energy between liquid and gas as directly observed, it has been calculated (Lukasik, 1957) that prefreezing anomalies in the specific heat should be readily detectable several degrees above $T_f$. In this elementary theory it is noteworthy, however, that the maximum effect of prefreezing at $T_f$ varies as $(\sigma_{SL})^{\frac{1}{4}}$. Such a steep dependence on this interface parameter obstructs any really close estimates of possible thermodynamic anomalies in melts near $T_f$, since the value of the controlling parameter $\sigma_{SL}$ is not accurately known and must be guessed at.

However in any case, a mere inversion of behaviour on either side of $T_f$ oversimplifies the situation. As already indicated in Chapter 12, it is necessary to give a much more specific description of structures of possible cooperative fluctuations, in both the solid and liquid phases, than can be characterized by the single somewhat indefinite parameter $\sigma_{SL}$. Because they are specific, structural features can be quite different in crystal and melt. In the crystals, if such a concept is valid at all, defect energies associated with 'liquid' fluctuations must depend to a marked degree on shape as well as extension of the region of the cooperative defect. These factors must then be accompanied by compression and tension parameters in the microregion (see Chapter 4) in a way not well represented by the conventional interfacial energy term $\sigma_{SL}$. For example, local correlation or its absence in the orientations of neighbours in molecular crystals could generate such a term (Deschamps, 1975). This mechanical difficulty in defining $\sigma_{SL}$ is much less marked for microcrystalline regions in the melt. As evidence, the elementary theory of cooperative fluctuations can indeed be used in a fairly satisfactory manner to calculate the concentration of microcrystalline regions which can serve as crystal nuclei in the melt (see Chapter 15).

An ingenious method for tracing the cooperative defects in crystalline and molten n-alkane alcohols uses the ESR spectra of small concentrations $(10^{-4}\,M)$ of a free radical molecule dissolved in either crystal or melt (Morantz and Thompson, 1970). Changes are claimed for the freedom of the —OH groups in crystals of the alcohols $C_nH_{2n+1}OH$. It is found that when $n \geqslant 8$, the OH groups are already 'free' in the crystals, but for shorter chains melting must first occur.

## 13.2 Crystallizable and anticrystalline clusters in relation to theories of cooperative fluctuations

In the melts, an important problem is to evaluate the concentration of cooperative fluctuations or clusters that differ measurably from the average. The greater free volume and lessened symmetry of packing of nearest neighbour atoms or molecules in a melt compared with the crystals may permit the formation of more than one type of cluster, with a diversity of structures. Crystal nuclei are only one form of cluster, privileged in the sense that continued addition of more units to the microregion would continue to fill space in a regular and complete way, since this is a basic property of crystal lattices. In some melts, non-crystallizable clusters containing an appreciable number of molecules may also be formed in rival arrangements with similar heat content and specific volumes as the domains or microregions that can act as crystal nuclei. Because the packing in anticrystalline clusters is not regular in three dimensions, any one such cluster cannot however continue to grow indefinitely without involving the inclusion of voids and formation of fronds. Usually these make up an increasing proportion of the total volume of the microdomain as the number of units attempting to pack closer into a cluster in an irregular way increases. For melts, the possibility to be examined in general terms is thus that around $T_f$ an appreciable fraction of their volume at any instant may be in the form of noncrystallizable clusters of limited size in addition to the ordered domains or microregions familiar in the conventional fluctuation theory of crystal nucleation in supercooled melts. Different kinds of clusters may be regarded as swimming in a more freely disordered, looser-packed fluid of quasigaseous structure. Overall make-up of a melt should in such cases be described as a conglomerate, with different fractions $\phi_1$, $\phi_2$, $\phi_3$ ... etc. of the total volume organized in different ways. This kind of description could supplement or make up for the deficiencies of elementary fluctuation theory when applied to a dense fluid containing non-spherical molecules.

Before discussing how clustering can affect prefreezing phenomena in specific melts derived from various kinds of crystals, two general comments can usefully be made. One of these refers to the widespread use of other 'quasicrystalline' models to describe the structure of melts. As stated in Chapter 11, in the present context, a *quasicrystalline* melt is defined quite precisely, as an arrangement of the units which can be referred to a specific crystal lattice, after allowing for the introduction of positional and orientational disordering, and some expansion of nominal lattice spacing, on melting. With this particular definition, melts cease to be describable as *quasicrystalline* when appreciable fractions of the total number of atoms or molecules in them take up rearrangements that can no longer be referred to the crysal lattice in the manner specified.

Examples of melts that are clearly not quasicrystalline have already been considered in preceding chapters and can serve to illustrate this point. Thus one instance is provided by melts of flexible molecules, such as long-chain paraffins. With these, the crystal usually selects only one rotational isomer, whereas the melt tolerates the presence of a whole variety in appreciable proportions. A second instance arises in certain ionic melts, in which 'association complexes' may be formed which cannot be described by reference to the parent lattice. A third instance refers to network melts. Though melting may occur with a jump of properties, the melt normally contains a diversity of network fragments more or less distorted after they have become detached from the crystal. A classic example is water (see below). The present chapter aims to review some general consequences of the formation of non-crystallizable clusters in different types of melts, as the freezing point is approached. In what follows, these will be termed *anticrystalline*.

## 13.3 Effects of cluster formation on enthalpy and specific volume

Whatever the structure of a cooperative fluctuation, to be of any significance it must be thermodynamically competitive with the crystal at $T_f$. *Anticrystalline* clusters can only occupy an important fraction of the total volume of a melt provided the decrease in enthalpy involved in their formation is of the same magnitude or even greater than for the crystalline microcluster (i.e. crystal nucleus) whose concentration can be approximately evaluated from the elementary theory of nucleus formation. In any viable cluster, whether crystallizable or anticrystalline, the packing in nearly every instance will be closer (in rare cases more open) than the overall average, to permit this reduction in enthalpy. The twin considerations about enthalpy and specific volume indicate that if the atoms or molecules in a melt are able to pack into anticrystalline clusters which are to be thermodynamically competitive with crystal nuclei, their progressive formation as the melt is cooled down is likely to give rise to anomalous changes in properties near to and below $T_f$. Such *prefreezing* anomalies may be evaluated in terms of the passing of a fraction $\phi_c$ of the total volume of a melt at any temperature into clusters, for example, of lower enthalpy. The extent of any prefreezing excess property $\Delta P_r$ is given by

$$\Delta P_r = \phi_c \, \delta p_r$$

where $\delta p_r$ is the difference per unit of structure. With regard to the heat content, it seems plausible to write

$$\Delta H_c \sim \phi_c H_f$$

so that the prefreezing anomaly in specific heat where $\Delta H_c$ is the excess heat content attributable to cluster formation, is given by

$$(C_p)_{\text{excess}} = \frac{d(\Delta H_c)}{dT} \sim H_f \cdot \frac{d\phi_c}{dT}$$

Experimentally, it is commonly observed that the specific heat of a melt rises on cooling towards $T_f$, which could be attributed to increasing cluster formation.

A second general comment about this model to describe prefreezing anomalies is that the packings in anticrystalline clusters, whilst their size is still very small, may be regarded as rival options to crystal lattice packing. No sharp transition from free melt to cluster is to be expected, but $d\phi_c/dT$ will normally increase progressively as the temperature falls even below $T_f$. This attracts attention to those properties of melts for which experimental methods of study can be readily extended into the supercooled range of temperatures without risk of spontaneous freezing. For example, viscosity measurements of various kinds frequently satisfy this experimental requirement whereas data of at least equal importance in studying prefreezing anomalies, such as volume or enthalpy parameters have seldom been extended to melts supercooled below $T_f$ because of heteronucleation of crystallization on the walls of most types of calorimeter or other container. Heat content data bear particularly close relationship to theories attributing prefreezing anomalies to cluster formation, in view of the general considerations outlined. As stated above, a diversity of melts show an upturn of specific heat on approaching $T_f$. No simple instances of a downturn appear to have been described. This probably indicates an enthalpy increase resulting from cluster formation, but extension of data for the melt down to temperatures even lower than $T_f$ is highly desirable. Examples of liquids with a fairly steep rise in specific heat on cooling down to $T_f$ include the following: $I_2$ (Carpenter and Hearle, 1937); n-$C_{18}H_{38}$ (Ubbelohde, 1938; van Hook and Silver, 1942); Pb (Douglas and Dever, 1954) (for this substance $C_p$ for the solid rises up to 7·03 cal (g atom)$^{-1}$ deg$^{-1}$ at $T_f$, while for the melt $C_p$ is 7·32 cal (g atom)$^{-1}$ deg$^{-1}$ and gradually falls); and Se (Grønvold, 1973). For Li and Na

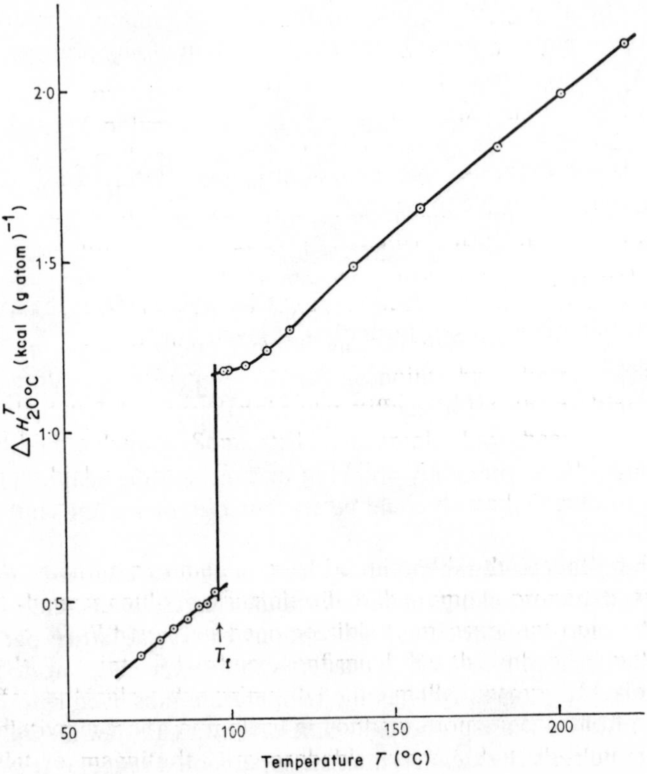


FIG. 13.1. Heat content increases of molten and solid lithium from 20°C

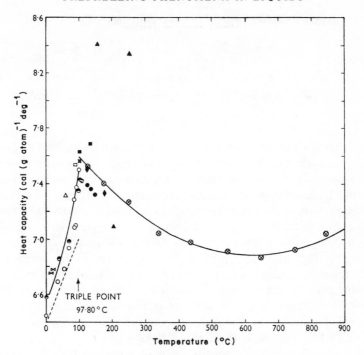

FIG. 13.2. Heat capacity of solid and molten sodium

(Schneider and Hilmer, 1956; see Ball, 1950), the additional specific heat of the melt may possibly pass through a maximum some $10°$ above $T_f$ (Figs. 13.1 and 13.2, and see original refs. noted above).

Vibration spectrum changes resulting from cluster formation have been proposed to predict clustering, even in liquid argon (Burton, 1969, 1970, 1971), but the specific heat near $T_f$ hardly shows any anomaly for this substance. Measurements of ultrasonic velocities and absorption, and evaluation of compressibilities are being made with increasing effectiveness in the search for reliable methods for evaluating structural and related prefreezing anomalies. Data for $o$-terphenyl and for diphenyl ether (D'Arrigo, 1975) correlate quite well with other information about these melts (p. 351). Data for molten Si and Ge point to only very gradual changes for structural anomalies in these melts above $T_f$ (Baidov and Gitis, 1970). Ultrasonic data for some polar organic molecules await supporting correlation with other methods for investigating prefreezing anomalies to provide evidence in the case of melts of methanol (Swamy and Narayana, 1974) and methyl cyclohexane (Rajagopal and Subrahmanyam, 1974).

Prefreezing effects have been studied systematically only for a restricted number of liquids, with rather erratic distribution of structural types.

## 13.4 The 'blocked volume' model for viscosity anomalies

In any theory of anomalous or excess properties, it is necessary to start with an adequate theory of 'normal' properties, from which excess values are to be computed. For many liquids, the quasicrystalline model can serve as a suitable norm. This indicates a viscosity–temperature relationship of the form

$$\log \eta = A_\eta + E_\eta/RT \qquad (13.1)$$

where $E_\eta$ has the properties of an activation energy. Equation (13.1) can be derived from a number of theoretical models (see Ubbelohde, 1964). In using (13.1) to calculate excess properties, values of $A_\eta$ and $E_\eta$ usually have to be determined from data obtained for the same liquid in a range of temperatures well above $T_f$, where the linear relationship is found to hold experimentally[*].

Once a suitable norm has been determined, 'ideal' values $\eta_i$ of viscosity can be calculated by extrapolation down to $T_f$, and in the supercooled region. It is reassuring to find that many 'normal' liquids are known for which the observed viscosity $\eta_{exp}$ does in fact remain close to $\eta_i$ right down to $T_f$. Other melts show more or less marked prefreezing anomalies, in the sense that $\eta_{exp}$ rises progressively above $\eta_i$ as $T$ is lowered. Examples of melts in which $\eta_{exp}$ falls *below* $\eta_i$ have not been instanced up to the present time (see however, Chapter 14 which describes liquid crystal behaviour).

To interpret this kind of viscosity anomaly with $\eta_{exp} > \eta_i$, which is illustrated by various examples in the following sections, the hypothesis has been put forward for various types of melts that part of the volume is blocked for flow, as a consequence of clusters formed in the melt, in increasing concentration as the temperature decreases to $T_f$ and falls below it. For interpreting the trend of viscosity with temperature the structure of such clusters need not be specified. It is only necessary to assume that all the molecules in the cluster are 'blocked', i.e. are compelled to move together in fluid flow at ordinary velocities. Under these circumstances the cluster behaves as a colloidal particle suspended in the fluid. As was first shown by Einstein, the presence of particles in suspension increases the viscosity above the ideal value $\eta_i$ by blocking flow. If the fraction $\phi$ of the total volume occupied by such particles does not exceed about 0.3, the actual viscosity $\eta_{exp}$ is given by (13.2) of which the first term was proposed by Einstein

$$\eta_{exp}/\eta_i = 1 + 2 \cdot 5\phi + 7\phi^2 \ldots \qquad (13.2)$$

---

[*] Some authors (e.g. Greenwood and Martin, 1952) have evaluated 'normal parameters' by applying standard error theory to fit the best straight line to *all* the data down to $T_f$ regardless of any more or less pronounced curvature in the experimental plot of $\log \eta$ against $T$. Such a normalization masks the kind of inquiry under discussion here. It inevitably introduces a break or singularity in $T_f$ when the smoothed values are subtracted from observed data over the whole supercooling range; this particular statistical procedure is thus erroneous and misleading.

(somewhat different numerical factors are found when the particle behaves as an ellipsoid rather than a sphere).

On this basis, the fraction $\phi$ can be calculated from observed deviations of melts from equation (13.1). Empirically it is then found that a logarithmic plot of $\log \phi$ vs. $1/T$ gives a straight line relationship in many cases (McLaughlin and Ubbelohde, 1958; Magill and Ubbelohde, 1958). Values of $\phi$ can often be correlated with the extent of interlocking or interbonding to be expected in the clusters. This is further discussed below.

## 13.5 Cluster formation and the glassy state

The hypothesis discussed in various parts of this book, that certain melts are far from quasicrystalline and contain appreciable concentrations of anticrystalline clusters, is of considerable interest in relation to other aspects of melting of such substances. Frequently, substances for which application of equation (13.2) indicates large volume fractions $\phi_c$ around $T_f$ also show abnormally large supercooling without crystallization and can readily be made to pass into the glassy state. By contrast, more normal melts (see Chapter 16) do not form glasses readily on cooling but nucleate spontaneously at about

$$T_N = 0 \cdot 8 T_f$$

if they have not already crystallized above this temperature owing to heteronucleation. Equation (13.2) becomes imprecise with $\phi$ greater than about $0 \cdot 3$ but qualitatively it seems plausible to suppose that in the melt various anticrystalline modes of atomic or molecular packing form clusters which compete more and more with the regular crystalline versions (crystal nuclei) also swimming in the disordered melt. On sufficient cooling, anticrystalline clusters may squeeze regular nuclei out completely. On still further cooling the melt then locks into a glass (Chapter 16). Examples of this behaviour have been collected (Tweer, Simmons, and Macedo, 1971). It should be added that the precursor rise in viscosity above values indicated by an Arrhenius equation, eventually leading to a glass on sufficient cooling, is widely observed experimentally. None of the different interpretations of this kind of temperature-dependence of the viscosity of glass-forming liquids have as yet achieved universal acceptance, since theoretical models are still semiempirical.

For some melts derived quantities such as the kinematic fluidity $\rho/\eta$ plotted as a function of specific volume may give more direct information about anomalies than the direct plot of $\eta$. Data are not always available through the melting temperature with sufficient accuracy to make this plot meaningful, however.

In a similar way, Batchinski's equation

$$\eta(V-b) = C$$

where $C$ and $b$ are parameters characteristic of a melt, can sometimes be used with advantage to reveal anomalies. Since $V$, the molar volume, is usually insensitive to these anomalies, the anomalous behaviour is shown by changes of the parameter $C$. Interpretation of anomalies is however much less direct than that given by the blocked volume theory.

A 'corrected' version of the Batchinski equation can be written in the form

$$(V - {}^0V) = A/\eta$$

where $V$ is the molar volume and ${}^0V$ is the molecular 'limiting volume', i.e. the value approached at infinite viscosity. $V - {}^0V$ can be regarded as a 'free volume', $1/\eta$ is the 'fluidity', and $A$ is a proportionality constant for any given molecule.

An amended equation is

$$V - C = A/\eta - B\eta$$

where $C$ is practically the same as the limiting volume ${}^0V$. Values of the molecular constants $A$ and $B$ have been collected, for example, for homologous series of n-alkyl derivatives (Bingham and Kinney, 1940). This two-parameter equation does not appear to have been tried out in the all-important prefreezing region of various melts that show anomalous fluidities. Its use is likely to diminish apparent anomalies, since it compensates for rapidly increasing $\eta$. However, the physical basis for this amended equation is not wholly clear.

### 13.5.1. *Centrifugal fields and clustering*

Interpretation of excess viscosity in a prefreezing range of temperatures in terms of the formation of clusters, probably anticrystalline in packing, would imply that there are measurable differences of *local* density as a result of these cooperative fluctuations in the melt. If so, application of centrifugal force under strictly isothermal conditions must result in migration of the denser fluctuations to regions of higher $g$ in the fluid. Not many techniques have been applied to test for consequences of migration in one-component systems; but with simple two-component systems in molten alloys (Kumar Singh and Sivaramakrishnan, 1967; Kumar and and Sivaramakrishnan, 1971) and in citric acid solutions (Mullin and Leci, 1969) evidence has been advanced in support of cluster migration in a gravitational field. Developments of such a technique could be rewarding.

## 13.6 Properties indicating prefreezing in melts of molecular crystals

In this group of crystals, which includes the great majority of organic substances, various physical properties have been applied in the study of prefreezing anomalies. Anomalous increases of viscosity have attracted particular attention, partly because of their theoretical interest for the structure of melts, and partly for technological reasons.

### 13.6.1. *Viscosity anomalies in polyphenyl melts*

Viscosity and other prefreezing studies on melts of polyphenyl hydrocarbons have the advantage that different molecular shapes can to a considerable extent be synthesized at will. Dipole and other stronger interaction forces are largely absent from these molecules. Any tendency to cluster formation may be expected to depend primarily on molecular configurations, which may either favour or obstruct interlocking. Thus the behaviour of a series of linear polyphenyls

(Andrews and Ubbelohde, 1955; Hu and Parsons, 1958) remains practically normal on approaching the freezing point of the melt. Data have also been extended to supercooled benzene. However by contrast, branched polyphenyls show prefreezing anomalies which become very prominent for some molecular structures such as $o$-terphenyl and tri-$\alpha$-naphthylbenzene (see Table 13.1 for the shapes of these molecular skeletons).

Consideration of the repulsion envelopes (Al Mahdi and Ubbelohde, 1953; Andrews and Ubbelohde, 1955) suggests that marked anomalies are associated with pronounced re-entrant portions on the molecules. Clustering in these melts may plausibly be attributed to the formation of interlocking close-packed assemblies. For these to behave as 'blocked' regions in viscous flow, it is necessary that the characteristic time $\tau_c$ for dispersion or build-up of such clusters remains large compared with the relaxation time $\tau_{vis}$ for normal viscous flow. On the conventional quasicrystalline theory of viscosity, $\tau_{vis}$ is the reciprocal of the 'hopping' frequency $\nu_a$ of the molecules and may be as low as $10^{-12}$ s. The condition $\tau_c \geqslant \tau_{vis}$ can be realized in principle for quite different forces leading to clustering, as discussed in the following sections.

For polyphenyl hydrocarbons, a useful index of the fraction of volume blocked for flow at $T_f$ is given by $\phi_f$ in Table 13.1 (see Magill and Ubbelohde, 1958; Ubbelohde, 1961; Binns and Squire, 1962) which illustrates how molecular shape influences this experimental flow parameter.

**Table 13.1.** *Volume fraction $\phi_f$ blocked for flow by clustering at $T_f$*

| Molecule | Shape | $T_f$ (°C) | $\phi_f$ |
|---|---|---|---|
| o-Terphenyl | | 55·5 | 0·58 |
| Triphenylene | | 198·1 | nil |
| 1,3,5-Tri-$\alpha$-naphthyl-benzene | | 197·3 | 0·40 |
| 1-Phenylnaphthalene | | $T_f$ unknown, probably yields glass below 40°C | 0·55 even at 25°C |
| 2-Phenylnaphthalene | | nucleates spontaneously 40°C below $T_f$ | 0·038 |

For these molecules, whose differences of packing in the melt can be attributed almost entirely to differences in repulsion envelopes, the contrast in prefreezing behaviour between various closely related pairs is particularly striking. Thus o-terphenyl and triphenylene give melts whose viscosity–temperature behaviour well above $T_f$ is practically identical (McLaughlin and Ubbelohde, 1958), whereas only the former shows any marked tendency to clustering at $T_f$ (Fig. 13.3). o-Terphenyl even passes into a glass if chilled fairly quickly. 1,3,5-Tri-$\alpha$-naphthylbenzene supercools even more readily to a glass (Chapter 16) through progressive interlocking of clusters (Magill and Ubbelohde, 1958). Again, melts of 1-phenyl-naphthalene show a far greater tendency to supercool than 2-phenylnaphthalene, and their viscosity anomaly indicates much more pronounced cluster formation.

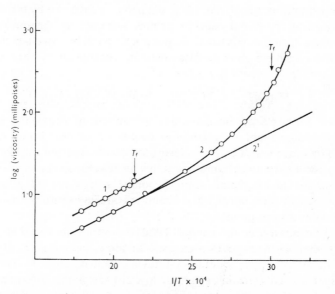

FIG. 13.3. Viscosity/temperature for o-terphenyl (2) and triphenylene (1)

### 13.6.2. Clustering as a result of association forces. Polar aromatic molecules

Anomalies in the viscosity and also in the dielectric properties have been reported for various aromatic molecules containing polar groups such as $-OCH_3$ and $-C_2H_5$. Related anomalies have also been reported in the velocity and the absorption of sound in melts of diphenyl ether, salol, menthol, azoxybenzene, benzophenone (Hunter, 1956), and various coordination compounds of $BF_3$ with carboxylic acids and esters (Greenwood and Martin, 1952).

For some polar molecules, the assumption of cluster formation in the melt is reinforced by evidence for their association, even in dilute solution. For example, polar derivatives of long-chain compounds ($C_9$—$C_{18}$) such as ketones, alcohols, acids, and soaps, form micelles even in fairly dilute solution in benzene (Pilpel, 1961). The extent of micelle formation will of course depend on the molecular 'solvent' environment, which in a melt may be more (or possibly less) favourable than in dilute solution. One way of testing this would be to investigate micelle formation with the long-chain hydrocarbon itself used as a solvent; such information does not appear to be yet available.

At the present stage of development of the themes discussed in this chapter, it is necessary to proceed by collecting and grouping examples of different kinds. When clustering is primarily due to the operation of specific intermolecular attractions, the effects can in principle be followed by the study of a variety of optical or other physical properties. For example, in dilute solutions in $CCl_4$ the clustering of molecules of phenol through hydrogen-bond formation leads to cascade formation of association complexes by repeated additions

$$C_6H_5(OH)+[C_6H_5OH]_n \;\rightleftharpoons\; [C_6H_5OH]_{n+1} \text{ etc.}$$

that can be followed by measurements of optical absorption. Again, the viscosity of ethyl alcohol is markedly enhanced by small additions (less than 1 mole %) of water which favours cluster formation (see Stuart, 1941). Such methods of study are unfortunately not always feasible in the melt, since the property under investigation often becomes insensitive as a means of differentiation between normal and clustering arrangements as the concentration of polar groups increases.

Results (Parthasarathy and Bindal, 1960) for ultrasonic velocities in melts of various phenolic derivatives are stated to show 'jumps' at $T_f$ despite the fact that the melts appear to supercool down to much below $T_f$. These melts are, however, too complex in behaviour to yield any simple inference.

### 13.6.3. *Prefreezing in melts of long-chain molecules*

In view of the special shape of long-chain molecules, it might be expected that distinctive clustering effects would be quite common in the prefreezing temperature range of their melts. When such molecules are rigid, cluster formation is in fact so marked that liquid crystals are frequently formed above $T_f$ (Chapter 14). However, with flexible molecules, such as n-paraffins, the many alternative configurations assumed in the melt (Chapter 7) apparently preclude marked formation of characteristic clusters. Microscope observations on the solidification of paraffin melts (Zocher and Machado, 1959) do, however, suggest that under suitable conditions partial alignment becomes appreciable near $T_f$ in the melt. 'Solidification' in a smectic phase has been reported.

### 13.6.4. *Rotational independence vs. clustering of molecules in melts*

The locking of molecules into an anticrystalline cluster usually blocks rotational independence of neighbours. The criterion of 'rotational freedom' in a melt can sometimes be used as a discriminating pointer to cluster formation, but it also includes much weaker correlations of orientation. Optical studies with sufficiently high resolving power often point to 'correlations' between neighbouring molecules in the melt, e.g. Raman spectra above $T_f$ in molten ethylene (Blumenfeld, Reddy, and Welsh, 1970).

A variety of optical studies give qualitative evidence of correlated orientation of axes of neighbouring molecules in molecular melts, without as yet leading to precise quantitative formulations. Related information has also been obtained by kinetic studies of viscosity, and from relaxation effects in dielectric measurements and NMR for benzene molecules substituted by the groups F, Br, CN, $NO_2$, and $CH_3$ (Ben-Reuven and Gershon, 1969). Whilst all these techniques support the general concept of hindered rotation in melts, especially in the prefreezing region, quantitative probability parameters expressing the falling off of correlation of molecular orientations with increasing remoteness from any one molecule chosen as reference are still wanting in most cases.

Similar comments and conclusions may be made about optical studies of thiophene derivatives (Kimelfeld Moskaleva, Mostovaya, Zhizkin, Litvinov, and Konyaeva, 1975). More detailed angular correlation functions have been sought from evidence from depolarized Rayleigh spectra of melts of benzene, acetone, and methyl iodide (Dill and Litovitz, 1975).

Mention may also be made of attempts to evaluate intermolecular correlations in melts of fairly simple polyatomic molecules from the half width of infrared bands (Kirov and Simova, 1973) and from cold neutron scattering (Trepadus, Rapenau, Padureanu, Pavlenov, and Novikov, 1974).

In light scattering, for a number of melts the intensity distribution of the 'wing' accompanying the scattered Rayleigh line changes in a significant way as the freezing point is approached. Although the detailed interpretation is complicated by uncertainty about the effects of hindered rotation on Rayleigh lines, it seems highly likely that the effects described are due to the growth in size and concentration of interlocking clusters as the melt approaches $T_f$ (Kastha, 1958). Such light scattering anomalies have been reported for pyridine, $\alpha$-picoline, $o$-cresol, $o$-chlorophenol, and $p$-xylene, in each of which molecular interactions are likely to be somewhat larger than conventional Van der Waals forces. For example, in $p$-xylene charge transfer effects may be postulated.

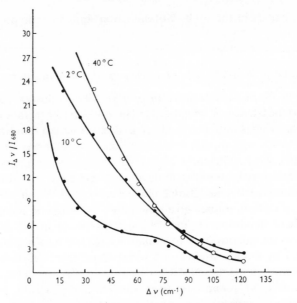

FIG. 13.4. Decrease in Rayleigh scattering on approaching $T_f$ for $o$-chlorophenol. $\Delta\nu$ measures the frequency difference to which the ratio $I_{\Delta\nu}/I_{680}$ refers. Reproduced by permission of the *Indian Journal of Physics*

Figures 13.4 and 13.5 ($o$-chlorophenol, pyridine) illustrate (Kastha, 1958) how the scattering near to the Rayleigh line ($680 \, \text{cm}^{-1}$ line for $o$-chlorophenol $654 \, \text{cm}^{-1}$ for pyridine), expressed by the ratios $I_{\Delta\nu}/I_{680}$ and $I_{\Delta\nu}/I_{654}$ respectively, decreases on approaching $T_f$. The effects suggest anomalous packing in clusters in the prefreezing region.

On the other hand it may be objected that studies of Rayleigh scattering of light from liquids containing various simple organic molecules such as $C_6H_6$, $C_6H_5Cl$, $C_6H_5Br$, $C_6H_5NO_2$, $C_6H_5CH_3$, $C_6H_4(CH_3)_2$, $(CH_3)_2CO$, $CS_2CHCl_3$, $(C_2H_5)_2O$, and $CH_3OH$ do not suggest the presence of marked oriented ordering (Shakhparonov, 1962). These experiments were, however, not extended into the prefreezing region of temperatures.

Changes of Raman spectra on melting are likely to be prominent particularly if the molecule can form strong hydrogen bonds which perturb molecular frequencies. Acetamide and m-phenyldiamine both give evidence from the Raman spectra of forms of association in the melts, different from those found under the more stringent conditions imposed by the crystal symmetry (Raskin and Secharev, 1959). For both these molecules, in the melts the number of lines is considerably reduced, compared with the crystals. For acetamide, whose crystals exist in $\alpha$ and $\beta$ modifications, the Raman lines of the melt suggest the presence of association clusters or domains related to

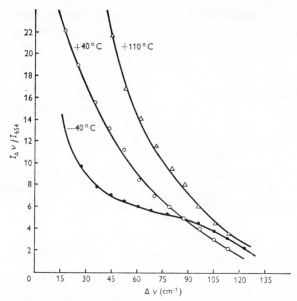

FIG. 13.5. Decrease in Rayleigh scattering on approaching $T_f$ for pyridine. Reproduced by permission of the *Indian Journal of Physics*

each of these polymorphs, though the spectra of both crystals are too complex to permit very detailed correlation with the melt.

A specific heat anomaly in liquid pyrrolidine may perhaps be attributed to changes in clustering associated with orientational disordering (see McCullough, Douslin, Hubbard, Todd, Messerly, Hossenlop, Frow, Dawson, and Waddington, 1959; also Hildebrand, Sinke, McDonald, Kramer, and Stull, 1959).

The heat capacity passes through a point of inflection which has been attributed to effects of hindrance to rotation in the melt. Transport studies on the melt do not however indicate any exceptionally prominent changes in the region where the heat-capacity trend with temperature is changing most rapidly (Hind, McLaughlin, and Ubbelohde, 1960). For this molecule, a small transition in the crystals at $T_c$ merges with the premelting region. Estimated values are:

$$T^0_f = 215.31 \text{ K}$$

$$T_c = 207 \cdot 14 \text{ K}$$

$$S_f = 9 \cdot 521 \text{ e.u.}$$

$$S_c = 0 \cdot 623 \text{ e.u.}$$

$$C_{p_s} = 23 \cdot 42 \text{ cal mole}^{-1} \text{ deg}^{-1}$$

$$C_{p_1} = 35 \cdot 73 \text{ cal mole}^{-1} \text{ deg}^{-1}$$

When the structures of the melt just above $T_f$ consists of clusters dispersed in a more uniformly disordered fluid, anomalies might perhaps also arise in the thermal conductance of the melt on approaching $T_f$, as well as in the supercooled melt. It seems likely (Horrocks and McLaughlin, 1963) that liquids conduct heat mainly through vibrational transfers. Formation of clusters which appreciably increase the scattering of phonons near $T_f$ would decrease the conductivity. Empirically, an effect of this kind has been reported for molten trinitrotoluene (Read and Lloyd, 1948); experimental uncertainties preclude any completely definitive attribution. Other authors (Ravich and Burtsev, 1961) dispute the reality of a drop in thermal conductance of trinitrotoluene in the prefreezing region but observe a minimum near $T_f$ which they attribute to the presence of a double layer of solid and liquid. Further, more precise experimentation is clearly desirable.

### 13.6.5. Correlation of neighbours in melts of inorganic molecules

Cyanogen halides CNCl and CNBr show a marked change in Raman spectra on passing from crystals to melts. It is estimated that up to 50% of the molecules associate into $n$-mers above $T_f$ (Pezolet and Savoie, 1971).

Raman and infrared spectra of molten $SiF_4$ suggest correlation of 'rotation' between neighbouring molecules in the melt, possibly owing to clustering (Besette, Capana, Fournier, and Savoie, 1970).

X-ray diffraction studies indicate 'short-range ordering' in molten $GeCl_4$ and $SnCl_4$ (Rutledge and Clayton, 1975).

Isotope effects in the self-diffusion of molten $PH_3$ and $PD_3$ point to fairly extended correlation of orientation between molecules in the melt (Krynicki, Sawyer, and Powles, 1975).

### 13.6.6. Surface effects in prefreezing of melts of polar molecules

On the basis that prefreezing anomalies may be attributed to rival possibilities for packing of the molecules into anticrystalline clusters, no particular anomaly would be anticipated in the free surfaces of melts in prefreezing regions of temperature. This is because in the free surfaces much stronger uncompensated forces impose an optimum packing little affected by modifications of arrangements in the interior of the melt. Although prefreezing anomalies have been reported in 'interior' properties such as the viscosity (Dodd and Hu, 1949) and dielectric constant (Dodd and Roberts, 1950) in phenyl ether, and other melts of polar molecules such as salol, menthol and azoxybenzene, it is thus not surprising that no anomaly is found in surface tension (Dodd, 1951).

13.6.7. *Roto-kinetic effects in liquids*

Differences in structure of melts near to $T_f$, and considerably above this temperature, are perhaps indicated by a fading out of the roto-kinetic effect (Borneas and Babutia, 1960, 1961). In benzene this fading-out temperature lies around 40°C and in water around 60°C. Theories of the effect are not well elaborated, but it seems to warrant further investigation. In brief, in this effect the surface tension of a melt in motion is stated to differ from that for the same melt at rest, in certain regions of temperature.

## 13.7 Prefreezing in water

The three-dimensional hydrogen-bonded networks formed by water molecules are so versatile that this unique substance provides experimental and theoretical material relating to several chapters of this book, see pp. 1, 34, 38, 282, 306, 334, 409 and 416.

Probably the most basic feature is that ice melts to a conglomerate with domains of several different structures, though, since ice is a network crystal, each of these domain structures may comprise a substantial proportion of network defects, and only relate approximately to any one specific crystal lattice as reference.

13.7.1. *Domain or cluster structure of molten ice*

A model for the conglomerate structure of water is outlined on p. 306. Other theoretical descriptions bear particular reference to selected aspects of this model. An approximate molecular size of domains with structures either like the polymorphs Ice I or Ice III is about 46 units (Shik Jhon, van Artsdalen, Grosh, and Eyring, 1967). The model proposed by Nemetty and Scheraga for water (p. 306) has been modified by replacing the 'intercluster fluid' by specific hydrogen-bonded assemblies of $H_2O$ molecules such as dimers or straight or branched polymers (Vand and Senior, 1965; see also Davis and Litovitz, 1965; Luck, 1967). Thermodynamic diffuse 'transitions' may even occur in some of these domains in water (Magat, 1935). For example, infrared absorption shows a peak anomaly around 36–38°C (Luck, 1965; Salama and Goring, 1966). More general infrared and Raman spectroscopic studies have also been reviewed (Hornung and Choppin, 1974). General theoretical discussions of the properties of water make variable allowance for any proposed conglomerate structure (Ives and Lemon, 1967; Drost Hansen, 1965a; Safford and Leung, 1971). Particularly striking changes are observed, as might be expected, in water supercooled below 0°C (Angell, Shuppert, and Tucker, 1973). A statistical thermodynamic treatment of a cluster model has been sufficiently refined so as to discriminate between $H_2O$ and $D_2O$ (Owicki, Lentz, Hagler, and Scheraga, 1975).

It is important to realize that such a flexible conglomerate structure as $H_2O$ can be very easily modified by wall effects, e.g. in biological systems or near various solid surfaces (Drost Hansen, 1965b, 1969) or by confining small droplets of water (1–5 $\mu$m) by relatively inert cell walls. This procedure modifies the anomalous specific heat $\Delta(C_p(\text{water}) - C_p(\text{ice}))$ of supercooled water, which begins to rise steeply around $-20°C$ using cell walls of n-heptane stiffened by sorbiton tristereate (Rasmussen, Mackenzie, Angell, and Tucker, 1973). In cells of methylcyclohexane, methylcyclopentane and n-heptane the temperature for spontaneous nucleation $T_N$ (Chapter 15) is $-38°C$ but increase of pressure to only 2 kbar lowers this to $-92°C$. This finding and other details suggest a shift of the local domain structure from that referable to Ice I to Ice III (Kanno, Speedy, and Angell, 1975).

## 13.8 Melts of sulphur, selenium, and tellurium

Melts of these atoms partake of network characteristics (Chapter 10) because of the strong covalent bonds which can be formed between the atoms. However no simple network behaviour pattern is followed, apparently because more than one kind of polyatomic assembly can be formed in the freedom of the melt, so that complicated cluster mixtures can arise. This is particularly notable for molten sulphur, in which eight-membered cyclic molecules predominate near $T_f$, changing to folded linear polymer assemblies at higher temperatures. The transformation passes through a critical lambda point around 197°C where many properties pass through 'peak' changes; most prominent is the flow viscosity which becomes extremely high (Patel and Borst, 1971). It is interesting to note that the maximum viscosity attained by molten sulphur is sharply reduced by addition of appropriate heteroatoms, which presumably interrupt (i.e. break) the chains of atoms in the polymeric form (Vezzoli, Dachille, and Roy, 1969). Information about the structure of molten selenium is still somewhat conflicting and dependent on the methods of observation used. The proportion of cyclic molecules $(\text{Se})_8$ at corresponding temperatures may be considerably smaller than in molten sulphur, but there are numerous long-chain or spiral molecules which behave as diradicals with two unpaired electrons. By way of illustration, the chain length of the diradicals is about 7200 at 230°C (Hamada, Yoshida, and Shirai, 1969; but see Waseda, Yokoyama, and Suzuki, 1974). Melts of tellurium, unlike those of sulphur or selenium, may not contain extended chains of atoms (Cabane and Friedel, 1971). A 3-covalent network model may serve to describe molten tellurium, but the lifetime of any bond must be quite short since its viscosity is not high (about 2 cP) (see also Tourand and Breuil, 1971; Hoyer, Thomas, and Wobst, 1975).

### 13.9 Melts of ionic crystals

Information about prefreezing in ionic melts is at present very uneven in quality. As discussed in Chapter 10, ionic network crystal lattices such as are formed by oxides of Si, B, Ge, and by related compounds incorporating cations, give rise to melts which show steep rises of viscosity on approaching $T_f$. This behaviour is well known for numerous molten silicates. As another example for albite (sodium aluminium silicate) the melting point $T_f$ is 1391 K and $\eta$ is still $10^8$ p at 1420 K, sinking to between $10^4$ and $10^5$ p at 1670 K (Bowen, 1915). Most probably at least part of this rise should be attributed to blocking of a fraction of the liquid volume, by cluster formation, though no detailed analysis of results on this basis has been published. A more usual view disregards any blocking of flow by clusters and points out that flow under external shearing stresses involves successive breaking and making of bonds between the atoms. This process could be associated with large activation energies; but a weakness of this as the sole explanation of the very high viscosities observed is that it would appear to indicate an activation energy for bond-breaking varying rapidly with temperature, which is not easy to justify.

In a manner closely analogous to the behaviour of the shear viscosity, electrical conductances of supercooled melts show a gentle curvature through $T_f$, without exhibiting any singularity at $T_f$, in the case of solutions of salts dissolved in molten supercooled electrolytes such as $N$-methylformamide and $N$-methylacetamide (French and Glover, 1955). The observed curvature could be attributed to progressive formation of anticrystalline clusters by the solvent molecules in such solutions, blocking part of their volume as their temperature is lowered towards and below $T_f$. Electrical conductances of supercooled melts of ionic compounds formed by complexing $BF_3$ with various acids such as $H(CH_3CO.OBF_3)$ are stated to show 'discontinuities' at $T_f$ (Greenwood and Martin, 1952) when $\log \sigma$ is plotted against $1/T$, but the statistical procedures used (see p. 348) to evaluate the data appear to need reconsideration.

Infrared absorption measurements on melts of hydroxides of lithium, sodium and potassium point to marked formation of association complexes in the melts (Wilmshurst, 1961). The temperature-dependence of this effect calls for further investigation before its relation to prefreezing anomalies generally can be made fully clear. Infrared and Raman studies point to an absence of free rotation of $(ClO_4)^-$ anions in molten $LiClO_4$, probably owing to clustering (Leong and James, 1969). Lack of freedom is also indicated by similar studies on melts of $ZnSO_4$ (Hester and Krishnan, 1968) and melts of lithium nitrate hydrate (Irish, Nelson, and Brooker, 1971).

The viscosity of supercooled melts of $NaClO_3$ points to eventual glass formation, probably preceded by cluster formation (Rapoport, 1967). The fact that the melting curves for $KNO_3$ (Babb, Chaney, and Owens, 1964;

Rapoport and Kennedy, 1965) and $KNO_2$ (Rapoport, 1966) both pass through maxima as the pressure is increased points to more than one mode of molecular packing in the melts, and thus to formation of clusters.

## 13.10  Molten metals

Information about possible conglomerate structures in molten metals, with consequent prefreezing effects, is still in a somewhat confused state. Thermodynamic properties such as the specific volume (Klemm, Spitzer, Lingenberg, and Junker, 1952) and the adiabatic compressibility (Kleppa, 1950) suggest some prolongation of structural changes at melting of 'semimetals' into the prefreezing region above $T_f$. Other thermodynamic data are referred to in Chapter 9 above.

### 13.10.1.  Structural measurements on molten metals

As discussed in Chapter 9, diffraction studies on molten metals often indicate an apparent uniformity of structure when conventional radial distribution functions are calculated. However alternative treatments that avoid smoothing out split peaks or similar diffraction intensity anomalies near $T_f$ reveal a conglomerate or 'hybrid' structure in a number of cases. General reasons for anticipating such a structure for metallic melts have been discussed in Chapter 9.

X-ray scattering from molten tin (Brozel, Handtmann, and Richter, 1962) at various temperatures has been interpreted in terms of the presence of two structures. One of these, whose domains are predominant near $T_f$, involves spherical close-packing, with a second covalent layer lattice structure gradually becoming more prominent as the temperature rises above $T_f$. However, the precision required in X-ray scattering experiments to detect anticrystalline clusters often exceeds that attained in much of the earlier data. Radial distribution functions suggest the possibility of anomalous prefreezing structures for melts such as Rb, Zn, Sn, and Sb (see Furukawa, 1962), though this method of representation is open to criticism for the present purpose.

Actual clusters of distorted crystallites or even anticrystalline regions whose diameter ranges between 50 Å and 150 Å have been proposed by a number of scientists using suitably refined methods of computations of the experimental data. The evidence collated by different workers forms a very substantial total; the original papers should be consulted for details, since different workers have concentrated on rather different aspects of these molten metals.

Conglomerate structures have been worked out for alkali metals as well as Ag, Au, Pb and Tl (Richter and Breitling, 1967c; Richter, 1975); for other metals see the references indicated: Cu, Bagchi (1972); Al, Bagchi (1972); Li,

Na, K, Bagchi (1972); Richter and Breitling (1968, 1970); Ag, Richter and Breitling (1970); Tl, Richter and Breitling (1970); Pb, Romanova and Mel'nik (1970), Bagchi (1972); Hg, Richter and Breitling (1967a; 1967b); Bi, Richter and Oehme (1967b).

In addition to the diffraction studies referred to, a number of less direct observations have led essentially to similar conclusions concerning the 'conglomerate' structure of many molten metals. Neutron scattering near $T_f$ indicates an overall similarity from the melt with scattering from the solid, and can be interpreted on the basis of 'solid-like' regions (clusters) of about 100 atoms diffusing past one another in the melt (Turberfield, 1962). Neutron scattering also reveals a 'correlation range' within which collective coupled vibrations exist in molten tin and in molten aluminium; these correspond with phonons in the crystals (Larson, 1964). Molten gallium clearly has an abnormal structure. Comparisons of positron annihilation in solid and liquid gallium (Gustafson and Mackintosh, 1963) make this clear, as do many other properties. Possibly dimers such as $(Ga)_2$ are present but this method of test is not very discriminating.

For some metals electron diffraction curves can be usefully extended well below $T_f$, since thin films can sometimes be more readily supercooled than metal in bulk. On this basis, molten tin has been examined at $T_f + 70°$, $T_f + 30°$, and $T_f - 10°$. At the lowest temperatures intensity peaks become progressively more pronounced. General correspondence with the crystal structure of white tin becomes progressively more marked as the temperature of the melt is lowered (Boiko, Palatnik, and Roukina, 1962).

### 13.10.2. *Viscosity and mass transport measurements in metal melts*

When a melt has a conglomerate structure, unless the rate of dispersal and reformation is sufficiently rapid, the presence of clusters will be revealed by transport measurements. For example, melts of the metals In, Sn, Zn, or Pb show coefficients of self-diffusion near $T_f$ that are greater than calculated (Nachtrieb, 1967). 'Coordinated motion' is also inferred from diffusion in molten alkali metals and molten salts (Jousset and Huntington, 1969). Experimental reports of steep increases of viscosity in the prefreezing region of certain metals are surprisingly controversial. Reports of rapid increases (Hopkins and Toye, 1950; Yao and Kondic, 1952; Jones and Bartlett, 1952, 1954; Toye and Jones, 1958; Budde, Fischer, Menz, and Sauerwald, 1961) conflict with other findings (Frost, 1954; Culpin, 1957; Gebhart and Kostlin, 1957). In the case of metals such as Sb or Bi, and especially Ga, viscosity measurements suggest that 'clusters' persist near $T_f$ when the solids are melted, but do not reform with equal ease in the melts which are supercooled from well above $T_f$. Unfortunately, any non-reversibility of this kind might well be promoted by heteroatomic impurities (Predel and Arpshofen, 1968) and be of chemical

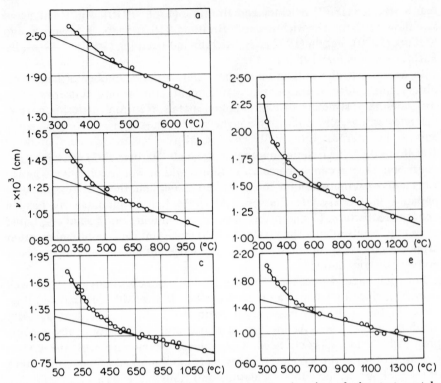

FIG. 13.6. Temperature curves of the kinematic viscosity of the test metals.
(a) Cadmium; (b) bismuth; (c) indium; (d) tin; (e) lead

rather than thermodynamic significance. This may help to explain the present rather controversial position with regard to this important prefreezing parameter. Viscosity measurements have also been cited as evidence for abnormal 'transitions' in molten aluminium (Lihl and Schwaiger, 1967) of 99·9% purity, as well as indicating clusters in molten Cd, Bi, In, Sn, and Pb (Lad'yanov, Arkharov, Novkhatsky, and Kisun'ko, 1972). The way in which viscosity might be used as a structure-sensitive property may be illustrated by plots for molten metals Cd, Bi, In, Sn, and Pb (Fig. 13.6).

Careful comparisons of a diversity of results on liquid Sn, Pb, Bi and K have confirmed the linear relationship $\log \eta = A + B/T$ except possibly near $T_f$ (Budde, Fischer, Menz and Sauerwald, 1961) with the following constants:

| Metal | $A \times 10^3$ | $B$ |
|-------|-----------------|------|
| Sn | 0·3320 | −2·3602 |
| Pb | 0·4915 | −2·3764 |
| Bi | 0·3170 | −2·3361 |
| K | 0·2523 | −2·0156 |

A sensitive test of cluster formation in a prefreezing region is to compare the viscosities of isotopic melts. Careful comparisons have been made between $^6$Li ($T_f = 180\cdot4°C$, $\eta_f = 4\cdot18 \pm 0\cdot05$ m.P.) and $^7$Li ($T_f = 180\cdot7°C$, $\eta_f = 6\cdot00 \pm 0\cdot05$ m.P.) (Ban, Randall, and Montgomery, 1962). The small difference in $T_f$ is not in accord with simple vibrational theories of melting (p. 82) and the ratio $\eta_7/\eta_6$ changes as the temperature sinks towards $T_f$. On any melting theories, the sensitive dependence on an isotopic mass can most readily be interpreted if transport clusters are formed whose size is a sensitive function of mass.

It is noteworthy that addition of potassium in dilute concentrations to mercury greatly enhances the viscosity, presumably owing to clustering of Hg atoms around each K atom (see Schulmann, 1962). Similar enhancements of viscosity are frequently found in binary mixtures of molten metals, when the electronegativity of the two components differs greatly. Persistence of clusters in such metallic melts is probable on many grounds (Bauer and Sauerwald, 1961; see Chapter 9). When the clusters that are postulated to explain this prefreezing effect involve covalent bonds, the time required for their aggregation and dispersion instead of being very short (see p. 80) might become comparable with or even greater than that of experimental observation. Prefreezing anomalies could then show a time-dependence even in conventional measurements (see observation on Fe, Schneider and Hilmer, 1956).

The fact (or claim) that the viscosity of melts of certain binary alloys 'follows' the properties of the solids (Yao and Kondic, 1952) must be interpreted with caution. When such properties are compared at temperatures in any way defined or selected with reference to the crystalline states, it is easy to introduce a hidden dependence on $T_f$, despite the absence of any real anomaly. However, when structure-sensitive properties are studied with precision, there appears to be little doubt that clusters 10 a.u. across or more are fairly common (Guntherodt, 1970).

Conflicting evidence from measurements of the viscosity of molten metals may in part be due to experimental errors. Particular attention must be paid to the method used by any author for establishing 'normal behaviour', since at present no general model for the liquid state is universally applicable. To illustrate this statement, viscosities determined down to temperatures (in °C) for Sn ($T_f - 0\cdot7$), Pb ($T_f - 11\cdot3$) Bi ($T_f - 4\cdot8$) and Al ($T_f - 1\cdot8$) using an oscillating sphere method (Rothwell, 1962) have been fitted to a calculated equation $\eta V^{\frac{1}{3}} = A \exp(-C/VT)$ where $V$ is the molar volume at $T$. Such a method of computation arbitrarily smoothes away any anomaly over the whole range down to the lowest temperature of measurement. Gradual departures from normality on approaching $T_f$ would not be apparent, and only a very steep increase could be detected in this way.

In another approach to measurements of prefreezing anomalies in molten metals the propagation of longitudinal sound in liquid Sn and Pb, using a pulse technique and a frequency of 5 Mc s$^{-1}$, has been measured at temperatures including supercooling down to below the freezing points. The observed velocities of sound are found to decrease linearly with temperature. When a root mean square straight line equation was fitted (Gordon, 1959), no departures from this line were observed on approaching $T_f$. Relaxation times $\tau$ should affect this method of measurement if of the order $\omega\tau \sim 1$ when $\omega \sim 10^7\,\text{s}^{-1}$ is the circular frequency of the imposed vibrations. If the viscosity becomes anomalously high, since the velocity $u$ of longitudinal waves in an isotropic medium is given by (see Pochapsky, 1951)

$$u^2\rho = 1/\beta + 4/3\mu \qquad (13.3)$$

where $\beta$ is the compressibility, $\mu$ the shear modulus, and $\rho$ the density, any rigidity of the liquid resulting from incomplete relaxation might be expected to affect $\mu$ and thus $u$. However, this argument (by the authors) only appears to be relevant if excess viscosities are due to some kind of delayed relaxation of shear in the liquid as a whole, i.e. regarded as a uniform structure. However, on a conglomerate model of a melt, if this contains regions blocked to viscous flow, as discussed above, it does not appear that anomalous viscosities necessarily determine $\mu$ and $u$ according to equation (13.3), which strictly applies only to a uniform medium. Clusters might merely increase scattering without having much effect on the velocity.

So far as pure metals are concerned, the extent of any prefreezing viscosity anomalies must be left open to question at the present. If they are genuine, the extent of liquid volume blocked by cluster formation can be readily calculated by the theory described above. Tentative calculations (McLaughlin and Ubbelohde, 1960) would indicate marked prefreezing clustering in molten tin and molten zinc. Using a different 'normal' relation for the viscosity/temperature dependence of fluids

$$\eta_i = V^{-1}(A + B/T)^3$$

similar evaluations of $\eta_i/\eta$ have been made in the prefreezing region on the basis of selected experimental data for the molten metals Sn, Pb, and Al. On this basis, aluminium also shows marked clustering in the range of temperatures between its melting point and about 695°C. At the melting point it is estimated that about 10% of the metal is associated into clusters, which contain about 100 atoms each (Mitra and Sharma, 1963).

Effects of small additions (up to 1 or 2 atom%) of other metals, such as Al, Fe, Pb, Ag, Cu, as impurities can lead to a steep decrease in viscosity of pure molten Zn or Sn. This has been attributed to the persistence of short-range order in these melts, even well above the melting point (Toye and Jones, 1958). Melts of Mg/Pb show marked prefreezing anomalies (Knappwost, 1959).

13.10.3. *Other properties of molten metals near* $T_f$

Brief mention may be made of other lines of investigation on molten metals, which have a bearing on possible conglomerate structures in the melts near $T_f$, but which have not attained any very decisive stage. Results reported for the influence of dissolved gases on melting points of various metals (Schofield, 1957; Forestier and Maurer, 1951) seem surprisingly large and call for confirmation.

Melts of IIB metals (Zn, Cd, Hg) lie on a reduced volume curve (McGonigal, 1962) different from that for IA and IIA metals and also other metals beyond IIB in the Periodic Table of elements. This may point to a fundamental difference in structure of the melts, either in their quasicrystalline packing, or possibly owing to the presence of different cluster types.

General theoretical discussion of the geometrical aspects of $n$-mer clusters of close-packed spheres (see Ginell, 1961) can only provide part of the background needed to elucidate the properties of molten metals. Nevertheless, they are of interest in examining the number of different forms of an $n$-mer of any given size. Spontaneous formation of 'crystallizable' clusters can be shown to be a rare event.

In molten Zn, and Al, it is claimed that heat transport across the solid–liquid interface is modified by the presence of clusters (Schaaber, 1952). Preservation of some kind of 'crystalline' structure in molten metals even well above $T_f$ has also been discussed in somewhat different terms (Borelius, 1953a,b; Borelius and Sandin, 1956, see Chapter 9.

## 13.11 Melts of semiconductors

According to the nature of the underlying process, any structural changes resulting from cluster formation in the prefreezing region will affect properties other than viscosity to very varying degrees. Striking changes of electrical resistivity provide one such property which has frequently been examined for metals (Chapter 9). For good conductors, the total change on melting as expressed by the ratio $\rho_L/\rho_S$ is not very large. Any anomalies resulting from cooperative fluctuations with slight differences of structure in the region of $T_f$ would be difficult to detect. In measuring this change the use of electrodes unfortunately precludes exploration of the supercooled melts below $T_f$, since it normally leads to heteronucleation (Chapter 15). However, with a number of semiconductors there is an increase in conductivity on melting; structural and other changes can also be considerably greater than for metals. This could make prefreezing anomalies easier to detect for such melts. A number of instances have been reported which apparently involve marked regions of prefreezing. For example, in melts of germanium it is claimed that the conductance rises to a maximum somewhat above $T_f$, and then falls as clusters are increasingly formed (Glazov and Chizhevskaya, 1962; but see Hamilton and Seidensticker, 1963; Busch and Tièche, 1963). Other measurements show that the electrical resistance of molten Te increases, but that it remains constant for molten Ge in the prefreezing region (Busch and Tièche, 1963; but see Blum, Mokrovskii, and Regel, 1952). Some authors report a density anomaly in the prefreezing region of molten tellurium, but this is disputed by others (Lucas and Urbain, 1962).

Analogous results have been reported with semiconductors that are crystal compounds such as AlSb, GaSb, InSb. In these experiments, electrical conductivities were observed by an electrodeless method from the angle of torsion in a rotating magnetic field, which in principle permits exploration of the supercooled region; spontaneous crystallization may, however, occur for other reasons. With gallium antimonide, the results were completely reversible even in the anomalous prefreezing regions, provided the melts were not heated too high above $T_f$. Strongly heated melts apparently failed to nucleate clusters during the time of observation, and gave supercooling without any drop in resistivity as $T_f$ was approached.

Crystal compounds whose structure is similar to that of zinc blende usually melt with shrinkage in volume and increase of coordination number. It has been reported that in melts of the compounds AlSb, GaSb, InSb, the viscosity may show a steep rise on approaching $T_f$. The activation free energy for viscous flow $G_\eta$ may pass through a minimum at a few tens of degrees above $T_f$, in the expression

$$G_\eta/RT = \ln(M\nu/Nh)$$

where $\eta$ is the kinematic viscosity, $M$ the molecular weight, and $N$ and $h$ have the usual meanings (Petrov and Glazov, 1958). Deviations are also found from Batchinski's formula in this temperature region. The viscosity-temperature relationship in melts of chalcogenide semiconductors points to a network structure, though it is difficult to identify regions of more specific structural anomaly.

The magnetic susceptibility as measured by the Faraday method (Glazov and Vertman, 1958) of some of the compounds cited in the preceding paragraph likewise shows a marked anomaly in the prefreezing region. A plausible interpretation might be that melting at first enhances metallic character through the breaking of covalent bonds. After maximum metallic behaviour has been reached the compound itself dissociates. When the temperature factor predominates an eventual fall is observed in magnetic susceptibility. However, like other prefreezing anomalies, these reported findings have been disputed. Thus it is claimed that electrical resistivities of liquid Ge and of InSb show no anomalous rise on approaching $T_f$, any more than the resistivity of molten Sn used as a standard (Hamilton and Seidensticker, 1963).

To sum up, in studying prefreezing anomalies care and insight are called for in selecting crystal types amenable to study and in determining which properties are sensitive to the kind of conglomerate structure which may or may not be formed. Unavoidably, the evidence recorded in this chapter is uneven in quality, and has in some cases only been included to draw attention to the problems to be considered from a new approach. There seems no doubt that, in arriving at models of the molten state near $T_f$ that are to be more realistic and more adaptable than the quasicrystalline model, advances are much to be desired both in physical techniques of measurement and in mathematical techniques for dealing with cooperative fluctuations and correlation of units into clusters.

## REFERENCES

A. A. K. Al Mahdi and A. R. Ubbelohde (1953) *Proc. Roy. Soc., Ser. A* **220**, 195.

J. N. Andrews and A. R. Ubbelohde (1955) *Proc. Roy. Soc., Ser. A* **228**, 435.

C. A. Angell, J. Shuppert, and J. C. Tucker (1973) *J. Phys. Chem.* **77**, 3092.

S. E. Babb, P. E. Chaney, and B. B. Owens (1964) *J. Chem. Phys.* **41**, 2210.

S. N. Bagchi (1972) *Acta Cryst.* **A28**, 560.

V. V. Baidov and M. B. Gitis (1970) *Sov. Phys. Semicond.* **4**, 825.

A. F. Ball (1950) *J. Res. Nat. Bur. Stand.* **45**, 23.

N. T. Ban, C. M. Randall, and D. J. Montgomery (1962) *Phys. Rev.* **128**, 6.

G. Bauer and F. Sauerwald (1961) *Wiss. Z. Martin Luther Univ. Halle Math. naturwiss.* **10**, No. 5, 1027.

A. Ben-Reuven and N. D. Gershon (1969) *J. Chem. Phys.* **51**, 893.

F. Bessette, A. Cabana, R. P. Fournier, and R. Savoie (1970) *Can. J. Chem.* **48**, 410.

F. C. Bingham and P. W. Kinney (1940) *J. Appl. Phys.* **11**, 192.

E. H. Binns and K. H. Squire (1962) *Trans. Faraday Soc.* **58**, 762.

A. I. Blum, N. D. Mokrovskii, and A. R. Regel (1952) *Izv. Akad. Nauk. SSSR, Ser. Fiz.* **16**, 139.

S. M. Blumfeld, S. P. Reddy, and A. L.,Welsh (1970) *Can. J. Phys.* **48**, 513.

B. T. Boiko, L. S. Palatnik, and N. I. Rodkina (1962) *Phys. Metals Metallography* (English trans.) **13**, 70.

G. Borelius (1953a) *Ark. Fys.* **6**, 20.

G. Borelius (1953b) *Ark. Fys.* **6**, 21.

G. Borelius and A. Sandin (1956) *Ark. Fys.* **10**, 187.

M. Borneas and I. Babutia (1960) *C.R. Acad. Sci., Paris* **250**, 1613.

M. Borneas and I. Babutia *Acta Phys. Pol.* **20**, 187.

N. L. Bowen (1915) *Am. J. Sci.* **39**, 175.

R. Brozel, D. Handtmann, and H. Richter (1962) *Naturwiss.* **49**, 129.

J. Budde, K. Fischer, W. Menz, and F. Sauerwald (1961) *Z. Phys. Chem. (Leipzig)* **218**, 100.

J. F. Burton (1969) *Chem. Phys. Lett.* **7**, 567.

J. F. Burton (1970) *J. Chem. Phys.* **52**, 345.

J. F. Burton (1971) *Nature, Lond.* **229**, 335.

G. Busch and Y. Tièche (1963) *Phys. Kond. Materie* **1**, 78.

B. Cabane and J. Friedel (1971) *J. Phys. (Paris)* **32**, 73.

L. G. Carpenter and T. F. Hearle (1937) *Phil. Mag.* **23**, 193.

N. F. Culpin (1957) *Proc. Phys. Soc., Lond.* **70B**, 1079.

G. D'Arrigo (1975) *J. Chem. Phys.* **63**, 61.

C. M. Davis and T. A. Litovitz (1965) *J. Chem. Phys.* **42**, 2563.

M. A. Deschamps (1975) *Chem. Phys.* **10**, 199.

J. F. Dill and T. A. Litovitz (1975) *J. Chem. Phys.* **62**, 3839.

C. Dodd (1951) *Proc. Phys. Soc., Lond.* **64**, 761.

C. Dodd and Hu Pak Mi (1949) *Proc. Phys. Soc., Lond.* **62B**, 454.

C. Dodd and G. N. Roberts (1950) *Proc. Phys. Soc., Lond.* **63B**, 814.

T. B. Douglas and J. L. Dever (1954) *J. Am. Chem. Soc.* **76**, 4824.

W. Drost Hansen (1965a) *1st Int. Symp. Water Desalination.*

W. Drost Hansen (1965b) *Ind. Eng. Chem.* **57**, 39.

W. Drost Hansen (1969) *Ind. Eng. Chem.* **61**, 11.

H. Forestier and J. Maurer (1951) *C.R. Acad. Sci., Paris* **232**, 1664.

C. M. French and K. H. Glover (1955) *Chem. Ind.* 252.

B. R. T. Frost (1954) *Progr. Met. Phys.* **5**, 96.

K. Furukawa (1962) *Rep. Progr. Phys.* **25**, 395.

E. Gebhart and K. Kostlin (1957) *Z. Metallk.* **48**, 636.

R. Ginell (1961) *J. Chem. Phys.* **34**, 992.

V. M. Glazov and S. N. Chizhevskaya (1962) *Sov. Phys. Solid State* **3**, 1964.

V. M. Glazov and A. A. Vertman (1958) *Proc. Acad. Sci. USSR* **123**, 805 (English).

R. B. Gordon (1959) *Acta Metall.* **7**, 1.

N. N. Greenwood and R. L. Martin (1952) *Proc. Roy. Soc., Ser. A* **215**, 46.

F. Grønvold (1973) *J. Chem. Thermodynamics* **5**, 525.

H. J. Guntherodt (1970) *Schweizer Archiv.* **36**, 179.

D. R. Gustafson and A. R. Mackintosh (1963) *Phys. Lett. (Netherlands)* **5**, 234.

S. Hamada, N. Yoshida, and T. Shirai (1969) *Bull. Chem. Soc. Japan* **42**, 1025.

D. R. Hamilton and R. G. Seidensticker (1963) *J. Appl. Phys.* **34**, 2697.

R. E. Hester and K. Krishnan (1968) *J. Chem. Phys.* **49**, 4356.

D. L. Hildebrand, G. C. Sinke, R. A. McDonald, W. R. Kramer, and D. R. Stull (1959) *J. Chem. Phys.* **31**, 650.

R. K. Hind, E. McLaughlin, and A. R. Ubbelohde (1960) *Trans. Faraday Soc.* **56**, 328, 331.

M. R. Hopkins and T. C. Toye (1950) *Proc. Phys. Soc., Lond.* **63B**, 773.

N. J. Hornung and G. R. Choppin (1974) *Appl. Spectrosc.* **8**, 149.

J. Horrocks and E. McLaughlin (1963) *Proc. Roy. Soc., Ser. A* **273**, 259.

W. Hoyer, E. Thomas, and M. Wobst (1975) *Z. Naturforsch.* **30A**, 1633.

P. M. Hu and R. W. Parsons (1958) *Proc. Phys. Soc., Lond.* **72**, 454.

A. N. Hunter (1956) *Proc. Phys. Soc., Lond.* **59B**, 965.

D. E. Irish, D. L. Nelson, and M. H. Brooker (1971) *J. Chem. Phys.* **54**, 654.

D. J. G. Ives and T. H. Lemon (1967) *Roy. Inst. Chem. Rev.* p. 62.

W. R. I. Jones and W. L. Bartlett (1952a) *J. Inst. Metals* **81**, 145.

W. R. I. Jones and W. L. Bartlett, *J. Inst. Metals* **83**, 59.

J. C. Jousset and H. B. Huntington (1969) *Phys. Status Solidi* **31**, 775.

H. Kanno, R. J. Speedy, and C. A. Angell (1975) *Science* **189**, 880.

G. S. Kastha (1958) *Indian J. Phys.* **32**, 473.

Ya. M. Kimelfeld, M. A. Moskaleva, G. N. Zhizkin, V. P. Litvinov, M. Mastovarayah, and I. P. Konyaeva (1975) *Opt. Spectrosk. USSR* **39**, 493.

N. Kirov and P. Simova (1973) *Izv. Fiz. Inst. ANEB (Bulgaria)* **24**, 115.

W. Klemm, H. Spitzer, W. Lingenberg, and J. H. Junker (1952) *Monatsh. Chem.* **83**, 629.

O. Kleppa (1950) *J. Chem. Phys.* **18**, 1331.

A. Knappwost (1959) *Z. Phys. Chem. (Frankfurt am Main)* **21**, 358.
K. Krynicki, D. W. Sawyer, and J. G. Powles (1975) *Proc. 18th Ampere Cong. Magnetic Resonance (1974) Amsterdam*, 511.
R. Kumer, M. Singh, and C. S. Sivaramakrishnan (1967) *Trans. Met. A.I.M.E.* **239**, 1219.
R. Kumer and C. S. Sivaramakrishnan (1971) *J. Mater. Sci.* **6**, 48.
V. L. Lad'yanov, V. I. Arkharov, J. A. Novkhotskii and V. Z. Kisun'ko (1972) *Fiz. Metal Metalloved.* **34**, 1060.
K. E. Larson (1964) *Symposium on inelastic scattering of neutrons, Bombay, 15–19 Dec. 1964.*
W. H. Leong and D. W. James (1969) *Austr. J. Chem.* **22**, 499.
F. Lihl and A. Schwaiger (1967) **58**, 777.
L. D. Lucas and G. Urbain (1962) *C.R. Acad. Sci., Paris* **255**, 3406.
W. Luck (1965) *Ber. Bunsenges. Phys. Chem.* **69**, 626.
W. Luck (1967) *Discuss. Faraday Soc.* **43**, 115.
S. J. Lukasik (1957) *J. Chem. Phys.* **27**, 523.
J. P. McCullough, D. R. Douslin, W. N. Hubbard, S. S. Todd, J. F. Messerly, I. A. Hosseniop, F. R. Frow, J. P. Dawson, and G. Waddington (1959) *J. Amer. Chem. Soc.* **81**, 5884.
P. J. McGonigal (1962) *J. Phys. Chem.* **66**, 1686.
E. McLaughlin and A. R. Ubbelohde (1958) *Trans. Faraday Soc.* **54**, 1804.
E. McLaughlin and A. R. Ubbelohde (1960) *Trans. Faraday Soc.* **56**, 988.
M. Magat (1935) *J. Phys. Radium* **6**, 179.
J. H. Magill and A. R. Ubbelohde (1958) *Trans. Faraday Soc.* **54**, 1811.
S. S. Mitra and L. P. Sharma (1963) *J. Chem. Phys.* **38**, 1287.
D. J. Morantz and R. C. Thompson (1970) *J. Phys. C.* **3**, 1335.
J. W. Mullin and C. L. Leci (1969) *Phil. Mag.* **19**, 1075.
N. H. Nachtrieb (1967) *Advan. Physics* **16**, 309.
J. C. Owicki, B. R. Lentz, A. T. Hagler, and H. A. Scheraga (1975) *J. Phys. Chem.* **79**, 2352.
S. Parthasarathy and V. N. Bindal (1960) *Indian J. Phys.* **34**, 272.
H. Patel and L. B. Borst (1971) *J. Chem. Phys.* **54**, 822.
D. A. Petrov and V. M. Glazov (1958) *Sov. Phys. Dokl.* **3**, 640.
M. Pezolet and R. Savoie (1971) *J. Chem. Phys.* **54**, 5266.
N. Pilpel (1961) *Trans. Faraday Soc.* **57**, 1426.
T. E. Pochapsky (1951) *Phys. Rev.* **84**, 553.
B. Predel and I. Arpshofen (1968) *Z. Naturforsch.* **23A**, 2052.
E. Rajagopal and S. V. Subrahmanyam (1974) *Pramana India* **2**, 312.
E. Rapoport (1966) *J. Chem. Phys.* **45**, 2721.
E. Rapoport (1967) private communication.
E. Rapoport and G. C. Kennedy (1965) *J. Phys. Chem. Solids* **26**, 1995.
Sh. Sh. Raskin and A. V. Secharev (1959) *Dokl. Akad. Nauk. SSSR* **128**, 67.
D. H. Rasmussen, A. P. Mackenzie, C. A. Angell, and J. C. Tucker (1973) *Science* **181**, 342.
G. B. Ravich and Yu. N. Burtsev (1961) *Bull. Acad. Aci. USSR Div. Chem. Sci.* (English trans.) November, p. 1951.
J. H. Read and D. M. G. Lloyd (1948) *Trans. Faraday Soc.* **44**, 721.
H. Richter and G. Breitling (1968) *Z. Naturforsch.* **23A**, 2063.
H. Richter and G. Breitling (1967a) *Z. Naturforsch.* **22A**, 658.
H. Richter and G. Breitling (1967b) *Z. Naturforsch.* **22A**, 655.
H. Richter and G. Breitling (1967c) *Advan. Phys.* **16**, 293.

H. Richter and G. Breitling (1970) *Z. Metallk.* **61,** 628.

H. Richter (1975) *Z. Naturforsch.* **30A,** 992.

H. Richter and H. Oehme (1967b) *Z. Naturforsch.* **22A,** 665.

E. Rothwell (1962) *J. Inst. Metals* **90,** 389.

C. T. Rutledge and G. T. Clayton (1975) *J. Chem. Phys.* **63,** 2211.

G. J. Safford and P. S. Leung (1971) *Techniques of Electrochemistry, Vol. II,* Wiley, New York.

C. Salama and D. A. G. Goring (1966) *J. Phys. Chem.* **70,** 3838.

O. Schaaber (1952) *Z. Metallk.* **43,** 251.

A. Schneider and O. Hilmer (1956) *Z. Anorg. Allg. Chem.* **286,** 97.

T. H. Schofield (1957) *J. Inst. Metals* **85,** 372.

H. Schulmann (1962) *Z. Anorg. Allg. Chem.* **317,** 204.

M. I. Shakparonov (1962) *Ukr. Fiz. Zh.* **7,** 782.

M. Shik Jhon, E. R. Van Artsdalen, J. Grosh, and H. Eyring (1967) *J. Chem. Phys.* **47,** 2231.

H. A. Stuart (1941) *Kolloid Z.* **96,** 149.

K. M. Swamy and K. L. Narayana (1974) *Z. Phys. Chem. (Frankfurt am Main )* **89,** 32.

G. Tourand and M. Breuil (1971) *J. Phys. Paris* **32,** 813.

T. C. Toye and E. R. Jones (1958) *Proc. Phys. Soc., Lond.* **71,** 88.

V. Trepadus, S. Rapeanu, I. Padureanu, Y. A. Parjenov, and A. G. Novikov (1974) *J. Chem. Phys.* **60,** 2832.

K. C. Turberfield (1962) *Proc. Phys. Soc., Lond.* **80,** 395.

H. Tweer, J. H. Simmons, and P. B. Macedo (1971) *J. Chem. Phys.* **54,** 1952.

A. R. Ubbelohde (1938) *Trans. Faraday Soc.* **34,** 292.

A. R. Ubbelohde (1961) *Pure Appl. Chem.* **2,** 263.

A. R. Ubbelohde (1964) *J. Chim. Phys.* **61,** 58.

V. Vand and W. A. Senior (1965) *J. Chem. Phys.* **43,** 1869, 1873, 1878.

A. Van Hook and L. Silver (1942) *J. Chem. Phys.* **10,** 686.

C. G. Vezzoli, F. Dachille, and R. Roy (1969) *J. Polym. Sci.* **7,** 1557.

Y. Waseda, K. Yokoyama, and K. Suzuki (1974) *Phys. Condens. Materie* **18,** 293.

J. K. Wilmshurst (1961) *J. Chem. Phys.* **35,** 1800.

T. P. Yao and V. Kondic (1952) *J. Inst. Metals* **81,** 17.

H. Zocker and R. D. Machado (1959) *Acta Cryst.* **12,** 122.

# 14. LIQUID CRYSTALS

## 14.1 Assemblies of partly ordered molecules in melts

The conglomerate structure suggested to explain the temperature variation of transport and other properties of certain melts involves clusters of molecules which are *anticrystalline*. Usually these clusters are far too small and irregular in other respects to have any very obvious influence on light scattering, though consequences of their presence may become apparent in certain observations on optical scattering from melts (p. 356), especially on supercooling below $T_f$.

Mesomorphic melts are special states of matter at temperatures intermediate between $T_f$ and a higher temperature $T_{clear}$ at which they become optically isotropic. Such 'liquid crystals' are conglomerate fluids containing quite large clusters or assemblies of partly oriented molecules. Mesomorphic melts of molecular crystals have recently been studied with much more attention paid to general physical and thermodynamic features, because of the technological possibilities they offer. Accordingly, lists of homologous compounds are now less often published with only the melting and transition temperatures, though, as discussed in the following sections important thermodynamic data still tend to be ignored. Useful general review articles include Chatelain (1954), Gray (1962), Usol'tseva and Chistyakov (1963), Sackmann and Demus (1969) Brown, Doane, and Neff (1970), Gray and Harrison (1971), Rapini (1973), Kelker and Hatz (1974), Stephen and Straley (1974), Chandrasekhar (1976) Mostly, thermodynamic aspects which are of particular interest for the present book are given only cursory treatment despite the enormous body of general physical observations on these materials. Formation of such optically anisotropic fluids at $T_f$ on melting certain types of crystals was first detected by microscope observations, using polarized light with crossed polarizer and analyser, which clearly reveal anisotropic regions in the fluids. Below $T_f$ crystals giving melts of liquid crystals behave in the same way as any other substances with similar molecules. Above $T_f$, optically oriented regions are observed in the melt over a range of temperatures, with an upper limit at a clearing temperature $T_{clear}$ above which the melt is optically isotropic.

Optical studies have revealed that assemblies of partly oriented molecules in molecular mesomorphic states can be grouped into three main types, which can be associated with three kinds of ordering, as follows.

(1) *Nematic liquid crystals* (see Fig. 14.1). The least restrictive ordering is found in this group which is the largest group. Between $T_f$ and $T_{clear}$ the melts

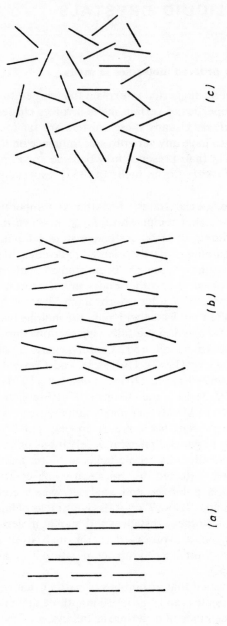

FIG. 14.1. Schematic representation of molecular order in the (a) crystalline, (b) nematic, and (c) isotropic phases. Reproduced with permission from Chandrasekhar, *Rep. Prog. Phys.* **39**, 615 (1976). Copyright by The Institute of Physics

contain bundles of molecules whose axes are approximately parallel, and fancifully described as nematic or worm-like.

(2) *Smectic liquid crystals* (see Fig. 14.2). In this, which is a more limited class, molecules are arranged end to end. Molecules in smectic melts have less freedom to move, and viscosities are markedly higher than for nematic melts. A subgroup distinction may be made between normal and tilted smectic structures.

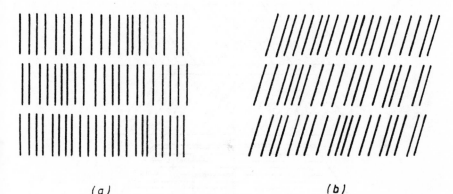

(a)    (b)

FIG. 14.2. (a) Normal, and (b) tilted smectic structures. Reproduced with permission from Chandrasekhar, *Rep. Prog. Phys.* **39,** 615 (1976). Copyright by The Institute of Physics

(3) *Cholesteric liquid crystals..* These are discussed below (see Fig. 14.3).

Until fairly recently, the exceptional ease of making optical studies using polarized light, especially under the microscope, has favoured the accumulation of much optical information about liquid crystals quite unsupported by studies of even other physical and especially of thermodynamic properties. However, research on such properties has gradually been building up. The present chapter is particularly concerned with thermodynamics in relation to the structure of these phases.

### 14.1.1. *Ionic mesophase melts*

Ordered melts of organic ionic liquids are discussed on p. 225. Though they resemble molecular liquid crystals in a number of ways, the stronger, longer-range forces in the ionic melts imposes behaviour for which there is no real analogy in molecular liquid crystals. Because of the long range of forces operating between the molecules, observations made by microscope-stage techniques may entrain serious perturbations of the interactions observed (e.g. Baum, Demus and Sackmann 1970; Demus, Sackmann and Seibert

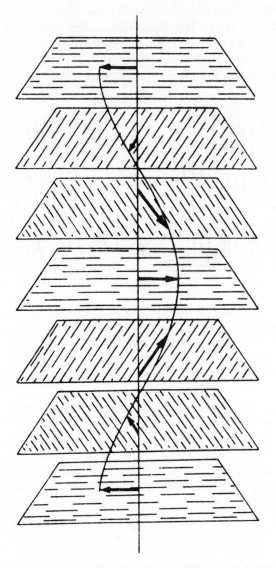

FIG. 14.3. The cholesteric liquid crystal: schematic representation of the helical structure. Reproduced with permission from Chandrasekhar, *Rep. Prog. Phys.* **39,** 615 (1976). Copyright by the Institute of Physics

1970). Thermodynamic data for ionic mesophase melts are best obtained on bulk material for this reason.

## 14.2 Melting in stages

With certain types of molecules, the progressive decrease of ordering found over a range of temperatures is best known in the sequence:

$$\text{crystal} \Big) \xrightarrow{T_f} \begin{array}{c} \text{smectic} \\ \text{liquid} \end{array} \Big) \xrightarrow{T_{c_1}} \begin{array}{c} \text{nematic} \\ \text{liquid} \end{array} \Big) \xrightarrow{T_{\text{clear}}} \begin{array}{c} \text{isotropic} \\ \text{liquid} \end{array}$$

### Table 14.1. Effects of constitution on liquid crystal transitions

($T_f$ = crystal → smectic, $T_{c_1}$ = smectic → nematic (or crystal → nematic if no smectic melt appears), $T_{c_2}$ = nematic → isotropic)

Series p-n-alcoxybenzoic acids (°C) $C_nH_{2n+1}OC_6H_4COOH$

| Substituent | $T_f$ | $T_{c_1}$ | $T_{c_2}$ |
|---|---|---|---|
| n = 1 | — | — | 184 |
| 2 | — | — | 196 |
| 3 | — | 145 | 154 |
| 4 | — | 147 | 160 |
| 5 | — | 124 | 151 |
| 6 | — | 105 | 153 |
| 7 | 92 | 98 | 146 |
| 8 | 101 | 108 | 147 |
| 9 | 94 | 117 | 143 |
| 10 | 97 | 122 | 142 |
| 12 | 95 | 129 | 137 |
| 16 | 85 | — | 132·5 |
| 18 | 102 | — | 131 |

Series 6 n-alcoxy-2-naphthoic acid
$R = C_nH_{2n+1}$

| | | | |
|---|---|---|---|
| n = 1 | — | 206 | 219 |
| 2 | — | 213 | 224 |
| 3 | — | 208 | 208·5 |
| 4 | — | 198 | 208·5 |
| 5 | — | 179·5 | 199 |
| 6 | — | 147 | 198·5 |
| 7 | — | 163 | 190 |
| 8 | — | 161·5 | 180 |
| 9 | (140) | 147·5 | 183·5 |
| 10 | 139 | 147 | 181 |
| 12 | 119 | 158·5 | 184 |
| 16 | 107 | — | 160·5 |
| 18 | 114 | — | 159 |

As with other types of crystals, for such substances, the disordering that takes place on melting occurs by stages. Clearly, unlike other examples of multiple mechanisms of melting, discussed in previous chapters, in liquid crystals some of the disordering processes demonstrably occur in the fluid state above $T_f$, since the persistence of strong correlations between the molecules can be observed from the physical properties. Illustrative examples of series of compounds which show such stages are recorded in Table 14.1. Very many other series have now been investigated.

Various qualitative rules have been proposed relating the structure of individual molecules with the appearance of mesophases. Thus substitution of halogen or nitro groups in the aromatic acid tends to lessen the 'existence range' of liquid crystals, presumably by interfering with packing of the molecules into anisotropic bundles. Strictly, however, both the crystal structure and the packing into clusters determine the existence range of any mesophase.

When the rigidity of the linear part of the molecules is increased, the range of liquid crystals existence is extended. Thus $p$-alcoxy derivatives of diphenyl in Table 14.1 may be compared with $p$-n-alcoxybenzoic acids in Table 14.2.

**Table 14.2. Series $4RO.C_6H_4—C_6H_4-COOH$ (4). Temperatures in $°C$;**
**$R = nC_rH_{2r+1}$**

| $r$ | $T_f$ | $T_{c_1}$ | $T_{c_2}$ |
|---|---|---|---|
| 1 | — | 258 | 300 |
| 2 | — | 256·5 | 301·5 |
| 3 | — | 260·0 | 287 |
| 4 | — | 234 | 284·5 |
| 5 | 227·5 | 229·5 | 275 |
| 6 | 213 | 243 | 272·5 |
| 7 | 194·5 | 251 | 265·5 |
| 8 | 183 | 255 | 264·5 |
| 9 | 176 | 256·5 | 258·5 |
| 10 | 172·5 | 256·5 | 257 |
| 12 | 165 | — | 252 |
| 16 | 151 | — | 241·5 |
| 18 | 150 | — | 238 |

### 14.2.1. *Cholesteric type of partially ordered melts*

A third (restricted) group of melts of molecules related to cholesterol, known as the cholesteric type, is referred to above and has also been detected by its optical microscopy. In these melts, orientation of the molecules in groups in the liquid is even more specialized (see De Vries, 1950) than in the nematic or even smectic types. Possibilities for attaining this arrangement depend on highly specific features of the molecular structure.

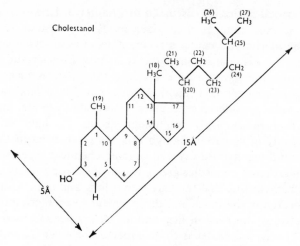

FIG. 14.4. Skeletal structure of cholestanol

Figure 14.4 illustrates the general skeleton of these molecules, whose repulsion envelopes must be plate-like if the cholesteric type of oriented melt is to be formed (Kast, 1955). Table 14.3 illustrates values of $T_f$ and $T_{clear}$ for a series of substituents.

*Table 14.3. Freezing and clearing points of some cholesteric-type melts*

Benzoates of sterine cholesterols (one double bond in the position indicated)

| | $T_f$ | $T_{clear}$ |
|---|---|---|
| | 137 | 155 |
| 5, 6 | 150 | 178 |
| 7, 8 | 158 | 176 |
| 8, 9 | 147 | 174 |
| 8, 14 | 115 | 140 |
| 14, 15 | 171 | no liquid crystals |

Benzoates of cholestadienols (two double bonds in the positions indicated)

| | $T_f$ | $T_c$ |
|---|---|---|
| 5, 6 : 7, 8 | 143 | 188 |
| 6, 7 : 8, 9 | 146 | 180 |
| 7, 8 : 14, 15 | 150 | no liquid crystals |
| 8, 9 : 24, 25 | 128 | 138 |
| 14, 15 : 24, 25 | 122 | no liquid crystals |

## 14.3 Some structural relationships in molecular anisotropic melts

The formation of ordered *fluids* above $T_f$ indicates that for these substances $T_f$ corresponds in certain ways with $T_{mech}$ as the temperature of fluidification of the entire crystal lattice, as discussed in Chapter 3. However, quite large entropy and volume changes take place at $T_f$ (see below). Evidently, in addition to the mechanical changes taking place, thermodynamic aspects of melting cannot be disregarded and the transition from crystal to ordered liquid is discontinuous. Furthermore, at $T_{mech}$ the lattice as a whole becomes fluid, whereas in conventional liquid crystals, mobility *within* each ordered bundle needs to be considered separately. Observations of melting by stages, with disordering mechanisms prominent above $T_f$, have stimulated a considerable body of research, though this has often been in the hands of microscopists and has mainly stressed organic chemical aspects. By far the largest known group of molecules giving rise to anisotropic melts are found to have as one characteristic that they are rather long and rigid. Usually several dipoles are located in the molecule; these appear to promote intermolecular coherence between the rods in any bundle. A gradual decrease of perfect parallelism between molecules is to be expected between $T_f$ and $T_{clear}$ in a liquid crystal on statistical thermodynamic grounds. For $p$-azoxyanisole, evaluations of the degree of ordering over the range of temperatures from 90°C to 130°C give fair concurrence between data based on proton resonance, diamagnetic susceptibility, refractive index, infrared spectrum, ultraviolet spectrum and theoretical calculations (see also Maier and Saupe, 1959). In a characteristic instance NMR studies in the region nematic → isotropic indicate about 8% of the molecules behaving as in the isotropic liquid (Grande, Limmer and Losche, 1975). Even at $T_{c_1}$, in the liquid crystal, marked tendency to parallelism of the molecular axes is found (Saupe and Maier, 1961). A jump in order at $T_{c_1}$ is predicted.

Gradual disordering of axes between $T_f$ and $T_{clear}$ is found when infrared spectra are studied (Maier and Englert, 1957). Observations of intensities of polarized and depolarized scattered light as a function of scattering angle can be conveniently made between $T_f$ and $T_{clear}$. These show a progressive decrease of 'coherence length' as the temperature is raised (Chu Bak and Lin, 1972), which may be regarded as a guide to cluster size.

Studies of effects of magnetic fields on states of order of azoxyphenol di-$p$-n-heptyl ether and azoxybenzoic acid di-$p$-ethyl ester show differences of behaviour in the smectic and nematic phases. In the smectic phase microordering is stated to be perfect (Weber, 1959). Similar studies have been made using proton magnetic resonance in the series of azoxyphenol di-$p$-n-alkylethers (Lippmann and Weber, 1957). For some molecular liquid crystals such as 4, 4'-di-n-pentyloxyazoxybenzene or 4, 4'-di-n-heptyloxyazoxybenzene measurements of the electrical conductivity have been utilized to follow the phase diagram (Heppke and Schneider, 1975).

Skilful organic synthesis has provided families of such molecules of related structures which permits a systematic investigation of the dependence of $T_f$ and $T_{clear}$ on molecular characteristics (e.g. Weygand and Gabler, 1938,

**Table 14.4. Melting and clearing temperatures of linear molecules with three phenyl groups (°C)**

| | $T_f$ | | | $T_{clear}$ | | |
|---|---|---|---|---|---|---|
| | I | II | Δ | I | II | Δ |
| H | 163·5 | 142 | 21·5 | — | | — |
| OH | 265 | 260·5 | 4·5 | — | | — |
| CH₃ | 189·5 | 191·5 | −2·0 | 264·5 | 264·0 | 0·5 |
| Cl | 180 | 203·5 | −23·5 | 288 | 289·5 | 1·5 |
| CH₃—O | 223 | 206 | 17 | 329·5 | 325 | 4·5 |
| C₂H₅—O | 198 | 201 | −3 | 322·5 | 324·5 | 2 |
| n-C₃H₇—O | 205·5 | 187·5 | 18 | 298 | 296·5 | 1·5 |
| i-C₃H₇—O | 196 | 192 | 4 | 233 | 228 | 5 |
| n-C₄H₉—O | 190 | 181 | 9 | 287 | 282 | 5 |

1939, 1941; Weygand and Lanzendorf, 1938; Kast, 1955; van der Veen, de Jeu, Groben, and Bover, 1972). Some typical classes of stuctures which melt to give rise to fluids containing optically oriented regions are listed in Table 14.4. Tables 14.1 and 14.2 record some trends in $T_f$ and $T_{clear}$ in homologous series which have been well investigated.

A further variant is to compare isomeric molecules. For example, molecular structures are readily prepared in which conjugation extends to three phenyl rings (Wiegand, 1957), with pairs of isomers (Fig. 14.5) as in I and II.

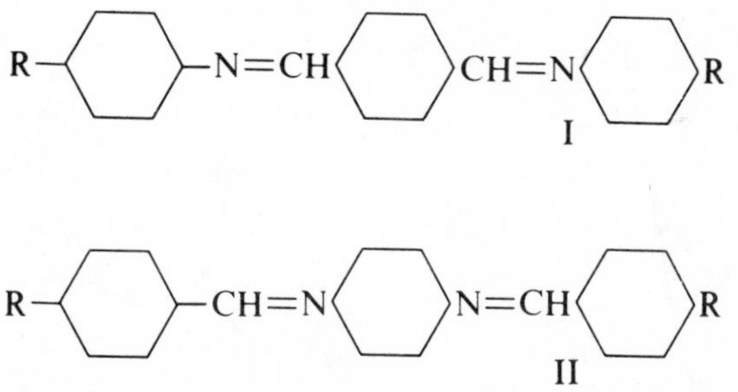

FIG. 14.5. Isomeric molecules with conjugation extending to three phenyl rings

It is noteworthy that I and II have quite large differences Δ in $T_f$ but practically the same clearing temperatures $T_{clear}$.

Obviously the melting point $T_f$ which will depend on precise close-packing in the *crystal lattices*, is considerably more sensitive to influences from nearest neighbour molecules than the nematic clearing point $T_{c_1}$. In a liquid crystal bundle, the molecules are parallelized but can undergo some relative movement lengthwise. Parallelism of the molecules on average appears to be destroyed at about the same clearing points $T_{clear}$ for both types I and II.

Similar behaviour is found for monoazomethines for which again $T_f$ is different but $T_{clear}$ practically the same. Similar studies of gradually decreasing order in a nematic liquid crystal have been made on melts formed by nonadiene 2–4 acid I, $C_4H_9(CH=CH)_2COOH$, for which $T_f = 23°C$, and $T_{c_1} = 53·5°C$, and by undecadiene 2–4 acid 1, $C_6H_{13}(CH=CH)_2COOH$, for which $T_f = 32°C$, and $T_{c_1} = 62·5°C$ (Maier and Markau, 1961).

When the homologous series is constituted by regular increases in the size of a normal alkyl group in the molecule, as is the case for many examples studied, it might be expected that convergence temperatures would be found

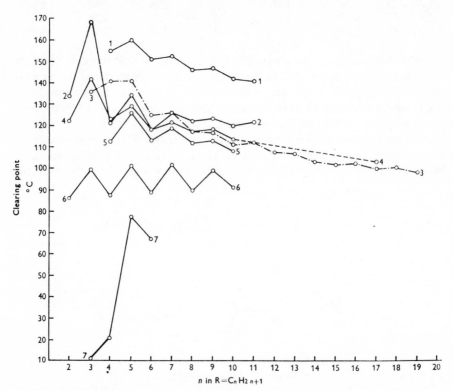

FIG. 14.6. Trends of clearing points in homologous series of molecules giving liquid crystals (Weygand and Gabler, 1941). (1) $p$-n-Alcoxybenzoic acids $R.O.C_6H_4COOH$; (2) $p$-n-alcoxyazobenzoles $R.O.C_6H_4.N=N.C_6H_4O.R$; (3)

$$\downarrow$$
$$O$$

phenetoleazophenolesters of n-fatty acids $R.COOC_6H_4N=N.C_6H_4OC_2H_5$; (4) $p$-n-alcoxybenzalphenetidene $R.OC_6H_4CH=N.C_6H_4.O.C_2H_5$; (5) $p$-n-alkyl benzoic acids $R.C_6H_4COOH$; (b) $p$-n-alcoxybenzal-1-aminonaphthalene-4-azobenzole $R.O.C_6H_4C=N.C_{10}H_6N=N.C_6H_5$; (7) $p$-iso-alcoxybenzal-1-aminonaphthalene-4-azobenzole

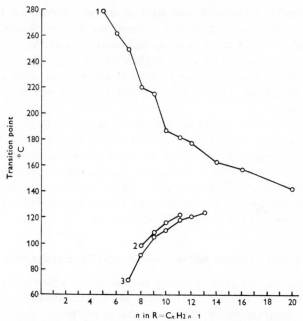

FIG. 14.7. Transition temperatures smectic–nematic in homologous series (Weygand and Gabler, 1941). (1) *p*-Azoxycinnamic acid esters of n-alcohols R.OOC.CH=CHC$_6$H$_4$N=NC$_6$H$_4$.CH=CH.COOR; (2) *p*-n-alcoxybenzoic

$$\downarrow$$
$$O$$

acids R.O.C$_6$H$_5$.COOH; (3) *p*-n-alcoxyazobenzenes R.O.C$_6$H$_4$N=N.C$_6$H$_4$.OR

$$\downarrow$$
$$O$$

for clearing points $T_{c_1}$ similar to the convergence effects for melting points $T_f$ of normal paraffins (Chapter 7). In general, $T_{c_1}$ does in fact approach asymptotically to a convergence value, but the shorter chains are associated with higher clearing temperatures. Unlike effects on $T_f$, the trend in $T_{c_1}$ with increasing molecular size is downwards. Figure 14.6 illustrates the fact that some alternation is found for the shorter chains, exactly as for $T_f$ but to a less pronounced degree. This suggests that in the nematic bundles the normal alkyl chains are in their stretched configurations.

For transitions smectic → nematic, trends may be either upwards or downwards a convergence limit, as illustrated in Fig. 14.7.

## 14.4 Ionic 'liquid crystals'

Most examples of mesophase melting have been concerned with molecular crystals, in which the presence of dipoles and hydrogen bonds, as well as the stretched out conformation of the molecules, favour their persistence in oriented bundles above the first melting point.

As explained in Chapter 8, discovery of means for ensuring the chemical stability of organic ionic melts has opened the way to molten states of an unusual kind, in which there is stong electrostriction of the organic parts of the molecule. A number of these organic melts pass though a mesophase which is optically anisotropic and whose stability range often extends over many tens of degrees (Table 8.36) before a second melting point to give isotropic melts.

By analogy, it is tempting to describe these mesophases as ionic liquid crystals, though for reasons indicated previously the use of the term 'liquid' may be misleading. The structures of organic ionic melts, when optically anisotropic, resembles that of the parent crystals, but is quite different from that of many molecular 'liquid crystals'. This structure is described on p. 224 above. Some of the thermodynamic and physical properties of anisotropic ionic melts are discussed in the following sections.

## 14.5 Disordering in stages above $T_f$—General Thermodynamic studies

With regard to the main purpose of this book, which aims to examine general relationships between structural and thermodynamic changes which accompany melting, some of the experiments which might yield most information about disordering in stages about $T_f$ have unfortunately been carried out with only a very limited range of substances. As typical of the smectic class, a readily available molecule is the ethyl ester of $p$-azoxybenzoic acid

$$C_2H_5COO-\langle\ \rangle-N{-}N-\langle\ \rangle-COOC_2H_5$$
$$\qquad\qquad\qquad\qquad\downarrow$$
$$\qquad\qquad\qquad\qquad O$$

for which

$$T_f = 114°C,\ T_{c_2} = 120°C$$

In the nematic class, much experimentation has been carried out with $p$-azoxyanisole

$$CH_3O-\langle\ \rangle-N{-}N-\langle\ \rangle-OCH_3$$
$$\qquad\qquad\qquad\qquad\downarrow$$
$$\qquad\qquad\qquad\qquad O$$

for which $T_f = 116°C$, and $T_{c_1} = 134°C$.

Measurements of the two principal thermodynamic quantities, the molar heat content and the molar volume, have been made on substances whose purity is not always well defined. Effects of pressure, which raises $T_{c_1}$ for nematic melts (Kast 1939) indicate that the isotropic liquid occupies a larger molar volume. A large difference between $T_{c_1}$ at constant pressure and constant volume is calculated for $p$-azoxyanisole (Chin and Neff, 1975). By differential calorimetry (Kreutzer, 1938), it has been shown for $p$-azoxy-anisole that

$$C_p \text{ (anisotropic)} = 0 \cdot 528 \pm 0 \cdot 009 \text{ cal g}^{-1} \text{ deg}^{-1},$$

$$C_p \text{ (isotropic)} = 0 \cdot 488 \pm 0 \cdot 009 \text{ cal g}^{-1} \text{ deg}^{-1},$$

$$\Delta H_{c_1} = 0 \cdot 46 \pm 0 \cdot 1 \text{ cal mole}^{-1},$$

where $\Delta H_{c_1}$ is the heat of transformation.

Correction of specific heats to constant volume using results for thermal expansion and compressibility leads to the values

For $p$-azoxyanisole-phenetole

$$\Delta C_v \text{ (anisotropic/isotropic)} \quad = 4 \cdot 2 \pm 0 \cdot 2 \text{ cal mole}^{-1} \text{ deg}^{-1},$$

$$\Delta H_c = 0 \cdot 91 \text{ cal mole}^{-1}$$

and for $p$-azoxyphenetole

$$\Delta C_v \text{ (anisotropic/isotropic)} = 7 \cdot 7 \text{ cal mole}^{-1} \text{ deg}^{-1},$$

$$\Delta H_{c_1} = 0 \cdot 90 \text{ cal mole}^{-1} \text{ deg}^{-1}$$

No simple structural difference of molecular arrangement in the melt has been proposed to explain this difference in behaviour between somewhat similar molecules. (Johnson, Porter, and Barrall, 1969).

## 14.6 Domain thermodynamics in liquid crystals

Considered from the general standpoint of the present book, the trans-formations discussed in mesomorphic states of matter are often physically spectacular, but thermodynamically not very prominent. This is revealed, for example, from precision studies of enthalpy and volume changes. These are often overlooked since the quantities involved may be small and difficult to evaluate compared with changes at the first melting point $T_f$. However, much interesting information is obtainable, for example when several parameters are measured at the same time as changes of specific volume, for $N$-($p$-hexyloxybenzalidene)-$p$-toluidine (Bahadur, Prakash, Tripathi, and Chandra, 1975). In particular, detailed calorimetry is becoming available for a number of examples. For each type of system, for molecular crystals the entropy change $S_f$ at the first melting point $T_f$ is notably larger than at subsequent transitions from one liquid crystal state to another, or from a mesomorphic to an isotropic liquid.

For example, for $N$-($p$-methoxylbenzylidene)-$p$-n-butylaniline) $S_f \sim$ 10·4 e.u., whereas $S_{c_1} \sim 0 \cdot 2$ e.u. for the transition from nematic to isotropic

(Shinoda, Maeda, and Enokido, 1974). Other calorimetric data (Johnson, Porter, and Barrall, 1969) confirm the general finding that the entropy change at the first melting point is many times larger than in subsequent disordering processes. A number of other examples leading to the same conclusion are listed (Chandrasekhar, 1976) in Table 14.5.

**Table 14.5. Melting transitions through mesophases**
**($^\circ$C and e.u. cal mole$^{-1}$ $^\circ$C$^{-1}$)**

| Molecule | $T_f$ | $S_f$ | $T_{c1}$ | $S_{c1}$ | $T_{c2}$ | $S_{c2}$ | $T_{c3}$ | $S_{c3}$ |
|---|---|---|---|---|---|---|---|---|
| p-Azoxyanisole | 118·2 (18·0) | | 135·9 (0·33) | | | | | |
| | | → nematic | | → isotropic | | | | |
| 4, 4'-Diheptyl | 74·4 (28·2) | | 95·4 (1·03) | | 124·2 (0·62) | | | |
| azoxybenzene | | → smectic C | | → nematic | | → isotropic | | |
| Terepthal-bis- | 113·0 (11·25) | | 144·1 (2·15) | | 172·5 (0) | | 199·6 | (0·15) |
| 4n-butylaniline | | → smectic B | | → smectic | | → smectic | | ↓ |

isotropic            236·5 (0·35)                    nematic
    ↑_____↑

Calorimetric data leading to similar conclusions about various other liquid crystal transitions have also been obtained (Arnold, 1966; Arnold and Roediger, 1966). The entropy changes for transitions above $T_f$ always involve smaller increases, sometimes much smaller than at the first melting point. Such data illustrate the fact that (at any rate for molecular crystals) melting to a mesomorphic fluid often involves far greater disordering than the final dispersal of anisotropic domains to give isotropic melts. In striking contrast, for ionic mesophases (Duruz and Ubbelohde, 1972) the two entropy changes lie closer together (see p. 226).

Whilst entropy and volume changes in mesomorphic transitions are not always easy to determine precisely, the shift of peak transition temperatures $T_{c1}$ on increasing the applied pressure can probably be more easily measured (Klement and Cohen, 1974; Feys and Kuss, 1974); Semenchenko Byankin, and Ya Baskakov, 1975; Chandrasekhar, 1976). Increase of pressure generally brings mesomorphism into greater prominence, because the isotropic phase is usually more open-packed than any mesomorphic intermediate, and also because potential barriers opposing free rotation in the melt are enhanced on compression. Raman studies point to free rotation in certain cases (Dvorjetski, Voltera, and Wiener-Avnear, 1975). It is notable that a negative volume discontinuity has been observed for one nematic–smectic A transition (Lin, Keyes, and Daniels, 1974) and that practically zero-pressure dependence has been found for some cholesteric transitions (Semenchenko Byankin, and Ya Baskakov, 1975). Other precursor effects have been established by optical studies; see Bulkin, Grunbaum and Santoro (1969), Afaras'eva Burlakov, and Zhizkin (1976).

It may be stressed that although domains in liquid crystals show extended long-range correlations of molecular orientations their domain structure is not *quasi-crystalline* in the strict sense used in this book. For this reason, a statistical model for orientational melting of crystals (see Pople and Karasz, Chapter 11) which is proposed from time to time (Chandrasekhar, Shashidar, and Tara, 1970) does not provide any very satisfactory treatment from the standpoint of thermodynamics. As stated above, the entropy changes found on first melting and on subsequent transformations to isotropic melts appear to be very unequal for molecular crystals. An alternative theory of translational and orientational melting has been proposed (Kobayashi, 1971). Original references (see above) should be consulted for other statistical models more particularly devised for transformations in mesomorphic domains.

### 14.6.1. *Two-component liquid crystals*

Effects of adding a second component as impurity has been studied, to test to what extent such impurities dissolve in the optically oriented bundles, or alternatively remain confined to the more open isotropic portions of the liquid surrounding the bundles. Impurities whose presence merely serves to dilute the amorphous environment depress the transition temperatures in a way presumably analogous to depressions of transition temperatures for any other 'phase' transformation. Calculation of the changes in heat content based on such observations are hindered by lack of information about the partition coefficient of the added impurity between the anisotropic and isotropic regions (see Kast, 1939). Furthermore, the freezing point depression equations to be used are not self-evident (see the discussion of effects on orientation resulting from solid solution, pp. 43, 100).

### 14.6.2. *Compound liquid crystals*

A very limited number of cases is known in which a two-component system actually forms liquid crystals more readily than either single species, suggesting that a 'compound packing' is formed in the melt, analogous with 'crystal compounds' whose melting temperature is at a maximum in relation to the composition of the binary system. As is well known, formation of crystal compounds occurs for many pairs of organic molecules, whose repulsion envelopes and attraction centres permit economical filling of space by regular sequences of the two components, packed so as to minimize potential energy.

Binary mixtures of molecules often show some difficulty of packing to form mixed liquid crystals. When the discrepancy in molecular repulsion envelopes is not too great, some depression of $T_{c_1}$ occurs for compositions intermediate between the two pure substances. Plots of clearing temperatures against composition resemble many ordinary freezing curves for solid solutions. Taken in pairs, the molecules $p$-azoxyanisole, $p$-acetoxybenzal, and $p$-phenetidine behave in this way (Dave and Lohar, 1959).

Conjugate properties favouring more regular compound-packing of two different molecules in parallelized bundles is much less well known. One reason may be that ordering within these bundles in liquid crystals is much less complete anyhow than in a crystal lattice regular in three dimensions, so that fitting together of appropriate segments of the dissimilar molecules tends to be less thermodynamically probable. Examples of molecules whose melts do not exhibit liquid crystal phenomena in one component systems but which do exhibit this additional ordering in mixture include (Dave and Lohar, 1959) are $p$-anisal $+ p$-phenetidine and $p$-nitrobenzal $+ p$-phenetidine in the range of compositions $21 \cdot 5$–$67 \cdot 5\%$ of the first component. Other aspects of mixed liquid crystals have been investigated by Dave and Dewar (1954) and McLaughlin, Shakespeare, and Ubbelohde, 1964.

## 14.7 Transport studies

In a number of cases transport properties such as viscosity and heat transfer have been measured on the melts from $T_f$ to above $T_{c_1}$. Marked anomalies are found in the liquid, but their complete interpretation still has not been made.

A mesomorphic fluid viewed at any moment of time contains 'bundles' of molecules with marked preferred orientations. Each 'bundle' is separated from neighbours by less strongly correlated molecules but the assembly has no permanency and represents a system of cooperative fluctuations. Neither theory nor experiment makes it clear in every case whether correlation within each bundle arises to a fairly steep maximum, or whether the modulation of properties in any liquid crystal is averaged less sharply.

### 14.7.1. *Flow properties in liquid crystals*

Flow properties of mesomorphic melts may be expected to show some unusual features associated with their domain structures (Porter and Johnson, 1963; McLaughlin, Shakespeare, and Ubbelohde, 1964; Porter, Barrall, and Johnson 1966; Ubbelohde, Michels, and Duruz, 1972; Michels and Ubbelohde, 1974). Smectic 'liquid crystals' may even show mechanical elasticity for stresses below a critical (low) value (Pershan and Prost, 1975). Cholesteric melts may show 'plug' flow (Helfrich, 1969). Unusual features of this kind are of considerable interest for the molecular interpretation of non-Newtonian flow, but they are probably too specialized to be useful as a general diagnostic indication of prefreezing structure in melts (Chapter 13). As previously discussed, hindered molecular rotation in melts can often lead to formation of *anticrystalline* clusters on approaching $T_f$; such clusters affect viscous flow but generally their structure may be only remotely related to the more ordered and often larger domains in liquid crystals.

A factor which can complicate such measurements is that the molecules, which are themselves oriented with respect to nearest neighbours in bundles in the fluid, can have these bundles subjected to further orientation as a whole fairly readily, by other force fields, particularly such as those that arise at the boundary walls of a container. Orienting magnetic or electrical fields can also be applied to the fluid as a whole (Kast, 1939, 1955). When the magnetic field is applied to the melts at right angles to the electric field, the anisotropy of dielectric polarizability of the molecules can be readily observed in favourable cases (see Carr, 1960). Wall effects are particularly striking, for example, at clean surfaces of single crystals, such as surfaces of freshly cleaved mica. In a typical nematic fluid, magnetic fields of the order of 1300–2000 gauss give practically complete orientation of the bundles in $p$-azoxyanisole; comparatively small gradients of temperature of the order of 2° per cm at 125°C have a correspondingly large disturbing effect, though the reasons are not fully understood (Stewart, Holland, and Raynolds, 1940). Even flow through a capillary can orient the bundles, so that the viscosity measured in this way may not show quite the same temperature-dependence as when a method is used, such as oscillation damping, which exerts much less marked orienting influence (Becherer and Kast, 1942). Smectic liquids, whose viscosity is very considerably greater than that of nematic liquids, are much less readily oriented by such adventitious means. Suggestive attempts to eliminate orientation effects by studying 'free' surfaces include measurements of the surface tension $\sigma_{LG}$ of azoxyanisole (Neumann, 1975).

### 14.8 Precursor effects in oriented melts

When the viscosity is measured by a capillary method the increase at the clearing point of $p$-azoxyanisole is quite sharp (Fig. 14.8). As already stated, flow viscosity is not a good general diagnostic property for mesomorphic fluids.

Information about the structure and size of the oriented groupings in the liquid crystal range of temperatures has generally been obtained only by rather indirect means. For $p$-azoxyanisole, bundles are estimated to contain approximately $10^6$ molecules (Stewart, Holland, and Reynolds, 1940). Formation of liquid crystals above $T_f$ becomes obvious for special classes of molecules mainly because of their optical properties, as is presumably connected with the comparatively large size of the bundles and their good orientation. The fairly sharp clearing points $T_{c_1}$ that are observed in these melts may likewise be attributed to the fact that the bundles are comparatively large; cooperative ordering is lost quite steeply as the temperature rises.

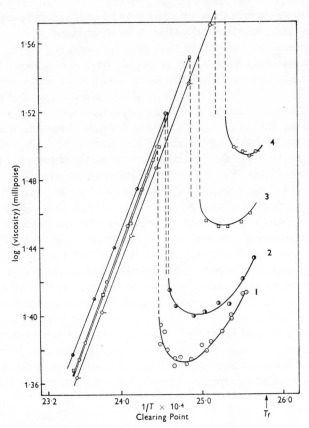

FIG. 14.8. Increase of viscosity at clearing point
of *p*-azoxyanisole. (1) 100% *p*-azoxyanisole; (2)
99·15% *p*-azoxyanisole, 1·85% phenanthrene; (3)
97·16% *p*-azoxyanisole, 2·84% phenanthrene; (4)
95·21% *p*-azoxyanisole, 4·79% phenanthrene

Difficulties in purification, and inadequacies in the methods of observation
most commonly used, leave it uncertain to what extent pretransition effects
precede discontinuous changes for the transformations crystal → liquid crystal, or liquid crystal → isotropic liquid. The order of magnitude of any precursor phenomena likely to become apparent can be estimated from the
elementary theory of cooperative fluctuations (Chapters 4 and 11). In its
present form this introduces an 'interfacial tension $\sigma_{12}$' between the two
'phases'. Unfortunately in many examples of pretransition phenomena $\sigma_{12}$
cannot be readily estimated from other sources independently. However,
structural considerations about the transformation liquid crystal → isotropic
liquid, suggest that $\sigma_{12}$ is likely to be particularly small. Important and quite

prominent pretransition effects may therefore be expected, though theories for their calculation may not be very realistic.

Precursor effects have been inferred on the basis of thermodynamic studies, such as volume measurements at different temperatures (Hoyer and Nolle, 1956; Porter and Johnson, 1963; McLaughlin, Shakespeare, and Ubbelohde, 1964). Even hysteresis has been reported in some cases (see Chapter 4), though effects of impurities in solutions may not have been fully characterized in this connection (Runyan and Nolle, 1957).

Other physical properties which yield evidence about pretransition phenomena include optical double refraction in streaming p-azoxyanisole (Tolstoy and Fedotov, 1947) and the anomalous propagation of ultrasonic waves (Curran, 1955; Gabrielle and Verdini, 1955; Hoyer and Nolle, 1956). Infrared measurements between $T_f$ and $T_{c_1}$ for nematic liquid crystals indicate gradual changes over as much as 25° (Bulkin, Grunbaum, and Santoro, 1969).

Incomplete ordering in the nematic phase of liquid crystals may also be deduced from dielectric studies (Maier and Meier, 1961). Nuclear magnetic resonance measurements (Weber, 1958) on p-azoxydipropyl phenol ethers confirm the findings for other members of the series

$$(C_nH_{2n+1})O.C_6H_4{-}N{=}N{-}C_6H_4.O(C_nH_{2n+1})$$
$$\downarrow$$
$$O$$

Tests on the 'rotational relaxation' of molecular axes of p-azoxyanisole have been made with the liquid crystals rotating in a strong magnetic field. There are differences of theoretical significance near $T_f$, but these have not been characterized sufficiently closely to give detailed information about changes of local ordering in the liquid crystal range (Lippmann, 1958).

### 14.8.1. Precursor phenomena above $T_{c_1}$

Nuclear magnetic resonance studies give particularly clear evidence of precursor phenomena (cluster formation) in the change from isotropic liquid to liquid crystal, for p-azoxyanisole, and for the benzoate, n-butyrate and propionate of cholesterol (Runyan and Nolle, 1957). Figure 14.9 illustrates this effect over about 1·5°C in the spin-lattice relaxation time of p-azoxy-anisole when this melt is cooled down to $T_{c_1}$.

At the transition from clear melt to liquid crystal, the resonance line broadens and shows 'shoulders' which change progressively, pointing to increasing broadening as the melt is supercooled to 102·0°C. ($T_f = 114·5°C$). Nuclear magnetic resonance data suggest that various rotational isomers of the molecule are present in the bundles, by virtue of the fact that it is not fully stretched. Similar decreases are found for the cholesterol compounds. Precursor effects have also been found from studies of optical scattering in

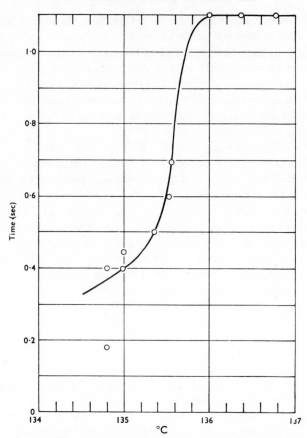

FIG. 14.9. NMR spin-lattice relaxation times near transition in *p*-azoxyanisole melts. Reproduced by permission of the American Institute of Physics

isotropic melts of ionic crystals on approaching the transition
isotropic → mesomorphic (Duruz and Ubbelohde, 1975).

## REFERENCES

N. I. Afaras'eva, V. M. Bulakov, and G. N. Zhizkin (1976) *J.E.T.P. Lett.* **23,** 461.
H. Arnold (1966) *Mol. Cryst.* **2,** 63.
H. Arnold and P. Roediger (1966) *Z. Phys. Chem. (Leipzig)* **231,** 3.
B. Bahadur, J. Prakash, K. Tripathi, and S. Chandra (1975) *Acustica* **33,** 217.
E. Baum, D. Demus, and H. Sackmann (1970) *Wiss. Z. Halle* **70,** 37.
G. Becherer and W. Kast (1942) *Ann. Phys.* **41,** 355.
G. H. Brown, J. W. Doane and V. D. Neff (1970) *Critical Reviews in Solid State Sciences* p. 303.

B. J. Bulkin, D. Grunbaum, and A. V. Santoro (1969) *J. Chem. Phys.* **51**, 160.

E. F. Carr (1960) *J. Chem. Phys.* **32**, 621.

S. Chandrasekhar (1976) *Rep. Progr. Phys.* **39**, 615.

S. Chandrasekhar, R. Shashidhar, and N. Tara (1970) *Mol. Cryst. Liquid, Cryst.* **10**, 337.

P. Chatelain (1954) *Bull. Soc. Fr. Mineral. Cristallogr.* **67**, 323.

J. C. Chin and V. D. Neff (1975) *Mol. Cryst. Liquid Cryst.* **31**, 69.

B. Chu, C. S. Bak, and F. L. Lin (1972) *Phys. Rev. Lett.* **28**, 1111.

D. R. Curran (1955) *J. Acoust. Soc. Am.* **27**, 997.

J. S. Dave and M. J. Dewar (1954) *J. Chem. Soc. London* 4616; *Proc. Nat. Acad. Sci., Wash.* **32A**, 105.

J. S. Dave and J. M. Lohar (1959) *Chem. Ind.* 597.

D. Demus, M. Sackmann, and K. Seipert (1970) *Wiss. Z. Halle* **19**, 47.

Hl. De Vries (1950) *Acta Cryst.* **4**, 219.

J. Duruz and A. R. Ubbelohde (1972) *Proc. Roy. Soc., Ser A* **330**, 1.

J. Duruz and A. R. Ubbelohde (1975) unpublished observations.

D. Dvorjeski, V. Volterra and E. Wiener–Avnear (1975) *Phys. Rev.* **12A**, 681.

M. Feyz and E. Kuss (1974) *Ber. Bunsenges, Phys. Chem.* **78**, 834.

L. Gabrielli and L. Verdini (1955) *Nuovo Cimento (Ser. 10)* **2**, 526.

S. Grande, S. Limmer, and A. Losche (1975) *Phys. Lett.* **54A**, 69.

G. W. Gray (1962) *Molecular Structures and Properties of Liquid Crystals*, Academic Press, London.

G. W. Gray and Brynmọr Jones (1953) *J. Chem. Soc. London* 4179.

G. W. Gray and K. J. Harrison (1971) *Symp. Faraday Soc.* No 5, p. 54.

W. Helfrich (1969) *Phis. Rev. Lett.* **23**, 372.

G. Heppke and F. Schneider (1975) *Z. Naturforsch.* **30A** 1640.

W. A. Hoyer and A. W. Nolle (1956) *J. Chem. Phys.* **24**, 803.

J. F. Johnson, R. S. Porter and E. M. Barrall (1969) *Mol. Cryst. Liquid Cryst.* **8**, 1.

W. Kast (1939) *Z. Electrochem.* **45**, 184.

W. Kast (1955) *Z. Angew. Chem.* **67**, 592.

H. Kelker and R. Hatz (1974) *Ber. Bunsenges. Phys. Chem.* **78**, 819.

W. Klement and L. H. Cohen (1974) *Mol. Cryst. Liquid Cryst.* **27**, 359.

K. K. Kobayashi (1971) *Mol. Crys. Liquid Cryst.* **13**, 137.

C. Kreutzer (1938) *Ann. Phys.* **33**, 192.

W. J. Lin, P. H. Keyes, and W. B. Daniels (1974) *Phys. Lett.* **49A**, 453.

H. Lippmann (1958) *Ann. Phys. (Leipzig)* **1**, 157.

H. Lippmann and K. H. Weber (1957) *Ann. Phys.* **20**, 265.

E. McLaughlin, M. A. Shakespeare, and A. R. Ubbelohde (1964) *Trans. Faraday Soc.* **60**, 25.

W. Maier and G. Englert (1957) *Z. Phys. Chem. (Frankfurt am Main)* **12**, 123.

W. Maier and K. Markau (1961) *Z. Phys. Chem, N.F.* **28**, 190.

W. Maier and G. Meier (1961) *Z. Naturforsch.* **162**, 262.

W. Maier and A. Saupe (1959) *Z. Naturforsch.* **14A**, 882.

H. J. Michels and A. R. Ubbelohde (1974) *Proc. Roy. Soc., Ser. A* **338**, 447.

A. W. Neumann (1975) *Mol. Cryst. Liquid Cryst.* **27**, 23.

P. S. Pershan and J. Prost (1975) *J. Appl. Phys.* **46**, 2343.

R. S. Porter and J. F. Johnson (1963) *J. Appl. Phys.* **34.** 51, 55.

R. S. Porter, E M. Barrell, and J. F. Johnson (1966) *J. Chem. Phys.* **45**, 1452.

A. Rapini (1973) in *Progress in Solid State Chemistry* **8**, 337 Properties and Applications of Liquid Crystals (Pergamon).

W. R. Runyan and A. W. Nolle (1957) *J. Chem. Phys.* **27**, 1081.

H. Sackmann and D. Demus (1969) *Fortsehritte Chem. Forsch.* **12,** 349.
A. Saupe and W. Maier (1961) *Z. Naturforsch.* **16A,** 816.
V. K. Semenchenko, V. M. Byankin, and V. Ya. Baskakov (1975) *Sov. Phys. Crystallogr.* (English Transl.) **20,** 111.
T. Shinoda, Y. Maeda, and H. Enokido (1974) *J. Chem. Thermodynamics* **6,** 921.
M. J. Stephen and J. P. Straley (1974) *Rev. Mod. Phys.* **46,** 617.
G. W. Stewart, D. O. Holland, and L. M. Reynolds (1940) *Phys. Rev.* **58,** 174.
N. A. Tolstoy and L. N. Fedotov (1947) *J. Exp. Theoret. Phys.* **17,** 564.
A. R. Ubbelohde, H. J. Michels, and J. J. Duruz (1972) *J. Phys. E.* **5,** 283.
V. A. Usol'tseva and L. G. Chistyakov (1963) *Russ. Chem. Rev.* **32,** 495.
J. van der Veen, W. H. de Jeu, A. H. Grobben and J. Boven (1972) *Mol. Cryst. Liquid Cryst.* **17,** 291.
K. H. Weber (1958) *Z. Naturforsch.* **13A,** 1098.
K. H. Weber (1959) *Z. Naturforsch.* **149,** 112.
G. Weygand and R. Gabler (1938) *Ber. Dt. Chem. Ges.* **71,** 2399.
G. Weygand and R. Gabler (1939) *Naturwiss.* **27,** 28.
G. Weygand and R. Gabler (1941) *Z. Phys. Chem.* **48B,** 148.
G. Weygand and W. Lanzendorf (1938) *J. Prakt. Chem.* **151,** 221.
C. Wiegand (1957) *Z. Naturforsch.* **12B,** 512.

# 15. RATE PROCESSES IN THE SOLID–LIQUID TRANSITION

## 15.1 Nucleation and crystal growth

Comment has already been made about the general observation that it is possible to supercool melts (and solutions) often by many tens of degrees, whereas few if any instances are known of the superheating of crystals. This can be interpreted on a structural basis, by recognizing that two quite different rate processes play leading but successive roles in the solid–liquid transition. Often these are described respectively as crystal nucleation from the melt and crystal growth.

Formal discussion of either of these rate processes encounter difficulties because they both involve transfer of molecules from one condensed state, i.e. the melt, to another condensed state, i.e. the crystal. In addition to special problems presented by the structures of these two states, the topology and structure of the melt–crystal interface, which is often assumed to have a controlling influence on rates, is seldom precisely known.

Nucleation kinetics is discussed in later sections. Classical discussions of kinetics of crystal growth usually concentrate on the crystal–melt interface, though in fact special features of all three states in the transfer of molecules may be relevant. Probably most neglected has been the structure of the 'free melt', i.e. at some distance from any interface with the crystals. This takes its simplest form for the case when melting involves crash injection of only one type of defect, so that the molten state is merely a highly defective version of a single, well-defined crystal lattice. Such ideally simple melts have been named 'quasicrystalline' in preceding chapters. More complex melts can however arise through the presence of microdomains derived from more than one crystal lattice, as well as from the presence of *anticrystalline* domains. Figure 10.1 (Chapter 10), which has already been discussed for network melting, is applicable generally.

Progressive advancement of crystal faces into a melt below $T_f$ is sometimes considered in terms of the successive location of one molecule after another at lattice sites of the interface, in so-called Kossel–Stranski kinetics. This approach involves detailed consideration about the 'roughness' of the crystal–melt interface and it serves to illustrate various fundamental growth features discussed below (see also Jackson, 1967). However, although growth of crystals from the vapour or from very dilute solutions may perhaps be given adequate description in terms of molecule by molecule location, it is not really suitable for melts. It can be argued that maximum rates in any such location

processes are often much too slow to account for observed rates of advancement of crystals into their melts (Peibst, 1967). Thus, velocities of an advancing crystallization front may range from values around $10^{-2}$ cm s$^{-1}$ in many
melts, up to even several m s$^{-1}$. To add one layer of molecules to 1 cm$^2$ of
interface involves locating about $10^{15}$ molecules and must be completed in
about $10^{-6}$ s if the crystal front as a whole is to advance at a speed of
$10^{-2}$ cm s$^{-1}$. To achieve this speed by individual location, each molecular
operation if it occurs singly needs to be completed within $10^{-21}$ s. However,
no periodic process in a crystal can occur with a frequency much faster than
the Debye crystal frequency; for a Debye temperature of 200 K the reciprocal
frequency is about $10^{-12}$ s. Apparently, $10^8$ atoms cm$^{-2}$ must thus be added
with cooperative simultaneity. Instead of operations of locating individual
molecules one at a time, elimination of excess defects would appear to occur
cooperatively. In melts of quasicrystalline structure this could perhaps take
place by movement of dislocations. Possibilities for melts of more complex
structure are briefly discussed by Peibst (1967).

Instead of focussing attention on a hypothetical crystal–melt interface one
molecule thick, it is however more appropriate to consider an interface zone
possibly several molecules thick whose thickness remains to be determined
from case to case, and which separates the fully stabilized crystal (with
equilibrium defect concentration $n_S/N$) from the free melt (with the much
higher equilibrium defect concentration $n_L/N$). Described in this way, the
interface zone is less defective than the free melt but more defective than the
stabilized crystal. Any gradient of defect concentration in the zone may
depend on how greatly the free melt differs in complexity from the hypothetical quasicrystalline version referred to above. When the presence of
anticrystalline clusters with fairly strong bonding needs to be considered (e.g.
in network melts, Chapter 10), their slow rate of break-up when entering the
transition zone may delay and sometimes even inhibit crystal growth altogether. Such vistas of crystal advancement involving 'cooperative' operations
for locating the molecules into the crystal lattice can be treated to give formal
kinetics resembling classical treatments (see Peibst, 1967), but they also show
that new interpretations must be given to parameters such as the often used
'interface energy' $\sigma_{SL}$. This warrants further scrutiny, particularly when an
interface extends over many molecular layers.

### 15.2  Classical kinetics at the solid–liquid interface

Crystal growth from a melt can proceed by transfer of molecules at an
interface between regions ordered in a crystalline lattice, and regions of the
more disordered melt, whenever the temperature gradient is such that the
solid is cooler than the melt. In principle the interface as defined thermodynamically is a surface which must be at $T_f$, but there is no *a priori* way of

determining over how many layers of molecules the transition zone will extend structurally, starting from layers in the melts whose structure disregards the presence of crystal (i.e. the free melt) and ending in crystal lattice whose properties disregard the presence of melt (i.e. the established crystal). As explained, for different types of crystal, particularly if the melts are not simply quasicrystalline different 'thicknesses' of interface region may be anticipated. Conditions must permit the dissipation of the heat of crystallization at a solid–liquid interface for growth to proceed. For many substances rate of crystallization from the melt is high, and the conditions must be selected somewhat narrowly to permit observations under reproducible conditions. So far as is known the growth process is reversed without perceptible hysteresis so long as both crystal and free melt remain present, when the temperature of the interface region is maintained slightly above $T_f$ instead of slightly below it. Even when the melt is absent initially, many investigations show that it is in fact difficult to superheat a crystal appreciably above its melting point; this implies that the mechanism of the process crystal $\rightarrow$ melt generally occurs at rates which are so rapid that under ordinary conditions any excess temperature of the melt in immediate contact with the solid remains very small. Even for melts of fairly complex structure, such as gallium, melting involves the injection of numerous defects as a first step into the crystal, and appears to be dislocation-assisted (Abbaschian and Ravitz, 1975). With very viscous melts appreciable superheating has however been claimed, for example up to 300°C for quartz and up to 40°C for cristobalite. Probably slow melting may be expected quite generally when melting involves a network-breaking mechanism (Chapter 10).

Simple formal kinetics of crystal growth may be briefly reviewed with the reminder that it may be difficult to find experimental examples that are as simple as the formal equations would suggest. In fact, these equations are better suited for solutions than for melts. Neglecting this consideration, the rate of advance of a crystal interface into a melt has been formulated in an elementary expression

$$U = (fD/\lambda)(1 - \exp(\Delta g/kT)) \qquad (15.1)$$

where the kinetic constant $D$ for interface motion in $cm^2\,s^{-1}$ is equated with the coefficient of self-diffusion in the melt, and the fraction of interface sites at which unit steps in the rate process can occur is approximately unity. Assuming the interface zone betwen free melt and established crystal is thin, the jump distance may be of the order of the crystal lattice spacing (for example, in crystallization of quartz the Si—O bond) and the 'driving free energy' is given by

$$\Delta g = H_f(1 - T/T_f) \qquad (15.2)$$

where $H_f$ is the molar heat of fusion. Assuming the Stokes–Einstein relation $D = kT(3\pi\lambda\eta)$, reasonable agreement is found between this expression and the microscopic observations. Observed rates tend to be higher than calculated, but, in the cases considered, catalysts, such as dissolved water or oxygen, may increase the rate process (Ainslie, Mackenzie, and Turnbull, 1961).

For a number of homologous n-paraffins ($C_4$—$C_8$) and n-alcohols ($C_2$—$C_6$) it is interesting to note that observed velocities of crystallization near $T_f$ prove to be of the same order of magnitude (Sackman and Sauerwald, 1950). Presumably, melts whose structure is far from quasicrystalline would have thicker interface zones and bring in operations which might require somewhat longer relaxation times, since the transfer of molecules from liquid to crystal must first permit molecular units to separate from clusters in the melt before they can be packed regularly one by one at the growing crystal interface. Abnormally difficult crystal growth has been interpreted formally in terms of activation free energies (see below) but these may not be meaningful. Unfortunately, experimental data are still very restricted. Most examples of crystal growth have been studied using solutions rather than melts. As yet, characteristic features of interface advancement in pure melts have been reported only in a few instances.

### 15.2.1. *Rates of advancement of the crystal–melt interface*

Solidification of a pure melt belongs to the class of rate processes known as interface reactions, but, as already stated, if the interface zone between free melt and established crystal is thick, complex situations can arise. Factors which may have been formally identified as modifying elementary expressions for growth rate include (i) the thermodynamic driving force or difference in free energy $\Delta g = G_L - G_S$ between solid and melt which may be appreciable even near $T_f$, (ii) the roughness or perfection of the crystal interface, with any corresponding modifications of the surface free energy, (iii) the mobility of molecules in the zone and at the interface as determined by the molecular shape and dimensions, and (iv) the disturbances resulting both from impurity molecules accumulating at the interface (i.e. growth poisons), and from thermal resistance in the dissipation of heat of crystallization. Various general discussions have been put forward (Jackson, 1967; Peibst, 1967; Kirkpatrick, 1975).

As stated above, crystal growth by cooperative elimination of defects from quasicrystalline melts has not been given any precise kinetic formulation; it appears to be highly likely from semiquantitative arguments. If ever individual location of molecules one by one may be assumed to operate, three distinct mechanisms have been proposed for interface advancement.

(1) Molecular rearrangement of the disordered liquid molecules may occur anywhere at the (rough) interface. For small supercooling the rate of

advancement $U = C_1 \Delta T$ where $C_1$ is determined by physical parameters of the system (Wilson, 1900; Frenkel, 1932). Melts with low heats of fusion such as quartz tend to have rougher interfaces than when $H_f$ is large (Kirkpatrick, 1975).

(2) Growth occurs only at molecular steps on a (perfectly smooth) surface. Lateral growth of each step is fast compared with the rate controlling process, which in this model is directly linked with the nucleation of new steps, that form 'pill boxes' of unimolecular height. The rate of advancement is controlled by the energy barrier for the formation of such a pill box and $U_2 = C_2 \exp(-C_3/\Delta T)$ (Volmer and Marder, 1931; Kaischew and Stranski, 1934). $C_3$ is an empirical constant which may be weakly dependent on $\Delta T$ and $C_2$ is determined by physical parameters of the system.

(3) The surface may be partially rough, for example owing to the presence of screw dislocations, one or more molecular diameters in height, intersecting it. During growth, such steps wind themselves into spirals without suppressing the central lattice disturbance. For an isolated spiral the growth law is $U = C_4(\Delta T)^2$ where $C_4$ is analogous with $C_1$ and $C_2$. Interaction between dislocations, or imperfect removal of heat of crystallization, can lead to a less rapid dependence on $\Delta T$. For a number of melts, approximately $U = C_4(\Delta T)^{1.7}$ cm s$^{-1}$ (Hillig and Turnbull, 1956). Experiments with zone-refined benzene ($T_f = 5.52 \pm 0.002°C$) give an average value of $C_4 = 0.108 \pm 0.008$ with an exponent $1.64 \pm 0.06$ for $\Delta T$, in good general agreement with the above. Rather higher growth rates have been observed in silver capillaries ($C_4 = 0.09$) in which the heat of crystallization was removed more effectively (Hudson, Hillig, and Strong, 1958). With benzene no favoured direction of crystal growth could be detected during the experiments.

In studying the growth of a crystal from its melt, the irregular packing of the units of structure at the interface is difficult to define experimentally; this is one limitation to the degree of refinement that can be given to theories of 'velocity of crystallization' at various crystal faces. Effects of clusters in the melt on the activation energy are discussed on pp. 349, 407.

Instead of attempting to describe the complex structural situation at a solid–melt interface whose thickness remains to be determined, more direct information can be obtained about a simpler but related problem which refers to the thin interface between condensed matter and vapour. Comparative rates of growth or evaporation of a liquid droplet or of a crystal from the vapour have been studied in the neighbourhood of $T_f$. For melts with fairly simple structure such as mercury, the accommodation coefficient for atoms striking the liquid surface from the vapour is effectively unity in the absence of impurities. Nevertheless, crystals of mercury can show rates of evaporation corresponding with accommodation coefficients higher than unity by 20–25%. This can most readily be attributed to a pronounced roughening of the nominal crystal faces by evaporation from preferred points (Neumann and Ramolla, 1961).

### 15.2.2. Dielectric charge accumulation

On crystallizing certain melts, charge build-up effects are observed whose origin is not finally elucidated. For example, electrical charges become trapped in the crystals (thermodielectric) on freezing naphthalene, possibly because polar impurities are squeezed out on freezing (Mascarenhas and Freitas, 1960).

### 15.2.3. Formation of stepped and spiral structures at interfaces

As with crystal growth from supersaturated solutions, crystal growth of certain types from melts supercooled below $T_f$ proceeds more rapidly at a built-in spiral defect, than at a uniform crystal face. Stepped and spiral structures can favour growth to a predominant degree in certain cases, but the wealth of information is much less than with solutions and will not be detailed here (see below).

## 15.3 Rhythmic growth from melts

Many examples of periodic precipitation of crystals from solution are known. Rhythmic effects in crystallization have also been reported for a number of melts containing polar molecules. However, substances for which such effects have been observed include many melts that must have been difficult to purify rigorously. Effects resulting from segregation of impurities are difficult to eliminate and may help to account for the periodicity of behaviour. For example, melts showing rhythmic growth include mannitol, ammonium lithoxanthate, ethyl tetracarboxylate, benzophenone, coumarin, sulphur, $p$-ethoxybenzylideneaminophenyl propionic acid, salol, methyl salol, methyl salicylic acid, $p$-toluonitrile, myristic acid, lauric acid, undecoic acid, decoic acid, antipyrine, sulphanol, malonamide, benzoin, benzonaphthol, santonin, $m$-diethylamine phenol, diaminonaphthalene, asparagine salts, benzoic acid, salicylic acid, hippuric acid, benzil, menthol, acetanilide, $m$-dinitrobenzene, piperonal, terpin hydrate, vanillin, phenyl benzoate, phenanthrene, cinnamic acid, thymol, and 3,5-dichloro-4-methyldiphenyl (Dippy, 1932). If effects of impurities may be discounted and the melts behave truly as one-component systems, possible origins of rhythmic growth are less easy to suggest. Possibly a film of melt is drawn over the surface of the crystals and does not immediately crystallize. This might be expected particularly when melts contain a large proportion of anticrystalline clusters (Chapter 13). These must disintegrate or dissociate before the individual molecules can be located afresh onto the surfaces at the correct crystal lattice sites. However a more probable explanation is that impurities such as $H_2O$ dissolved in the melts become segregated by the proces of crystallization, which thus introduce rhythmic fluctuations in concentration. In fact, this would imply that crystallization is being observed not from the pure melt but from a highly concentrated solution.

## 15.4 Dendritic growth

Instead of advancing more or less smoothly, a solid growing into a melt sometimes assumes striking dendritic or frond-like structures. Most studies on dendritic growth refer to supersaturated solutions. Conditions under which dendritic growth of crystals may occur in a supercooled melt, even in a one-component system such as a pure metal, have been discussed (Bolling and Tiller, 1961). Some anisotropy of surface free energy $\sigma_{SL}$ or of molecular fixation rates appears to be necessary before dendritic growth can occur, though physical conditions are not completely defined by current theory. Structural features such as the 'branching interval' of dendritic growth thus cannot as yet be predicted from other properties of the melts.

## 15.5 Spontaneous nucleation in melts

### 15.5.1. *Heteronucleation versus homonucleation*

One difficulty in the development of the study of nucleation of the transition melt → crystal has been that only very few nuclei are needed to transform large masses of melt at temperatures below $T_f$. A variety of solid impurities (heteronuclei) in very small concentration can often be found to start crystallization of entire continuous regions of a supercooled system; they can be quite difficult to eliminate. Their role is akin to that of heterogeneous catalysts, and should not be confused with homonucleation catalysts discussed below. Exclusion of such heteronuclei is essential in order to obtain significant information about spontaneous homonucleation of melts. When large extended volumes of melt are used, accidental heteronucleation is almost impossible to prevent. Currently, it is therefore usual to study homonucleation using a large number of supercooled separated droplets of the melt, in order to avoid practical interference from the small number of impurity nuclei which would be amply sufficient to act as catalysts for nucleation of large continuous regions. When heteronuclei are present in the melt in bulk, crystallization around them spreads with rapidity throughout the whole mass. However, if the melt is first subdivided into a very large number of separate droplets, catalytic nucleation followed by crystallization of a few of these, starting at impurity centres, does not influence the remaining droplets. It is interesting to note that when the calculated volume required for a nucleus (Defay and Dufour, 1962) is larger than the actual volume of an aerosol droplet, nucleation is hindered for purely volumetric reasons. However, normally the droplets are found to crystallize spontaneously by homonucleation at a critical supercooling limit $T_N$ lying within a quite narrow temperature interval for all the droplets, and reproducible in successive experiments. In one example, $T_N$ has been observed for metal droplets deposited on quartz slides freshly blown to prevent any surface nucleation (Turnbull, 1950). For other liquids $T_N$ has been evaluated using droplets as an aerosol suspension. Emulsions using organic oils have been used for droplets of low melting metals 5–10 $\mu$m in diameter (Rasmussen and Loper, 1975).

**Table 15.1. Maximum supercooling of metal droplets 50 m in size**

| Metal | $T_f$ (K) | $S_f$ (e.u.) | $\rho_N = T_N/T_f$ | $\sigma_{SL}$[a,b] (erg cm$^{-2}$) |
|-------|-----------|--------------|---------------------|------------------------------------|
| Hg    | 234·3     | 2·38         | 0·753               | 24·4                               |
| Ga    | 303       | 4·42         | 0·750               | 55·9                               |
| Sn    | 505·7     | 3·41         | 0·792               | 54·5                               |
| Bi    | 544       | 4·60         | 0·834               | 54·4                               |
| Pb    | 600·7     | 2·04         | 0·867               | 33·3                               |
| Sb    | 903       | 5·28         | 0·850               | 101                                |
| Al    | 931·7     | 2·74         | 0·860               | 93                                 |
| Ge    | 1231·7    | 4·94         | 0·816               | 181                                |
| Ag    | 1233·7    | 2·19         | 0·816               | 126                                |
| Au    | 1336      | 2·27         | 0·828               | 132                                |
| Cu    | 1356      | 2·29         | 0·826               | 177                                |
| Mn    | 1493      | 2·31         | 0·794               | 206                                |
| Ni    | 1725      | 2·43         | 0·815               | 255                                |
| Co    | 1763      | 2·08         | 0·813               | 234                                |
| Fe    | 1803      | 1·97         | 0·856               | 204                                |
| Pd    | 1828      | 2·25         | 0·818               | 209                                |
| Pt    | 2043      | 2·30         | 0·819               | 240                                |

[a] Calculated $\sigma_{SL}$ values are not very sensitive to the size of droplet assumed.
[b] For indium $\sigma_{SL}$ is estimated to be 101 dyne cm$^{-1}$ and an alternative estimate for Sn is 69 dyne cm$^{-1}$ (Herman and Curzon, 1974).

Direct observations of $T_N$ can be empirically correlated with $T_f$, in terms of the dimensionless ratio

$$T_N/T_f = \rho_N$$

Values are as recorded in Tables 15.1 and 15.2. To interpret these findings, a statistical theory of formation of crystal nuclei is required. Nucleation theories often bring in an interface surface energy $\sigma_{SL}$ which is further discussed below. In the present state of knowledge, $\sigma_{SL}$ is difficult to define independently and these calculations are best regarded as one means of

**Table 15.2 *Maximum supercooling of molten salts***

| Salt | $\rho_N = T_N/T_f$ | $\sigma_{SL}$ (erg cm$^{-2}$) |
|---|---|---|
| LiF | 0·79 | 181 |
| LiCl | 0·79 | 89 |
| LiBr | 0·76 | 46·6 |
| NaF | 0·78 | 179 |
| NaCl | 0·84 | 84·1 |
| NaBr | 0·84 | 71·5 |
| KCl | 0·84 | 65·6 |
| KBr | 0·83 | 57·1 |
| KI | 0·85 | 47·2 |
| RbCl | 0·84 | 55·7 |
| CsF | 0·86 | 60·2 |
| CsCl | 0·83 | 42·3 |
| CsBr | 0·82 | 51·7 |
| CsI | 0·78 | 54·7 |

defining $\sigma_{SL}$ in order to examine it in relation to other properties of crystal and melt. For example (Turnbull and Fisher, 1949; Turnbull, 1950), the number of nuclei $I$ formed s$^{-1}$ cm$^{-3}$ by spontaneous nucleation is calculated to be

$$I = n(kT/h)\exp-[\kappa\sigma_{SL}^3/(\Delta G_v)^2 kT]\exp[-\Delta G_a/kT] \qquad (15.3)$$

In this expression, $n$ is the number of atoms in the portion of melt considered, $\kappa$ is a shape factor, $\Delta G_v$ is the difference in free energy at $T$ per cm$^3$ of the solid and liquid phases, when both have infinite volume, and $\Delta G_a$ is the free energy of activation for transport of a molecule across the solid–liquid interface. Crystal growth studies indicate that $\Delta G_a$ is of the same order of magnitude as the activation required for viscous flow. In melts of normal metals, $\exp[-\Delta G_a/kT]$ is about $10^{-2}$. On this basis $I \sim 10^{33}\exp[-\kappa\sigma_{SL}^3(\Delta G_v)^2 kT]$ s$^{-1}$ cm$^{-3}$. In this expression, the temperature-dependence of $I$ is extremely steep. Measurable rates of nucleation in fact lie within a very narrow range of temperature, and a limiting value $T_N$ practically coincides with the temperatures above which frequencies are so low as to be unobservable in practice. If nuclei are to be taken to be spherical in shape, $\kappa = 16/3$. Making plausible assumptions about $\Delta G_v$ for metals (Turnbull, 1950), values of $\sigma_{SL}$ are as calculated in Table 15.1.

For liquid gallium (Bosio and Defrain, 1962), evidence has been advanced that an oxidation product, which it is difficult to remove completely, can lead to heteronucleation, so that the temperature of maximum supercooling of the melt is found to depend on the temperature to which it has previously been heated above $T_f$.

$T_f \sim 302\cdot8$ K, but with very careful removal of oxide (by dilute HCl) $T_N \sim 230$ K. The fact that for this melt the ratio $\rho_N = 0\cdot75$, i.e. somewhat below normal, points to abnormal complexities of structure in the melt, which is also indicated by other evidence (Chapter 8). These findings may however, be complicated by the fact (Defrain, 1960) that gallium crystallizes spontaneously in a metastable structure with $T_f \sim 256 \cdot 8$ K with a heat fusion of $9 \cdot 09$ cal g$^{-1}$ giving $S_f = 2 \cdot 47$ e.u.

Studies on the spontaneous freezing of aerosols of droplets condensed from salt vapours point to a remarkable uniformity of values for the ratio $T_N/T_f$, for many simple melts (Table 15.2) (Buckle and Ubbelohde, 1960, 1961). The empirical rule proposed by these authors is

$$T_N/T_f = 0 \cdot 8$$

For some halides of Group B metals, direct condensation vapour → crystal appears to take place more readily than the circuitous process vapour → liquid droplet → crystal. These compounds include salts of Cu, Bi and the halides of lead $PbF_2$, $PbCl_2$, and $PbI_2$ (Buckle and Hooker, 1962). In these salts, approximate numbers of ion pairs in the crystal nuclei may be plausibly estimated to range between 60 and 150.

For ice, $T_N/T_f = 0 \cdot 8$ (Mossop, 1955); for a number of organic liquids, values of $\sigma_{SL}$ have been estimated from the observed supercooling (Nordwall and Staveley, 1954; see Thomas and Staveley, 1952) as recorded in Table 15.3.

**Table 15.3. Maximum supercooling of some molecular liquids**

| Molecule | $T_f$ (K) | $T_N/T_f$ | $\sigma_{SL}$ (cal mole$^{-1}$) | $\sigma_{SL}/H_{vap}$ |
|---|---|---|---|---|
| BF$_3$ | 144·5 | 0·87 | 270 | 0·23 |
| Cyclopropane | 145·8 | 0·87 | 287 | 0·23 |
| Methyl bromide | 179·4 | 0·86 | 350 | 0·25 |
| CH$_3$NH$_2$ | 179·7 | 0·80 | 448 | 0·31 |
| SO$_2$ | 197·6 | 0·83 | 480 | 0·27 |
| CHCl$_3$ | 209·7 | 0·77 | 723 | 0·32 |
| Thiophene | 234·9 | 0·79 | 453 | 0·37 |

## 15.6 Melts which fail to nucleate spontaneously

Somewhat surprisingly, no spontaneous crystallization has been observed for a number of liquids, including krypton, $CO_2$, cyclopentane, n-heptane, toluene, pyridine, tetramethyl silicon, tetramethyl tin, $SiCl_4$, $C_2H_2Cl_2$, $CS_2$, and ethylene oxide. This problem of the failure to nucleate spontaneously is further discussed below (see 407). One difficulty in developing statistical mechanical theories of nucleation based on the concept of a surface energy $\sigma_{SL}$ of nuclei swimming in the melt is that the value of $\sigma_{SL}$ for microregions, of the size calculated by the theories described below, may not be the same as for the interface between a large area of the crystal surface in contact with its melt, since the lattice spacing in the (small) nucleus may differ from that in large crystals. When it is desirable to record this difference, the symbol $\sigma_{NL}$ should be used in place of the more general $\sigma_{SL}$. A minor addition to this real difficulty is that $\sigma_{NL}$ will depend on the indices of the crystal face.

Up to the present, no methods of evaluating $\sigma_{NL}$ by theory or experiment have been put forward, independent of the appearance of spontaneous nucleation. Even calculations of surface energies at the interface between two dissimilar crystal structures (Van der Merwe, 1963) introduces concepts of strain energy that could not readily be utilized when one of the structures is fluid melt. As indicated above, all that can be done, pending such essential cross-verification, is to evaluate $\sigma_{SL}$ from supercooling experiments, and to look for correlation with the structures of crystal and melt as derived by other means. Some welcome support for the above approach to theories of spontaneous nucleation is provided by the fact that values of $\sigma_{LG}$ calculated from the spontaneous nucleation of droplets of liquid, from the vapour, do agree with macroscopic values of $\sigma_{LG}$ observed directly (see Turnbull, 1950). As already stated, however, in a melt, the thickness of the interface zone surrounding a nucleus may be considerably greater than for the liquid–gas transition which makes the simple conventional treatment of nucleation not very suitable, particularly when the melts are not quasicrystalline.

## 15.7  Crystallization of anticrystalline melts through reconstructive changes

As already stated, growth from melts with quasicrystalline structure encounters kinetic hindrances mainly from the need to transpose the molecules so as to increase their ordering on passing from liquid to crystal. Even this process seems likely to be cooperative for the reasons given. However when the structure and texture of a melt differs in a more basic way from that in the crystal, both nucleation and crystal growth may require much more extensive reconstructive changes in the melt. Clusters or other molecular complexes in the melt have first to disaggregate so that the constituent molecules can be built in a more regular way into the crystal lattice. Molten sulphur and molten selenium are spectacular examples; many others will suggest themselves from discussions in the preceding chapters, but much more alert research is called for.

Reconstructive changes can modify both nucleation and growth of crystals:

(i) by introducing an activation free energy $\Delta G'$ for the jump of a molecule across the interface melt–finite crystal surface;

(ii) by introducing an activation free energy $\Delta G''$ for the jumping of a molecule across the interface nucleus (or cluster)–free melt.

Both (i) and (ii) could be modified by catalytic impurities, both positive and negative in their influence. For example, certain cations dissolved in silicate melts can constitute 'loose ends' to chains of silicate groups, and certain metal atom impurities might affect crystallization of molten sulphur.

Because of the need to introduce various parameters which cannot at present be calculated independently, mathematical theory as summarized in what follows is best regarded as only rather formal: Very thin interface layers are assumed, which is often not realistic.

Formal kinetics of nucleation in melts that are not quasicrystalline in structure have been discussed (Turnbull and Cohen, 1958). According to conventional nucleation theory as applied to crystals with simple units of structure (Turnbull and Fisher, 1949) the frequency $I$ of homogeneous nucleation, in the absence of heterogeneous particles, can be written

$$I = n\nu \exp[-(NW^* + \Delta G')/RT]$$

where $n$ is the number of atoms $cm^{-3}$, $\nu$ is the vibrational frequency at the interface nucleus–melt, $N$ is Avogadro's number, and $\Delta G'$ is as defined above. The 'thermodynamic' barrier $W^*$ is given by

$$W^* = \kappa \sigma_{NL}^{3} V_S^{2}/(\Delta G_{SL})^2$$

in which $\kappa$ is a numerical factor determined by the shape of the nucleus, $\sigma_{NL}$ is the interface tension nucleus–melt, $\Delta G_{SL}$ is the change in chemical potential per g atom in the macroscopic transformation melt → crystal (i.e. $\Delta G_{SL} = 0$ at $T_f$), and $V_S$ is the atomic volume of the crystal. This expression is of course subject to the uncertainties concerning the interface energy $\sigma_{NL}$ referred to above.

For a finite crystal, the linear velocity of growth of the interface is given by

$$U_c = a_0\nu \exp-[\Delta G''/RT][1 - \exp \Delta G/RT]$$

where $a_0$ is the spacing between units, and the other symbols are as above.

The usual estimation of the 'critical' supercooling temperature $T_N$ assumes this is attained when $I$ becomes $1\ cm^{-3}\ s^{-1}$. This leads to a value for the supercooling interval $\Delta T = T_f - T_N$ given by

$$\frac{\Delta T}{T_f} = \left\{ \frac{N\kappa \sigma_{NL}^{3} V_S^{2}}{H_f^{2}[RT_N \ln n\nu - \Delta G']} \right\}$$

Empirically, $\sigma_{NL} = \alpha H_f/N^{\frac{1}{3}} V_S^{\frac{2}{3}}$ where $\alpha$ is a number which ranges between $1/3$ and $1/2$ and $\beta = S_f/R$, where for simple liquids $\beta$ is of the order of unity. Then

$$\frac{\Delta T}{T_f} = \left\{ \frac{\kappa \alpha^{2} \beta RT_f}{RT_N \ln n\nu - \Delta G'} \right\}^{\frac{1}{2}}$$

Approximations (which seem reasonable) are: (a) no growth will be observed unless $U_C > 10^{-5}$ atom spacings s$^{-1}$ at $T_C$; (b) in the limiting case $\Delta G' = 0$, $\kappa = 16\pi/3n \sim 10^{23}$ atoms cm$^{-3}$, $\nu = 10^{12}$ s$^{-1}$. Thus $T_N$ will have a real value if $\alpha\beta^{\frac{1}{2}} < 0.9$; (c) $\alpha$ is very unlikely to exceed $0.5$ for simple molecules, and $\alpha\beta^{\frac{1}{2}}$ rarely exceeds $0.5$ for pure substances which do crystallize. Hence

$$\frac{\Delta T}{T_f} \sim \left[\frac{2\pi}{3} \frac{RT_f}{(80RT_N - \Delta G')}\right]^{\frac{1}{2}} \tag{15.4}$$

and $T_N$ will only have real values if

$$\Delta G' < 40RT_f$$

but not if $\alpha\beta^{\frac{1}{2}} > 0.9$. Actually, with the same approximations, $U_C$ will only exceed $10^{-5}$ atom spacings s$^{-1}$ if $\Delta G'$ or $\Delta G'' < 30$ as a limit above which no spontaneous nucleation is to be expected. When the empirical rule $T_N/T_f = 0.8$ is inserted into equation 15.4, finite but not unreasonable values must be attributed to $\Delta G'$.


### 15.7.1. Some experimental aspects of nucleation from anticrystalline melts

As has been indicated above (see Thomas and Staveley, 1952) certain liquids such as toluene fail to crystallize spontaneously and pass into a glass when sufficiently cooled. If the structure of such melts can still be regarded as quasicrystalline one possible empirical explanation is indicated in equation (15.4). Other causes of glass formation may however predominate if the melts are *anticrystalline* as indicated in previous chapters and in Chapter 16.

One useful consequence of a widely applicable general rule such as the limiting equation (Buckle and Ubbelohde, 1960, 1961) for the supercooling of melts

$$T_N/T_f \sim 0.8$$

is that it can serve to direct attention to melts whose behaviour is anomalous in not obeying this rule. Various sources of anomaly can in principle be foreseen. Thus if the bulk of the melt is largely associated into anticrystalline clusters, the equilibrium concentration of monomer may be reduced to such an extent that the sequence of the physicochemical aggregation equilibria, represented formally by the cascade of equilibrium equations

$$\ldots A_n + A \rightleftharpoons A_{n+1}$$
$$A_{n+1} + A \rightleftharpoons A_{n+2} \ldots \tag{15.4}$$

and so on, becomes extremely unfavourable for nucleus formation. This could push $T_N$ down to temperatures where rate processes have become too slow to permit easy crystallization. Conventional expressions analogous with equation (15.4) are discussed in Chapters 4 and 12, and also above, for the concentration of microregions of one phase dispersed in another phase of approximately the same free energy (i.e. Frenkel's 'heterofluctuations'). These expressions may in any case only be applied when the main bulk of the melt has a simple uniform structure, for example quasigaseous or quasi-crystalline. However, if appreciable concentrations of anticrystalline clusters are already present near $T_f$, these must compete with the crystallizable clusters which are the only crystallization nuclei. Presence and growth of anticrystalline clusters as the temperature cools would be expected to squeeze out normal nucleus formation progressively. This appears to explain the fact that those melts which readily pass into a glass, avoiding crystallization altogether, also generally already show marked prefreezing effects above $T_f$ (Chapter 13). A possible structure for such melts would appear to be a conglomerate, containing both crystallizable nuclei and anticrystalline clusters, swimming in less-ordered portions of the melt whose structure which can then be treated as a Van der Waals' condensate, which is a quasigaseous fluid. In the competitive series of polymerization equilibria of the general form

$$nA \; \rightleftharpoons \; A_n \quad \text{(anticrystalline clusters)}$$

$$nA \; \rightleftharpoons \; A'_n \quad \text{(any other polymer types)}$$

$$nA \; \rightleftharpoons \; a''_n \quad \text{(microcrystalline nuclei)}$$

only the process listed last can lead to a spontaneous conversion of the melt into an ordered crystal, at a rate determined by the kinetics of rearrangement of molecules from one cluster to the next.

Melts that are demonstrably anomalous with regard to homonucleation are only beginning to be identified. As one example, white phosphorus is a melt with exceptional properties in this respect. Despite the fact that $S_f$ is only 1·984 e.u., $T_N/T_f = 0·6$; there is no evidence of glass formation (Hildebrand and Rotariu, 1951). Its velocity of crystallization may exceed 2 m s$^{-1}$ (Powell, Gilman, and Hildebrand, 1951). It is not known whether this melt may perhaps owe its exceptional properties to an unusually low surface energy $\sigma_{SL}$ with respect to the solid, or to facile 'cooperative transfer' of molecules from melt to crystal. The melt has a high solvent power for mercury (Rotariu, Schramke, Gilman, and Hildebrand, 1951) which is itself unusual.

For n-alkanes, critical nucleation is unlikely to involve microregions that are of spherical symmetry. Models with cylindrical symmetry wtih $\sigma_{SL}$ dependent on chain length for the ends, and constant $\sigma_{SL}$ for the sides (about 6 erg cm$^{-2}$) have been proposed. The critical supercooling decreases as the chain length increases and shows odd/even variation, also found in other aspects of the melting of flexible molecules (Chapter 7).

For $C_{16}H_{34}$, $T_N/T_f = 0·95$, decreasing to the more usual value for $C_5H_{12}$ for which $T_N/T_f = 0·80$ (Uhlman, Kritchevsky, Straff, and Scherer, 1975).

## 15.8  Inverse crystallization temperatures of glasses

As inverse to spontaneous nucleation of crystallization at a temperature $T_N$ in supercooled melts, studies have been reported of the temperature $T_D$ of spontaneous crystallization of glasses, when these are warmed to progressively higher temperatures. For this purpose, vitreous states of matter are obtained by quenching vapours onto highly cooled surfaces, and determining at what temperature $T_D$ heat is evolved as a result of spontaneous crystallization (Nordwall and Staveley, 1956). It is interesting to compare $T_D$ with $T_N$ when both can be determined (see Table 15.4). Although values of $T_D$ are of obvious interest as a guide to nucleation kinetics, effects of heteronuclei cannot be so readily excluded in this technique as when aerosols of separated droplets of melt freeze spontaneously. This makes their interpretation less decisive. Furthermore, molecular arrangements in an assembly suddenly chilled may differ appreciably from those attained on more gradual cooling, particularly if the molecules are non-spherical, and may have to pack into one another during crash cooling without sufficient time for configurational adjustments.

**Table 15.4.** *Inverse crystallization temperatures $T_D$ of glasses*

| Molecule | $T_f$ | $T_N$ | $T_D$ $(K)$ |
|---|---|---|---|
| $H_2O$ | 273·2 | 233 | 105 |
| $D_2O$ | — | — | 140–160 |
| $NH_3$ | 195·5 | 155 | 92–95 |
| $CH_3Cl$ | 175·6 | 120 | 120–5 |
| $CH_3NH_2$ | 179·7 | 144 | 97–110 |
| $C_6H_6$ | 278·4 | 208 | not |
| $CH_3Br$ | 179·4 | 155 | 133 |
| Thiophene | 234·9 | 184 | 138 |
| Toluene | 178·2 | not | 123 |
| p-Xylene | 286·4 | not | 125 |
| Pyridine | 231·1 | not | 112 |
| $CS_2$ | 161·1 | not | not |
| $Sn(CH_3)_4$ | 218·2 | not | 138–139 |
| $Si(CH_3)_4$ | 174·1 | not | 119–23 |

## 15.9  Homonucleation catalysts

With the development of formal theories and of reliable experimental measurements to determine the critical formation of nuclei, it has become practicable to attempt catalysis of such transformations by the addition of impurities soluble in the melt. Negative catalysts have long been known in organic chemistry where they are known as 'nucleation inhibitors'; unfortunately, no systematic studies appear to be available. Positive catalysts for nucleation have likewise been long suspected, but only in quite general descriptive ways. Thus Fahrenheit first discovered the importance of dust for promoting crystallization of supercooled liquids (1724); generations of undergraduates struggling to crystallize their organic preparations have been told the legend about von Bayer's beard which is said to have carried crystal nuclei of all the solids ever made in his department, so that only one helpful shake was needed when the Master was standing by..

Systematic studies on homonucleation catalysts have been carried out using molten tin as the test substance (Glicksman and Childs, 1962). The extent to which this starting material can be supercooled was found to depend on previous heating above $T_f$ in a critical way. Most probably, the explanation is that heteronuclei originally present are dissolved when the melt is heated (to about $T_f + 12°C$) though the possibility of thermal decomposition of self-nuclei cannot be ruled out. With melts of tin 'sterilized' by heating above this temperature, attempts have been made to establish a sequence of metallic catalysts. Values in parentheses give the supercooling observed in °C for catalysts for the crystallization of 'sterile' melts of tin [Pt (4), Y (6, basal and prismatic faces), Ag (7), Y (8, basal face), TiC (15), MoS (17), Al (18)]. It is perhaps reassuring, in this as yet rather uncertain field of experimentation, that solids such as SiC, $Al_2O_3$, CoS, MgO and graphite did not catalyse nucleation.

Nucleation in supercooled melts of antimony under diverse conditions shows variations which likewise may be due to genuine catalytic effects (Mannchen and Hahn, 1958). However, owing to the use of bulk specimens the influence of growth factors upon these results cannot be excluded. Dependence on the degree of prior superheating is also reported. Mathematical interpretation of catalysed nucleation rates encounters difficulties in assumptions about the size and proportion of nuclei (Sundqvist, 1963; Glicksman, 1963). Presumably, if spontaneous crystallization is studied on a melt in bulk, catalysts can be far more effective, for reasons already discussed, when the melt is quasicrystalline than when it has a complex structure with a high proportion of anticrystalline clusters.

### 15.9.1. *The crystal–melt interface*

In view of the dominant role of the structure and thermodynamic properties of the material at the interface zone between crystal and melt, for both nucleation and growth, additional methods of studying this interface would be very welcome. Non-thermodynamic evidence for the change in molecular behaviour at the interface between solid and liquid is gradually being accumulated from various directions. One striking instance involves studies of the rate of chemical reactions such as free-radical recombination. Rates of polymerization resulting from the production of free radicals in various monomers, by irradiation, frequently prove to be much higher at the solid–

liquid interface than either in the bulk liquid phase somewhat above $T_f$, or in the bulk crystal somewhat below $T_f$. Similar enhanced reaction rates are found in the neighbourhood of transition temperatures $T_c$ (Chapter 4) in solid–solid transitions (Semenov, 1962). Although such enhanced reaction rates have not been fully elucidated, it seems likely that in a crystal environment (which may be assumed to be nearly free from defects) a molecule cannot rapidly dissipate excess energy of activation, because of its restricted mobility. However, this also hinders its ability to react chemically. In a melt the mobility is much greater but dissipation of excess energy occurs with much shorter relaxation times than in the crystal. At the crystal–melt interface, conditions appear to be highly favourable for the outward growth of polymer into the liquid phase; molecules in their activated state can migrate readily up to the interface, where polymerization can ensue. In crystalline benzene and crystalline diphenyl ether, enhanced effects appear to extend some way below $T_f$ (Magat, 1962) suggesting the intervention of cooperative defects in the crystals in a premelting region.

Other kinetic examples where activation energy does not become fully randomized at each reaction step appear to arise with some thermal decomposition processes in solids leading to detonation (Ubbelohde, 1955). The basic concept is that chemical energy liberated at a reaction step may not attain complete Maxwell–Boltzmann randomization before it is used again for activation processes that propagate further reaction, thus enhancing propagation from one layer of reactants to the next.

Corresponding non-random transfer of crystallization energy from one layer to the next may affect crystal advancement (if this requires activation). Further investigation of the interface zone between ordered crystal lattice and disordered melt is clearly desirable.

# REFERENCES

G. J. Abbaschian and S. F. Ravitz (1975) *J. Cryst. Growth* **28**, 16.
N. G. Ainslie, J. D. Mackenzie, and D. Turnbull (1961) *J. Phys. Chem.* **65**, 1718.
G. F. Bolling and W. A. Tiller (1961) *J. Appl. Phys.* **32**, 2587.
L. Bosio and A. Defrain (1962) *C.R. Acad. Sci., Paris* **254**, 1020.
E. R. Buckle and N. Hooker (1962) *Trans. Faraday Soc.* **58**, 1939.
E. R. Buckle and A. R. Ubbelohde (1960) *Proc. Roy. Soc., Ser. A* **259**, 325.
E. R. Buckle and A. R. Ubbelohde (1961) *Proc. Roy. Soc., Ser. A* **261**, 197.
R. Defay and L. Dufour (1962) *Inst. Roy. Met. Belg. Series* **B37**, 3.
A. Defrain (1960) *C.R. Acad Sci., Paris*, **250**, 483.
F. J. J. Dippy (1932) *J. Phys. Chem.* **36**, 2354.
D. C. Fahrenheit (1724) *Phil. Trans.* **39**, 78; *cf.* W. Ostwald (1903) *Lehrbuch der allg. Chemie Leipzig*, **2** ii, 740.
J. Frenkel (1932) *Phys. Z. Sowjet Union* **1**, 498.
M. E. Glicksman (1963) *Acta Metall.* **11**, 632.
M. E. Glicksman and W. J. Childs (1962) *Acta Metall.* **10**, 925.

R. P. Herman and A. E. Curzon (1974) *Can. J. Phys.* **52**, 923.

J. H. Hildebrand and G. J. Rotariu (1951) *J. Am. Chem. Soc.* **73**, 2524.

W. B. Hillig and D. Turnbull (1956) *J. Chem. Phys.* **24**, 914.

J. B. Hudson, W. B. Hillig, and R. M. Strong (1959) *J. Phys. Chem.* **63**, 1012.

K. A. Jackson (1967) *Current Concepts in Crystal Growth from the Melt. Progress Solid State Chem.* **4**, 53.

R. Kaischew and I. N. Stranski (1934) *Z. Phys. Chem.* **170A**, 295.

R. J. Kirkpatrick (1975) *Am. Mineral.* **60**, 798.

M. Magat (1962) *Pure Appl. Chem.* **5**, 496.

W. Mannchen and G. Hahn (1958) *Z. Elektrochem.* **62**, 926.

S. Mascarenhas and L. H. Freitas (1960) *J. Appl. Phys.* **31**, 1684.

S. C. Mossop (1955) *Proc. Phys. Soc., Lond.* **68**, 193.

K. Neumann and B. Ramolla (1961) *Z. Phys. Chem. (Frankfurt am Main)* **29**, 145.

H. J. Nordwall and L. A. K. Staveley (1954) *J. Chem. Soc., London* **224**.

H. J. Nordwall and L. A. K. Staveley (1956) *Trans. Faraday Soc.* **52**, 1207, 1061.

H. Peibst (1967) *Rev. Int. Hautes Temp. Refract.* **4**, 58.

R. E. Powell, T. S. Gilman, and J. H. Hildebrand (1951) *J. Am. Chem. Soc.* **73**, 2525.

D. H. Rasmussen and C. R. Loper (1975) *Acta Metall.* **23**, 1215.

G. J. Rotariu, E. Schramke, T. S. Gilman, and J. H. Hildebrand (1951) *J. Am. Chem. Soc.* **73**, 2527.

H. Sackman and F. Sauerwald (1950) *Z. Phys. Chem. (Leipzig)* **195**, 4.

N. N. Semenov (1962) *Pure Appl. Chem.* **5**, 496.

B. E. Sundqvist (1963) *Acta Metall.* **11**, 630.

D. G. Thomas and L. A. K. Staveley (1952) *J. Chem. Soc., London* 4569.

D. Turnbull (1950) *J. Appl. Phys.* **21**, 1022.

D. Turnbull and M. H. Cohen (1958) *J. Chem. Phys.* **29**, 1049.

D. Turnbull and J. C. Fisher (1949) *J. Chem. Phys.* **17**, 71.

A. R. Ubbelohde (1955) in *Chemistry of the Solid State*, ed. W. E. Garner, Butterworths, London.

D. R. Uhlman, O. Kritchevsky, R. Straff, and G. Scherer (1975) *J. Chem. Phys.* **62**, 4896.

J. H. van der Merwe (1963) *J. Appl. Phys.* **34**, 123.

M. Volmer and M. Marder (1931) *Z. Phys. Chem.* **154A**, 97.

H. A. Wilson (1900) *Phil. Mag.* **50**, 238.

# 16. MELTS AND GLASSES

## 16.1 Glasses as congealed melts

Because of their great technological importance, many investigations have been carried out on the structures and rates of crystallization of glasses. These have, in the main, been more *ad hoc* than fundamental in their approach. X-ray investigations show no long-range lattice correlation in glasses. One group of investigations leads to the view that glasses are structurally indistinguishable from fluid melts, though kinetically they only show creep phenomena (Plazek and Magill, 1966). Other researches lead to the suggestion that ultra-microcrystalline regions or domains of other enhanced ordering may be present in a variety of glasses (Mackenzie, 1960b).

Microregions with some kind of specific packing may be regarded as a congealed limit of conglomerate models for fluid melts (p. 343). Such models were suggested as a probable terminus for melts showing marked prefreezing anomalies, Chapter 13. It is just these melts that are particularly prone to pass into glasses on further cooling. Conglomerate models satisfy the requirements that there is no fundamental difference of structure or molecular packing between a melt and its glass. At the same time, they allow for the possibility of marked fluctuations of specific packing on passing from one microregion to the next in a melt or glass. In a way, fluctuations of specific ordered packing extending over small regions might be described as an ultra-microcrystalline structure. On the other hand, a conglomerate need not show any special preference for 'crystalline' microregions. High concentrations of alternative forms of specific close-packing which are anticrystalline would in fact hinder spontaneous crystallization as discussed on p. 349 and would facilitate smooth passage from a fluid to a congealed melt as a result of cooling. Glass formation in competition with the nucleation and growth of crystals has also been discussed by Turnbull (1969); the original paper should be consulted for details.

## 16.2 Viscosity criteria for congelation

Modern theories have developed two quite different approaches to glass formation. Thermodynamic considerations about the molecular origins of heat capacity and of thermal expansion have proved to be specially informative. In particular, they have substantiated the very general nature of the fluid → glass change for a great diversity of melts. As will be evident from earlier chapters of this book different types of melts must show significant differences of detail from one another in their fluid→glass relationships. Nevertheless, thermodynamic aspects of the change bring out fundamental relationships that are general for all types of melt. One convenient consequence is that useful experimentation can often be carried out on substances which form glasses at much more accessible temperatures $T_g$ than the historic silicates. Thermodynamic properties of glasses are discussed in later sections of this chapter. It is sometimes desirable to distinguish between glass temperatures, $T_g$, which can be traversed by experimental studies, from extrapolated glass temperatures, $T_g^*$, which are derived from theoretical calculations.

Viscosity criteria (which are technically very important) have been much longer studied. It was first shown by Tamman that on cooling the melts of silicate glasses a steep change in properties occurs around a temperature $\eta \geqslant 10^{13}$ poise. Though their structure was not known at the time, silicate melts are of the network type (Chapter 10). Subsequent experience has shown that a steep increase of viscosity on approaching their glass temperature is quite general with melts of other types which form glasses. This particular physical property cannot, however, be measured with sufficient precision in many cases to determine whether the limiting viscosity of about $10^{13}$ poise marks the fluid→glass change for every type of melt. Figure 16.1 illustrates the general form of semilogarithmic plots for silicates.

### 16.2.1. *Empirical correlations for viscosity of melts passing into glasses*

As might be expected from the form of Fig. 16.1, an Arrhenius type of plot

$$\log \eta = A + \exp(B/T)$$

brings the numerically large changes of this property into much more manageable form. Arrhenius plots are often approximately linear for various substances well above $T_g$. However, the molecular processes occurring in viscous flow of fluids cannot be as simple as linear Arrhenius equations would suggest, since the structure of any melt is irregular and must change as the liquid expands. For a very limited variety of molecules, such as linear polyphenyl hydrocarbons (Andrews and Ubbelohde, 1955) regular behaviour appears to be promoted by the molecular shape. Most melts may be expected to show much less conformable behaviour, however.

Many proposals to improve the Arrhenius plot have been made; an empirical Tamman–Vogel–Fulcher equation of the form

$$\log \eta = A + B/(T - T_0^*) \qquad (16.1)$$

can be made to give a good fit to experimental data on viscous flow, particularly not too close to $T_g^*$. Theoretical reasons for this behaviour have been discussed, (Weiler, Blaser, and Macedo, 1969). It is usual to emphasize relationships between viscosity and a 'free volume', since the repulsion

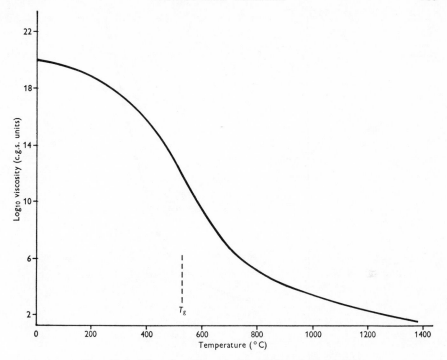

FIG. 16.1. Viscosity of a soda lime/silica glass up to $T_f$

envelope of each molecule occupies only a fraction of the total specific volume occupied by any melt at a given temperature and pressure. This may be a reasonable approximation for molecular liquids with rigid molecules whose envelopes can be considered approximately as ellipsoids. For other types of melts, except at high temperatures when the free volume is an appreciable fraction of the total, e.g. Cukierman, Lane, and Uhlmann (1973), it seems doubtful whether such a parameter can be regarded as the best to use. In the sections following it is referred to as $(T_0)_{relax}$, to distinguish it from an asymptotic $T_g$ derived from entropy considerations.

### 16.2.2 Types of glass from different types of melt

Emphasis on the unavoidable differences to be expected for melts of different types, discussed in earlier chapters of this book, gives support to the search for similitude relationships in viscosity/temperature plots within any type without expecting any similitudes to extend from one type to another. A successful example (Greet and Magill, 1967) is to plot log $\eta$ in terms of the reduced variable $T_f/T$ (see Figs. 16.2, 16.3 and 16.4). Such a plot brings molecules belonging to the same type of melt into remarkable congruence, and over extended ranges of temperature even close to $T_g^*$. It is noteworthy

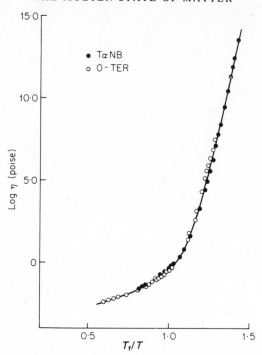

FIG. 16.2. Log $\eta$ vs. $T_f/T$ for a series of forked polyphenyl hydrocarbons. Reprinted with permission from Greet and Magill, *J. Phys. Chem.* **71**, 1746 (1967). Copyright by the American Chemical Society

that for different types of melts such as forked polyphenyl hydrocarbons (Fig. 16.2), aliphatic alcohols (Fig. 16.3), and metals (Fig. 16.4), there is remarkable congruence within any group but it does NOT extend to any other group. Even the slope of the limiting tangents for small $T_f/T$ differs (as it must do in view of the very different molecular interactions) for melts of different types. For network melts the congruence is less good, as might perhaps be expected (Fig. 16.5) in view of the structural peculiarities discussed (Chapter 10).

Water is shown to be unique on this plot (Fig. 16.3), again in line with previous discussions of its structure (p. 360), but it is perhaps noteworthy that no viscosity anomaly appears in the region of critical density. An indirect approach evaluates $T_g$ for amorphous water at $-137°C$, but this may differ very much from supercooled water as ordinarily obtained (Rasmussen and Mackenzie, 1971).

Other systematic studies of glass formation in relation to type of melt have more usually developed thermodynamic criteria as the main basis of investigation. Before discussing these, reference may usefully be added to viscosity

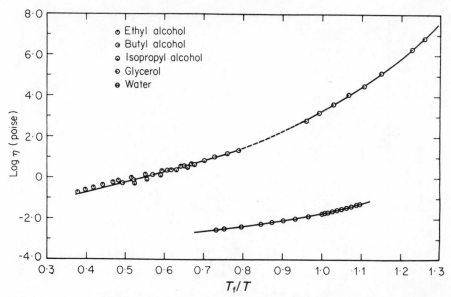

FIG. 16.3. Log $\eta$ vs. $T_f/T$ for a series of alcohols and water. Reprinted with permission from Greet and Magill, *J. Phys. Chem.* **71**, 1746 (1967). Copyright by the American Chemical Society

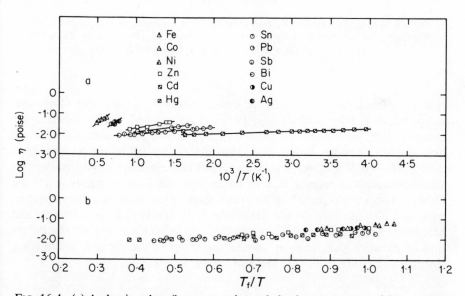

FIG. 16.4. (a) Arrhenius plots (log $\eta$ vs. reciprocal absolute temperature) for a variety of molten metals. (b) The same viscosity results of part (a) plotted with the reduced parameters $T_f/T$ as the abscissa. Reprinted with permission from Greet and Magill, *J. Phys. Chem.* **71**, 1746 (1967). Copyright by the American Chemical Society

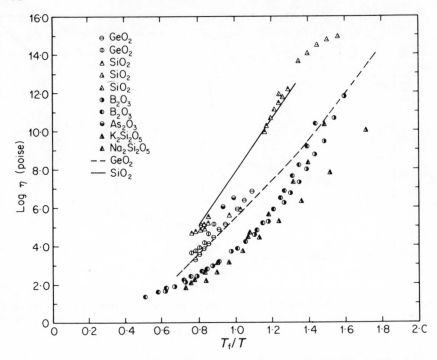

FIG. 16.5. Log $\eta$ vs. reduced parameter $T_f/T$ for a variety of inorganic oxides. Reprinted with permission from Greet and Magill, *J. Phys. Chem.* **71,** 1746 (1967). Copyright by the American Chemical Society

criteria applied to other aromatic glasses (Barlow, Lamb, and Matheson, 1966; Faerber, Kim, and Eyring, 1970). Viscosity and electrical relaxation measurements have also been applied to the study of glass formation in mixed nitrate melts (Rhodes, Smith, and Ubbelohde, 1967). The mixed ionic melt $0 \cdot 60 \ KNO_3 - 0 \cdot 40 \ Ca(NO_3)_2$ shows a remarkable tendency to vitrification. Its viscosity/temperature plot fails to agree with the Fulcher equation (16.1). Other network melts for which no 'free volume' model appears to be suitable, such as $B_2O_3$, likewise do not harmonize with (16.1) (Tweer, Laberge, and Macedo, 1971). Application of a possible spectroscopic probe for the detection of network structures in glasses may help to reduce this discrepancy (Ingram and Duffy, 1970; Angell and Wong, 1970).

## 16.3 Thermodynamic criteria for glass formation

Most of the historic glasses are, as stated, derived from network melts for which $T_g^*$ is inconveniently high, hindering precise measurements of various

thermodynamic parameters. However, for some typical glass-forming molecules $T_g$ falls in a convenient range for experimentation. For example,
measurements of heat capacity and of specific volume have been made, as
illustrated in Figs. 16.6, 16.7, 16.8 following pioneering experiments on the
specific heat of glycerol in supercooled liquid and crystalline states (Gibson
and Giauque, 1923). Many other physical properties of this substance have
now been studied. It is indicative of the generality of the change fluid → glass
that many of these properties also show critical changes in measurements
around the same $T_g$, as characterized from classical viscosity. In addition to
the specific heat, the refractive index, Fig. 16.9 (which is proportional to the
molar volume $V_m$ according to the Lorenz–Lorenz formula for molar refrac-
tivity $P$)

$$P = \frac{\mu^2 - 1}{\mu^2 + 2} \times V_m$$

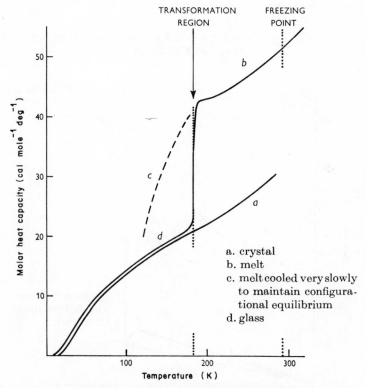

FIG. 16.6. Molar heat capacity of glycerol. Reproduced by permission of McGraw-Hill Book Company

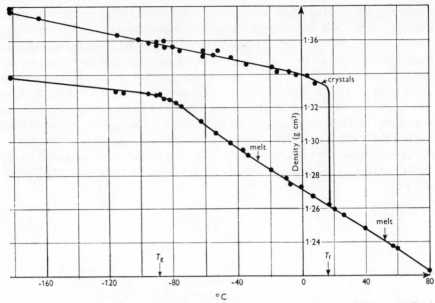

FIG. 16.7. Density of glycerol. Reproduced with permission from Schulz, *Kolloid Zeitschrift* **138,** 75 (1954)

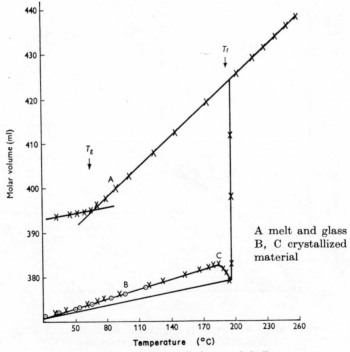

FIG. 16.8. Molar volume of tri-α-naphthylbenzene

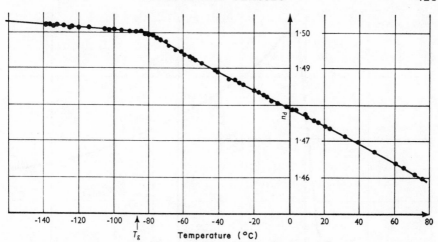

FIG. 16.9. Refractive index of supercooled molten and glassy glycerol.
Reproduced with permission from Schulz, *Kolloid Zeitschrift* **138**, 75 (1954)

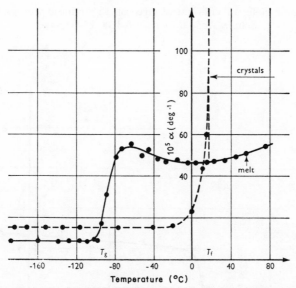

FIG. 16.10. Thermal expansion coefficient of glycerol.
Reproduced with permission from Schulz, *Kolloid
Zeitschrift* **138**, 75 (1954)

The thermal expansion coefficient, (Fig. 16.10), the static dielectric constant
$K$ (Fig. 16.11), and thermal conductivity (Fig. 16.12) all show critical changes
around the same experimental $T_g$. As indicated particularly clearly in Fig.
16.8, for supercooled melts of tri-$\alpha$-naphthylbenzene no discontinuous
change of properties is observed at $T_f$. Only if crystallization can be induced, is

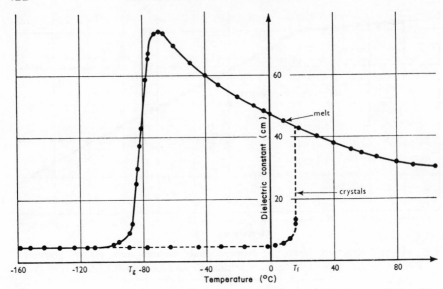

FIG. 16.11. Static dielectric constant of glycerol. Reproduced with permission from
Schulz, *Kolloid Zeitschrift* **138,** 75 (1954)

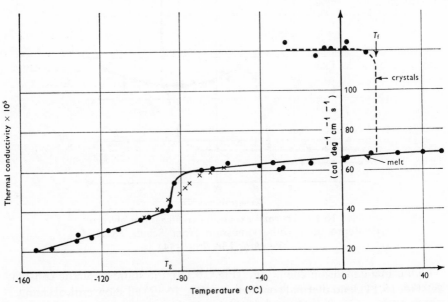

FIG. 16.12. Thermal conductivity of glycerol. Reproduced with permission from
Schulz, *Kollbid Zeitschrift* **138**, 75 (1954)

there a discontinuous change on passing from smelt to crystal. Below $T_f$ thermodynamic properties of the crystals take a normal course, quite distinct from that of the supercooled melts.

For some purposes it is useful to contrive model glasses which are free from hydrogen bonds and other strong local dipoles, unlike glycerol. A number of hydrocarbons have been deliberately synthesized with this aim in view. As stated previously Fig. 16.8 illustrates the transformation of a melt of tri-$\alpha$-naphthylbenzene into a glass (Magill and Ubbelohde, 1958). This shows many characteristic features of glass formation and is a conveniently available system. Glasses formed by cooling mixtures of various branched hydrocarbons have been proposed for use as media for 'trapped radicals'; their thermodynamic and mechanical behaviour has unfortunately not been described in any detail (Rosengren, 1962). Other melts with pronounced prefreezing anomalies have been discussed in Chapter 13 and many of these congeal into glasses.

## 16.4 Changes of configurational energy and entropy on forming a glass

A theory of glass formation which gives a reasonable account of all these experimental findings in terms of thermodynamics and structure generally stresses the role of configurational energy and entropy in the liquid. In a range of temperatures where changes of relative position of the atoms or molecules can still take place fairly rapidly in the melt, this adjusts its configurational energy by modifying these relative positions, to minimize its free energy at any temperature. So long as the relaxation time to attain configurational equilibrium of positions is not too large compared with the time of an experimental observation, this equilibrium is assured, and the assembly remains truly 'fluid'. In special cases, new mechanical tests must be devised to verify this (Ubbelohde, Duruz, and Michels, 1972; Michels and Ubbelohde, 1974).

Melts show changes in volume and heat capacity as the temperature changes, which can be regarded as roughly separable into configurational effects, and effects arising from thermal vibrations, i.e.

$$\frac{dV_{liq}}{dT} = \left(\frac{\partial V}{\partial T}\right)_{vib} + \left(\frac{\partial V}{\partial T}\right)_{config}$$

$$\frac{dV_{liq}}{dT} = \left(\frac{\partial H}{\partial T}\right)_{vib} + \left(\frac{\partial H}{\partial T}\right)_{config}$$

The importance of this formulation is that, for crystals, in which the nearest neighbour environment is usually very similar to that in the melt, the first term in each of these expressions is much the same as in the melt. However, the second term may be absent, or in any case much smaller in the solid than in the melt, depending on the crystal space group. On cooling down to the experimental $T_g$, the viscosity of the melt increases very steeply, either as a result of bond repairs to network defects (Chapter 10) or to the growth and interlocking of other kinds of anti-crystalline clusters (Chapter 13). Corresponding steep increases take place in the relaxation time for attaining configurational equilibrium. Below $T_g$, this relaxation time becomes so long, compared with ordinary experimental observations, that the positional configurations can be regarded as *frozen-in*, having arbitrary arrangements dependent on previous history of the system. By carrying out specific heat measurements in which each temperature is maintained steady very much longer than usual, the time available for reaching configurational equilibrium is somewhat lengthened. As illustrated in Fig. 16.6, curve C, some increase of $C_p$ above the 'frozen-in' value in the melt is then found at temperatures not too far from $T_g$. However, the rapidly increasing relaxation time as the temperature is further lowered soon reduces contributions from configurational changes, to negligible values. $T_g$ is thus somewhat dependent on the duration of observations; temperature-dependence of the effects is usually so steep that this effect of increased observational time is not marked. As a practical measurement, dilatometry can be much less hurried than calorimetry and for this reason it may be a more informative pointer to $T_g$, though less directly linked with the entropy.

## 16.5 Frozen-in configurational arrangements in glasses

On the basis of the evidence summarized above, a glass may be regarded as a melt whose configurational arrangement can no longer attain minimum free energy. Instead the configuration is maintained at an arbitrary value frozen-in around $T_g$. Plots of $C_p$ against ln $T$ for the crystal, melt, and glass (Jones and Simon, 1949) permit graphical evaluation of the excess entropy or disorder relative to the crystal. This becomes practically constant once $T_g$ is reached (see Fig. 16.13).

Despite the fact that different types of melt give rise to types of glass which differ in detailed molecular structure, on this view, a glass differs from a melt mainly through having arbitrary 'frozen-in' configurational disorder and entropy. Glasses thus do not represent states of thermodynamic equilibrium with respect to configurational changes of their molecular structure. On the other hand, vibrational energy is equilibrated in a glass, since its temperature can be made uniform. In fact, heat conduction ensures equilibrium for local distribution of energy for all modes that do not involve configurational changes.

Some authors regard such an arbitrary system as a thermodynamic 'anomaly'; this analogy with other transformations in the condensed state seems unwarranted, since it singles out a particular range of relaxation times

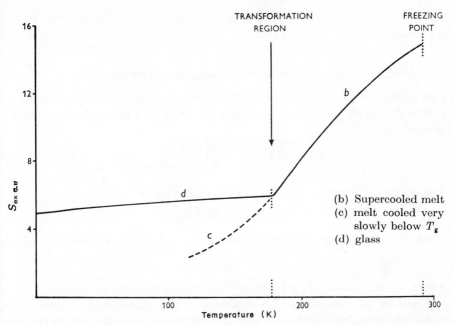

FIG. 16.13. Excess entropy of molten and glassy glycerol with respect to crystals.
Reproduced with permission from Jones and Simon, *Endeavour*, **8,** 177 (1949)

as of unique importance. Descriptive treatments and theoretical interpretations of glass formation discussed in the preceding sections are based on
the view that the glassy state of matter is not unique, but involves a degree of
arbitrariness unavoidably. From the standpoint of statistical thermodynamics,
the factor in the free energy which is associated with the configurations and
relative positions of the molecules becomes 'frozen-in' at a value dependent
on the past treatment of the material, and can then no longer be modified
unless the temperature is again raised above $T_g$. As stated, 'arbitrary' factors
are introduced in the overall partition function from 'frozen-in' equilibria.
Their occurrence in a variety of systems is indeed much more common than is
sometimes recognized in conventional texts on statistical thermodynamics,
(see discussions in Ubbelohde, 1952). Frozen-in factors do not introduce any
theoretical difficulties except in the region (in this instance around $T_g$) where
the factor can change with a relaxation time that may still be large but
comparable with that of the fluctuations whose operation alone guarantees
thermodynamic equilibrium. With regard to durations of experiments well
above the critical relaxation time for any type of variation affecting the overall
partition function, the relevant variable remains in complete equilibrium. Its
contribution to the free energy is to be minimized with all other variations
satisfying the conventional criteria for thermodynamic equilibrium. Well

below the critical relaxation temperature the factor is 'frozen-in' at an arbitrary value.

## 16.6 Compressibility changes accompanying glass formation

In liquids whose configurational entropy changes appreciably with volume, at least part of the adiabatic compressibility $\beta$ can be attributed to structural rearrangments, i.e.

$$\beta = \frac{1}{V}\left(\frac{\partial V}{\partial P}\right)_{\substack{\text{lattice} \\ \text{spacing}}} + \frac{1}{V}\left(\frac{\partial V}{\partial P}\right)_{\text{configurational}}$$

The velocity of sound $C$ is related to $\beta$ by the relation

$$\beta/(1+\tfrac{4}{3}\beta G_x) = 1/\rho u^2$$

where for many liquids the shear modulus $G_x$ can be neglected so that

$$\beta = 1/\rho u^2$$

where $\rho$ is the density. $T_g$ is frequency-dependent, but not sensitively.

When the velocity of sound is measured at increasingly high frequencies, conditions may be found such that the relaxation time for structural rearrangements exceeds $1/\nu$ where $\nu$ is the frequency of sound used. The

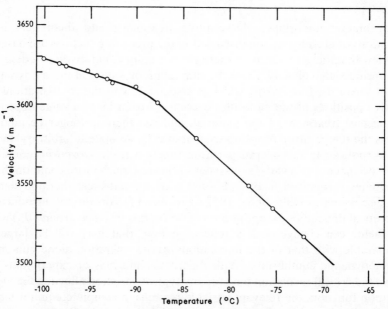

FIG. 16.14. Ultrasonic velocity (22 Mc) in molten and glassy glycerol. Reproduced by permission of the American Institute of Physics

compressibility decreases and $u$ increases quite steeply around this range of conditions. For practical reasons such relaxation effects are most conveniently studied keeping the frequency fixed and varying the temperature. Thus with glycerol, using sound of 22 Mc (Litovitz and Lyon, 1958), transition to a non-relaxing or glassy structure of the liquid is observed about 60°C above the normal $T_g$ transition temperature to a glass. For this liquid, as for others, the temperature coefficient of velocity of sound of this frequency changes quite steeply within 2–3°C of the critical range. Experiments of this kind illustrate the conclusion that glass formation essentially points to a freezing-in of configurational structure in a liquid (see Fig. 16.14).

## 16.7 Theoretical estimates of 'ideal' glass temperatures

As stated, the temperature variation of any dynamical property $P_r$ of a melt can be expressed with fair generality by means of an empirical Tamman–Vogel–Fulcher equation:

$$P_r = AT^{-\frac{1}{2}} \exp[B/(T - T_g^*)]$$

The parameters $A$ and $B$ as in the Arrhenius equation are assumed independent of temperature, which affects the property mainly through the limit term $(T_g^*)$. (The factor $T^{-\frac{1}{2}}$ is sometimes omitted but may 'straighten' a shallow curved plot.) In favourable cases, where more than one property can be measured, $(T_g^*)$ is approximately the same for all of them provided data are used not too near the experimentally observed $T_g$. For example, this applies to the electrical conductance, the shear viscosity, the bulk viscosity, or the diffusion coefficient for impurity ions, for melts of $Ca(NO_3)_2.4H_2O$ (Angell and Tucker, 1974a). Accordingly $(T_g^*)$ is often regarded as a convenient parameter marking the type of vitrified melt, and defining its 'ideal' glass temperature, i.e. $(T_g^*) = T_g$. (See, however, Weiler, Blaser, and Macedo, 1969.) In terms of fluctuation theory, the steep increase, e.g. in viscosity, of the melt approaching $T_g$ has been attributed to exceptionally large fluctuations around any critical temperature (Simmons, Macedo, Napolitano, and Haller, 1970). Alternatively, a growth of anticrystalline clustering on cooling (Magill and Ubbelohde, 1958; McLaughlin and Ubbelohde, 1958 see Chapter 13) would have a similar consequence without requiring any thermodynamic 'critical' phenomena.

To some extent, either form of relaxation theory is a rival interpretation of vitrification with theories based on configurational entropy or energy. Useful discrimination between relaxation theories and configurational entropy theories may perhaps be obtainable from experiments on the change of $T_g$ with applied pressure. The original should be consulted for details (Goldstein, 1973). The configurational entropy approach has been discussed in some detail particularly with regard to the vitrification of ionic melts (Angell and

Tucker, 1974a,b; for a more general discussion see Vogel, 1971). The main line of argument can be followed with reference to Fig. 16.6 and Fig. 16.13 and introduces what is sometimes termed the Kauzmann paradox (Kauzmann, 1948). At $T_f$, the specific heat $C_s$ of the crystal drops sharply, because of the loss of much ability to reduce its configurational entropy on further cooling. Accordingly, for the supercooled melt below $T_f$, $C_L > C_s$. The entropy of the melt $[S_L]_f$ is greater than that of the crystal $[S_c]_f$ at $T_f$, but it decreases faster below it. The paradox is that the entropy of the supercooled melt

$$[S_L]_f - \int_T^{T_f} C_L \, d(\ln T)$$

cannot fall below the entropy of the crystal

$$[S_c]_f - \int_T^{T_f} C_s \, d(\ln T)$$

which sets a lower limit temperature $(T_0^*)_{config}$ given by the requirement that

$$S_f + \int_{(T_0^*)_{config}}^{T_f} (C_s - C_L) \, d(\ln T)$$

must remain positive. The presumption is that the experimental $T_g$ must lie above any limit $(T_0^*)_{config}$ indicated by such reasoning.

In general, the difference $(C_L - C_s)$ will depend on the type of melt. It has been estimated (Angell and Tucker, 1974a) that, for ionic hydrate melts, $T_g^*/(T_g)_{config}$ lies between 1·07 and 1·3. This may be compared with ratios $T_g^*/(T_0)_{relax}$ where $(T_0)_{relax}$ is derived from kinetic data as stated above. For molecular melts $T_g^*/(T_0)_{relax}$ ranges between 1·2 and 1·4, and for high silicate network melts between 1·5 and 1·8 (Angell, 1968). For ionic melts $T_g^*/(T_0)_{relax}$ ranges between 1·05 and 1·15. Many other molten salts (Bartholomew and Lewek, 1970) and even metallic melts (e.g. Chen, 1974) as well as semiconductor melts (e.g. Haisty and Krebs, 1969) can be quenched into a glassy state by sufficiently rapid cooling, though the middle range of supercooled melts may be difficult to study because of their tendency to nucleate and grow crystals. The preceding discussion still applies but values of $T_g^*$ are not immediately comparable, as discussed in the next section

## 16.8 Vitrification in relation to molecular structures

For melts which show extensive prefreezing (Chapter 13) without spontaneous nucleation of crystallization, passage into the glassy state can usually be followed experimentally without much difficulty, especially in a critical range of temperature around $T_g$. Apparently the molecules in the melt become progressively incorporated into various kinds of anticrystalline clusters; this reduces the probability in a quasicrystalline melt of crystal nuclei being formed. Only these can build on additional molecules and continue their own pattern in space through crystal growth. Values of $T_g$ determined from the viscosity criterion or other methods of testing the freezing-in of configurational changes, such as the thermal expansion, are recorded in Tables 16.1 and 16.2 as obtained by direct observation.

Table 16.1. Glass-forming temperatures $T_g$ (Kauzmann, 1948; Magill and Ubbelohde, 1958)

| Molecule | $T_g$ | $T_f(K)$ |
|---|---|---|
| 3-Methylhexane | 80–90 | 153·7 |
| 2,3-Dimethylpentane | 80–85 | — |
| Ethanol | 90–96 | 155·8 |
| 1-Propanol | 86–100 | 146·0 |
| sec-Butyl alcohol | 100–115 | 184·0 |
| Propylene glycol | 150–165 | — |
| Glycerol | 180–190 | 291·0 |
| DL-Lactic acid | 195–206 | 291·0 |
| $Na_2S_2O_3.5H_2O$ | 231 | 232d |
| Sucrose | 340 | 459d |
| Glucose | 280–300 | 419·0 |
| Se | 302–308 | (453) |
| S | 244 | 392·3 |
| Tri-$\alpha$-naphthylbenzene | 350 | 470·0 |
| $B_2O_3$ | 470–530 | 567·0 |
| $SiO_2$ | 1500–2000 | ~1980·0 |

Table 16.2. Softening points $T_g$ (°C) for atactic polymers $-[CH_2.CHR]_n-$

| Substituent group R | $T_g$ |
|---|---|
| $-CH_3$ | -20 |
| $-OCH_3$ | -13 |
| $-OOC.CH_3$ | 0 |
| $-Cl$ | 81 |
| $-CN$ | 105 |
| $-C_6H_5$ | 100 |

On the other hand, for melts in which crystallization nucleates readily, gradual cooling usually fails to lead into the glassy state, before most of the melt has crystallized. As indicated, amorphous states can sometimes be frozen-in by extremely rapid chilling. Though it does not allow any detailed study of the onset of glass formation as a gradual process, it permits a kind of inverse approach to $T_g$ for such substances. Temperatures $T_D$ at which configurational changes can no longer be prevented from beginning to attain thermal equilibrium can then be observed by gradually heating the amorphous material, noting when the heat content or the volume (or appropriate non-thermal properties) point to a steep increase of molecular freedom. Some examples were recorded in Chapter 15, Table 15.4. Again, ice can be chilled into an amorphous state by condensing water vapour direct onto a pyrex surface chilled to 77 K. On heating such amorphous condensates $T_D$ appears to lie around 150 K (Ghormley, 1956). Systematic studies of inverse temperatures $T_D$ for a diversity of melts could help to throw valuable light on their structure.

As discussed in Chapter 13, at temperatures well above $T_g^*$, certain liquids show anomalously high viscosities; this can become prominent on approaching and traversing $T_f$. On the theory outlined in Chapter 13, for a number of polyphenyls, 'blocked' volumes through clustering attain a fraction comparable with 1/3. Such melts pass into glasses on further cooling. In melts of other types, viscosity anomalies build up more gradually, though they lead to glass formation eventually. Abnormally high 'activation energies' $E_\eta = B/R$ for viscous flow may appear to emerge when calculated from a non-linear relation between $\log \eta$ and $1/T$ using an Arrhenius type of equation

$$\log \eta = A + B/T$$

For example, with $BeF_2$, $E_\eta$ appears to be about 100 kcal mole$^{-1}$. Since the plot is curved, the attribution of an 'activation energy' for viscous flow is open to question. It is probably simpler to regard a fraction of the volume as blocked against flow. A network structure (Chapter 10) can be postulated for this melt (Mackenzie, 1960a,b) similar to that in molten quartz and molten $GeO_2$.

Groups or clusters are beginning to be identified in other melts to explain flow characteristics. Thus the phase diagram for the system $Na_2O-B_2O_3$ conforms reasonably well with that expected for a melt containing groups such as $B_4O_7^{2-}$, $B_8O_{13}^{2-}$ and $B_3O_{4.5}$ (Krogh-Moe, 1962). Indirect evidence supports the presence of such groupings in borate glasses. Developments of inorganic chemistry hold out promise of other novel melts, whose viscosity is exceptionally large and whose tendency to vitrification warrants investigation.

In polymers, the glass transition is associated with a 'softening point' but not with any clear cut changes of fluidity. However the freezing-in of local molecular movements, as discussed in relation to configurational freezing-in for conventional supercooled fluids above, can likewise occur in polymers and can result in significant changes of properties. Estimates of polymer softening points $T_g$ for a group of atactic polymers have been made (Goodman, 1967). Table 16.2.

## REFERENCES

J. N. Andrews and A. R. Ubbelohde (1955) *Proc. Roy. Soc.* **A228**, 435.
C. A. Angell (1968) *J. Am. Ceram. Soc.* **51**, 117, 126.

C. A. Angell and J. C. Tucker (1974a) *J. Phys. Chem.* **78,** 2, 78.
C. A. Angell and J. C. Tucker (1974b) in *Physical Chemistry of Process Metallurgy,* ed. Jeffes and Tait, Inst. Mining Metall. Pub., p. 207.
C. A. Angell and J. Wong (1970) *J. Chem. Phys.* **53,** 2053.
J. Barlow, J. A. Lamb, and A. J. Matheson (1966) *Proc. Roy. Soc., Ser. A* **292,** 322.
R. F. Bartholomew and S. S. Lewek (1970) *J. Am. Chem. Soc.* **53,** 3.
H. S. Chen (1974) *Acta Metall.* **22,** 1505.
M. Cukierman, J. W. Lane, and D. R. Uhlmann (1973) *J. Chem. Phys.* **59,** 3639.
G. L. Faerber, S. W. Kim, and H. Eyring (1970) *J. Phys. Chem.* **74,** 3510.
J. A. Ghormley (1956) *J. Chem. Phys.* **25,** 599.
Gibson and Giauque (1923) *J. Am. Chem. Soc.* **45,** 93; *cf.* G. N. Lewis and M. Randall *Thermodynamics,* McGraw-Hill, 1923, p. 444.
M. Goldstein (1973) *J. Phys. Chem.* **77,** 667.
I. Goodman (1967) *Roy. Inst. Chem. Lecture Series No. 3,* p. 47.
R. J. Greet and J. H. Magill (1967) *J. Chem. Phys.* **71,** 1746.
R. W. Haisty and H. Krebs (1969) *J. Non-Cryst. Solids I* **399,** 427.
M. D. Ingram and J. A. Duffy (1970) *J. Am. Chem. Soc.* **53,** 317.
G. Ö. Jones and F. Simon (1949) *Endeavour* **8,** 175.
W. Kauzmann (1948) *Chem. Rev.* **43,** 219.
J. Krogh-Moe (1962) *Phys. Chem. Glasses* **3,** 101.
T. A. Litovitz and T. Lyon (1958) *J. Acoust. Soc. Am.* **30,** 856.
J. D. Mackenzie (1960a) *J. Chem. Phys.* **32,** 1150.
J. D. Mackenzie (1960b) *Modern Aspects of the Vitreous State,* Butterworths, London.
J. H. Magill and A. R. Ubbelohde (1958) *Trans. Faraday Soc.* **54,** 1811.
E. McLaughlin and A. R. Ubbelohde (1958) *Trans. Faraday Soc.* **54,** 1804.
H. Michels and A. R. Ubbelohde (1974) *Proc. Roy. Soc., Ser. A* **338,** 447.
D. J. Plazek and J. H. Magill (1966) *J. Chem. Phys.* **45,** 3038.
D. H. Rasmussen and A. D. Mackenzie (1971) *J. Phys. Chem.* **76,** 967.
E. Rhodes, W. E. Smith, and A. R. Ubbelohde (1967) *Trans. Faraday Soc.* **63,** 1943.
K. Rosengren (1962) *Acta Chem. Scand.* **16,** 1421.
A. K. Schulz (1954) *Kolloid Z.* **138,** 75.
J. H. Simmons, P. B. Macedo, A. Napolitano, and W. K. Haller (1970) *Discuss. Faraday Soc.* **50,** 155.
D. Turnbull (1969) *Contemporary Physics* **10,** 473.
H. Tweer, N. Laberge, and P. B. Macedo (1971) *Am. Ceram. Soc.* **54,** 121.
A. R. Ubbelohde (1952) *Modern Thermodynamical Principles,* Oxford University Press.
A. R. Ubbelohde, J. J. Duruz, and H. Michels (1972) *J. Physics E. Scientific Instruments* **5,** 283.
W. Vogel (1971) *Structure and Crystallisation of Glasses,* Pergamon, p. 246. Headington Hall, Oxford.
R. Weiler, S. Blaser, and P. B. Macedo (1969) *J. Phys. Chem.* **73,** 4147.

# 17. MELTING AND CRYSTALLIZATION IN POLYMER SYSTEMS

## 17.1 Partly crystallized flexible macromolecules

In ordinary crystals, the molecule or unit of structure is many times smaller than the ordered system in which it is contained. In some polymers, on the other hand, the molecule can be so large that it extends over several crystalline regions, with portions passing through intermediate regions whose arrangement is disordered, as in liquids.

The possibility that parts of large molecule may be located in a crystalline environment and other parts in an amorphous environment clearly can lead to melting only in the case of macromolecules whose configuration is flexible. Many polymers either contain rigid macromolecules, or at any rate include numerous more or less rigid bonds between flexible macromolecules, with a breaking strength $D$ such that $D/k$ exceeds the temperature at which general pyrolytic breakdown occurs. With rigid macromolecules, or in polymers in which these cross-linking bonds are sufficiently numerous throughout the mass of material, no melting can of course be observed in the ordinary sense.

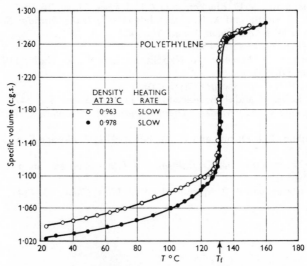

FIG. 17.1(a). Specific volume/temperature curve of samples of polyethylene. Reproduced with permission from Matsuoka and Aloisio, *J. Appl. Polym. Sci.* **40**, 116 (1960)

(For a discussion of thermodynamic consequences of cross-linking in polymer networks see Ziabicki and Klonowski, 1975a,b.) All the instances discussed in what follows involve considerable configurational flexibility of the macromolecules, and, when configurational flexibility can operate, various methods of investigation reveal phenomena akin to crystallization on cooling. Temperatures at which this occurs are determined by both entropy and energy factors, as for smaller molecules (see Dole and Wunderlich, 1959). Other methods of observation such as light scattering have also been used (Schaefer, 1975) to follow special situations.

It is mainly useful to consider investigations of the principal thermodynamic parameters, i.e. the heat content and specific volume. Figure 17.1 records typical studies of thermal properties (Matsuoka and Aloisio, 1960). It will be noted that the graphs resemble those for the behaviour of crystals of smaller molecules, but with premelting phenomena (Chapter 12) spread over an extremely wide range of temperatures. At the upper limit of this spread, a melting point $T_f$ is normally observed above which the polymer yields a fluid melt. However, it is interesting to note that even for macromolecules no

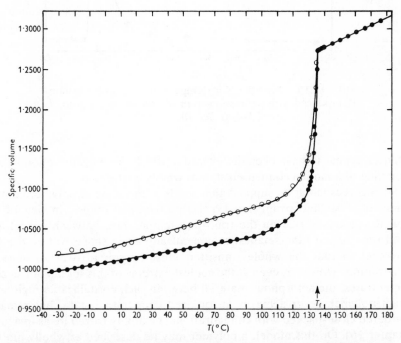

FIG. 17.1(b). Specific volume as a function of past history: (○○○) slowly cooled from $T_f$ to room temperature prior to fusion; ($\cdot\cdot\cdot$) crystallized 40 days at 130°C, then cooled to room temperature and melted

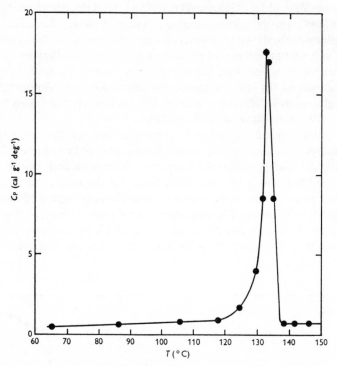

FIG. 17.1(c). Specific heat/temperature of polyethylene. Reproduced with permission from Matsuoka and Aloisio, *J. Appl. Polym. Sci.* **40,** 117 (1960)

instances appear to have been discovered in which the transformation from a crystalline to a molten condensed state is wholly continuous.

It seems reasonable to suppose that solids in this class of polymer contain regions of crystalline arrangement of varied sizes and shapes. In view of the influence of crystal size on the transition temperature between crystal and liquid (pp. 30 seq.) the smallest regions of any given shape will be the last to freeze, so that the whole transition must be spread over a range of temperatures. However, even at the coolest extreme of this spread of freezing temperatures, the amorphous material between such crystallizable regions in a high polymer never achieves an ordered cooperative state, but remains disordered close-packed with configurational flexibility frozen in, as in a glass (Chapter 16). On this model, a polymer may be described as wholly molten when it has lost all its crystalline regions; but it is never wholly crystalline throughout 100% of the space it occupies. There will thus be a unique peak melting point $(T_f)_{\text{lim}}$ but no unique freezing point. This statistical model is borne out by X-ray studies on polymers at various temperatures above and below their melting points. (For useful general reviews of the melting of polymers see Mandelkern, 1959; Goodman, 1967; Keith, 1974.)

Methods of determining the limiting melting point have to keep in view the gradualness of melting in high polymers as well as the dependence of melting behaviour on previous history. Penetrometer devices and measurement of other mechanical or physical changes (Edgar and Hill, 1952; Edgar and Ellery, 1952) are sometimes used to investigate melting, since these can be more convenient and more informative than measurements of thermodynamic properties. Microscope studies on melting of polymers are conveniently made on a microscope hot stage with polarizer and analyser crossed; in this case $T_f$ is taken as the temperature when the field becomes wholly dark. Figure 17.2 illustrates suggested changes of molecular configuration in a typical instance. With microscope hot stage studies the risk of interference from the containing walls must always be kept in mind, as was already noted for studies on liquid crystals.

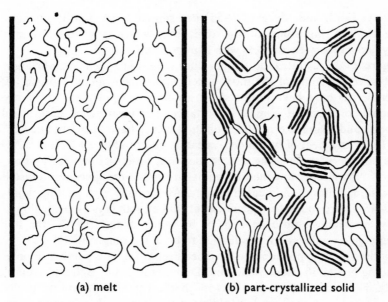

|   (a) melt   |   (b) part-crystallized solid   |

FIG. 17.2. Suggested changes of molecular configuration on melting polymers

## 17.2 Effects of past history on the melting of polymers

Studies on polychlorotrifluoroethylene (Hoffman and Weeks, 1962) indicate that the size of the crystalline regions of given material, and consequently the melting range, can be affected by the rate at which it was crystallized by cooling to various temperatures below the theoretical limiting melting point $(T_f)_{lim}$. Rates of crystallization of polyhexamethylene adipamide

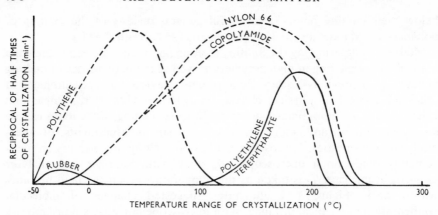

FIG. 17.3. Crystallization range of polymers. Reproduced by permission of the
Journal of Applied Chemistry

(Hartley, Lord, and Morgan, 1955) determined from density changes can
serve to illustrate some of the difficulties which any equilibrium thermodynamic theory encounters with high polymers.

Figure 17.3 illustrates the temperature ranges in which crystallization can
be observed in some of the most widely applied polymers (Morgan, 1954).
Fuller details have been obtained for a number of polymers where crystallization is of special technical importance. For unstretched natural rubber
(Roberts and Mandelkern, 1955) limiting values have been obtained of $T_f =
28 \pm 1°C$ with a heat of fusion per repeating unit of $15 \cdot 3 \pm 0 \cdot 5$ cal g$^{-1}$. As with
other polymers, the melting process depends on crystallite size and on past
history. Abnormally low melting points are found for specimens crystallized
at low temperatures and heated rapidly. Abnormally high melting points (up
to 35–45°C) are found for specimens of rubber with crystallites highly oriented by stretching before cooling, or by other means. At 0°C the limiting
percentage crystallization is estimated to be in the range 26–31%.

Multiple melting phenomena have been reported for a number of polymers
including Nylon 66 (Bell, Slade, and Dumbleton 1968; Bell and Dumbleton
1969). These depend on previous mechanical treatment and on prolonged
annealing and are attributed to folding of the macromolecule.

### 17.3 Melting and molecular structure

For polyethylene, statistical thermodynamic theory has been proposed to
correlate the melting range with the distribution of molecular weights of
normal chains, and the proportion of side-substituted chains (Richards,
1945). Good agreement is obtained when the temperature-dependence of the
fraction of crystallites in polyethylene is evaluated on the one hand by

calorimetric measurements of heats of fusion (Raine, Richards, and Ryder, 1945), and on the other hand by measurements of intensities of X-ray diffraction. Above 120°C, the whole system is amorphous so that this may be taken as the limiting melting point (Trillat, Barbezat, and Delalande, 1950).

An illustration of how the completion of melting can be determined from dilatometric measurements is given for natural rubber networks with various degrees of cross-linking in Fig. 17.4 (Roberts and Mandelkern, 1960). At the breaks in these plots, any expansion resulting from disordering of local crystallites on melting has been completed. Smooth correlations are found between the observed $T_f$ and the fraction of units cross-linked by vulcanization with sulphur.

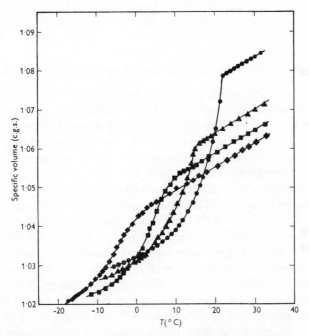

FIG. 17.4. Specific volume of rubber with different degrees of cross-linking by sulphur. Percentages of sulphur are: (....) 0.5; (▲▲▲) 1.0; (■■■) 1.5; (◆◆◆) 2.0. Reprinted with permission from Roberts and Mandelkern, *J. Am. Chem. Soc.* **82,** 1091 (1955). Copyright by the American Chemical Society

The synthesis of 'isotactic' polymers in which the successive segments have similar configurations has permitted interesting comparisons of heats and entropies of fusion. One procedure is to evaluate cryoscopic depressions of freezing points resulting from the addition of impurities. Experimentally the use of dilatometry to determine $T_f$ for various mixtures has the advantage that measurements can be made allowing plenty of time for equilibrium, which is not usually the case or even feasible when $T_f$ is determined either by hot stage microscopy with crossed Nichols or by calorimetry. On this basis (Danusso and Moraglio, 1959) values per unit of chain length ($-CH_2-CH_2-$ for polethylene) are given in Table 17.1.

*Table 17.1 Some melting parameters of polymers (per chain unit)*

| Polymer | $T_f$ (°C) | $H_f$ (cal mole$^{-1}$) | $S_f$ (e.u.) |
|---|---|---|---|
| Polyethylene | 138 | 1820 | 4·4 |
| Polypropylene (isotactic) | 176 | 2600 | 5·8 |
| Polystyrene (isotactic) | 240 | 2150 | 4·2 |

### 17.3.1. *Effects of the molecular units on the melting of macromolecules*

Just as the melting points of organic molecules can be broadly correlated with factors such as molecular shape, flexibility, and cohesive energy, the influence of these parameters can be discerned in broad general surveys of the melting points of polymers in relation to the units which constitute them (Bunn, 1955). Complicating factors have by no means been eliminated; the original discussion should be consulted.

As stated above, the melting of macromolecules differs in a fundamental way from the melting of conventional crystals; for these the size of the crystal always exceeds the size of the molecule, whereas for macromolecules the reverse is frequently the case. Regional melting may be anticipated for high polymers. In view of their technological importance, various systematic effects of molecular structure on melting points have been investigated in considerable detail. Increased molecular stiffness raises the melting point of a number of high polymers. Some striking examples have been recorded in the field of linear high polymers (Edgar and Hill, 1952), as a result of introducing $p$-phenylene groups in the chain. Comparisons are made in pairs in the Table 17.2.

Basically, melting parameters must be related by the equation $T_f = H_f/S_f$. Structural changes could in principle affect the jump in properties on melting, such as $H_f$ or $S_f$. When it has been possible to measure entropies of fusion as well as melting points, it has been confirmed that the enhanced melting point of the stiffer molecule is due primarily to its smaller entropy of fusion, i.e., the smaller relative increase in molecular configurations, see Table 17.3. (In this series, relative magnitudes are probably correct, but absolute values are obtained by indirect means and must thus be used with caution (Edgar and Ellery, 1952).

**Table 17.2. Effects of chain stiffening on melting macromolecules**

| Repeat sequence | $T_f(°C)$ |
|---|---|
| $-CH_2-CH_2-CH_2-$ | 135 |
| $-CH_2-\langle \text{ring} \rangle-CH_2-$ | 380 |
| $-CH_2O.CO(CH_2)_4.CO.OCH_2-$ | 50 |
| $-CH_2O.CO-\langle \text{ring} \rangle-CO.OCH_2-$ | 264 |
| $-(CH_2)_3NHCO(CH_2)_4CONH(CH_2)_3-$ | 264 |
| $-(CH_2)_3NHCO-\langle \text{ring} \rangle-CONH(CH_2)_3-$ | 440 |
| $-CO(CH_2)_4COO-$ | 85–98 |
| $-CO-\langle \text{ring} \rangle-CO.O-$ | 300 |

**Table 17.3. Melting parameters per repeat unit**

|  | $H_f$ (cal mole$^{-1}$) | $S_f$ (e.u.) |
|---|---|---|
| Polythylene terephthalate | 2200 | 4·0 |
| Polyethylene adipate | 3800 | 12·0 |
| Polyethylene sebacate | 3300 | 9·6 |

Some more general comparisons on the melting of common high polymers are recorded in Table 17.4.

### 17.3.2. Transformations in polymers below $T_f$

Just as simple crystals can exhibit several transition points in the solid state as their temperature is raised, polymers can in principle exhibit several types of disordering whose separate emergence at different peak temperatures, or whose coalescence within a melting range of temperatures, will depend on the internal environment of the ordered regions. This internal environment can be modified by chemical changes in the macromolecules, such as length of the primary chain, fraction and distribution of side-chain substituents, and fraction and location of cross-linking. It can also be modified by physical treatment of the polymer, for example by drawing or stretching, or by annealing or rapid quenching. Polyurethanes exhibit a number of characteristic effects of

**Table 17.4.** *General comparisons between densities and melting ranges of some well ordered macromolecules*

| Polymer | Repeat unit | Density theory | Expt. $(g/ml^{-1})$ | Melting range ($^\circ C$) |
|---------|-------------|----------------|---------------------|----------------------------|
| Polystyrene | $-CH_2-CH-$<br>$\qquad\quad\mid$<br>$\qquad\quad C_6H_5$ | 1·12 | 1·085 | 230 |
| Polypropylene | $-CH_2-CH-$<br>$\qquad\quad\mid$<br>$\qquad\quad CH_3$ | 0·938 | 0·92 | 158–170 |
| Polybutylene | $-CH_2-CH-$<br>$\qquad\quad\mid$<br>$\qquad\quad C_2H_5$ | 0·950 | 0·915 | 125–130 |
| Polypentene-1 | $-CH_2-CH-$<br>$\qquad\quad\mid$<br>$\qquad\quad C_3H_8$ | — | 0·87 | 75–80 |
| 1, 2-Polybutadiene | $-CH_2-CH-$<br>$\qquad\quad\mid$<br>$\qquad\quad CH=CH_2$ | — | 0·92 | 120 |
| Polyvinyl chloride | $-CH_2-CH-$<br>$\qquad\quad\mid$<br>$\qquad\quad Cl$ | 1·44 | 1·39–1·41 | 150 |

See Hendus, Schnell, Thurn, and Wolf (1959).

this kind (Kilian and Jenckel, 1959) illustrating how structural studies of melting mechanisms can be extended even to systems of high complexity.

### 17.3.3. *Degree of crystallization*

In view of the consideration that a polymer is not 100% crystalline, and unavoidably retains amorphous regions even down to the lowest temperatures of measurement, various physical methods for studying the proportion of crystalline order may be expected to give somewhat divergent estimates. Even moderate flow in a molten polymer can affect crystallization rates (Theil, 1975). In addition to the properties already referred to, other methods include observations (Hendus, Schnell, Thurn, and Wolf, 1959) on: (i) the freezing-in of dielectric polarization arising from orientation of permanent dipoles; (ii) a general decrease of mechanical moduli in the sense of decreasing rigidity with rising temperature in sensitive regions; (iii) anelastic processes in the polymers leading to temperatures of maximum damping of forced vibrations for any given frequency; (iv) changes in nuclear spin resonance corresponding with critical changes in local relative movement of portions of the macromolecules; and (v) scattering of X-rays by ordered regions.

Detailed comparisons have unfortunately seldom been worked out between these and other methods of estimation, for the same partially crystalline polymer at different temperatures. It seems abundantly clear however, and is important technically, that crystallization is dependent on the previous history of the specimen. Whatever the method of study, different rates of cooling are frequently observed to lead to the formation of varying proportions of crystalline ordered regions at the same temperature. This may be regarded as a consequence of the strains set up in the amorphous regions, which depend on the past history, as is further discussed below.

### 17.3.4. *Single crystals of macromolecules*

What has been stated above applies in general to 'polycrystalline' polymers in which cross-linking usually binds the macromolecule into a network. In favourable cases, minute single crystals of linear macromolecules can be grown (Keller, 1959; Hendus, Schnell, Thurn, and Wolf, 1959). A remarkable feature of such 'single' crystals is that their thickness may be only about 100 Å whereas the macromolecular chain extends over thousands of Ångstrom units. To achieve this unusual structure, in such 'single crystals' each macromolecule appears to pleat or fold back on itself after 'crystallization' with respect to neighbouring chains, allowing only a limited number of links between each pleat. Experimental evidence indicates the 'folding periodicity' is fairly uniform but depends to some extent on past history. Thus it can be increased by increasing the temperature at which the crystals separate from solution; prolonged tempering of the single crystals after separation also modifies the folding periodicity. To interpret these findings, theoretical attempts have been made to correlate the free energy of the crystals with the number of pleats in any stretch of the macromolecule. Vibrational contributions to the free energy of the crystal depend on the number of links between folds. Factors such as the intermolecular attraction and surface free energy also affect the free energy of the crystals, so that an overall minimum is to be expected (Peterlin and Fischer, 1960) corresponding roughly with the lengths of fold actually observed.

An alternative explanation of 'pleating' attributes this to a balance of opposing factors in the kinetics of crystallization, without postulating that any true equilibrium state of the crystal is reached (Frank and Tosi, 1961). Examples of 'pleating' depend on the polymer under observation as well as on the methods of study used; e.g., a paracrystalline structure is found for crystallized poly-($N$-vinyl carbazole) (Griffiths, 1975).

## 17.4 Changes of molecular association on melting

As in the case of smaller molecules, changes in the extent of association or dissociation may appear as a result of the greater freedom on passing to the more expanded melt, whereas these processes tend to be hindered by the influence of nearest neighbours in crystalline regions of the polymer. For example, a sharp infrared absorption peak appears at $6757\,\mathrm{cm}^{-1}$ in the spectra on melting polyamides. This has been attributed to the dissociation of hydrogen bonds in such melts (Glatt and Ellis, 1951) giving rise to 'free' NH groups; other interpretations of this difference between crystalline and molten states are however possible.

Catalytic methods for the production of 'crystalline' polymers in which the proportion of well ordered regions may be near 100% have been developed for a number of macromolecules. For these the specific volume is near to theoretical values for 100% crystalline macromolecules; the 'melting range' is comparatively narrow. Examples are included in Table 17.4.

Small differences between the infrared spectra of crystalline polystyrol and its melt have not yet been interpreted definitively (Hendus, Schnell, Thurn, and Wolf, 1959b).

## 17.5 Melting of macromolecules with mixed sequence of units

In a macromolecule, the repeat unit can frequently be varied by processes of 'copolymerization'. For example, polyethylene terephthalate can be copolymerized with polyethylene adipate or with polethylene sebacate. The melting point, $T_{\mathrm{f}}$ of such copolymers obeys Flory's relationship

$$\frac{1}{T_{\mathrm{f}}} = \frac{1}{{}^0T_{\mathrm{f}}} - \frac{R\ln X}{H_{\mathrm{f}}}$$

where $H_{\mathrm{f}}$ is the heat of fusion per repeating unit of (high melting) polymer, $X$ is the mole fraction of high melting units and ${}^0T_{\mathrm{f}}$ is the melting point in K for the pure high polymer (Edgar and Ellery, 1952). This expression is based on a model in which the one 'component' of the copolymerized macromolecule is regarded as crystallizable and the other as non-crystallizable (or any rate with a much lower $T_{\mathrm{f}}$) (Flory, 1947; see Flory and Mandelkern, 1956). A test in which the high melting component is decamethylene adipate, 'diluted' by copolymerization with various other groups, shows values of $T_{\mathrm{f}}$ agreeing within 1–2° with Flory's equation (Evans, Mighton, and Flory, 1947).

### 17.5.1. Melting of protein chain molecules

Melting of axially oriented crystalline polymers normally involves contraction, owing to the replacement of unique arrangements in the crystalline state by multiple configurations of a flexible chain in the melt. For protein macromolecules such as feather keratin, the melting point determined from the temperature at which contraction occurs is strikingly affected by the ionic environment as illustrated by Fig. 17.5 (see Mandelkern, Halpin, and Diorio, 1962; Flory, 1956). Changes in helix configurations are likely to be essentially related to phase transitions in biopolymers (Sture, Nordholm, and Rice,

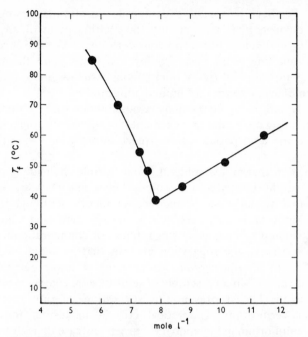

FIG. 17.5.  $T_f$ for feather keratin as a function of lithium bromide concentration. Reproduced with permission from Mandelkern, Halpin, and Diorio, *J. Polym. Sci.* **60,** 531 (1962)

1973). Details about these may perhaps be even less accounted for by thermodynamic considerations than for other polymers, though information is not yet very systematic.

## 17.6 Role of configurational entropy in the melting of macromolecules—glassy polymers

The conclusion, that flexible polymers should exhibit a limiting melting point above which the system behaves as a fluid, is reminiscent of the convergence temperature in homologous series of flexible molecules (Chapter 7), in which the configuration entropy changes become predominant at the series limit for melting. In polyethylene it has been estimated at $T_f$ that the configurational entropy of fusion is $1 \cdot 79 \pm 0 \cdot 15$ e.u. $mole^{-1}$ of $CH_2$. This term is about 33% less than the total entropy of fusion at constant pressure (Robertson 1969). The two kinds of solid are however not strictly comparable in every respect. Members of homologous series of flexible molecules are of a single molecular size; all of their melt changes into a crystal on freezing at $T_f$. In a polymer, a range of domains of flexible regions occurs bounded by rigid crossbonds. Further, compressive or tensile constraints arising from the amorphous environments may modify the melting point of any particular region in a high polymer. These two considerations make direct comparisons between the melting point observed for a polymer, and theoretical convergence melting temperatures of homologous series of pure crystals containing a single molecular species, of limited applicability.

With copolymers, even more extensive possibilities arise for the presence of regions of differing molecular arrangement. On sufficien⁺ cooling, calorimetry indicates that the amorphous regions freeze into a glas⸖ ⁻tate (Brown and Lowry, 1974).

The suggestion that an assembly of linear flexible chains may pass into an amorphous 'equilibrium' state on cooling (Gibbs and Di Marzio, 1958) has been developed by statistical thermodynamic theory. Although the eventual freezing of amorphous polymers into a 'glassy' state (see Chapter 16) is plausibly attributed to increasing sluggishness of configurational changes as the volume contracts, the suggestion made is that an actual 'second order' transformation to an *equilibrium* arrangement of lower entropy than the liquid may be reached in an assembly of semi-flexible chains, at temperatures not much below $T_g$. At present this suggestion has not received direct experimental verification. Theoretical criteria in testing for true thermodynamic equilibrium in high polymers are in any case difficult to apply with precision. Calorimetry even for polymers of fairly simple molecular composition such as gutta percha and polyurethane confirm the critical importance of past history of samples (Sochava and Smirnova, 1974).

## 17.7 Rates of nucleation and crystal growth in high polymers

Some effects of past history on the behaviour of polymers have been discussed above. Equilibrium (including configurational equilibrium) is not always uniquely determined at all temperatures. The rate at which segments on a flexible macromolecule form crystallization nuclei, and the rate at which these grow so that neighbouring segments assume long-range order, can be quite as variable as it is known to be for smaller molecules (Chapter 15). Some simplification of a very complicated situation arises from the fact that certain groups on a macromolecule are roughly equivalent, so that exchange of one for another does not greatly affect the tendency to crystallization (Bunn, 1954). For example, in polyvinyl alcohol with units of structure

$$-CH_2-\overset{\displaystyle H}{\underset{\displaystyle OH}{C}}-$$

the OH and H can be randomly placed on either side of the macromolecule, without impairing the tendency to crystallization whereas replacement of —OH by the considerably larger acetate group

$$
\begin{array}{c}
\text{O.CO.CH}_3 \\
| \\
-\text{C}- \\
| \\
\text{H}
\end{array}
$$

leads to an amorphous solid. Random replacement of a =CH$_2$ group (radius of repulsion 1.1 Å) by a =CF$_2$ group (radius of repulsion 1.35 Å) or by a

$$
\begin{array}{c}
\diagdown \quad \diagup \text{H} \\
\text{C} \\
\diagup \quad \diagdown \text{OH}
\end{array}
$$

group or a =C=O group (radius of repulsion 1.35 Å) does not prevent crystallization of a flexible macromolecule based on polythene, but random insertion of larger groups

$$
\begin{array}{ccc}
\diagdown \; \diagup \text{H} & & \diagdown \; \diagup \text{Cl} \\
\text{C} & \text{or} & \text{C} \\
\diagup \; \diagdown \text{Cl} & & \diagup \; \diagdown \text{Cl}
\end{array}
$$

such as those illustrated (repulsion radius 1.7 Å) inhibits crystallization. In polyvinyl chloride the crystallization observed is attributed to regular alternation of positions of H and Cl attached to the carbon chain. Broadly speaking, the factors involved are similar to whose which determine the limiting differences of molecular size, beyond which the formation of mixed crystals by copacking of two components becomes negligible (see Oldham and Ubbelohde, 1940).

A more speculative problem is to determine the critical supercooling $T_f - T_N$ (see Chapter 15) of any potentially crystallizable region in a polymer. It has been suggested (Bunn, 1954) that if the temperature of freezing-in of configurational entropy in a polymer $T_g$ lies much below $T_f$, nucleation is much more likely than when $T_g \sim T_f$. Although this seems plausible, independent means are not yet available for calculating $T_g$ and $T_N$ so as to elucidate the behaviour of supercooled polymers.

## 17.8 Melting of polymers under pressure

In principle, the effect of pressure in raising the freezing range of polymers is similar to its effect with crystalline materials (Chapter 2). The spread of fusion over a range of temperatures, and the occurrence of transitions below $T_f$ may however complicate the effects observed.

By way of illustration, for a specimen of natural rubber (Wood, Bekkedah, and Gibson, 1945), the melting point as determined by the disappearance of birefringence was raised from 36° to 70° by applying pressure of 1170 bar. Approximately $\Delta V_f = 0.0191$ cm$^3$ g$^{-1}$ at $T_f$ and $H_f = 16.2$ joule g$^{-1}$.

For isotactic polystyrene (Dedeurwaerder and Oth, 1959) use of the relationships

$$\Delta S_f = \Delta S_{vol} + \Delta S_{disorder} \tag{17.1}$$

$$\Delta S_{vol} = -\left(\frac{\alpha}{\beta}\right)\Delta V_f \tag{17.2}$$

where $\alpha$ is the coefficient of thermal expansion, and $\beta$ is the coefficient of compressibility, indicates

$$\Delta S_{vol} \sim 1.3 \text{ e.u.}$$

$$\Delta S_{disorder} \sim 2.6 \text{ e.u.}$$

(see Slater, 1939; Oriani, 1951). A more precise way of formulating this relationship is to calculate the entropy of melting at constant volume. This can be obtained from the relationship

$$(\Delta S_f)_v = (\Delta S_f)_p - \left(\frac{\partial S}{\partial V}\right)_T \Delta V_f \tag{17.3}$$

$$= (\Delta S_f)_p - \left(\frac{\partial P}{\partial T}\right)_v \Delta V_f \tag{17.4}$$

For polyethylene oxide (Malcolm and Ritchie, 1962) $\Delta V_f$ was calculated using known values of $\Delta S_f$ and the Clauius–Clapeyron equation

$$\frac{dT}{dP} = \frac{\Delta V_f}{\Delta S_f}$$

Measurements of the pressure coefficient

$$\left(\frac{\partial P}{\partial T}\right)_v$$

were made in a 'constant volume' thermometer apparatus. Values depend onlu slightly on molecular weight and yield $(\Delta S_f)_v = 1.41$ cal deg$^{-1}$ mole$^{-1}$ chain atoms or 4.22 cal deg$^{-1}$ mole$^{-1}$ repeating units. This value is in reasonable agreement with a model for melting (Starkweather and Boyd, 1960) of linear polymers which separates the entropy of fusion into two parts

$$(\Delta S_f)_v = \Delta S_{rot} + \Delta S_{dis}$$

in which $\Delta S_{rot}$ deals with configurational entropy (rotational isomerism) and $\Delta S_{dis}$ with a contribution from long-range disorder.

This treatment neglects any change of vibrational entropy on melting. However, the theory as a whole is only approximate. Its emphasis on entropy of orientation leads to a useful focussing of the problems involved.

For crystalline polythene (defect structure ($\sim2\%$) the temperature of crystallization at 4350 atm, has been estimated to be 227°C with $H_f = 69\cdot2$ cal $g^{-1}$ and with a specific volume $1\cdot001$ ml $g^{-1}$ (Wunderlich and Cormier, 1967).

# REFERENCES

J. P. Bell and J. H. Dumbleton (1969) *J. Polym. Sci.* **7**, 1033.

J. P. Bell, P. E. Slade, and J. H. Dumbleton (1968) *J. Polymer Sci.* **6**, 1773.

D. W. Brown and R. E. Lowry (1974) *J. Polym. Sci. Polym. Phys. Ed.* **12**, 1303.

C. W. Bunn (1954) *J. Appl. Phys.* **25**, 820.

C.·W. Bunn (1955) *J. Polym. Sci.* **16**, 323.

F. Danusso and G. Moraglio (1959) *Atti Accad. Naz Linei* **27**, 381.

R. Dedeurwaerder and J. F. M. Oth (1959) *J. Chim. Phys.* **56**, 940.

M. Dole and B. Wunderlich (1959) *Makromol. Chem.* **34**, 29.

O. B. Edgar and R. Ellery (1952) *J. Chem. Soc., London* 2633.

O. B. Edgar and R. Hill (1952) *J. Polym. Sci.* **8**, 1.

R. D. Evans, H. R. Mighton, and P. J. Flory (1947) *J. Chem. Phys.* **15**, 684.

P. J. Flory (1947) *J. Chem. Phys.* **15**, 685.

P. J. Flory (1956) *J. Am. Chem. Soc.* **78**, 5222.

P. J. Flory and L. Mandelkern (1956) *J. Polym. Sci.* **21**, 345.

F. C. Frank and M. Tosi (1961) *Proc. Roy. Soc., Ser. A* **263**, 323.

J. H. Gibbs and E. A. Di Marzio (1958) *J. Chem. Phys.* **28**, 373.

I. Goodman (1967) *Roy. Inst. Chem. Lectures* No. 3, 19.

L. Glatt and J. W. Ellis (1951) *J. Chem. Phys.* **19**, 449.

C. H. Griffiths (1975) *J. Polym. Sci. Polym. Phys. Ed.* **13**, 1167.

F. D. Hartley, F. W. Lord, and L. B. Morgan (1955) *Simp. Int. Chim. Macromol. Supp. Ric. Sci.* **25**, 3.

H. Hendus, G. Schnell, H. Thurn, and K. Wolf (1959) *Ergebn. Exakt. Naturwiss.* **31**, 220.

J. D. Hoffman and J. J. Weeks (1962) *J. Res. Nat. But. Stand.* **66A**, 13.

A. Keller (1959) *Makromol. Chem.* **34**, 1.

H. D. Keith (1974) *Metall. Trans.* **4**, 2747.

H. G. Kilian and E. Jenckel (1959) *Z. Elektrochem.* **63**, 951.

G. N. Malcolm and G. L. D. Ritchie (1962) *J. Phys. Chem.* **66**, 852.

L. Mandelkern (1959) *Rubb. Chem. Technol.* **32**, 1392.

L. Mandelkern, J. C. Halpin, and A. F. Diorio (1962) *J. Polymer Sci.* **60**, 31.

S. Matsuoka and J. Aloisio (1960) *J. Appl. Polym. Sci.* **4**, 116.

L. B. Morgan (1954) *J. Appl. Chem.* **4**, 160.

J. W. H. Oldham and A. R. Ubbelohde (1940) *Proc. Roy. Soc., Ser. A* **176**, 50.

R. A. Oriani (1951) *J. Chem. Phys.* **19**, 93.

A. Peterlin and E. W. Fischer (1960) *Z. Phys.* **159**, 272.

H. C. Raine, R. B. Richards, and H. Ryder (1945) *Trans. Faraday Soc.* **41**, 56.

R. B. Richards (1945) *Trans. Faraday Soc.* **41**, 127.

D. E. Roberts and L. Mandelkern (1955) *J. Am. Chem. Soc.* **77**, 781.

D. E. Roberts and L. Mandelkern (1960) *J. Am. Chem. Soc.* **82**, 1091.

R. E. Robertson (1969) *Macromolecules* **2**, 250.

D. W. Schaefer (1975) *Phys. Rev. Lett.* **35**, 1448.

J. C. Slater (1939) *Introduction to Chemical Physics* McGraw-Hill p. 261, New York and London.

I. V. Sochava, O. I. Smirnova, and A. A. Zhdanov (1974) *Sov. Phys. Solid State* **15**, 2001.

H. W. Starkweather, Jr and R. H. Boyd (1960) *J. Phys. Chem.* **64**, 410.

J. J. Trillat, S. Barbezat and A. Delalande (1950) *J. Chim Phys. Phys Chim. Biol.* **47**, 877.

M. H. Theil (1975) *J. Polym. Sci. Polym. Phys. Ed.* **13**, 1097.

L. A. Wood, N. Bekkedahl, and R. E. Gibson (1945) *J. Res. Nat. Bur. Stand.* **35**, 375.

E. Wunderlich and C. M. Cormier (1967) *J. Polymer Sci.* **5**, 987.

A. Ziabicki and W. Klonowski (1975a) *Rheologica Acta* **14**, 105.

A. Ziabicki and W. Klonowski (1975b) *Rheologica Acta* **14**, 113.

# INDEX

Adsorbed layers
  melting of, 32
Anticrystalline melts
  clusters, 343
  crystallization of, 405
  examples of, 3, 4, 121, 151, 170, 206, 253
Argon
  defect melting, 288
  vibrational melting, 72, 77
Autocomplexing
  in ionic melts, 193, 205
  of organic salts, 197
  optical studies, 210

Batchinski equation, 350

Calorimetry, 1, 9, 203
Carboxylates
  melting parameters of alkali salts, 222
Cell models of melts, 125
Characteristic frequency and melting
  at dislocation cores, 298
  Debye, 63, 68, 75
  Einstein, 62, 69, 72, 77
  electrical resistivity, 257
  helium, 71
  internal, 76
  phonon spectra, 63, 64, 65, 73
Clausius–Clapeyron equation, 11, 15, 184
  higher derivatives, 103
Coexistence of structures
  in metals, 238 *seq.*
  in rotator transitions, 134
  related transforms, 110
Cluster formation in melts
  centrifugal tests, 350
  cooperative fluctuations, 343
  effects on thermal conductance, 358
  in molten metals, 259, 362
  ionic melts, 205 *seq.*, 361
  molecular melts, 351
  paraffins, 354
  polyphenyls, 353
  precursor to glasses, 349

  rotational hindrance, 355
  semiconductors, 367
  water, 359
Compensation of electrostatic forces
  in ionic assemblies, 176
Compressibility
  change on glass formation, 426
  change on melting, 326, 330
  configurational terms, 61
Configurational entropy
  flexible molecules, 154, 222 *seq.*
  glass formation, 425
  macromolecules, 432
Continuous transitions
  solid/melt, 25 *seq.*, 293
  solid–solid, 102
Corresponding states
  applications to melting, 55
  in precursor effects, 316
  lattice energy parameter, 56
Cooperative defects and melting
  simple models, 126
  statistical theories, 296
Coordination numbers and melting
  hybrid structure of metals, 241
Critical transitions solid/liquid
  general, 25 *seq.*
  helium, 71
  quasi-crystalline melts, 293
Cryoscopy
  globular melts, 138
  ionic melts, 186, 204

Defects
  atom transfer, 215
  cooperative, 127, 296
  in ionic crystals, 173
  in metal crystals, 245
  positional energy, 173, 297
  valence switch, 216
Dislocation melting, 127, 296
Diffusion, 243
Disordering
  general theory, 54
  long-range correlations, 119

orientational, 3, 130, 202, 300
orientational correlations, 358
positional, 3, 126, 129
Dissociation
of ions on melting, 201
of salts dissolved in melts, 201
Domains
boundary activated sites, 115
melting at boundaries, 31
*see* Conglomerate liquids

Electrical resistivity changes on melting
ionic crystals, 192, 195, 199
metallic conductors, 256
semiconductors, 265
sub-lattices, 219
Electrostatic compensation
in ionic crystals and melts, 176, 205
Existence range of melts, 20, 245
Equilibrium
frozen-in states, 6, 92, 424
past history of macromolecules, 435
Entropy of melting
communal, 78, 284
configurational, 154, 443
globular molecules, 136
inert gases, 130
ionic crystals (B-type), 196
ionic crystals (inert gas type), 175, 178, 183
ionic crystals (polyatomic), 189, 203, 204
metals, 239, 240, 241
molecular crystals, 130, 131
polyphenyl molecules, 148
separation of contributions, 3, 131, 147, 155, 206, 237, 245, 257

Flexible molecules
configurational melting, 154
effects of stiffening macromolecules, 439
vibrational melting, 162
Free volume in melts
cell model, 286
halides, 202
nitrates, 202
Fusion
allene homologues, 146
at grain boundaries, 30
at high pressures, 14–16, 18, 19, 249

flexible macromolecules, 432
flexible molecules, 154
homologous series, 158, 160
in capillaries, 35
in gels, 37
in stages, 130
network breakdown, 275
of polymorphs, 49
of small regions, 30
of solid solutions, 43
on surfaces, 32
orientational, 300
paraphenyl homologues, 145
positional, 122 *seq.*
$P/T$ maxima, 17
rigid linear molecules, 143
rigid planar molecules, 148
sub-lattice, 218
tautomeric molecules, 168
vibrational, 61 *seq.*
volume change, 10, 11–13, 54

Glasses
compressibility, 426
configurational energy, 423
entropy, 419, 424
glass forming temperatures, 429
'ideal' glass temperatures, 427
inverse crystallization temperature, 409
Kauzmann paradox, 428
molecular types, 423
polymer types, 424
relation to cluster formation, 349
relation to melt viscosity, 414 *seq.*
thermal conduction, 83
Globular molecules
melting parameters, 136
rotational barriers, 141
Grotthus ionic conduction, 199
Growth at crystal/melt interface
block growth processes, 396
charge accumulation, 400
in polymer melts, 443
rhythmic, 400
spiral and stepped structures, 400

Hall effect in metals
changes on melting, 261
Helium
effects of pressure on melting, 46
melting of isotopes, 15, 46

vibrational melting, 70
High melting point solids, 20
Hybrid single crystals, 103
　acoustic detection, 111
　coexistence of related transforms, 110, 134
　storage of defect energy, 114
Hydrogen ions in crystals, 174
Hydrogen bond networks in melts, 305
　*see* Ice; Water
Hysteresis
　effects of impurities, 106, 113
　general theory for hybrid crystals, 108
　in orientational randomization, 93 *seq.*

Ice
　network domains in water, 359
　network structures in polymorphs, 382
　regelation, 34
Interface energy crystal/melt, 35, 244, 315, 396, 402, 411
　activation at interface, 115
Impurity effects on melting
　dissociating species, 42
　heterophase, 39
　not soluble in crystals, 39
　soluble in crystals, 43
Ionic crystals
　high valence ions, 178
　inert gas ions, 175
　ions of low polarizability, 177
　organic ions, 220, 229
　polyatomic ions, 200
　strongly polarizable ions, 192
　transport processes, 208
Isotope effects
　glass formation, 409
　melting points, 45, 47, 49, 69, 82, 97, 98, 129
　solid transformations, 96 *seq.*
　thermal conduction, 83
　transport, 80
　viscosity, 82, 365

Lambda phenomena in solids, 91
Lattice energy and melting, 57
Liquid state
　anticrystalline models, 3, 4, 121
　conglomerate models, 3, 120, 122, 275, 304, 344, 362
　domain topology, 2, 122, 254, 305, 385

existence range, 20–24
　principal types, 3, 125
　quasi-crystalline models, 3, 122, 294
　quasi-gaseous models, 1, 3
Long range correlation, 149, 355
Lindemann melting theory, 68
Lindemann parameters, 64
Liquid crystals
　compound formulation, 387
　homologous series, 377, 378, 379, 381, 382
　ionic, 4, 225, 375, 383
　molecular, 4, 373
　precursor effects, 389
　transport properties, 388

Magnetic susceptibility
　change on melting, 262
Melting (*see also* Fusion)
　helium, 70
　mechanical theories, 58, 65
　vibrational theories (one-phase), 61, 66
　vibrational theories (two phase), 72
Metallic state
　conglomerate melts, 247, 253
　effects of pressure on melting, 249
　melting parameters, 239, 241, 243
　transition from ionic, 217

Network melts
　elements, 278, 360
　general theory, 274
　halides, 280
　ionic, 197, 361
　metals, 248
　oxides, 278, 282
　semiconductors, 248, 267
　sensitive to impurities, 279
　viscosity, 277, 280 *seq.*
　water (*see* Ice), 282, 360
　zinc compounds, 198
Nucleation
　catalysts, 409
　failure, 404
　in anticrystalline melts, 395, 405, 407
　in capillaries, 38
　normal rules, 404

Optical absorption
　infrared: ionic melts, 213
　　organic melts, 361

Raman: inorganic melts, 358
  ionic melts, 217
  organic melts, 356
ultraviolet: ionic melts, 210
  organic melts, 168
Orientational randomization
  correlations in melts, 149
  effects of compression, 101, 133
  entropy increase due to, 147
  in ammonium salts, 90, 95
  in crystals below $T_f$, 90, 93
  in crystals with quasicylindrical molecules, 99
  in crystals with quasispherical molecules, 90
  in the theory of melting, 130, 300
  lubrication, 133
  of polyatomic ions, 201
  sensitiveness to impurities, 100

Phase transformations
  continuous, 86, 102
  in polymers, 439
  ionic crystals, 213
  of higher order, 101, 102
  persistence of crystal axes, 104
  solid/solid, 86
  vibrational, 73, 238
  volume changes, 95
Plastic crystals, 142, 151, 237
Polarizability of ions and melting, 181
Polyatomic ions
  disruption on melting, 215 seq.
Polymers
  degree of crystallization, 434
  effects of past history, 435
  melting, 4, 434
  melting of mixed sequence macromolecules, 442
  specific heat, 434
  thermal expansion, 432 seq.
Polymorphs
  melting points, 48
Phonons
  electrical resistance, 84
  fading at transitions, 66, 73, 87
  near $T_f$, 70
  thermal conduction, 68
Positional melting, 124
  and electrostatic compensation, 176
  a universal mechanism, 288

ionic crystals, 88
rigid molecules, 129
Positional randomization
  in phase transformations, 88, 217 seq.
Precursor effects
  ionic halide crystals, 182
  liquid crystals, 380, 389, 391
  premelting and prefreezing, 4, 10, 309
Pressure effects
  critical melting, 25 seq.
  liquid crystals, 386
  maxima on melting curves, 17, 249
  melting of polymers, 445
  on orientational randomization, 141, 142
  see also Clausius–Clapeyron equations; Simon equations
Premelting
  cooperative fluctuation, 315
  defect, 313
  effects of pressure, 184
  heterophase, 316
  homophase, 310, 320
  ionic crystals, 182 seq., 327
  Lennard-Jones model, 294
  liquid crystals, 389
  metals, 334
  molecular crystals, 321
  Mossbauer effects, 338
  n-paraffins, 321
  trivial two-phase, 310, 318
  vibrational, 312
Prefreezing
  ionic melts, 361
  Lennard-Jones model, 294
  liquid crystals, 391
  metals, 346, 362 seq.
  molten polyphenyls, 351
  non-inversion with premelting, 342
  polar organic molecules, 353 seq.
  water, 359

Quasicrystalline melts, 3, 122, 344
  change of reference structure on melting, 183
  dislocation model, 296, 299
  Lennard-Jones model, 288
  statistical theory, 288

Radial distribution functions, 119
  ionic melts, 180

metallic melts, 248, 253
molecular melts, 121
Rayleigh scattering by melts, 356
Rotation
    dipole rotation, 141
    effects of pressure, 133
    hindrance in clusters, 355
    in crystals, 130, 133–136
    mechanism of fusion, 300
    room to rotate in melt, 149
Rotational isomers
    melting of, 50
Resistivity changes on melting
    ionic crystals, 199, 214, 219, 230, 231
    metals, 256
Repulsion envelopes
    carboxylate anions, 223
    molecular crystals, 136
    nitrate anion, 223
    polyphenyls, 132
    rigid molecules, 132
    shrinking on melting ionic crystals, 177
Rotokinetic effect in melts, 359

Semiconductors
    melting of, 248, 265, 367
Simon equation
    general, 14
    ionic crystals, 185
Similitude theories, 125
    ionic crystals, 57, 172, 190
    metallic crystals, 241, 250
    molecular crystals, 134
    positional melting, 125
    structurally related molecules, 134
    viscosity/temperature plots, 416
Solid electrolytes, 89, 219
Solid solutions,
    lubrication of crystal transitions, 115
    lubrication of internal rotation, 100, 133
    melting theory, 43
    partition coefficient solid/melt, 44
Specific heat
    ionic melts, 183, 188, 204
    melts and crystals, 79
    metals, 239, 243, 245, 251
    polymers, 434
    prefreezing, 346
Silver salts
    melting, 175, 188, 205, 209
    phase transformations, 88, 219, 333

premelting, 329
Solid/solid transformations
    calorimetry, 9
    continuous, 102
    hysteresis, 10
    in metals, 249
    in polymers, 439
    isotope effects, 98
    nucleation effects, 105
    of higher order, 103
    persistence of crystal axes in cycles, 104
    vibrational, 238
    volume changes, 94
Sub-lattice melting, 174, 218 *seq.*
Sublimation, 20
Supercooling, 317, 395
    glasses, 349, 408, 424
    quasi-crystalline melts, 349, 404
Spiral growth, 400
Surface melting, 317
Superheating, 317
Surface energy
    interface surface energy of hybrid crys-
        tals, 106, 114, 247
    interface surface energy of melts, 30,
        247, 251, 315, 358, 359, 390, 410

Tautomeric molecules, 168
Thermal expansion
    and melting, 67, 293
    defect creation, 79, 294, 334
    vibrational, 67
Theories of melting
    mechanical, 58, 65
    thermodynamic, 3, 7
    vibrational, 61, 69
Thermal conduction
    crystal vs glass, 83
    effects of premelting, 338
    isotope effects in melts, 83
    solid/melt interface, 367
    vibrational transfer in melts, 358
Thermo-electric power
    molten metals, 260
Thermal transformations
    in microphases, 115
    in polymers, 439
Triple point ice, 8
Transport properties
    ionic melts, 209, 214, 219, 230, 231,
        361

metals, 80, 242, 363
molecular melts, 348, 350–355
network melts, 280, 281
Trends in melting parameters
 alternation in homologous series, 160
 liquid crystal series, 377, 378, 379, 381–383
 metals (quantum number), 252
 related molecules, 135, 136, 139, 144, 145, 146
 similitude systems, 127, 129, 172, 190, 241, 250, 416
 with molecular size, 136

Ultrasonic velocities
 and glass formation, 426
 at phase transformations, 87, 111
 in melts, 154
 in molten metals, 365

Viscosity
 Batchinski formulation, 350
 blocked volume model, 348
 criteria for glass formation, 414
 isotope effects, 82
 liquid crystals, 390
 molten metals, 81, 242, 363
 network melts, 280, 282
 polyphenyl molecules, 351, 416

prefreezing anomalies, 348, 349, 360, 363
 similitude plots, 415 *seq.*
 temperature dependence, 414
 vibrational theories, 80
Vibrational melting
 anharmonic line solid, 66
 correlations of parameters, 63 *seq.*, 75, 79
 flexible molecules, 162, 227
 Lindemann models, 62, 65, 69
Volume change on melting, 11
 benzene derivatives, 13
 ionic inert gas type, 175, 182
 metals, 239, 241
 molecular crystals, 121, 129, 131
 negative, 55, 203, 239, 241
 polyatomic ions, 203, 207, 227, 232
 polymethylene derivatives, 12
 polyphenyl molecules, 148
 tetrahalides, 135

Water
 conglomerate model, 305
 domain structure, 359
 entropy of fusion, 131
 reduced viscosity, 417

Zinc compounds, 198

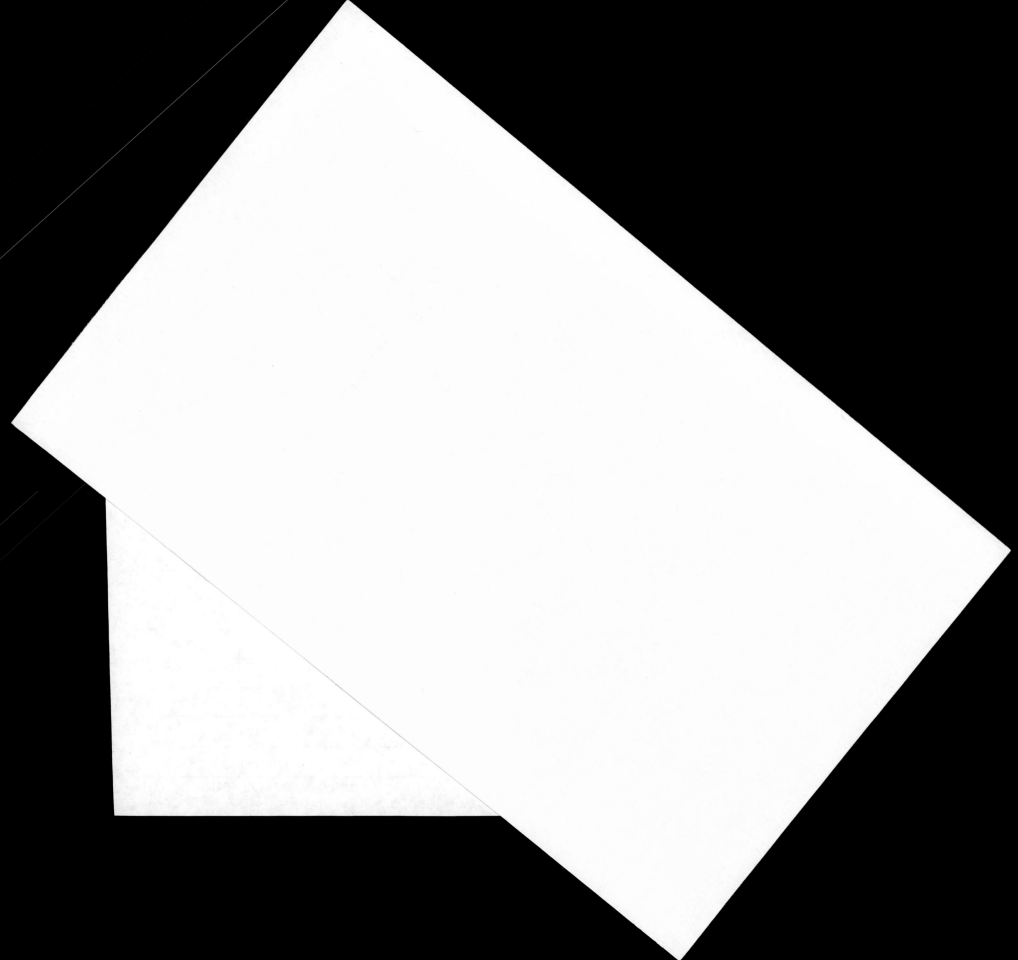